Advances in Natural and Technological Hazards Research

Volume 42

The book series entitled *Advances in Natural and Technological Hazards* is dedicated to serving the growing community of scholars, practitioners and policy makers concerned with the different scientific, socio-economic and political aspects of natural and technological hazards. The series aims to provide rapid, refereed publications of topical contributions about recent advances in natural and technological hazards research. Each volume is a thorough treatment of a specific topic of importance for proper management and mitigation practices and will shed light on the fundamental and applied aspects of natural and technological hazards.

Comments or suggestions for future volumes are welcomed.

More information about this series at http://www.springer.com/series/6362

Fabrice G. Renaud • Karen Sudmeier-Rieux
Marisol Estrella • Udo Nehren
Editors

Ecosystem-Based Disaster Risk Reduction and Adaptation in Practice

Editors

Fabrice G. Renaud
United Nations University Institute
  for Environment and Human
  Security (UNU-EHS)
Bonn, Germany

Karen Sudmeier-Rieux
Commission on Ecosystem Management
  International Union for Conservation
  of Nature
Gland, Switzerland

Marisol Estrella
Post-Conflict and Disaster Management
  Branch
United Nations Environment Programme
Geneva, Switzerland

Udo Nehren
Institute for Technology and Resources
  Management in the Tropics
  and Subtropics (ITT)
TH Köln, University of Applied Sciences
Köln, Germany

Disclaimer: The facts and opinions expressed in this work are those of the author(s) and not necessarily those of their organizations.

ISSN 1878-9897   ISSN 2213-6959   (electronic)
Advances in Natural and Technological Hazards Research
ISBN 978-3-319-43631-9   ISBN 978-3-319-43633-3   (eBook)
DOI 10.1007/978-3-319-43633-3

Library of Congress Control Number: 2016947057

Printed on acid-free paper

This Springer imprint is published by Springer Nature
The registered company is Springer International Publishing AG Switzerland

Foreword

Each year, disasters occupy headlines in different parts of the world. The frequency and impacts of disasters have increased in the last few decades due to human activities and exacerbated by a changing climate. Disasters such as earthquakes, landslides and floods cost human lives and affect economic growth of countries.

The Himalayas region, in particular, is highly vulnerable to disasters, while at the same time, it serves as a water tower and supports more than 1.5 billion people in the southern plains. Nepal as a country in the central Himalayas is listed in the top 20 most disaster-prone countries in the world, experiencing hundreds of disasters every year. The country faced tremendous loss due to the 7.8-magnitude earthquake that occurred in April 2015, killing more than 9000 people and several hundred thousand livestock and damaging numerous building, infrastructure and cultural sites.

We have long understood that taking care of our environment and working with nature constitute two of the best defences we have against natural hazards and climate change. Healthy, well-managed ecosystems can mitigate or prevent certain hazards, buffer against hazard impacts as well as enhance community resilience by providing basic needs and supporting livelihoods. But the potential role of harnessing nature and ecosystem-based solutions for reducing disaster and climate risks remains largely untapped globally.

I wish to mention one unique ecological zone referred to as the "Chure" in Nepal. Located between the southern Terai and northern midhills and with Himalayan mountains ranging from 300 to 1000 m above sea level, the Chure zone is experiencing rapid deforestation and rampant extraction of sand and gravel. Ecosystem degradation in the Chure has resulted in increased landslides, severe soil erosion and flash floods and has significantly reduced agricultural production in the region, known as the "breadbasket" of the nation. As head of state of the government of Nepal from 2008 to 2015, the government of Nepal initiated the "Chure Conservation Program" and established the President Chure Terai Madhesh Conservation Board, which was solely dedicated to manage the Chure region and reduce disaster risk through ecosystem management and climate-smart programmes. Drawing from these experiences, Nepal also has many lessons to share with the global community.

It is in this context that this book can make an important contribution. It was produced in a year when three major global policy agendas were negotiated, namely, the Sendai Framework for Disaster Risk Reduction (SFDRR), Sustainable Development Goals (SDGs) and the UNFCCC COP 21 Paris Agreement. Implementation of these agreements in countries and communities will demand integrated strategies. The book examines how improved ecosystem management helps to maximize the integration of disaster risk reduction, climate change adaptation and sustainable development. I appreciate the content of the book and congratulate authors and editors for their contribution in highlighting opportunities of working with nature to address our global challenges.

President of The Federal Democratic   Ram Baran Yadav
Republic of Nepal (2008–2015)
Kathamandu, Nepal

Acknowledgements

We are extremely grateful to the following experts who have volunteered their time and knowledge to review the chapters in this book. Alphabetically, our sincerest thanks go to Marwan Alraggad (Water, Energy and Environment Center (WEEC), University of Jordan, Jordan), James Blignaut (Beatus, Asset Research, Futureworks, SAEON, and Department of Economics, University of Pretoria, South Africa), James K. Boyce (University of Massachusetts Amherst, USA), Gernot Brodnig (International Union for the Conservation of Nature, Switzerland), John Connell (University of Sydney, Australia), Kathryn K. Davies (National Institute of Water and Atmospheric Research, New Zealand), Dolf de Groot (Wageningen University, the Netherlands), Nathalie Doswald (Biodiversity and Climate Change Consultant, Switzerland), Nigel Dudley (School of Geography, Planning and Environmental Management at the University of Queensland and Equilibrium Research, UK), Nadir Ahmed Elagib (Institute for Technology and Resources Management in the Tropics and Subtropics (ITT), TH Köln (University of Applied Sciences), Germany), Olivia E. Freeman (ASB Partnership for the Tropical Forest Margins, World Agroforestry Centre (ICRAF), Kenya), Christiane Grinda (Institute of Rescue Engineering and Civil Protection, TH Köln (University of Applied Sciences), Germany), Greg Guannel (The Nature Conservancy, USA), Johannes Hamhaber (Institute for Technology and Resources Management in the Tropics and Subtropics, TH Köln (University of Applied Sciences), Germany), Nivedita Haran (Government of Kerala, Thiruvananthapuram, India), James Hardcastle (International Union for Conservation of Nature, Switzerland), Srikantha Herath (United Nations University Institute for the Advanced Study of Sustainability, Tokyo, Japan), Dyah Rahmawati Hizbaron (Faculty of Geography, Universitas Gadjah Mada, Indonesia), Hiromu Ito (University of Tsukuba, Japan), Michel Jaboyedoff (Institute of Earth Science, University of Lausanne, Switzerland), Stacy Jupiter (Wildlife Conservation Society, Suva, Fiji), José Gabriel Vitória Levy (United Nations Development Programme, Mauritania), Ananda Mallawatantri (International Union for Conservation of Nature, Sri Lanka), Muh Aris Marfai (Faculty of Geography, Universitas Gadjah Mada, Indonesia), Maria

Luisa Martínez (Instituto de Ecología, A.C. Xalapa, Veracruz, Mexico), Brian G. McAdoo (Yale-NUS College, Singapore), Imen Meliane (7 Seas Environmental Strategies, Tunisia), Abdul Muhari (Directorate of Coastal and Ocean, Directorate General for Coasts, Marine and Small Islands, Ministry of Marine Affairs and Fisheries, Indonesia), Santhakumar Velappan Nair (Azim Premji University, Bangalore, India), Siddharth Narayan (National Center for Ecological Analysis and Synthesis (NCEAS), UC Santa Barbara, Santa Barbara, California, USA), Celia Norf (Institute of Rescue Engineering and Civil Protection, TH Köln (University of Applied Sciences), Germany), Pascal Peduzzi (GRID-Geneva, United Nations Environment Programme, Switzerland), Dinil Pushpalal (Graduate School of International Cultural Studies, Tohoku University, Japan), Borja G. Reguero (Department of Ocean Sciences, University of California-Santa Cruz, and The Nature Conservancy, USA), Jakob Rhyner (United Nations University Institute for Environment and Human Security, Bonn, Germany), Rebecca J. Sargisson (The University of Waikato, New Zealand), Dietmar Sattler (Institute of Geography, University of Leipzig, Germany), Zita Sebesvari (United Nations University Institute for Environment and Human Security, Bonn, Germany), Rajib Shaw (Graduate School of Global Environmental Studies, Kyoto University, Japan), Joerg Szarzynski (United Nations University Institute for Environment and Human Security, Bonn, Germany), Muralee Thummarukudy (Disaster Risk Reduction Programme Manager, UN Environment Programme, Switzerland), Paul Venton (International Development Consultant for Climate Change, Disasters and Environment, Boston, Massachusetts, USA), Gregory Verutes (Natural Capital Project, Stanford University, USA), Derek Vollmer (Future Cities Laboratory, Singapore-ETH Centre for Global Environmental Sustainability, Singapore), Vanja Westerberg (Altus Impact, Switzerland) and Pramaditya Wicaksono (Faculty of Geography, Universitas Gadjah Mada, Indonesia).

We are also grateful to Toa Loaiza from the Institute for Technology and Resources Management in the Tropics and Subtropics, TH Köln (University of Applied Sciences), Germany, for her invaluable editorial support during the preparation of this book.

Any errors found in this book are the sole responsibility of the chapter authors and of the editors.

Contents

Contributors

Sandra Alfonso is an architect and PhD student in geography at the University of Innsbruck, Austria. Her research is on the protection and sustainable management of urban ecosystems in Latin America with a particular focus on political ecology and ecosystem services.

Kalpana Ambastha is a research associate with Wetlands International South Asia and specializes in landscape-scale assessments of wetland catchments.

Yves Barthélemy is an expert in geo-information and tropical ecology. He is the manager of OBSCOM and a faculty member at Sciences-PO MPA in France. He has been working as a GIS consultant for UNEP for more than 10 years. Yves is currently based in Tanzania working on coastal sensitivity and vulnerability and developing capacity building programmes in GIS at the State University of Zanzibar.

Niloufar Bayani is a conservation biologist with a particular interest in coastal and marine issues. Since 2012, she has been working as a consultant with the United Nations Environment Programme in Geneva on ecosystem-based disaster risk reduction.

Michael W. Beck is the lead marine scientist for The Nature Conservancy and an adjunct professor in ocean sciences at the University of California-Santa Cruz, where he is based. Mike focuses on building coastal resilience in the interface between policy and practice to reduce risks to people, property and nature.

Rizaldi Boer is a staff member at the Department of Geophysics and Meteorology as well as the director of the Centre for Climate Risk and Opportunity Management (CCROM) of Bogor Agricultural University. His works involve work on both adaptation as well as mitigation for Southeast Asia. Key research areas are El Nino Southern Oscillation impacts on development, reduced emissions for defor-estation and forest degradation, peat lands and low carbon growth. He has authored key climate change-related documents, with emphasis of Indonesian National

Communications (1st and 2nd and working on the 3rd). He is a member of the IPCC.

Luc Boerboom is an assistant professor in the Department of Urban and Regional Planning and Geo-information Management at the Faculty of Geo-information Science and Earth Observation (ITC), the University of Twente, Enschede, the Netherlands. His research focus is on planning and decision-making concepts and methods.

Jil Bournazel recently graduated with an MSc in environmental management from the University of Nantes, France. Her main interest is climate change mitigation, with a particular focus on coastal ecosystem resilience.

Consuelo Castro is a professor in geography at the Pontifical Catholic University of Chile and an expert for coastal dune systems in Latin America. She recently published a book on the geography of coastal dunes in Chile.

Florie Chazarin is a geographer with interests in exploring links between forests, people and climate-related issues. She previously worked as a researcher for CIFOR (Center for International Forestry Research) based in Indonesia and Peru.

C. Gabriel David from Franzius Institute for Hydraulic, Estuarine and Coastal Engineering, Leibniz University of Hannover, Germany, works as a PhD student on coastal engineering and specializes on numerical modelling and simulation, ship wake phenomena as well as ecosystem-based coastal defence.

Huib de Vriend has served as director of science of Deltares and is emeritus professor of river engineering and eco-engineering at Delft University of Technology. His expertise is in river, estuarine and coastal dynamics, and he has led the first phase of the Building with Nature innovation programme in the Netherlands.

Mindert B. de Vries is a specialist on development of eco-engineering green solutions in relation to flood safety. Mr. de Vries has been involved in many advisory and research studies within the Netherlands and abroad. At present, he is coordinating the EU FP7 FAST project that links Earth observation to flood risk reduction effect of foreshores.

Febrina Desrianti has a background in rural sociology. Previously she worked for CIFOR (Center for International Forestry Research) and ICRAF (World Agroforestry Centre) in Indonesia.

Deepak Dhyani is a senior scientist in the field of biodiversity conservation with the Society for Conserving Planet and Life (COPAL), an NGO based in Srinagar Garhwal, Uttarakhand, India. His research focus is on awareness generation and capacity building for using underutilized wild edible resource of the Himalayas for landslide management and developing livelihood opportunities in the Western Himalayas.

Shalini Dhyani is a scientist and AcSIR assistant professor in the field of biodiversity conservation and management at the Environmental Impact and Risk Assessment Division of the National Environmental Engineering Research Institute (CSIR-NEERI) in Nagpur, India. Her research focus is on mainstreaming biodiversity in decision making and biodiversity-inclusive impact assessments with particular specialization in Himalayan forest ecosystems.

Bachir Diouf holds a PhD in geomorphology. He is a researcher and professor in the field of dynamic geomorphology at the Department of Geology, Cheikh Anta Diop University of Dakar in Senegal. His research focus is on coastal areas in Senegal and West Africa.

Houria Djoudi is a researcher in the field of adaptation to climate change at the Center for International Forestry Research (CIFOR). Her work focuses on adaptation to climate change, rural livelihoods, vulnerability and especially the role of forest ecosystems and trees in adaptation.

Luuk Dorren is a professor for natural hazards and GIS at the Bern University of Applied Sciences (Switzerland) and president of ecorisQ, the international association for natural hazards risk management. He specializes in quantitative risk analyses that account for the protective role of forests against natural hazards.

Lea Dünow is a technical advisor in the project "Prevention, Control and Monitoring of Bushfires in the Cerrado" in Brazil, for the German development corporation Deutsche Gesellschaft für Internationale Zusammenarbeit (GIZ) and AMBERO Consulting. Lea has work experience in Brazil, Germany, Mozambique and Serbia. Her main areas of interest are adaptation to climate change and mitigation, biodiversity, vulnerability and risk perception.

Lucy Emerton is director of environmental economics and finance at the Environment Management Group, based in Sri Lanka. Her work focuses on the economic valuation of biodiversity and ecosystem services and on the development of conservation incentives and financing mechanisms.

Marisol Estrella is the disaster risk reduction programme coordinator at the Post-conflict and Disaster Management Branch in the United Nations Environment Programme, based in Switzerland, where she works on global advocacy, partnerships, capacity building and field implementation.

Giacomo Fedele has a background in environmental science and is specialized in ecosystem-based approaches that seek to address climate change-related issues. He is currently a PhD student at AgroParisTech based in CIRAD (Agricultural

Research for Development), France, and he previously worked for CIFOR (Center for International Forestry Research) in Indonesia.

Maria Fernandez is a research fellow at the University of Massachusetts. Her training includes an MSc on natural resources conservation (National University, Costa Rica) and an MBA/MPPA on finance and international development policy (UMass Amherst). Maria is an Environmental Defense Fund Climate Corps fellow and a One Energy scholar.

Daniel A. Friess is an assistant professor in the Department of Geography, National University of Singapore. His lab investigates ecosystem services provided by mangrove forests, their stability under sea level rise and land cover change and payments for ecosystem services. For more information, please visit www. themangrovelab.com.

Naoya Furuta is a coordinator at IUCN Japan Liaison Office and a professor at the Institute of Regional Development, Taisho University, Tokyo, Japan, where he works on Eco-DRR and green infrastructure policy advocacy and research projects since the Great East Japan Earthquake in 2011.

Adi Gangga has a degree in forestry, and he worked with the Center for International Forestry Research (CIFOR) on a project studying the interactions between communities and forests in the context of climate vulnerability.

Zuzana V. Harmáčková is a researcher at the Department of the Human Dimensions of Global Change, CzechGlobe. She is focusing on modelling of ecosystem services and trade-offs, social-ecological system analysis and climate change adaptation.

Mark Huxham is a professor of environmental biology at Edinburgh Napier University, Edinburgh, Scotland. He works on coastal ecology and restoration, with a particular interest in mangrove forests. He is the founding director of the Association for Coastal Ecosystem Services, a charity which helps communities benefit from the conservation of their coastal resources.

Rohini Kamal is an economics PhD student at the University of Massachusetts Amherst and predoctoral fellow at the Global Economic Governance Initiative, Boston University. Her past work includes research at the Brac Development Institute in Bangladesh and at the Urban Climate Change Research Network, a project under NASA-GISS and Earth Institute, Columbia University.

Munish Kaushik leads the community-managed disaster risk reduction programme of Cordaid in India.

Johan Kieft is the head of the Green Economy Unit of the UN Office for REDD+ Coordination in Indonesia (UNORCID). His work involves all aspects of green economic development. He is working on green economy policy advisory services, providing technical advisory services on forest and peat fires, peat restoration and overall management of the Green Economy Unit. He has worked with senior ministerial officials on various climate change and sustainable development-related issues in countries such as Indonesia and Macedonia.

Julia Kloos is an agricultural economist with expertise in vulnerability and risk assessments and adaptation options with a focus on ecosystem-based approaches to adaptation and disaster risk reduction. Prior to her current engagement as scientific officer at the German Aerospace Center, she was associate academic officer in the Environmental Vulnerability and Ecosystem Services Section at the United Nations University Institute for Environment and Human Security in Bonn, Germany.

Bart Krol is a lecturer in the Department of Earth Systems Analysis and course director of applied Earth sciences at the Faculty of Geo-information Science and Earth Observation (ITC), University of Twente, Enschede, the Netherlands. His main focus is on natural hazard studies.

Ritesh Kumar is the conservation programme manager of Wetlands International South Asia and leads an interdisciplinary team on integrated management of water and wetlands.

Satish Kumar is a technical officer with Wetlands International South Asia and specializes in livelihood assessments for wetlands conservation programmes.

M. Priyantha Kumara is a senior lecturer at the Faculty of Fisheries and Marine Science, Ocean University of Sri Lanka. His research focus is on mangroves and coastal ecosystems.

Wolfram Lange is a research fellow at the Center for Rural Development of Humboldt-Universität zu Berlin. He works on sustainable urban development and natural resources management with a focus on ecosystem-based adaptation.

Bruno Locatelli is a research scientist with CIFOR (www.cifor.org) and CIRAD (www.cirad.fr). His background is in environmental sciences. His research interests relate to the relationships between forests and climate change, particularly the role of forest ecosystem services in adaptation.

Joan Looijen is a senior lecturer in the Natural Resources Department at the Faculty of Geo-information Science and Earth Observation (ITC), University of Twente, Enschede, the Netherlands. She concentrates on the use of spatial decision support tools for environmental impact assessment and strategic environmental assessment in a spatial planning context.

Eliška Krkoška Lorencová is a postdoctoral researcher at the Department of the Human Dimensions of Global Change at the Global Change Research Centre, Academy of Sciences of the Czech Republic (CzechGlobe). Her research focuses on climate change adaptation, perception of climate change and climate change impacts on ecosystem services.

Muh Aris Marfai is a professor in geography and vice dean of the Faculty of Geography, Universitas Gadjah Mada, Yogyakarta, Indonesia. His research focus is on coastal geomorphology and management, where he published several books and articles.

Jeffrey A. McNeely is a senior consultant for Thailand's Department of National Parks, Wildlife and Plant Conservation, a position he assumed following his retirement after 30 years at IUCN. As IUCN's chief scientist, he contributed to many aspects of biodiversity conservation, protected areas, agriculture, economics, climate change and disaster risk reduction.

Naoki Nakayama is a deputy director of the Office for Mainstreaming Biodiversity at the Japanese Ministry of the Environment, where he has been engaged in the international biodiversity policy making including policies related to Eco-DRR.

Udo Nehren is a senior researcher and lecturer in the field of ecosystem management at the Institute for Technology and Resources Management in the Tropics and Subtropics at TH Köln (University of Applied Sciences), Germany. His research focus is on tropical forests and coastal dune systems in Latin America and Southeast Asia.

Abdeljelil Niang is a researcher and lecturer in the field of dynamic geomorphology and natural hazards at the Department of Geography, Umm al-Qura University, Makkah, Kingdom of Saudi Arabia. His research focus is on monitoring the environmental changes in Mauritania and arid ecosystems.

Pranati Patnaik coordinated implementation of the Partners for Resilience programme during 2011–2013.

Carla Pesch is a teacher and researcher at the Zeeland University of Applied Sciences (HZ). She has been involved in research on implementation of Building with Nature in Dutch delta areas.

Christian Pirzer is a project manager and consultant at the independent research and consulting institute Endeva, based in Germany. Endeva's mission is to inspire and enable enterprise solutions for development challenges. Christian is leading the

Climate Change Adaptation Unit and explores how businesses can support low-income communities to manage climate risks and improve their climate resilience.

Claudia Raedig is a senior researcher and lecturer in biodiversity and connectivity conservation management at the Institute for Technology and Resources Management in the Tropics and Subtropics at TH Köln (University of Applied Sciences), Germany. Her current research focuses on tropical ecosystems in Latin America and Southeast Asia.

Fabrice G. Renaud is the head of the Environmental Vulnerability and Ecosystem Services Section at the United Nations University Institute for Environment and Human Security in Bonn, Germany, where he and his team are engaged in research and capacity development related to Eco-DRR, with a particular focus on coastal environments.

Simone Sandholz is assistant professor at the Institute of Geography, University of Innsbruck, Austria, where she is part of the working groups on development studies and sustainability science as well as natural hazards research. Her research focus is on urban areas of the Global South, in particular, on sustainable urban planning, culturally appropriate urban regeneration and risk management.

Junun Sartohadi is a professor in geography at Universitas Gadjah Mada, Yogyakarta, Indonesia. He is a specialist for disaster risk management, soil geomorphology and GIS applications in Indonesia and published several articles in international journals.

Anja Schelchen is a research associate at the Centre for Rural Development, Humboldt-Universität zu Berlin. She works on capacity development with focus on applied science and environment-related issues in Brazil and Africa.

Torsten Schlurmann is the head of Franzius Institute for Hydraulic, Estuarine and Coastal Engineering, Leibniz University of Hannover, Germany, where he and his team work on coastal engineering, nonlinear wave theories, time-frequency analysis methods, erosion and sedimentation processes, integrated coastal zone management (ICZM), marine environments and impacts, port and harbour design, scour phenomena and protection, offshore wind energy converters (OWEC) as well as risk and vulnerability assessment in coastal zones.

Nannina Schulz is a PhD student at Franzius Institute for Hydraulic, Estuarine and Coastal Engineering, Leibniz University of Hannover, Germany, and focuses on wave loads on offshore structures and particularly investigates the influence of marine growth on the structures by physical and numerical hydraulic modelling.

Massimiliano Schwarz is a forest engineer with a PhD on the contribution of root reinforcement in slope stability and currently works as scientific collaborator and lecturer in the group of mountain forests and natural hazards at the Bern University of Applied Sciences, Switzerland.

Satoquo Seino is an associate professor of Kyushu University, Japan. Her studies are on habitat restoration based on coastal ecological engineering. She has joined many national and local coastal environmental and disaster restoration policy formation and multi-stakeholder participation projects.

Ahmed Senhoury holds a PhD in geomorphology. He is a professor at the University of Nouakchott in Mauritania. A researcher in the field of marine and coastal environment, Mr. Senhoury had been the coordinator of a research component of the Regional Partnership for West Africa Coastal and Marine Conservation (PRCM). Since 2008, he is the director of PRCM.

Ipsita Sircar coordinates Partners for Resilience – India programme implementation. She works with Wetlands International South Asia on livelihood aspects of wetland management.

Talia Smith is a researcher at the Earth Institute of Columbia University in New York where she works on institutional policy issues of forest and peat fire management in Indonesia. Her research is also focused on disaster risk reduction, climate adaptation and sustainable development efforts across South and Southeast Asia.

Shiv Someshwar is the director for climate policy at the Center for Globalization and Sustainable Development (CGSD), Columbia University, and senior advisor for regional programmes at the International Research Institute for Climate and Society (IRI), Columbia University. He is an expert in climate and development policy. His focus has been on how policy makers and the institutions they represent deal with climate risk. His work covers amongst others fire risk management in Indonesia and adaptation financing in the Caribbean and Latin America.

Karen Sudmeier-Rieux is a senior researcher at the University of Lausanne, Institute of Earth Science, and lead of the Eco-DRR thematic group of the International Union for Conservation of Nature's Commission on Ecosystem Management. She is co-developer of the MOOC: "Disasters and Ecosystems: Resilience in a Changing Climate" and has published extensively on the topics of disaster resilience and community-based landslide risk reduction in Nepal.

Kazuhiko Takemoto is director of the United Nations University Institute for the Advanced Study of Sustainability (UNU-IAS) and a visiting senior research fellow of IR3S at the University of Tokyo, where he has been fully engaged in developing global environmental policies in climate change and biodiversity.

Kazuhiko Takeuchi is senior vice-rector of the United Nations University and director and professor of the Integrated Research System for Sustainability Science

(IR3S) at the University of Tokyo. His research focuses on establishing a sustainable and resilient society through a transdisciplinary approach.

Hiroaki Teshima is a Deputy Director of the National Park Division at the Ministry of the Environment, Japan, where he is engaged in national parks management policies including Eco-DRR.

Hoang Ho Dac Thai is the deputy director of the Institute of Natural Resources and Environmental Science at Hue University and has a strong track record in advisory services in forest landscape ecology, silvicultural treatments in tropical forests and restoration of coastal dune systems in Vietnam.

Yves-François Thomas holds a PhD in geomorphology. He is a senior researcher in the field of marine geomorphology and sedimentology at the Laboratory of Physical Geography of the National Centre for Scientific Investigation (CNRS, France). He is a specialist in the modelling of dynamics of beaches and dunes. He works today as an independent consultant.

Benjamin S. Thompson is a PhD candidate in the Department of Geography, National University of Singapore. His research interests centre on the conservation of coastal ecosystems, specifically mangrove forests and coastal fisheries. He is particularly interested in neoliberal approaches to conservation including payments for ecosystem services and ecotourism.

Nicholas Turner is a programme officer at the United Nations University Institute for the Advanced Study of Sustainability (UNU-IAS), where he supports the director in the management and development of the institute. His research interests lie in multilateral institutions, global governance, sustainable development and human rights.

David Vačkář is head of the Department of Human Dimensions of Global Change at the Global Change Research Centre, Academy of Sciences of the Czech Republic (CzechGlobe). His research focuses on multiple aspects of human-environment interactions, especially ecosystem services, climate change adaptation, anthropogenic transformation and environmental security.

Myra D. van der Meulen works as a researcher/consultant of marine ecology at Deltares. She is currently involved in a range of international projects, mainly for the European Union. She has been working on nature-based flood defence for several years now, including a project on developing a testing, monitoring and management plan for a soft dike.

Bregje K. van Wesenbeeck is an expert on nature-based flood risk management at Deltares in the Netherlands and at the Delft University of Technology. She is involved in multiple disaster risk reduction projects that aim at reducing flood risk in coastal areas in Indonesia, the Philippines and Colombia.

Cees van Westen is an associate professor in the Department of Earth Systems Analysis at the Faculty of Geo-information Science and Earth Observation (ITC), University of Twente, Enschede, the Netherlands. His research focus is on landslides and the use of spatial information for natural hazard and disaster risk management.

Marie-Jose Vervest is Wetlands International's programme head for community resilience. She leads Wetlands International's network engagement in programmes aimed at building community resilience using ecosystem-based approaches.

Marta Vicarelli is an assistant professor of economics at the University of Massachusetts Amherst. She is contributing author to the IPCC Fourth Assessment Report. Past appointments include research fellowships at the NASA Goddard Institute for Space Studies, the Harvard University Center for International Development and the Yale University Climate and Energy Institute.

Adam W. Whelchel is a senior scientist and practitioner immersed in the fields of ecosystem and urban resilience with The Nature Conservancy, where he and his team catalyse partnerships with strategic direction, build resilient communities and transform conservation around the globe.

Chapter 1
Developments and Opportunities for Ecosystem-Based Disaster Risk Reduction and Climate Change Adaptation

Fabrice G. Renaud, Udo Nehren, Karen Sudmeier-Rieux, and Marisol Estrella

Abstract In the past few years, many advances in terms of research, implementation and policies have taken place around the world with respect to understanding, capturing and facilitating the uptake of ecosystem-based approaches for disaster risk reduction (DRR) and climate change adaptation (CCA). We highlight some of these advances here, particularly for coastal (various hazards), riverine (floods), and mountain (landslides) environments. We also highlight that many international agreements reached in 2015 can facilitate the uptake of these approaches whereas ecosystem-based solutions can facilitate the achievement of many goals and targets related to DRR, CCA, and/or sustainable development enclosed in these agreements. Finally, the chapter provides an overview of the rest of the book.

1.1 Introduction

The role of ecosystems for disaster risk reduction (DRR), climate change adaptation (CCA) and development is increasingly recognised globally. In the short time since 2013 when the book "The role of ecosystems for disaster risk reduction" was

F.G. Renaud (✉)
United Nations University Institute for Environment and Human Security (UNU-EHS),
Platz der Vereinten Nationen 1, 53113 Bonn, Germany
e-mail: renaud@ehs.unu.edu

U. Nehren
Institute for Technology and Resources Management in the Tropics and Subtropics (ITT),
TH Köln, University of Applied Sciences, Köln, Germany

K. Sudmeier-Rieux
Commission on Ecosystem Management International Union for Conservation of Nature,
Gland, Switzerland

M. Estrella
Post-Conflict and Disaster Management Branch, United Nations Environment Programme,
Geneva, Switzerland

© Springer International Publishing Switzerland 2016
F.G. Renaud et al. (eds.), *Ecosystem-Based Disaster Risk Reduction and Adaptation in Practice*, Advances in Natural and Technological Hazards Research 42,
DOI 10.1007/978-3-319-43633-3_1

published (Renaud et al. 2013), tremendous developments have taken place in the field of ecosystem-based DRR (Eco-DRR) research, policies, and implementation on the ground. Some of these new insights were discussed at a workshop[1] co-organised, among others, by the Partnership for Environment and Disaster Risk Reduction (PEDRR), the Centers for Natural Resources and Development (CNRD), and the Indonesian Institute of Sciences (LIPI) in Bogor, Indonesia, in June 2014. The workshop focused on the role of ecosystems for disaster risk reduction and climate change adaptation (Eco-DRR/CCA) and had four main themes, namely (i) Evidence and economics of Eco-DRR/CCA; (ii) Decision making tools for Eco-DRR/CCA; (iii) Innovative institutional arrangements and policies for Eco-DRR/CCA; and (iv) Cutting edge scientific research and technical innovations on Eco-DRR/CCA. These themes were selected as they addressed some of the gaps that were identified in the first book (see Estrella et al. 2013) and now loosely provide the structure for this volume. Chapters written for this book emanate both from participants in the workshop and from invited authors.

2015 has been a critical year in terms of major global agreements and advancing international recognition of ecosystem-based approaches to DRR and CCA: first in March, the Sendai Framework for Disaster Risk Reduction (or SFDRR; UN 2015a) was approved in Sendai, Japan, replacing the Hyogo Framework for Action (UNISDR 2005). In September the UN General Assembly adopted the Sustainable Development Goals (or SDGs; UN 2015b). Finally in December, a new agreement to address climate change was reached in Paris (UNFCCC 2015). Ecosystems and ecosystem services are critical for helping achieve disaster risk reduction, sustainable development and climate change mitigation and adaptation, and this is now recognised by these agreements and others (Fig. 1.1).

In the last couple of decades, the number of concepts on the use of ecosystems for DRR, CCA and sustainable development has rapidly increased, and concepts such as Ecosystem-based Adaptation (EbA), Ecosystem-based Disaster Risk Reduction, Nature-based Solutions, Green Infrastructures, Working with Nature, and many more have emerged or been further developed (see Box 8.1 in van Wesenbeeck et al., Chap. 8). This recognition has facilitated increased implementation of Eco-DRR/EbA projects on the ground. Nonetheless, the variety of ecosystem-based concepts and definitions has generated some confusion, particularly for practitioners and policymakers.

With rapid progress made on concepts, policies, and implementation, it is perhaps time to take stock again on where we stand with respect to Eco-DRR/CCA. This is the purpose of this book, which was produced at a time when the three major global agreements mentioned above were being negotiated and agreed upon. In the next sections of this chapter, we will briefly discuss the concept of Eco-DRR/CCA, and show how in recent years the concept and other related ones have been promoted in research and practice. We will provide insights into some of the scientific advances related to coastal, riverine and forest ecosystems and their role

---

[1]http://pedrr.org/training/current-event/international-science-policy-workshop-bogor-2014/

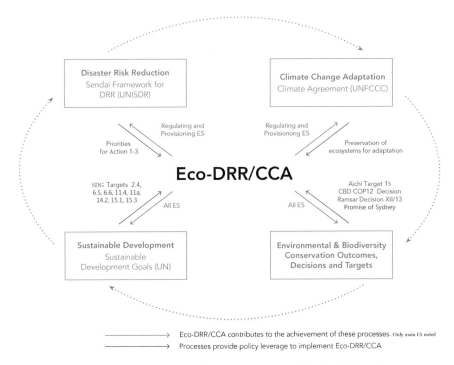

Fig. 1.1 Linkages between major international agreements and Eco-DRR/CCA. ES means ecosystem services

in disaster risk reduction and finally, present the structure and chapters of the book. Opportunities for the further development of Eco-DRR/CCA concepts and practice are discussed in more detail in the concluding chapter (Estrella et al. Chap. 24).

1.2   What Do We Mean by Eco-DRR/CCA?

Two key concepts feature in most of the chapters of this book: Eco-DRR and EbA. Definitions for each are given in Box 1.1. The two definitions are very similar (i.e. with a focus on ecosystem management, conservation and restoration for specific objectives and linking these to sustainable development), given that the Eco-DRR definition developed in 2013 drew on the existing definition of EbA which pre-dated it. One important difference is that one concept specifically addresses DRR and the other CCA. However, it can be easily argued that there are more similarities between the concepts than divergence, especially when addressing climate-related hazards (Doswald and Estrella 2015). Another key feature of both concepts, even if not spelled out explicitly in the definitions, is the fact that the approaches provide multiple benefits, beyond strictly DRR and CCA functions.

> **Box 1.1: Definitions of Eco-DRR and EbA**
> **Ecosystem-based Disaster Risk Reduction** (Eco-DRR) "is the sustainable management, conservation, and restoration of ecosystems to reduce disaster risk, with the aim of achieving sustainable and resilient development" (Estrella and Saalismaa 2013:30).
> **Ecosystem-based Adaptation** (EbA) "is the use of biodiversity and eco-system services as part of an overall adaptation strategy to help people to adapt to the adverse effects of climate change. Ecosystem-based adaptation uses the range of opportunities for the sustainable management, conservation, and restoration of ecosystems to provide services that enable people to adapt to the impacts of climate change. It aims to maintain and increase the resilience and reduce the vulnerability of ecosystems and people in the face of the adverse effects of climate change. Ecosystem-based adaptation is most appropriately integrated into broader adaptation and development strategies" (CBD 2009:41).

Although the definition of Eco-DRR does not include a reference to climate change, it was always considered that Eco-DRR could also contribute to climate change adaptation, as climate change is considered to be a risk amplifier now and in the future (Estrella and Saalismaa 2013). However, in this chapter, to be more explicit, we use the acronym Eco-DRR/CCA in order to emphasise that ecosystem-based approaches play a role for achieving *both* DRR and CCA. Therefore, we define Eco-DRR/CCA as: **"the sustainable management, conservation, and restoration of ecosystems to reduce disaster risk and adapt to the consequences of climate change, with the aim of achieving sustainable and resilient develop-ment"**. Although we use the term Eco-DRR/CCA in this chapter, authors of subsequent chapters have been given total freedom to elaborate on and use termi-nology that best describes their work.

1.3   Eco-DRR/CCA Gaining Steam Globally

Ecosystems for DRR and/or CCA have been advocated in many "commentaries" and "perspectives" of leading journals, particularly for coastal systems. For example, Barbier (2015) discussed in the journal *Nature* the feasibility of having three lines of coastal defenses: green and grey infrastructure as well as local stakeholders' engage-ment with a potential for application globally. This builds on an earlier perspective in *Science* where restoration of coastal ecosystems was considered a necessary step for long-term coastal adaptation (Barbier 2014). Again in *Science*, the case for "nature-based engineering solutions" in delta environments was made by Timmerman and Kirwan (2015), building on an earlier perspective in *Nature* encouraging a broader consideration of ecosystem-based, coastal defenses (Timmerman et al. 2013).

In *Nature Climate Change*, Cheong et al. (2013) discussed the role of ecological engineering for coastal adaptation. Finally, Martin and Watson (2016) made a general plea in *Nature Climate Change* for the consideration of ecosystems in adaptation to climate change. Furthermore, many scientific papers have been published on the topic during the last few years, some of them are reviewed in Sect. 1.4 of this chapter.

In addition to articles in scientific journals, many other publications related to ecosystem-based approaches have recently been published. Without intending to be exhaustive, a few can be mentioned. A recent example is a technical report by the European Environment Agency on Green Infrastructures as an option to mitigate climate change-related hazards, with a specific focus on landslides, avalanches, floods, and storm surges (EEA 2015). On the occasion of the 2014 World Parks Congress, the International Union for the Conservation of Nature (IUCN) published 18 case studies illustrating the interlinkages between protected areas and DRR and CCA (Murti and Buyck 2014). Ecoshape also showcased other examples such as oyster reefs to mitigate erosion, seabed landscaping to boost biodiversity, and more generally, the multiple benefits provided by Nature-based Solutions (De Vriend and Van Koningsveld 2012).

Technical and general guidelines are also increasingly being published. Examples include the role of protected areas for DRR (Dudley et al. 2015) which was released during the World Conference on Disaster Risk Reduction; the development of hybrid solutions for large scale coastal erosion control (Winterwerp et al. 2014); the use of mangroves (Spalding et al. 2014a) or natural and nature-based features (Bridges et al. 2015) for coastal protection and resilience; and comparisons of ecosystem-based and engineering solutions for coastal protection in Fiji (Rao et al. 2012).

In addition, and linked to the work leading to some of the publications above, many initiatives around the world have been developed that consider ecosystems as stand-alone solutions or as a component of hybrid solutions for DRR and/or CCA. Naming just a few and in no particular order: Living shorelines to restore America's estuaries[2]; the Building with Nature programme in Indonesia[3]; and the Coastal Resilience programme[4] (Beck et al. 2013).

As noted in the introduction, many positive developments have also taken place on the policy front. Ecosystems are mentioned as playing a critical role for DRR and CCA, a fact highlighted or reinforced in many recent international agreements. The role of ecosystems or of the environment features in numerous places in the Sendai Framework for Disaster Risk Reduction (SFDRR) (UN 2015a); they also play a critical role for many of the Sustainable Development Goals (SDGs) (UN 2015b); environmental or ecosystem integrity is mentioned in several places of the Paris Agreement (UNFCCC 2015); the Convention on Biological Diversity also puts an important emphasis on ecosystem-based solutions for CCA and DRR in

---

[2]https://www.estuaries.org/living-shorelines (accessed Oct 2015)

[3]http://www.ecoshape.nl/overview-bwn.html (accessed Oct 2015)

[4]http://coastalresilience.org/ (accessed Oct 2015)

a decision reached during the 12th Conference of the Parties (CBD 2014); and the Ramsar Convention on Wetlands adopted resolution XII.13 on "wetlands and disaster risk reduction" at its last Conference of the Parties in 2015 (Ramsar 2015). Figure 1.1 shows the possible linkages (the list is not exhaustive) between major international agreements and Eco-DRR/CCA.

There is clearly increasing interest in ecosystem-based solutions for DRR and CCA globally. In the next section, some recent scientific advances are further described for coastal protection, flood protection, and landslide risk reduction.

1.4 Progress on the Science Front

1.4.1 Coastal Ecosystems for Coastline Protection

Coastal social-ecological systems are exposed to various types of hazards (e.g. tropical cyclones, storm surges, tsunamis, flooding, erosion, sea-level rise) and are relatively vulnerable because of a variety of factors such as increasing population densities linked to urban expansion, and high levels of economic activities (e.g. Nicholls et al. 2008). As can be inferred from Sect. 1.3 of this chapter and in Chaps. 13, 14, 18, 19 and 20, many Eco-DRR/CCA activities have taken place or are being planned in coastal environments, particularly linked to the rehabilitation or conservation of coastal ecosystems, such as mangroves and sand dunes (Cunniff and Schwartz 2015; Gedan et al. 2011; Temmerman et al. 2013). Lacambra et al. (2013) provided a comprehensive review of the multiple roles of mangroves in terms of coastal protection. In the span of several years, many additional publications on the subject have emerged addressing the multiple dimensions regarding the role of coastal vegetation in buffering populations and infrastructures against hazards but also in providing other ecosystem services. Examples include reviews highlighting:

- the multiple benefits coastal ecosystems provide in the context of DRR such as reducing flooding and erosion, the ability of many ecosystems to self-repair or adapt to changing environmental factors, and the cost-effectiveness of ecosystem-based solutions (e.g. Spalding et al. 2014b);
- the critical role of coastal vegetation (e.g. mangroves, salt marshes, seagrasses) in terms of climate change mitigation (carbon sequestration) and adaptation (e.g. dissipation of wave energy, elevation of the land or the sea floor, sediment trapping, protection against coastal flooding and erosion) (Duarte et al. 2013). Mangroves, in particular, can store large amounts of carbon (Wicaksono et al. 2016), particularly below ground (Donato et al. 2011), and their destruction can lead to large emissions of carbon to the atmosphere (e.g. Murdiyarso et al. 2015);
- the reduction in height of wind and swell waves by mangroves (McIvor et al. 2012a, 2015);

- the linkages between mangrove presence and their ability to reduce storm surge peak water levels, flow speed and surge damage behind mangroves (McIvor et al. 2012b);
- the ability of mangroves, in many circumstances, to keep pace with local sea level rise (Duarte et al. 2013; McIvor et al. 2013) as long as there is a sustainable supply of sediment and organic matter (see also Alongi 2008). In addition, mangroves can migrate landward when facing e.g. rising sea levels but only if there are no obstacles behind them such as natural features or human infrastructure (Alongi 2008; Lovelock et al. 2015).

All these studies emphasise the fact that the cause-effect relationship between ecosystems and disaster risk reduction can be highly localised as multiple factors are at play when considering wave attenuation effects or increases in elevation of the land. Regarding the latter, Lovelock et al. (2015) noted that in 70 % of sites surveyed in the Indo-Pacific region, sea-level rise exceeded soil surface elevation gains. Nevertheless, based on these new insights and an increasing body of empirical evidence not reviewed here, several technical guidelines for experts and policymakers have been and are currently being developed (e.g. Spalding et al. 2014a; Dudley et al. 2015). Five chapters in this book discuss in varying details the role of coastal vegetation for DRR: Friess and Thompson (Chap. 4); van Wesenbeeck et al. (Chap. 8); Furuta and Seino (Chap. 13); Takeuchi et al. (Chap. 14); and David et al. (Chap. 20).

Another important type of ecosystem in the context of DRR are coastal dune systems (CDS) which provide a variety of ecosystem services, and in particular the physical buffer function that protects inland areas from coastal hazards such as tropical cyclones, storm surges, and coastal floods (Hettiarachchi et al. 2013). Coastal dunes can even prevent or at least mitigate tsunami impacts depending on the circumstances (Liu et al. 2005; Bhalla 2007; Mascarenhas and Jayakumar 2008). Furthermore, intact CDS control geomorphological processes such as coastal erosion (Prasetya 2007; Barbier et al. 2011) and mitigate effects of sea level rise and saltwater intrusion (Heslenfeld et al. 2004; Saye and Pye 2007). The effectiveness for hazard mitigation and long-term adaptation to climate change depends on the integrity of the protective ecosystem services. These are composed of the physical conditions, in particular height, width, shape and continuity (Gómez-Pina 2002; Takle et al. 2007; Thao et al. 2014), the ecological status (Nehren et al. in Chap. 18), and the dynamics of the CDS.

Despite their importance for coastal protection and CCA, losses and degradation of CDS are widespread phenomena around the globe, mainly triggered by urbanisation processes, overexploitation, mining, and tourism (Martínez et al. 2004). The growing global demand for sand and gravel (Peduzzi 2014) will most probably lead to intensified sand mining activities along beaches and shorelines in the near future, and further accelerate degradation processes in many coastal regions of the world – with severe consequences for biodiversity and the livelihoods and vulnerability of coastal communities.

In many mid-latitude countries, particularly in Europe and the USA, the problem has been recognized, and conservation and restoration measures for CDS have been established or are underway (Doody 2013). In these countries, current research related to DRR, CCA and ecosystem management of CDS focuses among others on mid- to long-term effects of climate change – in particular sea level rise and storm intensities – on morphology, species composition, and habitat losses of CDS (e.g. Feagin et al. 2005; Psuty and Silveira 2010; Prisco et al. 2013; Seabloom et al. 2013; Pakeman et al. 2015). Another research line deals with the protective services of CDS as well as conservation and restoration measures to maintain or restore these services (e.g. Feagin et al. 2010; Hanley et al. 2014; Sigren et al. 2014).

In tropical and subtropical countries, the databases on CDS and their role in coastal protection and adaptation are often very limited. Even though CDS of tropical and subtropical regions are frequently described as degraded (Moreno-Casasola et al. 2008), these assessments are often based on geographically restricted field studies and observations, so that inferences to larger areas are not possible. Due to the lack of ground-based data particularly in tropical and subtropical countries, there is as yet no global overview on the ecological status and change patterns of CDS. Considering the significance of CDS for coastal protection, climate change adaptation and biodiversity conservation, there is an urgent need to foster research and action with respect to the status and management of CDS in developing and emerging countries, where livelihoods of coastal dwellers are most affected. Furthermore, in-depth research on the protective and other ecosystem services of CDS are needed for a more targeted implementation of conservation, restoration and sustainable use measures. Finally, policymakers need to be convinced that in many cases the short-term benefits of sand dune exploitation are associated with higher costs for coastal protection in the long run. This requires an improved database on the socio-ecological system including the valuation of ecosystem services of CDS.

1.4.2   Riverine Ecosystems for Floods Protection

Flooding is the hazard that causes the majority of disasters and economic losses. Between 1994 and 2013, floods accounted for 43 % of all recorded events and affected nearly 2.5 billion people (EM-DAT 2015). In addition to higher concentrations of populations in flood plains, more extreme precipitation is one of the hazards likely to become more frequent due to climate change (IPCC 2014). Reducing flooding can be very costly, and mitigation measures range from high-technology structural engineered flood defenses around densely populated areas, to non-structural measures such as early warning systems or floodplain zoning (Senhoury et al. Chap. 19). Along with increasing numbers of flooding events, high economic losses and the uncertainty that flood defenses are inadequate to protect against increasing flood risk, a shift is occurring to consider more integrated flood risk management, including natural flood defenses (Bubeck et al. 2015; Day et al. 2007; van Wesenbeeck et al. 2014; van Staveren et al. 2013, van Wesenbeeck, Chap. 8). These include wetlands, lakes and rivers which have been restored to make "room for water" and can retain water in upper catchments and provide space for excess water (Bubeck et al. 2015). The importance of Integrated Water Resources Management (IWRM) which considers water management issues in watersheds and river basins was especially highlighted in the SFDRR.

However, the uptake of integrated approaches varies considerably among countries, depending on the frequency of flooding events and the public demand and support for certain types of flood risk management (Bubeck et al. 2015). The major floods which struck Europe between 1998 and 2004 led to several important European Union directives, including the Water Framework Directive (EC 2000) and the Flood Directive (EC 2007). The Water Framework Directive, in particular, is one of the few directives with a dual ecological and DRR component while promoting an integrated approach to water and drought risk management. It points to the need to achieve a balance between ecological requirements and the need for drought measures and flood defense based on good ecological science (Sudmeier-Rieux 2013). As a result of these two directives, a number of countries, notably the Netherlands, U.K., Germany, Belgium and France developed programmes, which promoted the use of wetlands, rivers and other natural spaces as reservoirs for excess water. One example is the Netherlands' "Room for the River", a €2.3 billion programme which was conceived to create more space for the rivers while improving flood protection, recreation possibilities and improved environmental quality of rivers in the country. According to the main government agency overseeing the project, in addition to flood protection, any extra space created for the rivers will also remain permanently available for this purpose and for other recreational and ecological functions (Dutch Ministry of Water Management, Transport and Public Works 2012). Although not part of the EU but following this paradigm shift in flood management, the Government of Switzerland's third Rhone River Correction programme is a 30-year initiative which will allow to control potential flood damages, re-establish and strengthen biological functions of the river and maintain social and economic priorities along the upper catchment of the Rhone River (between the town of Brig and the mouth of river in the Canton of Vallis) (Pahl-Wostl et al. 2006).

Global estimates of inland (freshwater) wetlands vary between 5.3 and 9.5 million km^2 but are also considered underestimated (Russi et al. 2013). The Economics of Ecosystem and Biodiversity (TEEB) report on water and wetlands (Russi et al. 2013) has estimated that inland wetlands (floodplains, swamps, marshes and peatlands) provided regulating services of 23,018 USD/hectare/year and a total of 44,597 USD/hectare/year. This value does not consider the many non-monetary values that wetlands provide, such as aesthetics, rich biodiversity, educational, cultural and recreational ecosystem services.

The core of the new flood risk management paradigm is a recognition of ecosystem services in attenuating flooding, which needs to be based on a careful scientific analysis of the linkages between wetlands and flooding (Janssen et al. 2014; van Wesenbeeck et al. 2014). According to van Eijk et al. (2013), river basins are highly dynamic systems, and the periodic rise and fall of floodwaters is a normal pulsing feature in the river landscape. The role of wetlands in regulating floods is far from universal and will depend on the scale of the flood event, the size and health status of the wetlands, its location in a river basin and local climate. Depending on the study, wetlands can both contribute to flood reduction and increase it (van Eijk et al. 2013). This points to a wide heterogeneity of ecosystem services related to flood attenuation, which requires more localised expertise and study. Thus according to the situation:

- Peatlands, wet grasslands and other wetlands can store water and release it slowly, reducing the speed and volume of runoff after heavy rainfall or snowmelt in springtime (Brouwer and van Elk 2004; Javaheri and Babbar-Sebens 2014)
- Marshes, lakes and floodplains release wet season flows slowly during drought periods and can contribute to recharging ground water (Maltby 2009; Wilson et al. 2010)

However despite their many benefits, wetlands face severe pressures especially due to land conversion, development of dams, eutrophication and pollution due to intensification of agriculture. In Europe, 80 % of wetlands have disappeared over the past 75 years, as compared to 50 % in North America (van Verhoeven 2013). In 2012, 28 % of 127 governments reporting to the Ramsar Convention stated that their wetlands had deteriorated, while only 19 % reported any improvements (Russi et al. 2013).

1.4.3   Protection Forests for Landslide Risk Reduction

From the geological and geomorphological viewpoint, landslides can be principally considered natural phenomena, which are usually triggered by rainfall or earthquakes. However, human interference, such as road construction, quarrying, deforestation, agricultural practices in mountainous terrain, can contribute to or aggravate their destructive forces (Dolidon et al. 2009; Walker and Shiels 2013). Another important root cause for landslides is the change of the vegetation cover (Papathoma-Koehle and Glade 2013). To mitigate in particular the risk of shallow landslides (i.e. with a depth of 2–10 m), conservation and restoration of vegetation (e.g. from grasses with deep roots to mountain forests) are recommended, often combined with engineered structures such as fences and debris flow barriers (Dietrich et al. 1998; Wehrli and Dorren 2013).

The effectiveness of protection forest depends on various factors, such as the hazard type, the geological and topographical setting, the location of the forest, its tree composition and dynamics, as well as management aspects (Wehrli and Dorren 2013). There are many experiences with respect to the creation and maintenance of protection forests particularly in Europe and the US, where protection forests are not only used for landslide risk reduction, but also as buffers against rockfall, avalanches, debris flows, flooding and erosion (Brang et al. 2006). A prominent example is found in the Swiss Alps, where protection forests are a main component of the national disaster risk reduction programme, and the Government spends over USD 120 million per year on the management of its protective forests (Wehrli and Dorren 2013). However, the planning process takes a time span of 50–100 years and requires public willingness to contribute to the forests' maintenance. On the other hand, Wehrli and Dorren (2013) point out that the creation and maintenance of protection forest cost 5–10 times less than structurally engineered structures over time.

Current research on protection forests is concentrated in Europe, North America, Australasia, and Japan and focuses among others on the ideal composition of tree species to maximise the degree of protection. Models that take into account the structural diversity and species composition include parameters that have a major impact on slope stabilisation, such as root density, root tensile strength, and root orientation (Danjon et al. 2008; Mao et al. 2012; Preti 2013). These models build on studies on root systems of different tree species in various environments (e.g. Schmid and Kazda 2001, 2002; Roering et al. 2003; Bischetti et al. 2005, 2009; Mattia et al. 2005; De Baets et al. 2008; Abdi et al. 2009) and works on root characteristics (Stokes et al. 2009). Other models include the effects of vegetation, reinforcement and hydrological changes (Greenwood 2006), forest structure (Kokutse et al. 2006) and hydro-mechanical effects of different vegetation types (González-Ollauri and Mickovski 2014). Important research along these lines include the impact of successional stages and plant density for landslide control (Cammeraat et al. 2005; Pohl et al. 2009; Loades et al. 2010), management aspects of protection forests (Dorren et al. 2004; Schönenberger et al. 2005; Brang et al. 2006; Runyan and D'Odorico 2014; Basher et al. 2015), and geomorphologically-controlled variations of ecological conditions on root reinforcement (Hales et al. 2009). A quantitative tool developed to determine the slope stabilising effect of protection forests in Switzerland is presented by Dorren and Schwarz (Chap. 11).

Within the last years, the potential of protection forests for landslide risk reduction has also been recognised in developing countries and emerging economies, and several projects have been implemented, often together with local communities. In this context, Anderson et al. (2014: 128) stress the implementation challenges of community-based landslide risk reduction measures in developing countries and point out "the need for disaster risk reduction researchers and practitioners to develop future environmental scenarios as the basis for modeling landslide triggers in vulnerable communities."

For landslide-affected areas in Asia and the Pacific, the FAO (2013) published a report that provides a good overview of the affected regions and shows strategies for effective risk management, with a focus on protection forests and land management practices. For Dolakha District in central-eastern Nepal, Jaquet et al. (2013) analysed landslides trends and demonstrated that proper management of community forests significantly contributes to slope stabilisation and thereby reduces the risk of shallow landslides. For China, there are also some studies that focus on floristic and vegetational aspects, in particular the root systems of different forest types (Genet et al. 2010; Ji et al. 2012).

Also in Latin America as well as in Sub-Saharan Africa, the role of forests and good agricultural management including slope terracing, agroforestry, and silvopastoral systems for landslide and flood prevention has become increasingly recognised. However, the number of scientific publications, in particular with respect to ground-based data, is still limited. Among the few publications that exist are those by Anderson et al. (2011) on community-based landslide risk reduction in the Eastern Caribbean; Lange et al. (Chap. 21 in this book) on risk perception for participatory ecosystem-based adaptation to climate change in the Atlantic Forest of Rio de Janeiro State; Lange et al. (2016) on the potential of ecosystem-based measures for landslide risk reduction in the city of Rio de Janeiro; and some studies that have been conducted on landslides in the Mt. Elgon area (Bintoora 2015).

The Eco-DRR/CCA advances reviewed above for coastal, floodplain and mountain environments show the increase interests of the scientific and practitioner communities on the concept. However, much more knowledge remains to be generated to fully understand the role ecosystems can play in mitigating hazards of different magnitudes and frequencies and in helping societies adapt to climate change. This could be further facilitated in the future by the recognition of the role of ecosystems for DRR, CCA and development in major international agreements (Fig. 1.1). Further advances, practical examples, and suggestions for the way forward for Eco-DRR/CCA are presented in the following chapters of the book.

1.5 Structure of the Book

This book comprises 24 chapters divided into four main sections as well as an overall introduction (this Chapter) and an overall conclusion by Estrella et al. (Chap. 24) which summarizes the main points developed throughout the book, and discusses emerging issues related to the four themes mentioned earlier in this chapter.

Part I, entitled "Economic approaches and tools for Eco-DRR/CCA" is composed of four chapters, which examine how best to capture, from an economic perspective, the multiple benefits generated by Eco-DRR approaches. Emerton et al. (Chap. 2) present and discuss a conceptual framework for the integration of ecosystem values in development planning in the context of climate change. Applications of the framework are presented for coastal areas in Kenya and Sri Lanka. Vicarelli et al. (Chap. 3) make the case for the consideration of cost-benefit analyses for Eco-DRR and EbA projects, by providing a detailed review of best practices and providing examples from case studies. Friess and Thompson (Chap. 4) discuss the concept of Payment for Ecosystem Services for mangroves in the context of DRR, outlining some of the pre-requisites that are necessary for these types of schemes to work efficiently. Finally, Harmáčková et al. (Chap. 5) present a case study in the Czech Republic where participatory scenario building, GIS modelling and economic evaluation were used to analyze economic costs and benefits of adaptation scenarios.

Part II of the book entitled "Decision-making tools for Eco-DRR/CCA" comprises seven chapters. Whelchel and Beck (Chap. 6) provide, through the analysis of case studies, lessons learned and recommendations related to decision support tools and approaches for Eco-DRR and EbA. In Chap. 7, Krol et al. provide an overview of the use of geo-information tools for Eco-DRR and how they can be used to compare different DRR options. The decision support tool RiskChanges is also presented. Van Wesenbeeck et al. (Chap. 8) present approaches which could better integrate the role of ecosystems in coastal flood risk management engineering projects and, by doing so, provide additional incentives for coastal engineers to consider ecosystem-based solutions for coastal flood management. Kloos and Renaud (Chap. 9) review ecosystem-based approaches for drought risk reduction, with a focus on Sub-Saharan Africa. The chapter also presents some criteria to determine when approaches can be considered ecosystem-based. In Chap. 10, Bayani and Barthélemy show how the Integrated Valuation of Ecosystem Services and Tradeoff (InVEST) tool can be used to assess ecosystems and disaster risk in data-scarce environments, with examples from Haiti and the Democratic Republic of the Congo. In their chapter, Dorren and Schwarz (Chap. 11) present a quantitative tool called SlideforNET which was developed to determine the slope stabilising effect of protection forests in Switzerland. In the last chapter of Part II, Kumar et al. (Chap. 12) describe a cluster approach used for disaster risk reduction planning in the Mahanadi Delta, India.

Part III of the book entitled "Innovative institutional arrangements and policies for Eco-DRR/CAA" is composed of five chapters. The first two chapters (Furuta and Seino; Takeuchi et al.) address the integration (or lack thereof) of ecosystem-based approaches in the rebuilding process in the aftermath of the 2011 Great East Japan Earthquake (GEJE). In both chapters, the debates and policies enacted after this disaster are discussed in detail. Furuta and Seino (Chap. 13) also describe the role that ecosystems played during the GEJE. In addition to the GEJE case study, Takeuchi et al. (Chap. 14) showcase the multiple benefits of Eco-DRR activities in two other regions of the world, Ghana and Myanmar. Sandholz (Chap. 15) addresses urban disaster risk reduction through the example of Kathmandu Valley in Nepal and illustrates how unplanned urban development, political instability and the non-enforcement of existing policies and laws constitute hurdles to the integration of ecosystem-based approaches in DRR. Kieft et al. (Chap. 16) discuss anticipatory management of peat fires in Indonesia and the integration of the concept into existing procedures of fire prevention and into spatial and development planning. The early warning system "Fire Risk System" is also presented. Finally, McNeely (Chap. 17) argues for the greater consideration of protected areas in national strategies linked to CCA and DRR and proposes various management approaches for protected areas in this context.

Part IV "Research and Innovation" has six chapters. Nehren et al. (Chap. 18) highlight the importance of coastal dune systems for DRR through case studies from three countries: Vietnam, Indonesia, and Chile. They also suggest indicators for assessing the degradation of coastal dune systems and for assessing ecosystem services. In Chap. 20, David et al. elaborate on the perspectives of coastal engineers on ecosystem-based coastal protection measures and highlight the multiple benefits as well as the limitations of "low-regret measures", such as green belts, coir fibers, and porous submerged structures. Senhoury et al. (Chap. 19) present an assessment of flood risk for Nouakchott, Mauritania, and highlight, among other things, the importance of preserving and restoring the coastal dune belt that can protect the city. Lange et al. (Chap. 21) present research results from a case study area in Brazil that focused on perception analysis to determine how to more effectively promote local community participation in Eco-DRR and EbA activities; the hazards considered in this chapter are landslides, mudslides and floods. Dhyani and Dhyani (Chap. 22) also address land degradation, but this time from the Indian Himalayas' perspective, and discuss the important role of forests for DRR, and critically, for improving local livelihoods. They show in detail the complex interactions between society and their natural environment and discuss the role that fodder banks can play in supporting livelihoods and ecosystems. Last but not least, Fedele et al. (-Chap. 23) discuss the role of forest ecosystems for livelihoods when disasters strike in Indonesia. Through an analysis of ecosystem services, they emphasise the roles that forests play in reducing the vulnerability of communities exposed to various hazards.

With this second book volume, we hope to spark ongoing dialogue, research and practice that advance global understanding and, most importantly, applications of ecosystem-based solutions for disaster risk reduction and climate change adaptation.

References

Abdi E, Majnounian B, Rahimi H, Zobeiri M (2009) Distribution and tensile strength of Hornbeam (Carpinus betulus) roots growing on slopes of Caspian Forests, Iran. J For Res 20(2):105–110

Alongi DM (2008) Mangrove forests: resilience, protection from tsunamis, and responses to global climate change. Estuar Coast Shelf Sci 76:1–13

Anderson MG, Holcombe E, Blake JR et al (2011) Reducing landslide risk in communities: evidence from the Eastern Caribbean. Appl Geogr 31:590–599

Anderson MG, Holcombe E, Holm-Nielsen N, Della Monica R (2014) What are the emerging challenges for community-based landslide risk reduction in developing countries? Nat Hazard Rev 15(2):128–139

Barbier EB (2014) A global strategy for protecting vulnerable coastal populations. Science 345 (6202):1250–1251

Barbier EB (2015) Hurricane Katrina's lessons for the world. Nature 524:285–287

Barbier EB, Hacker SD, Kennedy C et al (2011) The value of estuarine and coastal ecosystem services. Ecol Monogr 81(2):169–193

Basher L, Harrison D, Phillips C, Marden M (2015) What do we need for a risk management approach to steepland plantation forests in erodible terrain? NZ J For 60(2):7–10

Beck MW, Gilmer B, Ferdaña Z et al (2013) Increasing the resilience of human and natural communities to coastal hazards: supporting decisions in New York and Connecticut. In:

Renaud FG, Sudmeier-Rieux K, Estrella M (eds) The role of ecosystems in disaster risk reduction. UNU Press, Tokyo, pp 140–163

Bhalla RS (2007) Do bio-shields affect tsunami inundation? Curr Sci 93:831–833

Bintoora AK (2015) Initiatives to combat landslides, floods and effects of climate change in Mt Elgon Region. In: Murti R, Buyck C (eds) Safe havens: protected areas for disaster risk reduction and climate change adaptation. International Union for Conservation of Nature, Gland, pp 132–148

Bischetti GB, Chiaradia EA, Simonato T, Speziali B, Vitali B, Vullo P, Zocco A (2005) Root strength and root area ratio of forest species in Lombardy (Northern Italy). Plant Soil 278:11–22

Bischetti GB, Chiaradia EA, Epis T, Morlotti E (2009) Root cohesion of forest species in the Italian Alps. Plant Soil 324:71–89

Brang P, Schönenberger W, Frehner M et al (2006) Management of protection forests in the European Alps: an overview. For Snow Landsc Res 80(1):23–44

Bridges TS, Wagner PW, Burks-Copes KA et al (2015) Use of natural and nature-based features (NNBF) for coastal resilience. Engineer Research and Development Center, ERDC SR15-1, US Army Corps of Engineers

Brouwer R, van Elk R (2004) Integrated ecological, economic and social impact assessment of alternative flood control policies in the Netherlands. Ecol Econ 50:1–21

Bubeck P, Kreibich H, Penning-Rowsell EC et al (2015) Explaining differences in flood management approaches in Europe and in the USA – a comparative analysis. J Flood Risk Manag. doi:10.1111/jfr3.12151

Cammeraat E, van Beek R, Kooijman A (2005) Vegetation succession and its consequences for slope stability in SE Spain. Plant Soil 278:135–147

CBD (2009) Connecting biodiversity and climate change mitigation and adaptation: report of the second ad-hoc technical expert group on biodiversity and climate change. CBD Technical Series No. 41, Secretariat of the Convention on Biological Diversity, Montreal, 126p

CBD (2014) Decision adopted by the conference of the parties to the convention on biological diversity at its twelfth meeting. XII/20. Biodiversity and climate change and disaster risk reduction. UNEP/CBD/COP/DEC/XII/20, 17 October 2015

Cheong S-M, Silliman B, Wong PP et al (2013) Coastal adaptation with ecological engineering. Nat Clim Chang 3:787–791

Cunniff S, Schwartz A (2015) Performance of natural infrastructure and nature-based measures as coastal risk reduction features. Environmental Defense Fund, 35 pp

Danjon F, Barker DH, Drexhage M, Stokes A (2008) Using three-dimensional plant root architecture in models of shallow-slope stability. Ann Bot 101:1281–1293

Day JW, Boesch DF, Clairain EJ et al (2007) Restoration of the Mississippi Delta: lessons from Hurricanes Katrina and Rita. Science 315:1679–1684

De Baets S, Poesen J, Reubens B et al (2008) Root tensile strength and root distribution of typical Mediterranean plant species and their contribution to soil shear strength. Plant Soil 305:207–226

De Vriend HJ, Van Koningsveld M (2012) Building with nature: thinking, acting and interacting differently. EcoShape, Building with Nature, Dordrecht

Dietrich WE, Real de Asua R, Coyle1 J et al (1998) A validation study of the shallow slope stability model, SHALSTAB, in forested lands of Northern California. StillwaterEcosystem. Watershed & Riverine Sciences, Berkley, USA

Dolidon N, Hofer T, Jansky L, Sidle R (2009) Watershed and forest management for landslide risk reduction. In: Sassa K, Canuti P (eds) Landslides – disaster risk reduction. Springer, Heidelberg, pp 633–649

Donato DC, Kauffman JB, Murdiyarso D et al (2011) Mangroves among the most carbon-rich forests in the tropics. Nat Geosci 4:293–297. doi:10.1038/NGEO1123

Doody JP (2013) Sand dune conservation, management and restoration. Springer, Dordrecht

Dorren LKA, Berger F, Imeson AC et al (2004) Integrity, stability and management of protection forests in the European Alps. For Ecol Manag 195:165–176

Doswald N, Estrella M (2015) Promoting ecosystems for disaster risk reduction and climate change adaptation: opportunities for integration. Discussion paper, United Nations Environment Programme, Geneva, 48p

Duarte CM, Losada IJ, Hendriks IE et al (2013) The role of coastal plant communities for climate change mitigation and adaptation. Nat Clim Chang 3:961–968

Dudley N, Buyck C, Furuta N et al (2015) Protected areas as tools for disaster risk reduction, A handbook for practitioners. MOEJ and IUCN, Tokyo/Gland, 44p

Dutch Ministry of Water Management, Transport and Public Works (2012) Dutch water program – room for the river. Dutch Ministry of Water Management, Transport and Public Works, The Hague

EEA (2015) Exploring nature based solution. The role of green infrastructure in mitigating the impacts of weather- and climate change-related natural hazards. EEA Technical Report No 12/2015, European Environment Agency, Copenhagen, 61p. Available at http://www.eea. europa.eu/publications/exploring-nature-based-solutions-2014. Accessed Nov 2015

EM-DAT (2015) The human cost of natural disasters 2015, a global perspective. Centre for Research on the Epidemiology of Disasters (CRED) Université Catholique de Louvain 58pp

Estrella M, Saalismaa N (2013) Ecosystem-based disaster risk reduction (Eco-DRR): an overview. In: Renaud FG, Sudmeier-Rieux K, Estrella M (eds) The role of ecosystems in disaster risk reduction. UNU Press, Tokyo, pp 26–54

Estrella M, Renaud FG, Sudmeier-Rieux K (2013) Opportunities, challenges and future perspectives for ecosystem-based disaster risk reduction. In: Renaud FG, Sudmeier-Rieux K, Estrella M (eds) The role of ecosystems in disaster risk reduction. UNU Press, Tokyo, pp 437–456

European Commission (EC) (2000) Directive 2000/60/EC of the European Parliament and of the Council of 23 October 2000 establishing a framework for community action in the field of water policy (Directive 2000/60/EC). European Commission Brussels, Belgium

European Commission (EC) (2007) Directive 2007/60/EC of the European Parliament and of the Council of 23 October 2007 on the assessment and management of flood risks. (Directive 2007/ 60/EC). European Commission Brussels Belgium

FAO (2013) Forests and landslides: the role of trees and forests in the prevention of landslides and rehabilitation of landslide-affected areas in Asia, 2nd edn. Forbes K, Broadhead J (eds) Regional office for Asia and the Pacific, Bangkok

Feagin RA, Sherman DJ, Grant WE (2005) Coastal erosion, global sea-level rise, and the loss of sand dune plant habitats. Front Ecol Environ 3:359–364

Feagin RA, Mukherjee N, Shanker K et al (2010) Shelter from the storm? Use and misuse of coastal vegetation bioshields for managing natural disasters. Conserv Lett 3:1–11

Gedan KB, Kirwan ML, Wolanski E et al (2011) The present and future role of coastal wetland vegetation in protecting shorelines: answering recent challenges to the paradigm. Clim Chang 106:7–29

Genet M, Stokes A, Fourcaud T, Norris JE (2010) The influence of plant diversity on slope stability in a moist evergreen deciduous forest. Ecol Eng 36:265–275

Gómez-Pina G (2002) Sand dune management problems and techniques, Spain. J Coast Res 36:325–332

González-Ollauri A, Mickovski SB (2014) Integrated model for the hydro-mechanical effects of vegetation against shallow landslides. Int J Environ Qual 13:37–61

Greenwood JR (2006) Slip4ex – a program for routine slope stability analysis to include the effects of vegetation, reinforcement and hydrological changes. Geotech Geol Eng 24:449–465

Hales TC, Ford CR, Hwang T et al (2009) Topographic and ecologic controls on root reinforcement. J Geophys Res Earth Surf 114:17

Hanley ME, Hoggard SPG, Simmonds DJ et al (2014) Shifting sands? Coastal protection by sand banks, beaches and dunes. Coast Eng 87:136–146

Heslenfeld P, Jungerius PD, Klijn JA (2004) European coastal dunes: ecological values, threats, opportunities and policy development. In: Martínez ML, Psuty NP (eds) Coastal dunes: ecology and conservation. Springer, Berlin/Heidelberg

Hettiarachchi SSL, Samarawickrama SP, Fernando HJS et al (2013) Investigating the performance of coastal ecosystems for hazard mitigation. In: Renaud FG, Sudmeier-Rieux K, Estrella M (eds) The role of ecosystems for disaster risk reduction. United Nations University Press, Tokyo

IPCC (2014) Synthesis report. Contribution of working groups I, II and III to the fifth assessment report of the intergovernmental panel on climate change Core Writing Team, Pachauri RK, Meyer LA (eds). Geneva, Switzerland, 151 pp

Janssen SKH, Van Tatenhove JPM, Otter HO, Mol APJ (2014) Greening flood protection—an interactive knowledge arrangement perspective. J Environ Policy Plan 17(3). doi:10.1080/1523908X.2014.947921

Jaquet S, Sudmeier-Rieux K, Derron M-H, Jaboyedoff M (2013) Forest cover and landslide trends: a case study from Dolakha District in central-eastern Nepal, 1992–2009. In: Renaud FG, Sudmeier-Rieux K, Estrella M (eds) The role of ecosystems in disaster risk reduction. UNU Press, Tokyo, pp 343–367

Javaheri A, Babbar-Sebens M (2014) On comparison of peak flow reductions, flood inundation maps, and velocity maps in evaluating effects of restored wetlands on channel flooding. Ecol Eng 73:132–145

Ji J, Kokutse NK, Genet M et al (2012) Effect of spatial variation of tree root characteristics on slope stability. A case study on Black Locust (Robinia pseudoacacia) and (Platycladus orientalis) stands on the Loess Plateau, China. Catena 92:139–154

Kokutse N, Fourcaud T, Kokou K et al (2006) 3D numerical modelling and analysis of forest structure on hill slopes stability. In: Marui H, Marutani T, Watanabe N et al (eds) Interpraevent 2006: disaster mitigation of debris flows, slope failures and landslides, September 25–27, Niigata, Japan. Universal Academy Press, Tokyo, pp 561–567

Lacambra C, Friess DA, Spencer T, Möller I (2013) Bioshields: mangrove ecosystems as resilient natural coastal defences. In: Renaud FG, Sudmeier-Rieux K, Estrella M (eds) The role of ecosystems in disaster risk reduction. UNU Press, Tokyo, pp 82–108

Lange W, Sandholz S, Nehren U (2016) Strengthening urban resilience through nature: the potential of ecosystem-based measures for reduction of landslide risk in Rio de Janeiro. Lincoln Institute of Land Policy, Cambridge, MA, in press

Liu PL-F, Lynett P, Fernando H et al (2005) Observations by the international tsunami survey team in Sri Lanka. Science 308:1595

Loades KW, Bengough AG, Bransby MF, Hallett PD (2010) Planting density influence on fibrous root reinforcement of soils. Ecol Eng 36:276–284

Lovelock CE, Cahoon DR, Friess DA et al (2015) The vulnerability of Indo-Pacific mangrove forests to sea-level rise. Nature. doi:10.1038/nature15538

Maltby E (2009) The Changing Wetland Paradigm. In: Maltby E, Barker T (eds) The wetlands handbook. Wiley-Blackwell, Oxford. doi:10.1002/9781444315813.ch1

Mao Z, Saint-André L, Genet M et al (2012) Engineering ecological protection against landslides in diverse mountain forests: choosing cohesion models. Ecol Eng 45:55–69

Martin TG, Watson JEM (2016) Intact ecosystems provide best defence against climate change. Nat Clim Chang 6:122–124

Martínez ML, Maun MA, Psuty NP (2004) The fragility and conservation of the World's coastal dunes: geomorphological, ecological and socioeconomic perspectives. In: Martínez ML, Psuty NP (eds) Coastal dunes. Springer, Berlin/Heidelberg

Mascarenhas A, Jayakumar S (2008) An environmental perspective of the post-tsunami scenario along the coast of Tamil Nadu, India: role of sand dunes and forests. J Environ Manag 89:24–34

Mattia C, Bischetti GB, Gentile F (2005) Biotechnical characteristics of root systems of typical Mediterranean species. Plant Soil 278:23–32

McIvor AL, Möller I, Spencer T, Spalding M (2012a) Reduction of wind and swell waves by mangroves. Natural coastal protection series: report 1. Cambridge Coastal Research Unit working paper 40. The Nature Conservancy and Wetlands International, 27p

McIvor AL, Spencer T, Möller I, Spalding M (2012b) Storm surge reduction by mangroves. Natural coastal protection series: report 2. Cambridge Coastal Research Unit working paper 41. The Nature Conservancy and Wetlands International, 35p

McIvor AL, Spencer T, Möller I, Spalding M (2013) The response of mangrove soil surface elevation to sea level rise. Natural coastal protection series: report 3. Cambridge Coastal Research Unit working paper 42. The Nature Conservancy and Wetlands International, 59p

McIvor A, Spencer T, Spalding M et al (2015) Mangroves, tropical cyclones, and coastal hazard risk reduction. In: Ellis J, Sherman DJ (eds) Coastal and marine hazards, risks, and disasters. Elsevier, Amsterdam, pp 403–429

Moreno-Casasola P, Martínez LM, Castillo-Campos G (2008) Designing ecosystems in degraded tropical coastal dunes. Ecoscience 15(1):44–52

Murdiyarso D, Purbopuspito J, Kauffman JB et al (2015) The potential of Indonesian mangrove forests for global climate change mitigation. Nat Clim Chang. doi:10.1038/NCLIMATE2734

Murti R, Buyck C (eds) (2014) Safe heavens: protected areas for disaster risk reduction and climate change adaptation. IUCN, Gland, Switzerland 168p. Available at https://portals.iucn.org/library/node/44887. Accessed Nov 2015

Nicholls RJ, Wong PP, Burkett V et al (2008) Climate change and coastal vulnerability assessment: scenarios for integrated assessments. Sustain Sci 3:89–102

Pahl-Wostl C, Berkamp G, Cross K (2006) Adaptive management of upland rivers facing global change: general insights and specific considerations for the Rhone basin. Rosenberg International Forum on Water Policy, Banff

Pakeman RJ, Alexander J, Beaton J et al (2015) Species composition of coastal dune vegetation in Scotland has proved resistant to climate change over a third of a century. Glob Chang Biol 21 (10):3738–3747

Papathoma-Koehle M, Glade T (2013) The role of vegetation cover change in landslide hazard and risk. In: Renaud FG, Sudmeier-Rieux K, Estrella M (eds) The role of ecosystems in disaster risk reduction. UNU Press, Tokyo, pp 293–320

Peduzzi P (2014) Sand, rarer than one thinks. UNEP Global Environmental Alert Service (GEAS) Thematic focus: ecosystem management, environmental governance, resource efficiency. Available at: http://www.unep.org/pdf/UNEP_GEAS_March_2014.pdf. Accessed Mar 2016

Pohl M, Alig D, Körner C, Rixen C (2009) Higher plant density enhances soil stability in disturbed alpine ecosystem. Plant Soil 324:91–102

Prasetya GS (2007) The role of coastal forest and trees in combating coastal erosion. In: Braatz S, Fortuna S, Broadhead J, Leslie R (eds) Coastal protection in the aftermath of the Indian Ocean Tsunami: what role for forests and trees? FAO, Bangkok

Preti F (2013) Forest protection and protection forest: tree root degradation over hydrological shallow landslides triggering. Ecol Eng 61(1):633–645

Prisco I, Carboni M, Acosta ATR (2013) The fate of threatened coastal dune habitats in Italy under climate change scenarios. PLoS ONE 8(7), e68850. doi:10.1371/journal.pone.0068850

Psuty NP, Silveira TM (2010) Global climate change: an opportunity for coastal dunes? J Coast Conserv 14:153–160. doi:10.1007/s11852-010-0089-0

Ramsar (2015) Resolution XII.13 on wetlands and disaster risk reduction. Resolution adopted at the 12th meeting of the conference of the parties to the convention on Wetlands Punta del Este Uruguay 1–9 June 2015. Available at http://www.ramsar.org/sites/default/files/documents/library/cop12_res13_drr_e_0.pdf. Accessed Jan 2016

Rao NS, Carruthers TJB, Anderson P et al (2012) A comparative analysis of ecosystem–based adaptation and engineering options for Lami Town, Fiji. A synthesis report by the Secretariat of the Pacific Regional Environment Programme, 28p. Available at https://www.sprep.org/attachments/Publications/Lami_Town_EbA_Synthesis.pdf. Accessed Nov 2015

Renaud FG, Sudmeier-Rieux K, Estrella M (eds) (2013) The role of ecosystems in disaster risk reduction. UNU Press, Tokyo

Roering JJ, Schmidt KM, Stock JD et al (2003) Shallow landsliding, root reinforcement, and the spatial distribution of trees in the Oregon Coast Range. Can Geotech J 40:237–253

Runyan CW, D'Odorico P (2014) Bistable dynamics between forest removal and landslide occurrence. Water Resour Res 50:1112–1130

Russi D, ten Brink P, Farmer A et al (2013) The economics of ecosystems and biodiversity for water and wetlands. IEEP London and Brussels Ramsar Secretariat, Gland, 84 pp

Saye SE, Pye K (2007) Implications of sea level rise for coastal dune habitat conservation in Wales UK. J Coast Conserv 11:31–52

Schmid I, Kazda M (2001) Vertical distribution and radial growth of coarse roots in pure and mixed stands of Fagus sylvatica and Picea abies. Can J For Res 31:539–548

Schmid I, Kazda M (2002) Root distribution of Norway spruce in monospecific and mixed stands on different soils. For Ecol Manag 159:37–47

Schönenberger W, Noack A, Thee P (2005) Effect of timber removal from windthrow slopes on the risk of snow avalanches and rockfall. For Ecol Manag 213:197–208

Seabloom EW, Ruggiero P, Hacker SD et al (2013) Invasive grasses, climate change, and exposure to storm-wave overtopping in coastal dune ecosystems. Glob Chang Biol 19(3):824–832

Sigren JM, Figlus J, Armitage AR (2014) Coastal sand dunes and dune vegetation: restoration, erosion, and storm protection. Shore Beach 82:5–12

Spalding M, McIvor A, Tonneijck FH et al (2014a) Mangroves for coastal defence. Guidelines for coastal managers and policy makers. Wetlands International and The Nature Conservancy, 42p. Available at http://www.wetlands.org/Portals/0/publications/Book/Mangroves%20for% 20Coastal%20Defence_A%20Decisionmakers%20Guide_Web%20Version.pdf.   Accessed Nov 2015

Spalding MD, McIvor AL, Beck MW et al (2014b) Coastal ecosystems: a critical element of risk reduction. Conserv Lett 7(3):293–301

Stokes A, Atger C, Bengough AG et al (2009) Desirable plant root traits for protecting natural and engineered slopes against landslides. Plant Soil 324:1–30

Sudmeier-Rieux K (2013) Ecosystem approach to disaster risk reduction. Basic concepts and recommendations to governments, with a special focus on Europe. European and Mediterranean Major Hazards Agreement (EUR-OPA) Council of Europe, 31 pp

Takle ES, Chen T-C, Wu X (2007) Protection from wind and salt. In: Braatz S, Fortuna S, Broadhead J, Leslie R (eds) Coastal protection in the aftermath of the Indian Ocean Tsunami: what role for forests and trees? FAO, Bangkok

Temmerman S, Meire P, Bouma TJ et al (2013) Ecosystem-based coastal defence in the face of global change. Nature 504:79–83

Thao ND, Takagi H, Esteban M (eds) (2014) Coastal disasters and climate change in Vietnam. Elsevier, London/Waltham

Timmerman S, Kirwan ML (2015) Building land with a rising sea. Science 349(6248):588–589

Timmerman S, Meire P, Bouma TJ et al (2013) Ecosystem-based coastal defence in the face of global change. Nature 504:79–83

UN (2015a) Sendai framework for disaster risk reduction 2015–2030. United Nations. Available at http://www.preventionweb.net/files/43291_sendaiframeworkfordrren.pdf. Accessed Oct 2015

UN (2015b) Sustainable development goals. Available at https://sustainabledevelopment.un.org/? menu = 1300. Accessed Oct 2015

UNFCCC (2015) Adoption of the Paris agreement. Draft decision -/CP.21. FCCC/CP/2015/L.9/ Rev.1. Available at http://unfccc.int/resource/docs/2015/cop21/eng/l09r01.pdf. Accessed Jan 2016

UNISDR (2005) Hyogo Framework for action 2005–2015: building the resilience of nations and communities to disaster. United Nations International Strategy for Disaster Reduction, 23p

van Eijk P, Baker C, Gaspirc R, Kumar R (2013) Good flood, bad flood: maintaining dynamic river basins for community resilience. In: Renaud F, Sudmeier-Rieux K, Estrella M (eds) The role of ecosystems in disaster risk reduction. UNU Press, Tokyo, pp 221–247 van Staveren MF, Warner JF, van Tatenhove JPM, Wester P (2013) Let's bring in the floods: de-poldering in the Netherlands as a strategy for long-term delta survival? Water Int 39 (5):686–700, http://dx.doi.org/10.1080/02508060.2014.957510

van Verhoeven JTA (2013) Wetlands in Europe: perspectives for restoration of a lost paradise. Ecology and biodiversity. Ecol Eng 66:6–9

van Wesenbeeck BK, Mulder JPM, Marchand M et al (2014) Damming deltas: a practice of the past? Towards nature-based flood defenses. Estuar Coast Shelf Sci 140:1–6

Walker LR, Shiels AB (2013) Chapter 6 Living with landslides for landslide ecology. USDA National Wildlife Research Center – Staff Publications. Paper 1637. Available at http://digitalcommons.unl.edu/icwdm_usdanwrc/1637. Accessed Mar 2016

Wehrli A, Dorren L (2013) Protection forests: a key factor in integrated risk management in the Alps. In: Renaud FG, Sudmeier-Rieux K, Estrella M (eds) The role of ecosystems in disaster risk reduction. UNU Press, Tokyo, pp 343–415

Wicaksono P, Danoedoro P, Hartono, Nehren U (2016) Mangrove biomass carbon stock mapping of the Karimunjawa Islands using multispectral remote sensing. Int J Remote Sens 37(1):26–52

Wilson L, Wilson J, Holden J et al (2010) Recovery of water tables in Welsh blanket bog after drain blocking: discharge rates, time scales and the influence of local conditions. J Hydrol 391:377–386

Winterwerp H, van Wesenbeeck B, van Dalfsen J et al (2014) A sustainable solution for massive coastal erosion in Central Java. Towards Regional Scale Application of Hybrid Engineering. Discussion Paper, Deltares and Wetlands International, 45p. Available at http://www.wetlands.org/LinkClick.aspx?fileticket = rv2jbvHx%2BHw%3D&tabid = 56. Accessed Nov 2012

Part I
Economic Approaches and Tools for Eco-DRR/CCA

Chapter 2
Valuing Ecosystems as an Economic Part of Climate-Compatible Development Infrastructure in Coastal Zones of Kenya & Sri Lanka

Lucy Emerton, Mark Huxham, Jil Bournazel, and M. Priyantha Kumara

Abstract Even though 'green' options for addressing the impacts of climate change have gained in currency over recent years, they are yet to be fully mainstreamed into development policy and practice. One important reason is the lack of economic evidence as to why investing in ecosystems offers a cost-effective, equitable and sustainable means of securing climate adaptation, disaster risk reduction and other development co-benefits. This chapter presents a conceptual framework for integrating ecosystem values into climate-compatible development planning. Case studies from coastal areas of Kenya and Sri Lanka illustrate how such an approach can be applied in practice to make the economic and business case for ecosystem-based measures. It is argued that, rather than posing 'grey' and 'green' options as being necessarily in opposition to each other or as mutually incompatible, from an economic perspective both should be seen as being part and parcel of the same basic infrastructure that is required to deliver essential development services in the face of climate change.

Keywords Climate-compatible development • Coastal ecosystems • Economic valuation • Mangroves

L. Emerton (✉)
Environment Management Group, Colombo, Sri Lanka
e-mail: lucy@environment-group.org

M. Huxham
School of Life, Sport and Social Sciences, Edinburgh Napier University, Edinburgh, UK
e-mail: m.huxham@napier.ac.uk

J. Bournazel
Independent Consultant, Edinburgh, UK
e-mail: jil.bournazel@gmail.com

M.P. Kumara
Ocean University of Sri Lanka, Tangalle, Sri Lanka
e-mail: kumarampp@yahoo.com

© Springer International Publishing Switzerland 2016
F.G. Renaud et al. (eds.), *Ecosystem-Based Disaster Risk Reduction and Adaptation in Practice*, Advances in Natural and Technological Hazards Research 42,
DOI 10.1007/978-3-319-43633-3_2

2.1   Introduction

Several authors have noted that, even though ecosystem-based approaches are gaining in popularity, they are for the large part yet to be fully mainstreamed into development decision-making as compared to more conventional 'grey' measures (ProAct Network 2008; UN Global Compact et al. 2011; Renaud et al. 2013). It is argued that a major reason for this omission is the lack of economic evidence as to why investing in ecosystems offers a cost-effective, equitable and sustainable means of securing climate adaptation and disaster risk reduction benefits (Colls et al. 2009; UNEP 2011; Munroe et al. 2012). Intensifying competition over scarce private and public investment funds, coupled with increasing demands from shareholders and taxpayers for information about how their money has been spent, means that the need to demonstrate cost effectiveness and value for money is becoming an ever-more pressing concern (Tompkins et al. 2013; Ferrario et al. 2014). While figures are readily available on the benefits of hard engineering or built infrastructure options, and are routinely used to guide and report on investment decisions, much less information is on hand about the potential gains associated with investing in green disaster risk reduction and adaptation measures.

This chapter describes how economic valuation can assist in communicating the advantages of ecosystem-based options for climate-compatible development (CCD) in coastal areas. It contends that, rather than posing 'grey' and 'green' investments as being necessarily in opposition to each other, or as mutually incompatible, both should be seen as being part and parcel of the same basic economic infrastructure that is required to deliver essential development, adaptation and disaster risk reduction services. In turn, if CCD is to reach its full potential, decision-makers must be equipped with the tools and information that will enable them to explicitly recognise the economic values associated with ecosystem services, factor them into investment calculations, and develop policy instruments and management approaches which will better capture and harness them in support of climate adaptation and disaster risk reduction. This requires a shift in the way in which land use and development trade-offs are conceptualised and evaluated — moving from a paradigm which undervalues ecosystem services to approaches which count and invest in them as an economic part of climate-compatible development infrastructure.

2.2   The Economic Value of Coastal Ecosystem Services

On the face of it, coastal planners and decision-makers would seem to be well aware of the value of natural resources. Such figures are accorded a prominent role in most national economic statistics and indicators, and in the development decisions they inform. For example, a compilation of country-level trade accounts indicates fish to be the most valuable agricultural commodity on world markets: recorded export earnings are now worth more than coffee, cocoa, sugar and tea combined (OECD 2008). Sea fisheries, alone, are documented to generate income in excess of USD 80 billion a year, provide for around 35 million jobs and support the livelihoods of more than 300 million people (Beaudoin and Pendleton 2012).

While these kinds of statistics suggest that it is hardly a novel insight that coastal natural resources make a major contribution to local, national and even global economies, there has long been a tendency to conceptualise ecosystem values only in terms of the commodities that are traded in formal markets, such as fisheries, timber, minerals or tourism (Emerton 2006). This definition however remains an incomplete one, because it excludes the host of other goods and services that coastal ecosystems generate. In particular, the economic values associated with subsistence-level and non-market production and consumption and with the protection and regulation of natural and human systems – arguably those which are of the most importance to adaptation, disaster risk reduction and climate-compatible development – tend to be largely left out of the equation. Almost half the global population are thought to depend on marine and coastal biodiversity in some way for their basic livelihoods (SCBD 2009). In Myanmar, for example, the food, fuel, construction materials and medicinal products obtained from natural ecosystems contributes around 83 % of per capita GDP for rural populations in the coastal zone (Emerton 2014c). Meanwhile, at least 100 million people, worldwide, benefit in economic terms from the disaster risk reduction services provided by coral reefs or would incur hazard mitigation and adaptation costs should these ecosystems be degraded (Ferrario et al. 2014). Up to three times this number are thought to be vulnerable to other climate-related effects in coastal areas (ProAct Network 2008).

The economic significance of these largely uncounted ecosystem services is substantial, and often far outweighs that of the direct physical products that are obtained from coastal lands and resources (Agardy et al. 2005; UNEP-WCMC 2006; Barbier et al. 2011; Shepard et al. 2011). Recent work carried out in India and Thailand, for example, finds that mangrove coastline protection and stabilization services are worth around USD 10,000/ha/year (Das 2009; Das and Crépin 2013; Barbier et al. 2011). Similarly, the protection afforded by natural ecosystems against waves, storm surges and other extreme weather events in Indonesia, Malaysia and Singapore has been calculated at just under USD 200,000 per km of coastline (MPP-EAS 1999). In Belize, coral reefs and mangroves help to reduce beach erosion and wave-induced damages to coastal property by up to USD 250 million a year, a value that translates to more than a quarter of national GDP (Cooper et al. 2008). In Sri Lanka coastal wetlands provide flood control and water purification functions to a value in excess of USD 2500 per hectare (Emerton and Kekulandala 2003), while mangrove storm protection services were assessed to be almost USD 800,000/ha/year just before the 2004 Indian Ocean tsunami (Batagoda 2003).

Not only do these figures make the point that the value of coastal ecosystems extends far beyond that which is conventionally included in the calculations that inform development decisions, but they also serve to demonstrate that managing ecosystems for their services is frequently a far cheaper and more cost-effective option than employing artificial technologies or taking remedial or mitigative measures when these essential functions are lost (ProAct Network 2008; Haisfield et al. 2010; Beck and Shepard 2012; Sudmeier-Rieux 2013; Temmerman et al. 2013; Spalding et al. 2014). Every dollar invested in coastal ecosystem-based mitigation is, for example, estimated to reduce the US taxpayer burden by USD 4 in terms of avoided costs, losses and damages from storm-surge effects and other natural hazards (MMC 2005). In southern Vietnam, the restoration of 12,000 ha of mangroves has saved an estimated USD 7.3 million/year in dyke maintenance, a figure that is more than six and a half times the costs of planting (Powell et al. 2010). On the west coast of Sri Lanka, long-term climate adaptation benefits and costs saved were found to be more than twice as high as the costs of conserving coastal and estuarine ecosystems (De Mel and Weerathunge 2011).

2.3 How Undervaluation Poses a Problem for Development Decision-Making

Despite these impressive figures, coastal ecosystem undervaluation remains a persistent problem. For the most part, calculations of the relative returns to different land, resource and investment choices simply do not factor in such costs and benefits. A review of past patterns of coastal development would reinforce the observation that decision makers have perceived there to be few economic benefits associated with the conservation of natural ecosystems, and few economic costs attached to their degradation and loss. The net result is that even though substantial amounts of public and private investment funds have been ploughed into establishing the built infrastructure that is required to stimulate and sustain economic development processes in coastal zones, much less attention has been paid to maintaining (or even improving) the natural capital base that underpins and protects them.

As a consequence, investments in CCD infrastructure in coastal zones tend to continue to be heavily skewed towards those hard engineering and built infrastructure options for which a monetary return can easily be calculated (Ferrario et al. 2014). Most of the cost-benefit analyses that are applied to investigate the relative desirability of different investment choices simply do not take environmental values into account (Chadburn et al. 2013; Shreve and Kelman 2014; also see Vicarelli et al. Chap. 3). The small number of cases where economic methods are used to assess ecosystem-based approaches for adaptation and disaster risk reduction tend to confine themselves to direct, physical costs and benefits – thus underestimating massively the gains and value-added that can be secured as compared to, or in combination with, 'hard' and 'grey' infrastructure options. There remain very few real-world instances where broader ecosystem values and development co-benefits are factored into calculations (also see Vicarelli et al. Chap. 3).

The effects of undervaluation are also manifested at the policy level. Across the world, there is a long history of economic policies which aim to stimulate production and growth having also hastened the process of resource and habitat, degradation and discouraged ecosystem-based investments. In coastal zones, a wide variety of tax breaks and fiscal inducements, often combined with low or non-existent environmental penalties and fines, provide a powerful incentive to intensify resource exploitation and modify and reclaim natural habitats for more 'productive' commercial uses. One obvious example is fisheries subsidies, estimated to be worth between USD 30–34 billion a year worldwide (MRAG 2009), which have led to a massive expansion in the capacity of fishing fleets and resulted in the over-exploitation (and in some cases collapse) of fish stocks (UNEP 2004). Another well-known case is the generous tax breaks, import duty exemptions, export credits and preferential loans offered to shrimp farming in many countries (Primavera 1997; Bailly and Willmann 2001).

The policy distortions and perverse incentives that result from ecosystem under-valuation mean that prevailing prices and market opportunities in many countries mean that it frequently remains more profitable for people to engage in economic activities that degrade ecosystems – even if the costs and losses that arise for other groups, or to the economy as a whole, outweigh the immediate gains to the land or resource user that is causing the damage (Mohammed 2012). The loss of potential economic benefits in the global fishery due to subsidy-driven fish stock depletion and over-capacity is for example estimated to cost more USD 50 billion per year (World Bank and FAO 2009). At the local level, work carried out in the Togean Islands in Indonesia shows that while the costs associated with the loss of ecosystem services caused by commercial logging and agriculture in coastal areas outweigh the income they generate by a factor of more than four, it is still more profitable for households and businesses to clear and reclaim coastal habitats than to engage in other more sustainable land and resource uses (Cannon 1999; Emerton 2009). Similarly, in Sri Lanka, it is possible to gain high market returns from clearing mangroves for shrimp farming; however, if the costs and negative externalities associated with ecosystem service loss were factored into prices and markets, shrimp farming would cease to be a financially viable land use option (Gunawardena and Rowan 2005).

In many ways undervaluation can thus be seen to have encouraged a negative investment process in coastal areas, whereby ecosystems have been destroyed, degraded and converted in the course of expanding the built environment, stimulating particular sectors or production activities, or even while attempting to take action to reduce the risk of disasters and protect against the effects of climate change (Emerton 2014a; also see Freiss and Thompson, Chap. 4). If ecosystems have no value, then such decisions would be perfectly rational ones from both a financial and an economic point of view. In a similar vein, should there be low or zero costs attached to ecosystem degradation and depletion, then there would be no particular economic advantage to be gained from considering green adaptation and disaster risk reduction measures. This is however clearly not the case. The problem is not so much that ecosystems have no economic importance, but rather that this value is poorly understood, rarely expressed in numerical or monetary terms, and as a result is frequently omitted from decision-making. A pressing question then becomes: how can we better articulate the economic opportunities, value-added and costs avoided that are associated with adopting ecosystem-based approaches, and integrate this information into climate-compatible development planning?

2.4 Frameworks for Identifying and Demonstrating Ecosystem Values

Over the last two decades, a set of useful (and increasingly widely-used) economic methods and tools have been developed which help to overcome the problems associated with ecosystem undervaluation. The concept of total economic value has now emerged as one of the most commonly-applied frameworks for identifying and categorising ecosystem values. This represents a move away from the very narrow definition of benefits that economists traditionally applied, which saw the value of ecosystems only in terms of raw materials, physical products and traded commodities. Total economic value also encompasses subsistence and non-market values, ecological functions and non-use benefits (Fig. 2.1) – in other words, the full gamut of provisioning, regulating, supporting and cultural services that ecosystems generate (Millennium Ecosystem Assessment 2005). Looking at the total economic value of an ecosystem essentially involves considering its full range of characteristics as an integrated system — its resource stocks or assets, flows of environmental services, and the attributes of the system as a whole.

The question of how to ascribe values to ecosystem services has long posed something of a challenge to economists. The easiest and most straightforward way, and the method used conventionally, is to look at their market price: what they cost to buy or are worth to sell. However, as ecosystem services very often have no

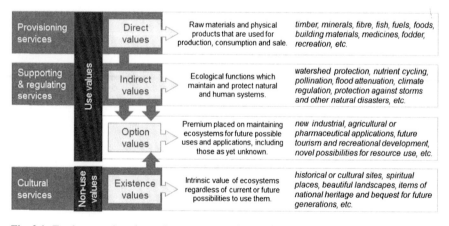

Fig. 2.1 Total economic value and ecosystem services (Adapted from Emerton 2006, 2014a)

market price (or are subject to market prices which are highly distorted), these techniques obviously only have very limited application. Parallel to the advances made in defining and conceptualising the economic value of ecosystem services, techniques for quantifying ecosystem values in monetary terms have also moved forward substantially over the last 20 years or so.

Today a suite of methods is available for valuing ecosystem services that cannot be calculated accurately via the use of market prices, including in coastal environments (see, for example, van Beukering et al. 2007; UNEP-WCMC 2011; Wattage 2011; Beaudoin and Pendleton 2012). Applying these methods basically requires carrying out three interrelated steps: characterising the change in ecosystem structure, functions, and processes that gives rise to changes in ecosystem service(s); tracing how these changes influence the quantities and qualities of ecosystem service flows to people; and using valuation to assess and articulate changes in human wellbeing that result from the change in ecosystem services (see Barbier et al. 2011).

These methodological developments enable a wide range of formerly unvalued or undervalued coastal ecosystem goods and services to be expressed in monetary terms, and – in principle at least – incorporated into the calculations that are used to inform development decisions. Ecosystem valuation has for some time been a relatively well-accepted and widely-used component of environmental and biodiversity conservation research and planning. For example, a large volume of studies now exists on the economic value of coastal ecosystems, covering most major habitat types and many regions of the world. Yet, although it can in theory provide a powerful tool for placing ecosystem-based options on the agenda of development planners and decision-makers, the use of ecosystem valuation techniques in climate adaptation and disaster risk reduction still remains in its infancy and as yet there have only been a small number of real-world applications (also see Harmáčková et al. Chap. 5; Clark et al. 2012; Naumann et al. 2011; Chadburn et al. 2013; Rao et al. 2013; Emerton 2014b; Shreve and Kelman 2014).

The following sections illustrate how economic valuation approaches were applied to generate information which could be used to assist in making the case for integrating ecosystem-based options for CCD into coastal zone planning in Puttalam Lagoon, Sri Lanka and the Kwale coastline, Kenya. The objective was to demonstrate to national and local decision-makers and budget-holders the potential gains from green CCD strategies as well as the costs and losses associated with failing to factor ecosystems into coastal development planning. The studies focussed on assessing the costs, benefits and trade-offs associated with investing in mangrove rehabilitation and conservation as a means of strengthening climate adaptation and disaster risk reduction, at the same time as generating other development co-benefits for coastal populations.

2.5   Weighing up the Opportunity Costs of Land Use Change in Puttalam Lagoon, Sri Lanka

Puttalam Lagoon is located on the north-west coast of Sri Lanka, and covers a surface area of some 33 km^2. It connects to the open sea at the northern end, and is separated from the Indian Ocean on the west by a narrow strip of sand dunes and long sandy beaches. Mangroves are currently estimated to cover between 700 and 1000 ha of the lagoon's inner shoreline (Weragodatenna 2010; Kumara 2014; Bournazel et al. 2015). Tidal flats, seagrass beds, salt marshes, dry monsoon forest, coastal scrub jungles and dry thorny scrublands are also found (Kumara and Jayatissa 2013). On the eastern and southern fringes, large tracts of land have been converted to agriculture and aquaculture, including around 1500 ha of crop-land, a similar area of salterns, several thousand hectares of coconut plantations and at least 1000 ha of mainly small-scale shrimp ponds (Weragodatenna 2010; Bournazel et al. 2015). Some 45,000 households or 185,000 people live in the administrative divisions abutting the lagoon.

The expansion of shrimp farming in Sri Lanka over the past three decades has dramatically changed the coastal landscape, and Puttalam has experienced some of the most destructive development in the country. It is estimated that a third or more of mangrove cover in the lagoon has been lost since the early 1990s (Bournazel et al. 2015). Meanwhile, problems with disease meant that many shrimp farms performed poorly in financial terms, leading to their being abandoned after a relatively short time, leaving denuded and unproductive landscapes (Dahdouh-Guebas et al. 2002; Westers 2012). The conversion of mangroves to aquaculture ponds and their subsequent abandonment pose potentially serious risks to develop-ment in the Puttalam Lagoon area, in terms of negative effects on local livelihoods and increased vulnerability to the impacts of climate change.

Ecosystem-based approaches have been proposed as a means of addressing the problems associated with environmental degradation, and are also being mooted as a way of strengthening the livelihoods and adaptive capacity of local communities. These are envisaged to be based around the restoration of mangroves in abandoned shrimp farms, and the promotion of environmentally sustainable aquaculture tech-nologies and practices among functioning and developing enterprises. However, in order to have traction with local and national decision-makers (especially those in the fisheries and agricultural sectors that exert such a heavy influence on land use change patterns), the economic rationale for these green CCD options needs to be made explicit. There is still a widespread belief that mangroves and other natural habitats comprise 'uneconomic' areas, or land 'taken out' of production. There has to date been little recognition among decision-makers of the far-reaching economic costs, losses and damages that can result from the modification and conversion of coastal environments.

Against this backdrop, the valuation study assessed the trade-offs associated with alternative land use development options, with a view to demonstrating the opportunity costs of mangrove conversion in terms of climate compatible development benefits foregone.

First of all it was necessary to identify the main ecosystem services and economic processes associated with the mangroves in Puttalam Lagoon, and select the techniques that would be used to value them. Seven services relating to climate adaptation, mitigation and associated livelihood benefits were identified as being of key importance: fuelwood, timber, non-wood/non-fish products, protection against saline intrusion, water quality regulation, carbon sequestration and avoided emissions, and provision of breeding and nursery habitat for fisheries[1] (Emerton 2014b). As is so commonly the case in ecosystem valuation, conventional market price techniques only had limited applicability: for valuing wood and non-wood/non-fish products (using local farmgate prices) and carbon storage services (via the prevailing voluntary forest carbon price for Asia). The protective functions of mangroves were valued based on the replacement costs of installing and operating wastewater treatment facilities which would bring the quality of water being discharged into the lagoon to a commensurate level (for water quality regulation services), and the expenditures on alternative drinking water sources in order to mitigate or avert the effects of surface water contamination (for protection against saline intrusion). The role of mangroves in maintaining nursery populations and habitat for commercially-important fish species was assessed by tracing effects on the productivity and catch of near shore and lagoon fisheries.

The resulting analysis indicated that the 731 ha of mangroves in Puttalam Lagoon are currently providing ecosystem services worth some USD 2.8–3.0 million a year, or between USD 3800–4100 per hectare (Table 2.1).

The second step was to examine the economic impacts of mangrove degradation and loss. The analysis covered the period 1992–2012, for which mapping, land use change analysis and carbon modelling had been carried out (see Bournazel et al. 2015). This period registered a net loss of some 934 ha of mangroves, most of which were converted to shrimp farms, salt pans, coconut plantations and other agricultural land uses (Fig. 2.2).

The extent to which, or ways in which, human populations utilise or depend on mangroves in Sri Lanka is not the same today as it was 20 years ago. Thus, in addition to considering the impact of changes in mangrove area on ecosystem values, the economic model also accounted for the ways in which the real price or value of ecosystem services had altered over time. This involved tracking the considerable socio-economic changes which have occurred in and around Puttalam Lagoon since 1992 (these are well-documented in the literature: see, for example, IUCN 2012; Kumara and Jayatissa 2013). Factors such as shifts in demography and settlement patterns, fluctuations in production and demand, changes in the relative

---

[1]It should be noted that two 'classic' mangrove ecosystem services do not appear in Puttalam Lagoon: protection against shoreline erosion and extreme weather events. This is due to the fact that the sheltered lagoon/estuary system is not exposed directly to the sea, and mangroves are found only on the inside shores, not on the coastline abutting the Indian Ocean.

Table 2.1 Current value of mangrove ecosystem services in Puttalam Lagoon

Ecosystem services		Total value (USD '000)	Unit value (USD/ha)
Provisioning	Wood products	367.4	506
	Non-wood/non-fish products	121.3	167
Regulating	Support to fisheries productivity	1757.5	2421
	Water quality regulation	553.3	762
	Protection against saline intrusion	192.3	265
	Carbon sequestration & avoided emissions	183.9	217
Total		2808.4–2991.9	3832–4121

Note: Individual ecosystem service values cannot simply be summed to give a total, as this would result in double-counting. As some services are partially or wholly mutually exclusive, a range of values is given. Water quality regulation services are applied only to those mangrove areas which protect major freshwater inflows into the lagoon

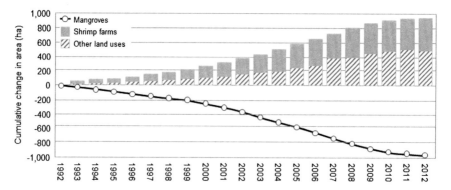

Fig. 2.2 Land use change in mangrove areas of Puttalam Lagoon 1992–2012 (Based on data presented in Bournazel et al 2015)

scarcity or abundance of natural resources, varying human dependence on (and preference for) mangrove products, and the price and availability of substitutes have all affected ecosystem service values.

The resulting analysis showed that the loss of mangroves in Puttalam Lagoon has been accompanied by a progressive decline in the value of ecosystem services[2] (Fig. 2.3). Overall, the value of mangrove services today is around USD 4 million lower than that which was available in 1992. This is even though in many cases the per hectare value of mangrove regulating services has actually shown a steady

---

[2]There are two exceptions to these general trends – carbon and fisheries. The slight improvement in carbon sequestration and avoided emissions values after 2007 is accounted for by the slowed pace of mangrove conversion. The dip in fisheries productivity values in 2005 and 2006 can be attributed to the sharp drop and then slow recovery of fish catch resulting from the impacts of the 2004 Indian Ocean tsunami.

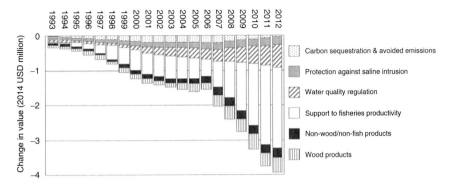

Fig. 2.3 Puttalam Lagoon: change in mangrove ecosystem service values 1992–2012 (Based on data presented in Emerton 2014b)

Table 2.2 Financial and economic impacts of land use change in Puttalam Lagoon 1992–2012 (based on data presented in Emerton 2014b)

	Value (USD mill)
Added income from shrimp farms	12.43
Added income from salterns	0.04
Added income from coconut plantations	0.17
Total added income from mangrove conversion	12.63
Foregone value of mangrove ecosystem services	−32.29
Net economic gain/loss from land use change	−19.66

increase over time, mainly due to population growth, and the intensification of settlement and industry in the area.

The third and final stage of the economic analysis entailed comparing the ecosystem values foregone due to the loss of mangrove habitats with the additional income and revenues earned from their conversion to other land uses. It was important to carry out this comparison, so as to consider the full opportunity costs and economic impacts of alternative land use, investment and development choices. The three land uses which together account for the vast majority of mangrove conversion since 1992 were considered – shrimp farms, coconut plantations and salterns. Per hectare budgets were developed for establishing, developing and maintaining each of these enterprises (including the restoration and rehabilitation of mangrove cover), and applied to the annual land use change figures. This showed that in total, mangrove ecosystem services worth USD 32.29 million (with a net present value (NPV) of USD 9.5 million) were lost between 1992 and 2012 (Table 2.2). This figure is around two and a half times more than the income earned from the shrimp farms, coconut plantations and salterns that were established on cleared mangrove land.

2.6 Articulating the Economic Gains from Ecosystem-Based Approaches on the Kwale Coastline, Kenya

Even though Kenya's mangroves generate a wide range of economically valuable goods and services to surrounding populations, they are being rapidly depleted, degraded and converted. Mangrove cover in 2010 was estimated at just over 45,000 ha, representing a reduction of 18 % from that recorded in 1985 (Kirui et al. 2013). The southern portion of the coastline has witnessed some of the highest rates of loss, driven largely by rapid population growth, escalating resource demands and intensifying settlements, infrastructure and industry (Rideout et al. 2013); if current trends continue, it is likely that mangroves may soon disappear altogether at many southern sites (Huxham et al. 2015).

There is presently a great deal of debate about the relative merits of different development approaches in Kenya's coastal zone. In the face of growing concerns about the vulnerability of the local population and economy to the effects of climate change, CCD has been gaining ground. Current development plans specify an ambitious (and costly) array of investments and activities aiming to protect and climate-proof coastal settlements and infrastructure, and strengthen the resilience and adaptive capacity of local livelihoods and production systems. Yet there remains very little information about the potential gains and relative cost-effectiveness of ecosystem-based approaches. As a result, green CCD options have to date been accorded only a minor role in public investment programmes. In an attempt to fill these gaps in evidence, the valuation study aimed to demonstrate the gains and value-added from investing in mangrove rehabilitation and conservation as a core component of climate-compatible development in the coastal zone. It focused on Kwale County on the southern Kenyan coast, which stretches approximately 90 km south from Mombasa to the border with Tanzania. The study area covered the four main mangrove areas of Mwache, Gazi, Funzi, and Vanga, which together contain just under 5600 ha of mangroves and around 22,000 people or 4500 households.

The study followed a process similar to that outlined above for Puttalam Lagoon. This first of all calculated the current baseline value of mangrove ecosystem services, moved on to assess the economic consequences of ecosystem change, and then articulated the value-added and costs-avoided that might be gained from integrating ecosystem-based approaches into CCD planning. Ten mangrove ecosystem services were valued: honey, fuelwood, timber, protection against shoreline erosion, defence against extreme weather events, carbon sequestration, nursery habitat for fisheries, tourism, research, and cultural practices. A variety of market and non-market valuation techniques were applied (see Huxham 2013 for further details). These included looking at mitigative and avertive expenditures on coastal defence structures (for protection against coastal erosion services), replacement costs of building and maintaining seawalls for storm and wave protection (for shelter against extreme weather services), and effects on fisheries production (for nursery habitat services).

Table 2.3 Current value of mangrove ecosystem services on the Kwale coastline (based on data presented in Emerton 2014b)

Ecosystem services		Total value (USD '000)	Unit value (USD/ha)
Provisioning	Timber, fuelwood & honey	1148.1	206
	Capture fisheries (finfish)	609.0	109
	Capture fisheries (crustaceans)	716.2	129
Regulating	Protection against coastal erosion	2196.5	395
	Protection against extreme weather events	192.5	35
	Carbon sequestration	1397.3	251
	Tourism, education & research	228.6	41
Total		5747.5–6488.1	1033–1166

Note: Individual ecosystem service values cannot simply be summed to give a total, as this would result in double-counting. As some services are partially or wholly mutually exclusive, a range of values is given

The calculations suggested that the services generated by Kwale's mangroves are currently worth between USD 5.75–6.5 million a year, or around USD 1100 per hectare (Table 2.3). It is worth pointing out that coastal protection services (including climate mitigation, erosion control and defence against extreme weather events) dominate these figures. Together they are worth more than one and a half times as much as the direct income from the provisioning services – forest and fisheries products – that economic value estimates would conventionally be confined to.

Two possible development and ecosystem futures were then modelled: business as usual (BAU) and ecosystem-based climate-compatible development (CCD). These reflected qualitative storyline scenarios developed by local stakeholders (including representatives from government, NGOs, communities and regulatory bodies), which laid out alternative visions for future land use and development along the Kwale coastline over the next 20 years (see King and Nap 2013; Huxham et al. 2015). In brief, BAU was depicted as entailing the gradual loss of mangrove cover and degradation of remaining forests, decline in fisheries resources, increasing coastal vulnerability and poverty, while CCD emphasised ecosystem conservation and sustainable management resulting in healthy mangroves supporting improved local livelihoods and enhanced resilience. Quantitative risk mapping and modelling of forest cover was also carried out (see Huxham et al. 2015), informed by the stakeholder scenarios and assuming the continuation of key risk factors and past trends in mangrove forest loss (Kirui et al. 2013; Rideout et al. 2013).

The risk mapping and land use change projections suggested that a 43 % loss of mangroves would occur over the next 20 years under BAU (with 100 % loss at the most vulnerable site, Mwache). Under the CCD scenario, forest cover was predicted to expand by 8, 7, 9 and 13 % in Funzi, Gazi, Mwache and Vanga, respectively (see Huxham et al. 2015). Because most of the area cleared of mangroves over the past 25 years has been left unused, mangrove restoration would generally not entail opportunity costs. Thus, unlike in the Puttalam study, no comparison was made between mangroves and the value of alternative land uses. However, as was the case for Puttalam, the economic model allowed for changes in the real value of ecosystem services, according to likely future trends in resource demands, user numbers and relative dependency on mangrove goods and services. Running the scenario analysis indicated a progressive decline in mangrove values over the next 20 years under BAU and a sustained increase in ecosystem benefits under CCD, yielding total values of USD 95 million and USD 156 million respectively, or NPVs of USD 43 million and USD 61 million (Fig. 2.4, Table 2.4).

Using these figures, it was then possible to portray the economic implications of the two coastal development alternatives for the Kwale coastline. Should BAU continue, the economic model indicated that ecosystem services worth more than USD 41 million will be lost over the next 20 years as compared to those that would have been available had the area and quality of mangroves remained at current levels (Table 2.5). In contrast, the CCD scenario stands to generate economic gains

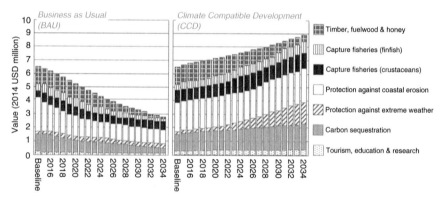

Fig. 2.4 Kwale coastline change in mangrove ecosystem service values under Business as Usual (BAU) and Climate Compatible Development (CCD) scenarios 2014–34 (Based on data presented in Emerton 2014b)

Table 2.4 Kwale coastline value of mangrove ecosystem services under BAU and CCD 2014–34 (based on data presented in Emerton 2014b)

	BAU value (US mill)		CCD value (US mill)	
	Total	NPV@10%	Total	NPV@10%
Timber, fuelwood & honey	16.17	8.01	19.82	9.20
Capture fisheries (finfish)	9.25	4.20	13.78	5.45
Capture fisheries (crustaceans)	12.92	5.43	19.68	7.42
Protection against coastal erosion	28.36	12.94	50.24	19.72
Protection against extreme weather	5.11	1.81	14.29	4.14
Carbon sequestration	20.10	9.02	33.55	13.04
Tourism, education & research	3.07	1.44	5.13	2.03
Total	94.98	42.85	156.48	61.01

Table 2.5 Incremental costs and benefits of BAU and CCD scenarios for the Kwale coastline 2014–34 (based on data presented in Emerton 2014b)

	Total (USD mill)	NPV@10% (USD mill)
Costs incurred by BAU over the baseline	−41.27	−12.38
Value-added incurred by CCD over the baseline	20.23	5.77
Value-added incurred by CCD over BAU	61.50	18.16

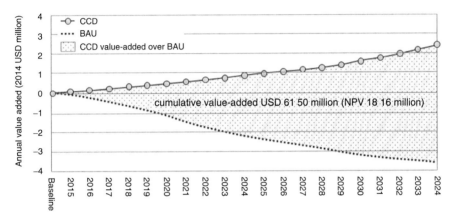

Fig. 2.5 Kwale coastline CCD value-added over BAU 2014–34 (Based on data presented in Emerton 2014b)

of more than USD 20 million as compared to the baseline. Adding these two figures together indicates the potential value-added and costs avoided of shifting from business as usual to an ecosystem-based climate-compatible development model would be in excess of USD 60 million (Fig. 2.5). This is, in effect, the return to investing in ecosystem-based CCD measures (or, conversely, the cost of policy inaction as regards mangrove conservation and rehabilitation). By the year 2034, mangrove ecosystem services will be generating values worth just under USD 10 million a year under a CCD scenario (almost 40% more than they are worth today), as compared to under USD 3 million under BAU (less than half of today's value).

2.7 Encouraging Investments in Ecosystems as Climate-Compatible Development Infrastructure

Case studies from Sri Lanka and Kenya have been used to illustrate the ways in which economic valuation can serve to articulate both the gains from investing in ecosystems as a key component of climate-compatible development infrastructure, and the losses that can result from not doing so. On the one hand, the findings from Puttalam Lagoon demonstrate clearly the cost of omitting natural ecosystems from land use development planning. Since 1992, the conversion of mangrove habitats to seemingly more 'productive' or 'profitable' uses has cost the local economy more than USD 31 million in foregone benefits. These losses amount to a sum that is more than twice as high as the income earned by shrimp farming and the other land uses that replaced mangroves. In the Kwale case study, ecosystem-based CCD options were shown to offer the potential to secure an additional USD 20 million of adaptation, disaster risk reduction and other livelihood benefits over the next 20 years as compared to those that would have been available if the area and quality of mangroves remains at current levels. This is more than one and a half times as much as the gains that would realised under a continuation of a 'business as usual' model for coastal development.

These kinds of approaches thus offer a means of generating potentially powerful – and usually much-needed – evidence and data about the economic opportunities, cost savings and avoided losses associated with ecosystem-based approaches. The resulting figures make the important point that green options have value not just because they provide a cost-effective way of securing climate adaptation and disaster risk reduction gains, but also due to the considerable development co-benefits that they generate in terms of value-added and costs avoided to other economic sectors and processes. The implication is that, from an economic perspective, ecosystems should be treated, counted and invested in as an integral part of climate-compatible development infrastructure — as a stock of facilities, services and equipment which are needed for the economy to function, grow, adapt and maintain its resilience in the fact of climate change and other hazards (Emerton 2006, 2014a).

It is, nevertheless, important to underline that valuation is not an end in itself. While a lack of economic evidence may act as a major constraint to ecosystem-based approaches being fully mainstreamed into development decision-making, the story does not end with generating strikingly large figures on costs and benefits. Even if information on ecosystem values is a necessary condition for increasing the budgetary and policy priority given to green adaptation and disaster risk reduction, by itself it is rarely sufficient. In both the Sri Lanka and Kenya cases, considerable further work was required to develop and deliver a communication strategy and set of messages which would prove convincing to coastal decision-makers. Equally importantly, however much ecosystem services are demonstrated to be worth in theory, and however convinced decision-makers are that it is in the public or private interest to invest in them, this has little meaning unless it translates into real-world changes in the way in which policies are formulated and decisions are made, and is reflected in the prices and profits that people face as they choose between alternative land, resource and investment options. Ultimately, it is those who manage, use and impact on natural ecosystems on a day to day basis who must be willing – and economically able – to invest in their continued upkeep and maintenance.

Yet, for the most part, a better understanding, and more accurate quantification, of the economic benefits of ecosystem conservation (and economic costs of ecosystem degradation and loss) is still reflected weakly in the policies, markets and prices which actually drive people's economic behaviour. The Sri Lanka case, in particular, illustrates that what might be the most profitable or desirable or beneficial land, resource or investment choice from the perspective of the wider economy is not necessarily the one which has the most immediate appeal to landholders in coastal areas. Converting land to aquaculture and agriculture makes more financial sense to local landholders than sustainably using and managing mangroves. Shrimp farming, coconut farming and salt production all generate higher cash returns and more immediate sources of earnings for the landholder – even if (as is the case for shrimp farming) this income cannot be sustained over the long-term, or imposes significant negative impacts and externalities on other groups and sectors. The bottom line is that there remain few economic incentives for landholders to maintain mangroves on their land.

The key challenge then becomes one of moving beyond merely articulating the value of ecosystem services for adaptation and disaster risk reduction, to identifying where there are needs and niches to capture these values as concrete incentives and finance for ecosystem management. The application of ecosystem valuation tools and approaches does not just involve estimating and demonstrating ecosystem service values, but also seeking solutions using economically informed policy and management instruments (TEEB 2008, 2010). The aim is to help to change the economic conditions and circumstances that cause people users to convert or degrade ecosystems in the course of their economic activities, and instead set in place the economic opportunities and rewards which will encourage, enable and motivate the investments and actions that are required for continued maintenance of valuable 'natural' climate compatible development infrastructure.

Acknowledgement This chapter presents the findings of research conducted under the iCoast project 'understanding the fiscal and regulatory mechanisms necessary to achieve CCD in the coastal zone'. The project was carried out by Edinburgh Napier University, LTS International, Birmingham University, Ruhuna University in Sri Lanka and Kenya Marine and Fisheries Institute, in collaboration with Ecometrica. It was funded by the UK Department for International Development (DFID) and the Netherlands Directorate-General for International Cooperation (DGIS) under the Climate & Development Knowledge Network (CDKN), for the benefit of developing countries. However, the views expressed and information contained in this chapter are not necessarily those of or endorsed by DFID, DGIS or the entities managing the delivery of the Climate and Development Knowledge Network, nor the project's implementing institutions, which can accept no responsibility or liability for such views, completeness or accuracy of the information or for any reliance placed on them.

References

Agardy T, Alder J, Dayton P et al (2005) Coastal systems. In: Hassan R, Scholes R, Ash, N (eds) Ecosystems and human well-being: current state and trends: findings of the condition and trends working group, Island Press, Washington, DC

Bailly D, Willmann R (2001) Promoting sustainable aquaculture through economic and other incentives. In: Subasinghe R, Phillips M, Bueno P et al (eds) Aquaculture in the third millennium. Technical proceedings of the conference on aquaculture in the third millennium, Bangkok, Food and Agriculture Organisation of the United Nations (FAO), Rome

Barbier E, Hacker S, Kennedy C et al (2011) The value of estuarine and coastal ecosystem services. Ecol Monogr 81(2):169–193

Batagoda, B (2003) The economic valuation of alternative uses of mangrove forests in Sri Lanka, UNEP/Global programme of action for the protection of the Marine Environment from Land-based Activities, The Hague

Beaudoin Y, Pendleton L (2012) Why value the oceans? A discussion paper. TEEB Office, United Nations Environment Programme, Geneva

Beck M, Shepard C (2012) Coastal habitats and risk reduction. In: World risk report. Published in collaboration with the UN University Institute for Environment and Human Security (UNU-EHS) and The Nature Conservancy (TNC), Alliance Development Works (Bündnis Entwicklung Hilft), Berlin

Bournazel J, Kumara M, Jayatissa L et al (2015) The impacts of shrimp farming on land-use and carbon storage around Puttalam Lagoon, Sri Lanka. Ocean Coast Manag 113:18–28

Cannon J (1999) Participatory workshop for the economic valuation of natural resources in the Togean Islands, Central Sulawesi. NRM/EPIQ, Jakarta

Chadburn O, Anderson C, Cabot Venton C, Selby S (2013) Applying cost benefit analysis at a community level: a review of its use for community based climate and disaster risk management. Oxfam Research Reports June 2013, Oxfam International and Tearfund, London

Clark S, Grossman T, Przyuski N et al (2012) Ecosystem-based adaptation to climate change: a cost-benefit analysis. Report prepared for Conservation International by Bren School of Environmental Science & Management, University of California at Santa Barbara

Colls A, Ash N, Ikkala N (2009) Ecosystem-based adaptation: a natural response to climate change. International Union for Conservation of Nature (IUCN), Gland

Cooper E, Burke L, Bood N (2008) Coastal capital: economic contribution of coral reefs and mangroves to Belize. World Resources Institute, Washington, DC

Dahdouh-Guebas F, Zetterstrom T, Ronnback P et al (2002) Recent changes in land-use in the Pambala-Chilaw Lagoon complex (Sri Lanka) investigated using remote sensing and GIS: conservation of mangroves vs. development of shrimp farming. Environ Dev Sustain 4:185–200

Das S (2009) Can mangroves minimize property loss during big storms? An analysis of house damage due to the super cyclone in Orissa. South Asia Network for Development and Environmental Economics (SANDEE), Kathmandu

Das S, Crépin A (2013) Mangroves can provide protection against wind damage during storms. Estuar Coast Shelf Sci 134:98–107

De Mel M, Weerathunge C (2011) Valuation of ecosystem services of Maha Oya. Environmental Foundation, Colombo

Emerton L (2006) Counting coastal ecosystems as an economic part of development infrastructure. Ecosystems and Livelihoods Group Asia, International Union for the Conservation of Nature (IUCN), Colombo

Emerton L (2009) Investing in natural infrastructure: the economic value of Indonesia's marine protected areas and coastal ecosystems. Coral Triangle Initiative, The Nature Conservancy. (TNC), Denpasar

Emerton L (2014a) Using valuation to make the case for economic incentives: promoting investments in marine and coastal ecosystems as development infrastructure. In: Essam M (ed) Economic incentives for marine and coastal conservation: prospects, challenges and policy implications. Earthscan, London

Emerton L (2014b) Valuing and investing in ecosystems as development infrastructure: economic analysis of options for climate-compatible development in coastal zones of Kenya & Sri Lanka. Technical report to iCoast project, LTS International, Penicuik

Emerton L (2014c) Assessing, demonstrating and capturing the economic value of marine & coastal ecosystem services in the Bay of Bengal Large Marine Ecosystem. BOBLME-2014-

Socioec-02, Bay of Bengal Large Marine Ecosystem project, Food and Agriculture Organisation of the United Nations (FAO), Phuket

Emerton L, Kekulandala B (2003) The economic value of Muthurajawela wetland, Sri Lanka. International Union for the Conservation of Nature (IUCN) Regional Environmental Economics Programme and Sri Lanka Country Programme, Colombo

Ferrario F, Beck M, Storlazzi C et al (2014) The effectiveness of coral reefs for coastal hazard risk reduction and adaptation. Nat Commun. doi:10.1038/ncomms4794

Gunawardena M, Rowan J (2005) Economic valuation of a mangrove ecosystem threatened by shrimp aquaculture in Sri Lanka. Environ Manag 36(4):535–550

Haisfield K, Fox H, Yen S et al (2010) An ounce of prevention: cost-effectiveness of coral reef rehabilitation relative to enforcement. Conserv Lett 3(2010):243–250

Huxham M (2013) Economic valuation of mangrove ecosystem services in Southern Kenya. Report prepared for iCoast Project, School of Life, Sport and Social Sciences, Napier University, Edinburgh

Huxham M, Emerton L, Kairo J et al (2015) Envisioning a better future for Kenyan mangroves: comparing business as usual with climate compatible development. J Environ Manag 157:168–181

IUCN (2012) An Environmental and Fisheries Profile of the Puttalam Lagoon System. Regional Fisheries Livelihoods Programme for South and Southeast Asia (GCP/RAS/237/SPA) Field Project Document 2011/LKA/CM/06. International Union for Conservation of Nature (IUCN), Sri Lanka Country Office, Colombo

King L, Nap G (2013) What the future holds: visioning as a tool to develop climate compatible development scenarios: tourism as a case study. iCoast Policy Brief, LTS International, Penicuik

Kirui K, Kairo J, Bosire J et al (2013) Mapping of mangrove forest land cover change along the Kenya coastline using Landsat imagery. Ocean Coast Manag 83:19–24

Kumara M (2014) Field notes on the returns to shrimp farming and other land uses in mangrove areas of Puttalam Lagoon. Report prepared for iCoast Project, Ocean University of Sri Lanka, Tangalle

Kumara M, Jayatissa L (2013) Case study template for Puttalam Lagoon. Report prepared for iCoast Project, Ocean University of Sri Lanka, Tangalle

Millennium Ecosystem Assessment (2005) Ecosystems and human well-being: synthesis. Island Press, Washington, DC

MMC (2005) Natural hazard mitigation saves: an independent study to assess the future savings from mitigation activities. Findings, conclusions, and recommendations, vol 1. Multihazard Mitigation Council, National Institute of Building Sciences, Washington, DC

Mohammed E (2012) Payments for coastal and marine ecosystem services: prospects and principles. International Institute for Environment and Development (IIED), London

MPP-EAS (1999) Total economic valuation: marine and coastal resources in the Straits of Malacca. MPP-EAS technical report no. 24/PEMSEA technical report no. 2, GEF/UNDP/IMO Regional Programme for the Prevention and Management of Marine Pollution in the East Asian Seas (MPP-EAS)/Partnerships in Environmental Management for the Seas of East Asia (PEMSEA), Quezon City

MRAG (2009) Fisheries and subsidies. Policy brief 9, Marine Resources Assessment Group Ltd. (MRAG) and UK Department for International Development (DFID) London

Munroe R, Doswald N, Roe D et al (2012) Does EbA work? A review of the evidence on the effectiveness of ecosystem-based approaches to adaptation. IIED, BirdLife International, Cambridge University and UNEP-WCMC, Cambridge

Naumann S, Anzaldua G, Berry P et al (2011) Assessment of the potential of ecosystem-based approaches to climate change adaptation and mitigation in Europe. Final report to the European Commission, DG Environment, Contract no. 070307/2010/580412/SER/B2, Ecologic Institute and Environmental Change Institute, Oxford University Centre for the Environment

OECD (2008) Pro-poor growth and natural resources: the economics and politics. Development Cooperation Directorate, Development Assistance Committee, OECD, Paris

Powell N, Osbeck M, Sinh Bach Tan, Vu Canh Toan (2010) World resources report case study: mangrove restoration and rehabilitation for climate change adaptation in Vietnam. World Resources Report, Washington DC

Primavera J (1997) Socio-economic impacts of shrimp culture. Aquac Res 28:815–827

ProAct Network (2008) The role of environmental management and eco-engineering in disaster risk reduction and climate adaptation. ProAct Network, Nyon

Rao N, Carruthers T, Anderson P et al (2013) An economic analysis of ecosystem-based adaptation and engineering options for climate change adaptation in Lami Town, Republic of the Fiji Islands. Secretariat of the Pacific Regional Environment Programme (SPREP), Apia

Renaud F, Sudmeier-Rieux K, Estrella M (2013) The role of ecosystems in disaster risk reduction. United Nations University Press, Tokyo

Rideout A, Joshi N, Viergever K et al (2013) Making predictions of mangrove deforestation: a comparison of two methods in Kenya. Glob Chang Biol 19:3493–501

SCBD (2009) Biodiversity for development and poverty alleviation. Secretariat of the Convention on Biological Diversity, Montreal

Shepard C, Crain C, Beck M (2011) The protective role of coastal marshes: a systematic review and meta-analysis. PLoS ONE 6(11), e27374. doi:10.1371/journal.pone.0027374

Shreve C, Kelman I (2014) Does mitigation save? Reviewing cost-benefit analyses of disaster risk reduction. Int J Disaster Risk Reduct 10:213–235

Spalding M, Ruffo S, Lacambra C et al (2014) The role of ecosystems in coastal protection: adapting to climate change and coastal hazards. Ocean Coast Manag 90:50–57

Sudmeier-Rieux K (2013) Ecosystem approach to DRR: basic concepts and recommendations to governments, with a special focus on Europe

TEEB (2008) The economics of ecosystems and biodiversity: an interim report. European Commission, Brussels

TEEB (2010) The economics of ecosystems and biodiversity: mainstreaming the economics of nature. A synthesis of the approach, conclusions and recommendations of TEEB. United Nations Environment Programme (UNEP), Nairobi

Temmerman S, Meire P, Bouma T et al (2013) Ecosystem-based coastal defence in the face of global change. Nature 504(7478):79–83

Tompkins E, Mensah A, King L et al (2013) An investigation of the evidence of benefits from climate compatible development. Centre for Climate Change Economics and Policy working paper no. 124, London School of Economics and Sustainability Research Institute Paper No. 44, University of Leeds

UN Global Compact, UNEP, Oxfam, WRI. (2011) Adapting for a green economy: companies, communities and climate change. A caring for climate report by the United Nations Global Compact, United Nations Environment Programme (UNEP), Oxfam, and World Resources Institute (WRI), UN Global Compact Office, New York

UNEP (2004) Analyzing the resource impact of fisheries subsidies: a matrix approach. Economics and Trade Branch, United Nations Environment Programme (UNEP), Geneva

UNEP (2011) Making the case for ecosystem-based adaptation: building resilience to climate change. United Nations Environment Programme (UNEP), Nairobi

UNEP-WCMC (2006) In the front line: shoreline protection and other ecosystem services from mangroves and coral reefs. World Conservation Monitoring Centre, Cambridge

UNEP-WCMC (2011) Marine and coastal ecosystem services: valuation methods and their application. UNEP-WCMC Biodiversity series no. 33, UNEP World Conservation Monitoring Centre, Cambridge

van Beukering P, Brander L, Tompkins E, McKenzie E (2007) Valuing the environment in small Islands: an environmental economics toolkit. Joint Nature Conservation Committee, Peterborough

Wattage P (2011) Valuation of ecosystem services in coastal ecosystems: Asian and European perspectives. Ecosystem Services Economics (ESE) working paper series no. 8, Division of Environmental Policy Implementation, United Nations Environment Programme (UNEP), Nairobi

Weragodatenna D (2010) An atlas of the Puttalam Lagoon area. IUCN – International Union for Conservation of Nature, Colombo

Westers T (2012) Assessing sustainability of smallholder shrimp farms in Sri Lanka. Thesis submitted to the Faculty of Graduate Studies in partial fulfilment of the requirements for the degree of Master of Science, Faculty of Veterinary Medicine, Calgary, Alberta

World Bank and FAO (2009) The sunken billions – the economic justification for fisheries reform. The World Bank, Washington, DC

Chapter 3
Cost Benefit Analysis for Ecosystem-Based Disaster Risk Reduction Interventions: A Review of Best Practices and Existing Studies

Marta Vicarelli, Rohini Kamal, and Maria Fernandez

Abstract Cost Benefit Analysis (CBA) is underutilised in assessing Ecosystem-based Disaster Risk Reduction (Eco-DRR) interventions, the protocols used are not always rigourous and the analytical framework is unclear. However, CBAs which follow best practices could be extremely beneficial and helpful to policy makers in establishing priorities for Eco-DRR interventions. A robust and systematic economic analytical approach might be useful, if not necessary, to justify large upfront investments and promote the implementation of this type of risk reduction intervention at an even broader scale. Identifying a common core of best practices for CBA applied to Eco-DRR would also increase comparability between studies, reproducibility of assessments, and facilitate much needed external review. The purpose of this chapter is to (i) outline the fundamental principles and best practices of rigourous cost-benefit analysis (CBA) applied to ecosystem-based adaptation (EbA) and (Eco-DRR) interventions; (ii) review existing studies; and – based on this review of past work – (iii) outline the possible areas of improvement to strengthen future CBAs of Eco-DRR projects.

Keywords Cost-benefit analysis • Program evaluation • Best practices • Review

M. Vicarelli (✉)
Department of Economics, University of Massachusetts, Amherst, MA, USA

Center for Public Policy Analysis, University of Massachusetts, Amherst, MA, USA
e-mail: mvicarelli@econs.umass.edu

R. Kamal
Department of Economics, University of Massachusetts, Amherst, MA, USA

M. Fernandez
Center for Public Policy Analysis, University of Massachusetts, Amherst, MA, USA

Isenberg School of Management, University of Massachusetts, Amherst, MA, USA

© Springer International Publishing Switzerland 2016   45
F.G. Renaud et al. (eds.), *Ecosystem-Based Disaster Risk Reduction and Adaptation in Practice*, Advances in Natural and Technological Hazards Research 42,
DOI 10.1007/978-3-319-43633-3_3

3.1 Introduction

The purpose of this chapter is to (i) outline the fundamental principles and best practices of rigourous cost-benefit analysis (CBA) applied to ecosystem-based adaptation (EbA) and ecosystem-based disaster risk reduction (Eco-DRR) interventions; (ii) review existing studies; and – based on this review of past work – (iii) outline the possible areas of improvement to strengthen future CBAs of Eco-DRR projects. The motivation behind this chapter is that CBA is underutilised in assessing Eco-DRR interventions, the protocols used are not always rigorous and the analytical framework is unclear. However, CBAs which follow best practices could be extremely beneficial and helpful to policy makers in establishing priorities for Eco-DRR interventions (also see Vackar et al., Chap. 5).

There is a growing literature on Eco-DRR interventions, however existing studies usually highlight the environmental and socio-economic benefits of such interventions but they seldom discuss long-term costs and benefits in economic terms (also see Emerton et al, Chap. 2). And, in particular, social welfare implications at the community or household level are hardly examined. In fact, to our knowledge, most studies are qualitative and very few apply a systematic methodological framework for performing an economic assessment of long-term costs and benefits (IFRC 2011).

Cost benefit analysis (CBA) has its limitations and it should not be considered as either necessary or sufficient for making policy decisions. However, it can provide a very useful framework for consistently organizing disparate information. It can help to compare and choose between policies and programmes. And it can also be useful to retrospectively assessing existing interventions. In general, by illuminating the tradeoffs involved in making different kinds of social investments, CBA may inform policy decisions and improve their outcomes (Arrow et al. 1996; Goulder and Stavins 2002).

The abundant qualitative assessments of Eco-DRR interventions seem to show that benefits are disproportionally larger than implementation costs, especially in light of current climate change projections (IPCC 2014). Yet, the uptake of Eco-DRR approaches is slow and one reason is that their net benefits are undervalued (Renaud et al. 2013). A robust and systematic economic analytical approach might be useful, if not necessary, to justify large upfront investments and promote the implementation of this type of risk reduction intervention at an even broader scale. Identifying a common core of best practices for CBA applied to Eco-DRR would also increase comparability between studies, reproducibility of assessments, and facilitate much needed external review. This chapter attempts to identify a core of best practices to implement CBA to Eco-DRR projects, including data requirements and collection strategies.

The remainder of this study is organised in three sections. The first section outlines the fundamentals of CBA, presents best practices to effectively perform CBA of Eco-DRR interventions, and discusses optimal data collection strategies and protocols. A box at the end of the section proposes a didactic exercise: it superimposes a CBA framework to an existing Eco-DRR project in Bangladesh. The second section reviews in detail existing Eco-DRR studies applying CBA, and discusses strengths and limitations of their methodological approach. The final section concludes and provides recommendations.

3.2   Economic Analytical Framework for Assessing Long-Term Costs and Benefits of Eco-DRR Interventions.

The rigourous assessment of Eco-DRR policies and programmes presents many challenges due to the interdisciplinary nature of the subject. Standard protocols and best practices for CBA (Arrow et al. 1996; Goulder and Stavins 2002) are at the core of our analytical framework, informed by the ecology and economics literature on valuation of environmental goods and services (Heal et al. 2005; Costanza et al. 2014). The rich development economics literature on experimental design has also provided invaluable insights in elaborating data collection strategies, especially for community-level studies (e.g. Duflo et al. 2006).

3.2.1   Fundamentals of Cost Benefit Analysis

In evaluating policies or projects, economists will recommend the most *efficient* one: the project for which the net benefits (i.e. NB, the difference between total benefits and total costs) are maximized. Projects and policies have a stream of costs and benefits over time. When an analysis includes multiple time periods, economists will speak of *dynamic efficiency*. Assessing efficiency in a dynamic setting requires *discounting*, a mathematical operation that translates the stream of costs and benefits into a single monetary value, the present value (PV). The current value of a future cost or benefit (PV) is obtained by discounting the future sums of money to equivalent current sums, using the following formula:

$$PV = \frac{Future\,value\,in\,t^{th}\,period}{(1 + discount\,rate)^t}$$

where t indicates the time periods (i.e. *year 0, 1, 2, 3, . . ., t* etc.) and the *discount rate*, or annual rate of return on the investment. Suppose a project involves benefits and costs over a time span from the present moment (time 0) to T years from now. Let B_t and C_t be, respectively, the benefit and cost t years from now. The present value of net benefits (PVNB, sometimes also indicated as *net present value* NPV) is calculated using this formula:

$$PVNB = \sum_{t=0}^{T} (B_t - C_t)/(1 + r)^t$$

In a dynamic setting (i.e. multiple periods), an efficient policy maximises the present value of net benefits to society (PVNB). Some studies also report the *benefit cost ratio* (BCR): the ratio of the present values of benefits and costs.

There are three sources of uncertainty in the PVNB estimate: the choice of the discount rate r, the choice of time horizon (T), and the very estimates of future costs and benefits. The best practice is to perform *sensitivity tests* in estimating PVNB using a range of values for the discount rate, time horizon, and future benefits and costs.

Discount rates adopted usually range from 0 to 10 %, which is the approximate marginal pretax rate of return on an average investment in the private sector. The choice of discount rate has important *intergenerational equity* implications: a large discount rate will discount more (i.e. value less) benefits and costs of future generations (Kelman 1991; Stern 2007). This point is further expanded below.

When PVNB is positive (i.e. BCR is bigger than 1), the benefit to the winners is larger than the losses to the losers. Theoretically, after compensation from winners to losers, the policy would yield what economists call a 'Pareto Improvement': some individuals would be better off and no individual would be worse off. The PVNB analysis is based on the Potential Pareto Improvement (PPI) criterion (Goulder and Stavins 2002). However, the compensation may not be possible and may not actually take place. Indeed, the PPI criterion, or PVNB efficiency analysis, does not take into account *distributional or social (intra-generational) equity* considerations. The best practice is therefore to always complement CBA with evaluation criteria that incorporate equity analysis and identify distributional consequences (Kelman 1991; Arrow et al. 1996).

3.2.2 Cost Benefit Analysis for Eco-DRR Intervention: Best Practices

In their seminal paper on "The Role of Cost Benefit Analysis in Environmental, Health and Safety Regulations" (1996), Kenneth Arrow and ten other prominent fellow economists outlined the best practices of CBA, and stressed the importance of adopting a common set of economic assumptions to increase the feasibility of comparisons across analyses. This section outlines the key assumptions and best practices that make CBA a useful framework to organise disparate information in assessing and/or comparing Eco-DRR interventions.

(i) Define the *scope*, or objective of the analysis. CBA may be forward-looking, when comparing and choosing between policies or projects to be implemented, or retrospective (backward-looking), to evaluate policies already in place.

(ii) Identify the *geographic scale* of the study, and the *population affected.* Benefits and costs may differ across different spatial scales. Accurately identifying stakeholders and communities affected directly and/or indirectly by the intervention is necessary to providing the best estimates of socio-economic costs and benefits. Moreover the choice of data collection strategies depends on geographic scale and stakeholders' analysis. Community-level projects always require high-resolution data collection strategies.

(iii) The net benefits are always measured with respect to a *reference baseline scenario.* In assessing an ecosystem restoration project, the baseline scenario would be the state of no intervention. Only a clear definition of the reference baseline scenario allows a systematic comparison between the conditions before and after a project. Another type of Eco-DRR assessment may focus on the costs and benefits of preserving an existing ecosystem. In these cases, the baseline scenario would be the absence (or partial degradation) of the ecosystem.

Many Eco-DRR projects include complementary *concurrent interventions* aimed at mitigating risk (e.g. community awareness/preparedness programs) or generating new revenues (e.g. sustainable land management, livelihood diversification, ecotourism). The definition of the scope of the intervention and baseline scenario is crucial. The program evaluation of an Eco-DRR intervention should isolate the Eco-DRR component from possible confounding factors to avoid estimation errors. A biased CBA could assign a large PVNB to an expensive reforestation program when, in fact, the community vulnerability was effectively reduced thanks to a less costly disaster preparedness plan. For this reason, the baseline scenario and scope of the intervention should emphasise the distinction between core Eco-DRR interventions and corollary activities.

(iv) Set the *time horizon* of the appraisal and distinguish between: (a) the duration of the project implementation, and (b) the longevity of its net benefits. They may differ. Monitoring of projects tends to correlate with donors' project cycles; however costs and benefits may extend beyond the implementation of the project. Neglecting the real extent of benefits over time would compromise the validity of the analysis. Hence, the study should (i) ensure that for every time period in the project timeline all costs and benefits are included, (ii) provide a best estimate of the longevity of Eco-DRR costs and benefits (even beyond the project implementation), and (iii) clarify the assumptions adopted to estimate it. This exercise may require taking into account forecasts of frequency and intensity of hazard events.

(v) Best estimates of *costs and benefits* should always be presented along with a description of the uncertainties. Arrow et al. (1996) argue that government agencies (or international agencies) should establish a set of default values for typical benefits and costs, and develop a standard format for presenting results. In fact, there are no such default standards for CBA. Moreover, Eco-DRR studies may strongly differ depending on the context, hazards and ecosystems. Nevertheless, for comparability purposes, it is useful to identify categories of costs and benefits to be systematically included in the analyses. Below, we propose a core set of costs and benefits to be complemented on a case-by-case basis by context-specific data:

Standard implementation costs for ecosystem restoration/preservation projects include basic expenditures: planning, training, awareness-building in the communities, project development and maintenance, data collection (monitoring), and evaluation. Possible opportunity costs of not using the area for other revenue-producing activities may also be included.[1]

Standard economic benefits are estimated in terms of savings in operation or maintenance costs of existing physical infrastructure (with respect to the baseline scenario); as well as damage costs avoided to agro-ecosystems, private assets, and public property (buildings, infrastructure).

Environmental net benefits, such as environmental goods and ecosystem services, are a crucial component in the assessment of Eco-DRR interventions. And yet, they are often underestimated or not included. Estimates of the value of global ecosystem services have progressively evolved and are publicly available since 1997 (Costanza et al. 1997, 2014). The most recent effort to update global estimates has been promoted by the United Nations Environment Programme's global initiative: *The Economics of Ecosystems and Biodiversity* (TEEB 2010a, b). De Groot et al. (2012) estimate the value of ecosystem services in monetary units provided by 10 main biomes (open oceans, coral reefs, coastal systems, coastal wetlands, inland wetlands, lakes, tropical forests, temperate forests, woodlands, and grasslands) based on local case studies across the world.[2] Many research groups are currently working on better understanding, modelling, valuation, and management of ecosystem services and natural capital. Their efforts are monitored and coordinated by emerging regional, national, and global networks, like the Ecosystem Services Partnership (ESP) (Costanza et al. 2014). New free online tools like the Natural Capital Project's InVEST allow users to "quantify natural capital in biophysical, socio-economic and other dimensions, to visualise the benefits delivered today and in the future, to assess the tradeoffs associated with alternative choices, and to integrate conservation and human development aims".[3] These tools and models enable dynamic CBA analysis by taking into account how ecosystem services may change over time.

---

[1]This is appropriate only if the cost has not been included as benefit/revenue in the baseline scenario.

[2]These studies covered a large number of ecosystems, types of landscapes, different definitions of services, different areas, different levels of scale, time and complexity and different valuation methods. In total, approximately 320 publications were screened and more than 1350 data-points from over 300 case study locations were stored in the Ecosystem Services Value Database (ESVD). Available via http://www.fsd.nl/esp/80763/5/0/50. A selection of 665 of these value data points were used for the analysis, values were expressed in terms of 2007 'International' $/ha/ year, i.e. translated into US$ values on the basis of Purchasing Power Parity (PPP) (de Groot et al. 2012).

[3]Natural Capital Project (http://www.naturalcapitalproject.org/) is a partnership between Stanford University, the University of Minnesota, The Nature Conservancy and the World Wildlife Fund. Together they have created InVEST ('Integrated Valuation of Ecosystem Services and Tradeoffs'), a free and open-source software suite. Other tools include ARIES: http://www. ariesonline.org/about/intro.html and ArcSWAT: http://swat.tamu.edu/software/arcswat/

Additional *socioeconomic costs and benefits* may depend on the scale of the project:

- In *large-scale projects*, socioeconomic costs and benefits may include changes in agricultural production, revenues (associated to direct/indirect business interruption), property values, relief expenditure or government aid, morbidity, and mortality.
- *Small-scale, community-level projects* may include the above indicators but also require more high-resolution data collection efforts, such as interviews, focus groups and household surveys. These practices, increasingly popular in CBAs for community-level disaster risk reduction (DRR) projects for collecting high resolution indicators (Venton 2010; Shreeve and Kelman 2014) are also applicable to Eco-DRR community-level projects (Table 3.1), including: household income; food and non-food expenditures; household assets (e.g. property, livestock); value of crop-yields; level of credit and insurance payouts (when applicable); unemployment; and human capital, i.e. household health expenditures (due to injury and diseases), school attendance, school achievements, and child-labor.[4]

Unfortunately reliable data are not always available but best estimates should always be provided. Lastly, *intangible (non-quantifiable) costs and benefits* are usually excluded from the quantitative analysis but should nevertheless be highlighted if they may significantly affect the final results of the CBA. For instance, the possible degradation of the cultural heritage of indigenous groups in risk-prone areas may be impossible to quantify but it may carry an important weight in the final decision about the implementation of a project. Multi Criteria Decision Making (MCDM) strategies, also known as Multi Criteria Analysis (MCA), or Multi-Criteria Evaluation (MCE), may complement – or even provide an alternative to – CBAs in assessing intangible costs and benefits. MCDMs do not evaluate public expense *efficiency* (i.e highest net present value*)*, but only its *effectiveness* towards achievable goals on the basis of other criteria (e.g. human safety, human rights) (Boyce 2000). Indeed, MCDM indicators are not defined in monetary terms, their construction is usually based on ranking, weighting, and scoring of qualitative impact categories and criteria. MCDMs take into account multiple stakeholders and multiple decision-making criteria.

(vi) *Socio-economic data collection methodologies* should be thoroughly documented and clarified in the final report. Reliable and complete information on data sources and collection protocols are indispensable to ensuring reproducibility of the study, effectiveness of external review, and comparability between studies. This is particularly important for variables subject to high levels of uncertainty, such as the value of environmental goods and services. When socioeconomic

---

[4]To our knowledge, existing assessments of community-level Eco-DRR projects have seldom included socio-economic impacts at the household or community level, and when this happens data is self-reported, collected retroactively and analysis are mostly qualitative.

Table 3.1 Households and communities offer two different levels of data resolution and different types of socio-economic variables can be collected. Drawing from standard development economics protocols, this table summarizes, at both the households and community levels, the types of socioeconomic variables that can be collected, the required collection methodologies, and the respondents

Unit of observation	Variables	Data collection method	Respondent
Household	ECONOMIC STATUS		
	Income		
	Assets: land, livestock		
	Savings		
	Crop-yields		
	Agricultural revenues		
	Credit and insurance		
	Employment status	1. Focus groups	
	Soil productivity		1. Members of the community
	Water quality	2. Panel survey:	2. Head of household
	MIGRATION	Baseline survey	3. Individual household members (in detailed surveys)
	Temporary-permanent	Follow-up survey(s)	
	Remittances		
	HAZARDS		
	Occurrence		
	Observed damages		
	Government aid received		
	HUMAN CAPITAL		
	Health:		
	Mortality		
	Morbidity		
	Health expenditures		
	Children:		
	Child labor		
	School attendance		
	School Achievements		

(continued)

Table 3.1 (continued)

Unit of observation	Variables	Data collection method	Respondent
Community	ECONOMICS		
	Employment level		
	Property values		
	MIGRATION	Panel survey:	
	Temporary-permanent	Baseline survey	Community leader(s)
	Remittances	Follow-up survey(s)	
	HAZARDS		
	Occurrence		
	Observed damages		
	Relief expenditure		
	Government aid received		
	HUMAN CAPITAL		
	Health:		
	Mortality		
	Morbidity		

data are collected through surveys or interviews, copies of the questionnaires, as well as the survey schedule and the survey protocols should be provided for external reviews to check for possible data collection bias or contamination of the results. A broader discussion and more recommendations about data collection strategies are presented in Sect. 1.3.

(vii) Given uncertainties in identifying the correct *discount rate r*, it is appropriate to use a range of rates (Arrow et al. 1996). The choice of the optimal discount rate is an ongoing debate.

Since 1992, the US federal Office of Management and Budget (OMB) has recommended a 7 % real discount rate for the analysis of federal programs, noting that "this rate approximates the marginal pretax rate of return on an average investment in the private sector" (OMB 1992). If the value of PVNB is positive, the project will yield a higher return than the market interest rate. This rate may be appropriate for projects that exclusively affect the allocation of capital, but does not seem appropriate for projects that affect private consumption.

Arrow et al. (1996) discussed the choice of discount rate and noted that the rate at which costs and benefits are discounted will generally not equal the rate of return on private investments. It should reflect how individuals trade off future consumption. The *Social Rate of Time Preference* is often adopted in response to this concern. This is the rate at which society is ready to substitute present for future consumption. In the United States, the federal opportunity cost of capital (rate on treasury bonds) is generally used as a proxy. For instance, the National Oceanic and Atmospheric Administration (NOAA) has adopted a 3 % discount rate as a proxy for the social rate of time preference for discounting interim service losses and restoration gains when scaling compensatory restoration (NOAA 1999).

Some economists and ethicists, concerned by *intergenerational equity* issues, recommend an even lower discount rate. The higher the discount rate, the lower is the weight of future costs and benefits in the analysis. A zero discount rate would attribute the same weight to present and future generations and prevent present generations from ignoring or underestimating the long-term social impacts of present day decisions. The idea behind choosing a 0 discount rate is that present generations have the moral obligation to protect the interests of future generations, because these future recipients of benefits are yet unborn and cannot express their own (future) preferences (Kelman 1991; NOAA 1999; Stern 2007). In conclusion, due to intergenerational distribution and equity concerns, it is very important to always clarify the assumptions behind the choice of discount rate.

(viii) Perform *sensitivity analysis* to test sensitivity of PVNB to minor changes in numeric inputs. Tests require using a range of values for: discount rate, time horizon T, estimates for costs and benefits with a high degree of uncertainty. Sensitivity tests are becoming a standard protocol in numerous organisations (e.g. US-OMB 2003).[5] Non-linearities and threshold effects associated with ecosystem benefits should be explored in the sensitivity tests (Perrings and Pearce 1994; Folke et al. 2004). Lastly, considerations about the impacts of climate change, in terms of probabilistic climate scenarios, addressing potentially catastrophic outcomes, are also recommended to further increase the reliability of sensitivity analysis (Weitzman 2009).

(ix) The *CBA final report* should always indicate the results of the sensitivity tests and the assumptions used. It is good practice to highlight the highest, intermediate, and lowest value of the PVNB. And estimates should always be presented along descriptions of the uncertainties (Arrow et al. 1996; Goulder and Stavins 2002; Graham 2008). Policy makers should have less confidence in studies where the sign of the PVNB is highly sensitive to the discount rate or to small changes in future benefits and costs (Arrow et al. 1996; Goulder and Stavins 2002).

(x) CBA assesses the efficiency of policies or programs. However, a good policy analysis should also identify and discuss important *distributional consequences*, namely, intra- and inter-generational equity considerations about the impacts of the program on subgroups of the population. Social Impact Assessments (SIA) and equity analysis techniques are synergistic with Environmental Justice Analysis (EJA) and may provide helpful insights on equity considerations as well as possible health, demographic, and market changes (NOAA 1994, 2007; EPA 2014).

---

[5]The U.S. Office of Management and Budget (US-OMB) published a formal guidance document that outlines the US Government's standards for regulatory analysis, especially CBA (Circular A-4, September 2003), stressing the necessity "to provide a sensitivity analysis to reveal whether, and to what extent, the results of the analysis are sensitive to plausible changes in the main assumptions and numeric inputs".

(xi) *Peer reviews* of economic analysis should be used for projects with potentially large environmental and/or socioeconomic impacts. "The more external review the cost benefit analyses receives, the better they are likely to be" (Arrow et al. 1996, p. 222). Trans-disciplinary in nature with costs and benefits affecting both environmental and human capital, CBA applied to Eco-DRR projects may be very complex and should require external reviews by experts from different disciplines, in both the natural and social sciences.

3.2.3   Socioeconomic Data Collection Strategies for Retrospective CBA

Existing CBA studies can be divided in two categories: (i) forward looking assessment aimed at comparing and choosing between policies; and (ii) retrospective backward looking studies assessing the role of ecosystems (natural or managed) in responding to hazards that have already occurred. The latter category represents a form of *programme evaluation*. Arrow et al. (1996) stress the importance of periodically carrying out *retrospective assessments* of selected regulatory impact analyses to inform future policies and programmes.

Collecting data during the Eco-DRR project implementation should become a universally accepted best practice, and data collection should be part of the original project design. It allows to perform retrospective assessments, and to indirectly or directly inform future policies or programs. This practice may also have important positive implications for the development of the project itself: continuous monitoring allows intermediate programme assessments, useful to iteratively improve and calibrate the intervention's efficiency over time, which in turn maximises returns to investment of donors' resources.

In the past twenty years, development economists have refined programme evaluation frameworks to rigorously assess the socioeconomic impacts of projects or programmes, especially in developing countries through experimental and quasi-experimental techniques (Gerber and Green 2012; Duflo et al. 2006; Glennerster and Takavarasha 2013). Best practices in programme evaluation are now standardised and provide excellent tools for data collection strategies and analytical framework design, especially to analyse household and community-level data. A new project is evaluated in its *pilot phase*, before expanding the project at full scale. Any expansion is conditional on the positive outcome of the pilot phase evaluation. Data collection strategies and evaluation protocols are always included in the original project design. This enables project managers to collect a baseline database before the project starts, maximising efficiency in data collection and measurement standards.[6]

---

[6]In our literature review (Sect. 3.2.1) we have not found any study that adopts standard program evaluation techniques. The only studies that attempted to create a baseline dataset using a survey, did so only after the completion of the project, by asking respondents to report their best recollection.

A rigourous CBA programmme evaluation of the pilot project has many bene-fits: (i) more abundant and reliable data, tailored to assessing to real scope and impacts of the project; (ii) even when no extreme event occurs during the pilot phase, it provides insights on programme performance for recalibrating investments and improving implementation protocols during the project expansion; (iii) it enables reliable data gathering to better understand the socio-economic implica-tions of Eco-DRR projects and inform future project design.

Programme evaluation may be strengthened by *randomisation*: during the pilot phase, two comparable subsets of the intervention area are identified as *treatment* and *control samples*. The evaluation compares net benefits between the intervention area (treatment) and baseline scenario (control). Ravallion (2012) contributed to the programme evaluation debate by highlighting the possible limitations of the randomisation process: the ability to randomise the distribution of treatment may limit the type of questions and programmes analysed.

The design of the survey to assess the programme (project) is usually preceded and informed by focus groups interviews, carried out to identify areas of concern for communities and households. The final *panel survey* includes baseline and follow-up surveys with the same respondents. Households and communities offer two different levels of data resolution and different types of socio-economic vari-ables can be collected. Drawing from standard development economics protocols, Table 3.1 summarises, at the households and community level, the types of socio-economic variables that can be collected, the required collection methodologies, and the suitable respondents. The best practice should be for survey data points to have spatial coordinates to be spatially merged with environmental data. Spatial analysis techniques are an indispensable tool for evaluating Eco-DRR interven-tions. They allow combining spatial data on demographics, socioeconomics, geo-morphology, hazard impacts, land-use features (e.g. forest cover), and evolution in ecosystems services.[7]

Lastly, after the pilot evaluation and before expanding the project to the entire area, benefit and cost estimates obtained during the pilot can be used to build reliable CBA estimates for a forward-looking study covering the entire intervention area.

Box: Case Study: Bangladesh Afforestation Program*

This box proposes a didactic exercise: we superimpose the CBA framework and guidelines outlined in Sect. 3.1 to show its usefulness to an existing Eco-DRR project in Bangladesh. Implemented in flood-prone areas of Bangladesh from 2009 to 2013, the *Community Based Adaptation to Climate Change through Coastal Afforestation* programme (CBACC-CA) focuses on mangrove restoration to foster climate adaptation and risk mitigation. No

(continued)

---

[7]For instance, the InVest model provides spatial maps of ecosystems services scenarios.

formal programme evaluation was performed for this project. We deconstruct the CBACC-CA programme to outline key components of a retrospective CBA framework with emphasis on spatial data integration. This programme represents a useful case study due to the magnitude of disaster risk in Bangladesh, the large size of population exposed, the nexus between risk and poverty as well as vast DRR investments, potential socio-economic impacts and public policy implications. Indeed, evaluations of programmes of this magnitude would be useful for informing future interventions.

- *Motivation:* Bangladesh's climate history and future projections make managing disaster risks a national priority. The risk of floods, and consequent loss of life and property is one of the highest risk factors in Bangladesh, due the significant proportion of its population living in low-lying coastal zones and flood plains (IPCC 2014). Climate change is projected to exacerbate floods linked to extreme rainfall events, rising sea level, and tropical storms; thus increasing exposure and vulnerability of the growing population (Lichter et al. 2011; Mimura 2013; IPCC 2014).
- *Scope:* Mangrove afforestation is designed to reduce risk exposure and improve climate adaptation. Mangroves act as a natural barrier protecting the lives and property of coastal communities from storms and cyclones, flooding, and coastal soil erosion (Menéndez and Priego 1994, IPCC 2014).
- *Geographic Scale* Project sites include multiple coastal areas in the Chittagong, Noakhali, Bhola, and Patuakhali districts (Fig. 3.1a). These districts are densely populated**, (Fig. 3.1b), and with pockets of extreme poverty (Fig. 3.1c) in cyclone affected areas, prone to floods (Fig. 3.1d). The coastal region is predominantly agrarian, and the primary sources of rural livelihoods of the local communities are agriculture, fisheries, forestry, and livestock (Ahammad et al. 2013). Rural poverty is projected to increase due to land degradation associated to climate change.
- *Baseline scenario:* The mangrove cover is rapidly declining with over 50 % of coastal areas underutilised due to exposure to soil salinity and tidal inundation (CBACC-CA 2008 Project Document).
- *Concurrent interventions:* (i) community agro-ecosystem practices to improve livelihoods; (ii) capacity building of local communities, and government officials; (iii) national policy review in light of project development; and (iv) knowledge dissemination.
- *Time Horizon(s):* CBA sensitivity tests may span multiple time horizons***. The project period is 2009–2013, and it takes about 25 years for mangrove forests to reach maturity. However, benefits can be measured well after the 25 years horizon (Harvey 2007).

(continued)

- *Costs and benefits:* Expenditures associated to the programme implementation are well documented and amount to about 5.4 million USD from the
  Government of Bangladesh (GoB) the United Nations Development Program (UNDP), and the Global Environment Facility (GEF). The project
  was implemented by UNDP, and jointly monitored by the GoB and the
  GEF (Government of Bangladesh and UNDP 2008). However, benefits are
  not measured in monetary terms. Programme documents do not report the
  value of benefits in terms of avoided risk or provided ecosystem goods and
  services. They indicate the number of trees planted (i.e. 8,500). Tools such
  as InVEST could help in creating a catalogue of benefits associated to the
  status of ecosystem services over time, using spatially-defined scenarios.
  The use of *spatial analysis* tools allows combining multiple dimensions:
  geography, exposure to risk, demographics, and socio-economic variables,
  including poverty levels (Fig. 3.1a).
- *Socio-economic data collection:* No data were collected on the impacts of
  mangrove afforestation on livelihoods. In projects of this size, the best
  strategy would be to develop a pilot phase, targeting a subset of the total
  region, with treatment and control areas to collect baseline and follow-up
  surveys. The pilot may include multiple treatment categories, in this case:
  areas treated with mangrove afforestation, and areas where agro-
  ecosystem practices are also implemented. This type of analytical framework allows to rigourously assessing the benefits of mangrove afforestation as compared with the benefits of additional complementary practices.
  *Abundant public information in CBACC-CA project documents and
  research papers informs our analysis (CBACC-CA Project Document
  2008; Annual Progress Reports 2010, 2012, 2013; Ahammad
  et al. 2013), and personal communications by Doctor Paramesh Nandy,
  UNDP Project Manager.
  **According to 2011 census data the population in these districts corre-
  sponds to 34,843,751 people (2011 Bangladesh census data).
  ***Sensitivity tests should include a range of interest rates, as previously
  discussed in Sect. 3.2.2.

The original CBACC-CA programme measured the increase in income associated to the complementary agro-ecological practices on a subset of households, by
collecting baseline and follow-up measurements. However, from a statistical perspective, without a control group, this change in income cannot be statistically
attributed to the intervention, as it could be associated to other regional changes that
have not been monitored.

The maps below emphasise the importance of spatial data analysis and planning
for assessing the intervention in its multi-dimensional complexity: status of ecosystem services under alternative scenarios and their socio-economic implications,
including disaster risk reduction.

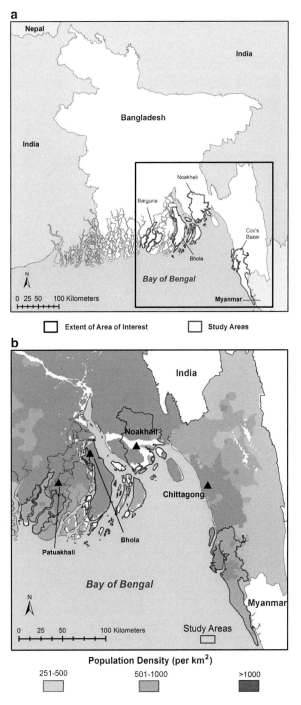

Fig. 3.1 (**a**) Bangladesh: study areas (Source: Ahammad et al. 2013; GADM 2015); (**b**) Population Density 2010 (Source: Center for International Earth Science Information Network 2011; Balk et al. 2006); (**c**) Proportion of Population below the Lower Poverty Line. Poverty Head Count Rates estimated using the Small Area Estimates (SAE) technique. Lower poverty line corresponds to the extreme poor household whose total expenditures are equal to the food poverty line using the Cost of Basic Needs method (Source: The World Bank et al. (2010): Report of the Household Income & Expenditure Survey (HIES) 2010); (**d**) Storm surge magnitude and extent of cyclone affected areas (Source: SPARRSO 2010)

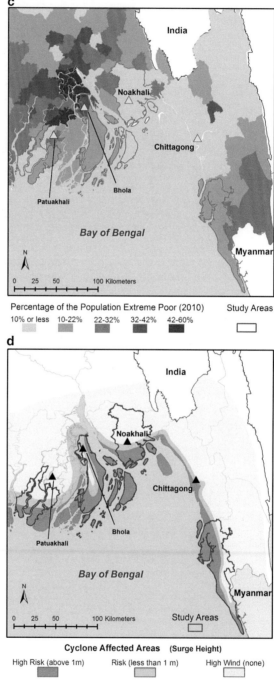

Fig. 3.1 (continued)

3.3   Review of Literature on the Economics of Eco-DRR

There is a growing literature attempting to examine the role of ecosystems in disaster risk reduction from an economic viewpoint. Many studies adopt a qualitative approach (Robledo et al. 2004; Farber et al. 2006; Ingram et al. 2006; Costanza et al. 2008; Feagin et al. 2010). Quantitative studies have diverse methodologies, including: empirical models (Kathiresan and Rajendran 2005; Bebi et al. 2009); regression analysis (Bradshaw et al. 2007; Peduzzi 2010); and predictive models (Brouwer and Van Ek 2004). Their analytical frameworks also differ widely for the variables examined: some studies focus only on physical impacts of disasters; others examine socioeconomic variables, too. The large variety of methodologies and assumptions make it difficult to systematically and effectively compare studies and results. An ongoing study review of the current literature on the economics of Eco-DRR by Vicarelli (2016) found that only twenty-two Eco-DRR quantitative studies perform some form of economic assessment of the protective services of ecosystems; and of these only 15 perform CBA (Table 3.2). The remaining studies use replacement cost methods (e.g. Sathirathai and Barbier 2001; Dorren and Berger 2012), or quantification of losses avoided (e.g. Batker et al. 2010, IFRC 2011). Among the CBA studies only nine are peer-reviewed articles, seven are reports or book chapters (e.g. Hoang Tri et al. 1998; Sathirathai and Barbier 2001; White and Rorick 2010; IFRC 2011) (Table 3.3). In this section we examine the fifteen Eco-DRR CBA studies identified by Vicarelli (2016) and compare their methodologies using the framework outlined in section one (Table 3.2).

3.3.1   Scope and Purpose of the Analysis

Assessments are performed with different purposes: some retrospectively evaluate the protective role of ecosystems (n = 4), others compare future scenarios (n = 12).

The majority of *retrospective CBAs* evaluate programs (n = 3) and only one study retrospectively evaluates the protective role of a natural ecosystem (i.e. mangroves) in response to past shocks (Barbier 2007). Among the three programme evaluations only one study was peer-reviewed (Walton et al. 2006); the remaining two were conducted by NGOs to evaluate their own programmes:, Mercy Corps Nepal (White and Rorick 2010) and the International Federation of the Red Cross and Red Crescent Societies (IFRC 2011). The IFRC approach is a model for policy recommendations: it performed a retrospective CBA of a 2-year Eco-DRR programme, and used the results in a forward looking CBA, building net benefits projections until the year 2025, to assess whether the continuation of the programme was an efficient strategy.

Table 3.2 Summary of existing studies performing CBA simulations or CBA of Eco-DRR projects, sorted by year of publication (Source: Vicarelli 2016)

Year	Authors	Peer reviewed	Metastudy	Scale	Location	Ecosystem	Hazard	Assessment of existing project	Timeframe of benefits forward looking (ex-ante) or retrospective study (expost)	Duration (Time horizon)	Discount rate(s) used
1997	Kramer et al. 1997	Yes	No	Regional	Madagascar	Riparian and mountain tropical forest	Floods	No	exante	46 years	10, 15 %
1998	Dedeurwaerdere (1998)	No	No	National	Pampanga province, Phillippines	Rainforest	Flood, lahar (mudflows)	Yes	exante	3-years (River channel improvement); 4-years (bamboo plantation); 30-years (rainforestation farming	20 %
1998	Hoang Tri et al. (1998)	Yes	No	Local/community based	Northern Vietnam	Mangroves	Tropical cyclones	No	exante	25 years	3, 6, 10 %
2001	Sathirathai and Barbier (2001)	Yes	No	Local/community based	Tha Po, Thailand	Mangroves	Storms, tsunamis, water pollution	No	ex-ante	5-20 years	10, 12, 15 %
2004	Brouwer and Van Ek (2004)	Yes	No	National	Netherlands	Floodplains	Floods	Yes	ex-ante	100 years	4 %
2006	Dorren and Berger (2012)	No	Yes	Local/community based	Alps	Subalpine forest	Avalanches, drought	No	ex-ante	100 years (the lifetime of avalanche barrier)	–
2006	Walton et al. (2006)	Yes	No	Local/community based	Panay Island, Philippiness	Mangrove	Erosion and extreme storms	Yes	ex-post	15 years	–

2007	Barbier (2007)	Yes	No	National	Thailand	Mangrove	Storms	No	ex-post	9-10 years (1996–2004)	10, 12, 15 %
2007	Ming et al. (2007)	Yes	No	Local/community based	Momoge National Reserve, China	Wetlands soils	Flood	No	exante	Not discussed	–
2010	Batker et al. (2010)	No	No	Local/community based	USA	Mississippi River Delta	Storms, hurricanes, erosion	No	ex-ante	100 years	0, 2, 3.5, 5 %
2010	Chen et al. (2010)	Yes	No	Local/community based	Xiamen, China	Coastal	Flooding/Storms	No	ex-ante	70 years	4.5 %
2010	White and Rorick (2010)	No	No	Local/community based	Kailali, Nepal	Riparian	Flood, erosion	Yes	ex-post (evaluation of a 2 year project)	10 years	12 %
2011	IFRC (2011)	No	No	Local/community based	Vietnam	Mangrove	Hydro meterorological	Yes	ex-ante (using results of a program evaluation conducted in 2011)	31 years (1994–2025)	7.23 %
2012	Dorren and Berger (2012)	No	No	Local/community based	Swiss Alps	Mountain Forest	"Gravitational hazard": rock-falls, ava-lanches, land-slides, flooding/debris flow	Yes	ex-ante	Not discussed	–
2013	Pernetta et al. (2013)	Yes	Yes	Regional	South China Sea	Emphasis on Man-groves, but also Seagrass, Coral Reefs and Coastal Wetlands	Implicitly, storms	No	ex-ante	5 years (dura-tion of project)	4 %

Table 3.3 Categories of benefits valued in monetary terms

Benefits of Eco-DRR intervention		
ECOSYSTEM USE VALUE		
DIRECT VALUE		
Marketable goods	Food, timber and other raw materials	Chen et al. (2010), Batker et al. (2010), White and Rorick (2010), Barbier (2007), Walton et al. (2006), Sathirathai and Barbier (2001) and Hoang Tri et al. (1998)
	Increase yield in aquaculture	IFRC (2011)
	Mangrove-fisheries linkage	Pernetta et al. (2013), Barbier (2007) and Sathirathai and Barbier (2001)
Marketable service	Tourism	Pernetta et al. (2013), Chen et al. (2010) and Walton et al. (2006)
Non-marketable services	Water supply	Batker et al. (2010)
	Water quality	Batker et al. (2010)
	Genetic resources	Chen et al. (2010)
INDIRECT VALUE		
Hazard protection	Reduction in expected storm damage avoidedeconomic losses in agriculture, infrastructure, properties	IFRC (2011), Batker et al. (2010), White and Rorick (2010), Barbier (2007), Brouwer and Van Ek (2004), Dedeurwaerdere (1998) and Kramer et al. (1997)
	Avoided cost of building and maintaining an engineering solution of comparable protective potential	Dorren and Berger (2012), Brouwer and Van Ek (2004), Ming et al. (2007) and Hoang Tri et al. (1998)
	Water flow regulation, flood protection	Batker et al. (2010) and Kramer et al. (1997)
Additional services	Waste treatment	Chen et al. (2010) and Walton et al. (2006)
	Carbon sequestration	Pernetta et al. (2013), IFRC (2011) and Batker et al. (2010)
	Biodiversity maintenance	Chen et al. (2010)
	Habitat refugium	Batker et al. (2010)
	Nutrient regulation	Batker et al. (2010)
	Erosion control, sediment retention	Chen et al. (2010)
ECOSYSTEM NON-USE VALUE		
Aesthetic, cultural, spiritual		Pernetta et al. (2013), Batker et al. (2010), Chen et al. (2010) and Brouwer and Van Ek (2004)

The rest of the studies include only *forward-looking projections*. Three compare possible programs or policies (Kramer et al. 1997; Brouwer and Van Ek 2004; Batker et al. 2010); eleven simulate future responses of ecosystems to hazards and apply CBA to the results showing the net benefits of Eco-DRR solutions (Chen et al. 2010; e.g. Dorren and Berger 2012).

3.3.2 Geographic Scale

Several studies evaluate costs and benefits at the national-regional scale (Kramer et al. 1997; Dedeurwaerdere 1998; Brouwer and Van Ek 2004; Barbier 2007; Batker et al. 2010), others perform community-level (Hoang Tri et al. 1998; Pernetta et al. 2013) and household-level analyses (Sathirathai and Barbier 2001; Walton et al. 2006; White and Rorick 2010; IFRC 2011).

Most studies (n = 10) were conducted in Asia (China, Nepal, Philippines, Thailand, and Vietnam), the remaining studies in Africa (n = 1, Madagascar), Europe (n = 3, French Alps and The Netherlands), and in North America (n = 1, US).

3.3.3 Ecosystems and Hazards Analyzed

Six studies focus on mangroves (e.g. Granek and Ruttenberg 2007) and two on coastal ecosystems in general. One study explores the role of coral reefs and sea grass beds (Pernetta et al. 2013); one examines river-delta systems (Kremer et al. 1997; Batker et al. 2010). Forest-ecosystems studies focus on mountain forests (n = 2) (Dorren 2006; Dorren and Berger 2012) and tropical rainforest (n = 1) (Dedeurwaerdere 1998). The remaining studies examine wetlands (Ming et al. 2007) floodplains (Brouwer and Van Ek 2004), and riparian ecosystems (White and Rorick 2010).

Hydro-meteorological hazards (i.e. tropical cyclones, hurricanes, extreme storms, and storm surges) are the most studied (n = 15), including floods (n = 6) and drought (n = 2). Other hazards studied include gravitational hazards (i.e. landslides, avalanches and rockfalls) (n = 2), and tsunamis (n = 1).

3.3.4 Time Horizon and Longevity (Duration) of Costs and Benefits

The duration of a project may differ, and in general be shorter than, the longevity of its costs and benefits; and benefits usually last longer. Most studies do not discuss at all this difference in defining the time horizon of the analysis (T). In some cases the value of T is not indicated (Ming et al. 2007). In other cases the choice seems arbitrary, and no motivation is provided (T = 70 years in Chen et al. 2010). Batker et al. (2010) arbitrarily set T = 100 years arguing that long time horizons accentuate the differences between scenarios. Less arbitrary choices include: duration of project implementation (T = 5 years in Walton et al. 2006; Pernetta et al. 2013); time before complete deforestation of a forested ecosystem (T = 46 years in Kramer et al. 1997); data-availability in retrospective studies (T = 8 years in Barbier 2007);

or life-cycle of ecosystems, such as forest-avalanche-barriers ($T = 100$ years, Dorren 2006) and mangrove forest ($T = 31$ years in IFRC 2011).

3.3.5   Baseline

The baseline is seldom discussed explicitly. The default baseline is usually implicitly assumed to be 'business as usual' (be it conservation or degradation) and in some retrospective studies "pre-hazard conditions". Baseline socio-economic data were never collected before the project implementation.

3.3.6   Model Specification

Most studies adopt a simplified version of CBA. Their analysis is not structured as a yearly stream of costs and benefits over a number of years; the net benefits are estimated only at the beginning and at the end of the program before calculating the PVNB. Only three studies include series of net benefits over time, by introducing variables that account for change: deforestation rates (Kramer et al. 1997), wetland loss trends (Batker et al. 2010) or risk and frequency of disasters (IFRC 2011).

3.3.7   Concurrent Programs and Confounding Effects

Eco-DRR interventions are usually combined with community awareness, early warning systems, and livelihood improvement programs. Costs and benefits of ecosystem restoration or preservation should be rigourously disentangled from costs and benefits of complementary programs. None of the studies discusses these possible confounding effects, with the exception of IFRC (2011), and White and Rorick (2010). No study attempts an in-depth CBA of all the separate concurrent programs (following standard program evaluation techniques).

3.3.8   Benefits and Costs Estimates

None of the studies attempt to provide a complete catalog of costs and benefits of a given Eco-DRR policy or program. In general authors include only some costs and benefits in the CBA until they are able to show that net benefits associated to the Eco-DRR approach are positive or higher than in alternative scenarios.

Table 3.3 presents an overview of the main categories of monetised benefits included in the CBAs: (a) ecosystem goods and services; (b) post-disaster physical losses avoided (e.g. infrastructure; agricultural or industrial productive activities); (c) hypothetical ecosystem replacement cost (the ecosystem value is estimated as the cost of building an artificial structure of comparable protective characteristics[8]).

Marketable environmental goods extracted from ecosystems are more likely to be included than non-marketable goods and services due to the large uncertainty associated to their estimation. Authors often argue that the PVNB would be higher had they included benefits to social or environmental capital that could not be monetised. Unfortunately this translates into a chronic underestimation of benefits of Eco-DRR projects, making them less competitive against development projects (Barbier 2007), or traditional structural solutions (Brouwer and Van Ek 2004).

Costs are usually associated with expenditures: cost of the intervention (e.g. Hoang Tri et al. 1998; IFRC 2011; Pernetta et al. 2013), cost of land reclamation projects (Chen et al. 2010), cost of incentives and indemnizations for loss of land property returned to wetlands or other natural ecosystems (Brouwer and Van Ek 2004), and administrative costs per project beneficiary (IFRC 2011). Some studies also include opportunity costs of conservation versus alternative land use (e.g. shrimp farming) (Sathirathai and Barbier 2001; Barbier 2007).

An important finding is that the value of human life is never measured and rarely discussed: Hoang Tri et al. (1998), White and Rorick (2010), and IFRC (2011) explicitly chose not to include it. The impacts of Eco-DRR on health are never economically quantified. White and Rorick (2010) collected data on health and analysed them qualitatively.

3.3.9 Socio-economic Data Collection Strategy

Methodologies adopted to collect socioeconomic data vary across studies and depend on the scale of the project. Some community-based projects use question-naires to elicit the perceived value of the protective role of an ecosystem, or to calculate the economic value of ecosystems' marketable goods (Sathirathai and Barbier 2001; Walton et al. 2006). Other studies use focus groups and detailed household surveys to collect in-depth data on the socio-economic and environmental costs and benefits of a restoration intervention (White and Rorick 2010; IFRC 2011). White and Rorick's (2010) detailed dataset includes: savings; soil productivity; water quality; flood income losses (annual crop production, belongings lost or damaged); and any health and education costs associated to floods.

None of the studies use experimental design with treatment and control groups to assess the benefits of the intervention. Besides the lack of *control group*, a major limitation is that the data is always collected *a posteriori* (after the project implementation): there is no baseline, or it is built by asking respondents to report their

---

[8]The cost of avoided maintenance of such structures is also considered in some studies. However, this method underestimates the benefits provided by the goods and services of the ecosystem.

best recollection. Data reliability is further compromised by its *self-reported*, and often *qualitative* nature (e.g. respondents are asked if income has improved but not by how much, or if they consider the intervention to have had positive effects).

3.3.10   Discount Rate

Four studies do not report discount rates (e.g. Walton et al. 2006; Ming et al. 2007) or do not perform any discounting. When used, discount rates range between 0 % and 20 %. Six studies use only one rate (e.g. Chen et al. 2010; Pernetta et al. 2013;). Five studies use a range of two to four discount rates (e.g. Hoang Tri et al. 1998; Barbier 2007; Batker et al. 2010). Most studies do not motivate their choice. Brouwer and Van Ek (2004) apply the Dutch Treasury rate (4 %). Kramer et al. (1997) chose 10 % because it is often used by multilateral organizations, testing also 15 % to prioritize current generations in deep poverty.

3.3.11   Benefit Cost Ratio (BCR), and Present Value of net Benefits (PVNB)

All studies report the PVNB (sometimes indicated as *net present value, NPV*), and some studies report also the BCR. IFRC (2011) presents two BCR values for each community: including or neglecting the ecological benefits of carbon sequestration (which are very high). Other studies do not report a BCR, but include elements that would make it possible to measure it for different policy options (Permetta et al. 1993; Sathirathai and Barbier 2001; Barbier 2007; Dorren and Berger 2012).

3.3.12   Uncertainty of Estimates and Sensitivity Analysis

Most studies provide a description of the uncertainty associated to their cost and benefit estimates. Authors sometimes perform sensitivity tests with a range of discount rates (Hoang Tri et al. 1998; Sathirathai and Barbier 2001; Barbier 2007; Batker et al. 2010). Others perform more complex analyses. White and Rorick (2010) test three discount rates, two different time horizons for the duration of project benefits, and calculate BCR for two net-benefit estimates (best estimates, and 20 % lower benefits). Kramer et al. (1997) perform sensitivity tests to account for the uncertainty key random variables (i.e. deforestation rate, decrease in storm flow, percent land in paddy, net returns per hectare, exchange rate, and discount rate).

3.3.13   *Distributional Issues*

Some studies qualitatively address distributional and social equity considerations (e.g. Sathirathai and Barbier 2001) but none of the study performs a quantitative equity analysis. IFRC 2011 uses interviews and surveys to study distributional and welfare implications that cannot be detected with CBA.

3.4   Conclusions and Recommendations

Drawing from the extensive literature on CBA methods, this chapter outlined a core of indispensable best practices to perform rigorous CBAs for Eco-DRR. We have then used this theoretical framework to examine the fifteen existing Eco-DRR CBA studies identified by Vicarelli (2016) and compared their methodologies. We found that none of the studies reviewed followed every one of the best practices that we identified; indeed their methodological approaches are rather diverse and the assumptions adopted are not always explained. Our final goal is to identify possible areas of improvement in current practices and provide recommendations to strengthen and make more consistent the analytical framework used in CBA for Eco-DRR. The adoption of a more robust framework may be useful to consistently organise disparate information, compare and choose between Eco-DRR policies and programmes, and retrospectively assess existing interventions. In this section we summarize our key findings and provide recommendations.

Our literature review suggests that studies should be more rigourous in defining analytical boundaries (i.e. geographic scale and time horizons), initial conditions (i.e. baseline), and model specifications (i.e. longevity of benefits and discount rates). A strong limitation of some studies is that instead of estimating a dynamic series of costs and benefits they take static snapshots before and after the intervention, which may lead to underestimation of ecosystem services.

In order to avoid estimation errors in assessing the role of ecosystems in DRR, the effect of concurrent complementary programs (e.g. micro-finance, disaster awareness/preparedness training) should be systematically disentangled from the effect of the ecosystem component. Yet, in most studies, this concern is not discussed and this procedure ignored. Other observed methodological weaknesses include absence of sensitivity tests, and discussion about distributional issues in reporting the results.

For retrospective studies in developing countries, especially community-level interventions, changes in welfare conditions and development status should be addressed and best estimates always provided. A major challenge is represented by the lack of reliable data. Most studies make estimates before the project is implemented, other studies attempt retrospective estimates using self-reported data, which is not very reliable. The most effective way to collect data is to include data collection strategies in the design of the intervention itself. However, a careful literature review suggests this is not the standard practice. Rigorous protocols for data collection are indispensable toward more reliable estimates of costs and benefits. None of the studies in our review collected ex-ante baseline data (before the beginning of the intervention) and follow-up surveys. Baseline data were collected only ex-post (in two studies). We recommend that retrospective studies always include data collection strategies in the original project design.

Forward-looking CBAs carry more uncertainties than retrospective ones. Peer-reviewed retrospective CBAs allow for more planning and better data quality. Following rigourous program evaluation protocols, including data collection strategies incorporated in the original program design, retrospective CBAs are useful to: (i) collect more abundant good quality data; (ii) strengthen our understanding of long-term costs and benefits; (iii) build a global database of cost-benefit estimates and international standards to efficiently compare results across analyses; and lastly (iv) calculate more reliable forward-looking estimations of costs and benefits.

A retrospective assessment can provide extremely useful and cost-effective when used as mid-project evaluation prior to full-scale expansion of the intervention. Data collected in the retrospective CBA can provide reliable estimates for the forward-looking CBA associated to the full project. Moreover, in large resource-intensive projects, retrospective assessments may allow for iterative calibration of the resources invested, and progressive optimisation of the intervention toward the most efficient outcome. Eco-DRR interventions should adopt this approach as best practice.

Many of the studies reviewed were not peer-reviewed. Our last but important recommendation is that a peer-review of economic analysis should become the standard for projects with potentially large socio-economic impacts and/or with potentially irreversible impacts on ecosystems. External reviews by multi-disciplinary teams would be the best practice.

References

Ahammad R, Nandy P, Husnain P (2013) Unlocking ecosystem based adaptation opportunities in coastal Bangladesh. J Coast Conserv 17(4):833–840

Arrow K, Cropper M, Eads G et al (1996) Is there a role for benefit-cost analysis in environmental, health, and safety regulation? Science 272:221–222

Balk DL, Deichmann U, Yetman G, Pozzi F, Hay SI, Nelson A (2006) Determining global population distribution: methods, applications and data. Adv Parasitol 62:119–156

Barbier EB (2007) Valuing ecosystem services as productive inputs. Econ Policy 22(49):177–229

Batker DP, de la Torre R, Costanza P et al (2010) Gaining ground – wetlands, hurricanes and the economy: the value of restoring the Mississippi river delta. Earth Economics, Tacoma

Bebi P, Kulakowski D, Rixen C (2009) Snow avalanche disturbances in forest ecosystems—State of research and implications for management. For Ecol Manag 257(9):1883–1892

Boyce J (2000) Let them eat risk: wealth, rights, and disaster vulnerability. Disasters 24:254–261

Bradshaw CJ, Sodhi NS, Peh KSH (2007) Global evidence that deforestation amplifies flood risk and severity in the developing world. Glob Chang Biol 13(11):2379–2395

Brouwer R, Van Ek R (2004) Integrated ecological, economic and social impact assessment of alternative flood control policies in the Netherlands. Ecol Econ 50(1):1–21

CBACC-CA (2008) Community based adaptation to climate change through costal afforestation project document 2008. Bangladesh Government, UNDP, GEF, Swiss Agency for Development and Cooperation. Kingdom of the Netherlands. Available via http://www.cbacc-coastalaffor.org.bd/index.php?option=com_content&view=article&id=1320&Itemid=688. Accessed 13 Oct 2015

CBACC-CA (2010) Community based adaptation to climate change through costal afforestation annual report 2010. Bangladesh Government, UNDP, GEF, Swiss Agency for Development and Cooperation. Kingdom of the Netherlands. Available via http://www.cbacc-coastalaffor.org.bd/index.php?option=com_content&view=article&id=1320&Itemid=688. Accessed 13 Oct 2015

CBACC-CA (2012) Community based adaptation to climate change through costal afforestation annual progress report 2012. Bangladesh Government, UNDP, GEF, Swiss Agency for Development and Cooperation. Kingdom of the Netherlands. Available via http://www.cbacc-coastalaffor.org.bd/index.php?option=com_content&view=article&id=1320&Itemid=688. Accessed 13 Oct 2015

CBACC-CA (2013) Community based adaptation to climate change through costal afforestation annual progress report 2013. Bangladesh Government, UNDP, GEF, Swiss Agency for Development and Cooperation. Kingdom of the Netherlands. Available via http://www.cbacc-coastalaffor.org.bd/index.php?option=com_content&view=article&id=1320&Itemid=688. Accessed 13 Oct 2015

Center for International Earth Science Information Network (2011) Global Rural–Urban Mapping Project, Version 1 (GRUMPv1): Population Count Grid. Palisades, NY: NASA Socioeconomic Data and Applications Center (SEDAC). CIESIN – Columbia University, International Food Policy Research Institute – IFPRI, The World Bank, and Centro Internacional de Agricultura Tropical – CIAT. http://dx.doi.org/10.7927/H4VT1Q1H. Accessed 12 Jan 2014

Chen W, Zhang L, Lu C et al (2010) Estimating the ecosystem service losses from proposed land reclamation projects: a case study in Xiamen. Ecol Econ 69(12):2549–2556

Costanza R, d'Arge R, de Groot R et al (1997) The value of the world's ecosystem services and natural capital. Nature 387:253–260

Costanza R, Pérez-Maqueo O, Martinez ML et al (2008) The value of coastal wetlands for hurricane protection. AMBIO J Human Environ 37(4):241–248

Costanza R, de Groot R, Sutton P et al (2014) Changes in the global value of ecosystem services. Glob Environ Chang 26:152–158

de Groot RS, Brander L, van der Ploeg S et al (2012) Global estimates of the value of ecosystems and their services in monetary units. Ecosyst Serv 1:50–61

Dedeurwaerdere A (1998) Cost-benefit analysis for natural disaster management – a case-study in the Philippines. CRED, Brussels

Dorren L (2006) Managing ecosystems for disaster reduction services: examples from around the world. In: International disaster risk reduction conference, Davos

Dorren LKA, Berger F (2012) Integrating forests in the analysis and management of rockfall risks: experiences from research and practice in the Alps. In: Landslides and engineered slopes: protecting society through improved understanding. Taylor and Francis Group, London, pp 117–127

Duflo E, Glennerster R, Kremer M (2006) Using randomization in Development Economics Research: a toolkit. Available via http://www.povertyactionlab.org/sites/default/files/documents/Using%20Randomization%20in%20Development%20Economics.pdf. Accessed 20 Jan 2015

EPA (2014) Social impact assessment analysis. Available via http://www.epa.gov/sustainability/analytics/social-impact.htm. Accessed 10 Oct 15

Farber S, Costanza R, Childers DL et al (2006) Linking ecology and economics for ecosystem management. Bioscience 56(2):121–133

Feagin RA, Mukherjee N, Shanker K et al (2010) Shelter from the storm? Use and misuse of coastal vegetation bioshields for managing natural disasters. Conserv Lett 3(1):1–11

Folke C, Carpenter S, Walker B et al (2004) Regime shifts, resilience, and biodiversity in ecosystem management. Annu Rev Ecol Evol Syst 35:557–581

GADM (2015) Database of global administrative areas. Available via http://www.gadm.org/. Accessed 27 Oct 2015

Gerber A, Green D (2012) Field experiments: design, analysis and interpretation. Norton WW & Company, Inc, New York

Glennerster R, Takavarasha K (2013) Running randomized evaluations: a practical guide. Princeton University Press, Princeton

Goulder L, Stavins R (2002) Discounting: an eye on the future. Nature 419:673–674

Government of Bangladesh, UNDP (2008) CBACCC project report. Available via UNDP

Graham J (2008) The evolving regulatory role of the U.S. Office of Management and Budget. Rev Environ Econ Policy 1(2):171–191

Granek EF, Ruttenberg BI (2007) Protective capacity of mangroves during tropical storms: a case study from 'Wilma' and 'Gamma' in Belize. Mar Ecol Prog Ser 343:101–105

Harvey N (2007) Global change and integrated coastal management: the Asia-Pacific region., Springer Netherlands

Heal GM, Barbier EB, Boyle KJ et al (2005) Valuing ecosystem services: toward better environmental decision making. The National Academies Press, Washington, DC

Hoang Tri N, Adger WN, Kelly PM (1998) Natural resource management in mitigating climate impacts: the example of mangrove restoration in Vietnam. Glob Environ Chang 8(1):49–61

IFRC (2011) Breaking the waves. Impact analysis of coastal afforestation for disaster risk reduction in VietNam. IFRC (International Federation of Red Cross and Red Crescent Societies), Geneva

Ingram JC, Franco G, Rio CRD et al (2006) Post-disaster recovery dilemmas: challenges in balancing short-term and long-term needs for vulnerability reduction. Environ Sci Pol 9 (7):607–613

IPCC (2014) Fifth assessment report Intergovernmental Panel on Climate Change working group II

Kathiresan K, Rajendran N (2005) Coastal mangrove forests mitigated tsunami. Estuar Coast Shelf Sci 65(3):601–606

Kelman S (1991) Cost-Benefit analysis: an ethical critique -with replies. AEI J Govt Soc Reg 5:33–40

Kramer RA, Richter DD, Pattanayak S, Sharma NP (1997) Ecological and economic analysis of watershed protection in Eastern Madagascar. J Environ Manag 49(3):277–295

Lichter M, Vafeidis A, Nicholls R et al (2011) Exploring data-related uncertainties in analyses of land area and population in the 'low-elevation coastal zone' (LECZ). J Coast Res 27:757–68

Menéndez L, Priego A (1994) Los Manglares de Cuba: Ecología. En el Ecosistema de Manglar en América Latina y la Cuenca del Caribe: su manejo y conservación. Rosenstiel School of Marine and Atmospheric Sc., Univ. Miami, Fla. & The Tinker Found, New York, pp 64–75

Mimura N (2013) Sea-level rise caused by climate change and its implications for society. Proc Jpn Acad Ser B Phys Biol Sci 7:281

Ming J, Xian-Guo L, Lin-Shu X et al (2007) Flood mitigation benefit of wetland soil—a case study in Momoge National Nature Reserve in China. Ecol Econ 61(2):217–223

NOAA (1994) Guidelines and principles for Social Impact Assessment. http://www.nmfs.noaa.gov/sfa/social_impact_guide.htm#sectIV. Accessed 10 Oct 15

NOAA (1999) Discounting and the treatment of uncertainty in natural resource damage assessment: Technical paper 99–1. National Oceanic and Atmospheric Administration. Silver Spring, MD. Available via http://www.darp.noaa.gov/library/pdf/discpdf2.pdf

NOAA/NMFS Council Operational Guidelines—Fishery Management Process (2007) Guidelines for the assessment of the social impact of Fishery Management Actions.http://www.nmfs.noaa.gov/sfa/reg_svcs/NMFSI_01-111-02.pdf. Accessed 10 Oct 15

Peduzzi P (2010) Landslides and vegetation cover in the 2005 North Pakistan earthquake: a GIS and statistical quantitative approach. Nat Hazards Earth Syst Sci 10(4):623–640

Pernetta JC, Ongb JO, Oyardo Padillac NE, Rahimd KA, Chinh NT (2013) Determining regionally applicable economic values for coastal habitats and their use in evaluating the cost effectiveness of regional conservation actions: the example of mangroves, in the South China Sea. Ocean Coast Manag 85:177–185

Perrings C, Pearce D (1994) Threshold effects and incentives for the conservation of biodiversity. Environ Resour Econ 4(1):13–28

Ravallion M (2012) Fighting poverty one experiment at a time: a review essay on Abhijit Banerjee and Esther Duflo, poor economics. J Econ Lit 50:103–114

Renaud F, Sudmeier K, Estrella M (eds) (2013) The role of ecosystems in disaster risk reduction. United Nations University Press, Tokyo

Robledo C, Fischler M, Patiño A (2004) Increasing the resilience of hillside communities in Bolivia: has vulnerability to climate change been reduced as a result of previous sustainable development cooperation? Mt Res Dev 24(1):14–18

Sathirathai S, Barbier EB (2001) Valuing mangrove conservation in southern Thailand. Contemp Econ Policy 19(2):109–122

Shreeve CM, Kelman I (2014) Does mitigation save? Reviewing cost-benefit analyses of disaster risk reduction. Int J Disaster Risk Reduct 10:213–235

SPARRSO (2010) Bangladesh Space Research and Remote Sensing Organization. http://www.sparrso.gov.bd/x/. Accessed 26 Oct 2015

Stern N (2007) The economics of climate change: the Stern review. Cambridge University press, Cambridge

TEEB Foundations (2010a) The economics of ecosystems and biodiversity: ecological and economic foundations. Earthscan, London/Washington

TEEB Synthesis (2010b) Mainstreaming the economics of nature: a synthesis of the approach, conclusions and recommendations of TEEB. Earthscan, London/Washington

The World Bank, World Food Program, Bangladesh Bureau of Statistics (2010) Report on the household income and expenditure survey 2010. Available at: http://catalog.ihsn.org/index.php/catalog/2257/download/36931. Accessed 26 Oct 2015

US-OMB (1992) Office of management and budget circular no.A-94. Revised Guidelines and Discount Rates for Benefit-Cost Analysis of Federal Programs October 29. Available via https://www.whitehouse.gov/omb/circulars_a094#8

US-OMB (2003) US Government's standards for regulatory analysis. U.S. Office of Management and Budget. Circular A-4, September 2003

Venton CC (2010) Cost benefit analysis for community based climate and disaster risk management: synthesis report. Oxfam – America; Tearfund

Vicarelli M (2016) Review of eco-DRR studies performing economic valuations. Working paper Depratment of Economics University of Massachusetts Amherst

Walton M, Samonte-Tan GPB, Primavera JH, Edwards-Jones G, Le vay L (2006) Are mangroves worth replanting? The direct economic benefits of a community-based reforestation project. Environ Conserv 33:335–343

Weitzman ML (2009) On modeling and interpreting the economics of catastrophic climate change. Rev Econ Stat 91(1):1–19

White BA, Rorick MM (2010) Cost-benefit analysis for community-based disaster risk reduction in Kailali, Nepal. Mercy Corps Nepal, Lalitpur

Chapter 4
Mangrove Payments for Ecosystem Services (PES): A Viable Funding Mechanism for Disaster Risk Reduction?

Daniel A. Friess and Benjamin S. Thompson

Abstract Mangrove forests provide a multitude of ecosystem services, many of which contribute to Disaster Risk Reduction (DRR) along tropical coastlines. In the face of rapid deforestation, Payments for Ecosystem Services (PES) schemes such as Reducing Emissions from Deforestation and Forest Degradation (REDD+) has been heralded as a potential avenue for financing conservation, although PES schemes remain in an embryonic state for mangroves. Several challenges must be overcome if mangrove PES is to advance. Firstly, challenges exist in quantifying multiple ecosystem services, especially those that contribute to DRR, such as wave attenuation and the control of coastal erosion. Secondly, the permanence of quantified ecosystem services is a central tenet of PES, but is not guaranteed in the dynamic coastal zone. Mangroves are affected by multiple stressors related to natural hazards and climate change, which are often outside of the control of a PES site manager. This will necessitate Financial Risk Management strategies, which are not commonly used in coastal PES, and introduces a number of management challenges. Finally, and most importantly, PES generally requires the clear identification and pairing of separate service providers and service users, who can potentially overlap in the context of DRR. This chapter reviews and discusses these emerging issues, and proposes potential solutions to contribute to the more effective implementation of mangrove PES. Ultimately however, difficulties in pairing separate and discreet service providers and users may render PES for DRR unfeasible in some settings, and we may need to continue traditional modes of DRR finance such as insurance and donor support.

Keywords Blue carbon • Coastal erosion • Permanence • REDD+ • Valuation • Wave attenuation • Wetland

D.A. Friess (✉) • B.S. Thompson
Department of Geography, National University of Singapore, 1 Arts Link, 117570 Singapore, Singapore
e-mail: dan.friess@nus.edu.sg; benjamin.thompson@u.nus.edu

© Springer International Publishing Switzerland 2016
F.G. Renaud et al. (eds.), *Ecosystem-Based Disaster Risk Reduction and Adaptation in Practice*, Advances in Natural and Technological Hazards Research 42, DOI 10.1007/978-3-319-43633-3_4

4.1   Introduction

With high population densities in the coastal zone, hundreds of millions of people are currently exposed to coastal hazards such as storms, cyclones and sea level rise. Over 789,000 people were killed by tropical cyclones between 1970 and 2009 alone (EM-DAT 2011). The number of people exposed to coastal hazards is expected to rise substantially according to future climate change scenarios. In particular, future increases in both cyclone frequency/intensity and coastal population densities could lead to an extra 149.3 million people in the tropics being vulnerable to coastal hazards, with 90 % of exposed people found in Asia (Peduzzi et al. 2012). Based on some sea level rise forecasts, the population exposed to a 1-in-a-100 year flood event is projected to increase to 350 million people by 2050 (Jongman et al. 2012). Reducing vulnerability to threats such as sea level rise will require increasing the height of coastal defences by up to 1 m across the globe (Hunter et al. 2013).

Due to the future expense of more hard coastal defences, attention has turned to potential ecological engineering solutions (see van Wesenbeeck et al., Chap. 8 and David et al., Chap. 20). Coastal mangrove forests are an important halophytic vegetated ecosystem found throughout the tropics and subtropics. Mangroves provide a multitude of ecosystem services, tentatively valued at US$239 to US$4185 per hectare in Southeast Asia (Brander et al. 2012). These ecosystem services provide a range of biophysical and ecological benefits, and include fisheries, timber, pollutant assimilation, carbon storage, and DRR services such as hydrodynamic energy attenuation and shoreline erosion control (Barbier et al. 2011; Lacambra et al. 2013). Despite their importance, mangroves are experiencing rapid and sustained decline globally due to deforestation for new land uses such as aquaculture, agriculture and urban development (UNEP 2014). Deforestation is resulting in the loss and possible extinction of mangrove vegetation species (Polidoro et al. 2010), and is reducing the provision of ecosystem services upon which hundreds of millions of people depend across the tropics.

Mangroves – like many forested ecosystems – have been managed and conserved under traditional government-led protected area approaches. However, recent years have seen a move towards neoliberal conservation instruments that attempt to balance conflicts between conservation and economic growth priorities. Payments for Ecosystem Services (PES) is one instrument with which to balance such conflicts, and is broadly defined as "voluntary transactions between service users and service providers that are conditional on agreed rules of natural resource management" (Wunder 2015). While serious issues relating to social equity, governance and commodification exist with PES (Phelps et al. 2010; Pascual et al. 2014), this instrument has been touted as, "probably the most promising innovation in conservation since Rio 1992[1]" (Wunder and Wertz-Kanounnikoff 2009:576). PES has several key tenets that must be satisfied (Fig. 4.1).

---

[1]The 1992 Earth Summit in Rio de Janeiro, Brazil, instigated the Convention on Biological Diversity (CBD) and United Nations Framework Convention on Climate Change (UNFCCC), which later spawned the Kyoto Protocol.

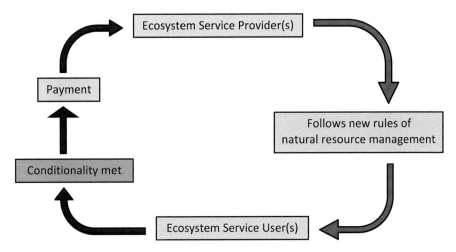

Fig. 4.1 The key tenets of PES showing: ecosystem service flows (*green arrows*), payment flows (*red arrows*), key players (*yellow boxes*), key voluntary transactions (*blue boxes*), and key criteria (*lilac box*) (Based on Wunder (2015))

PES is a concept that has been discussed for decades, with Costa Rica's adoption of PES at a national scale in 1997 viewed as a key moment that instigated new research and policy directions (Chaudhary et al. 2015). Focusing on the tropics, PES schemes that pay for stored carbon, such as Reducing Emissions from Deforestation and Forest Degradation (REDD+) have been discussed at the international level for almost a decade, with several operational schemes now in place throughout the tropics and a large number in the proposal stage. Explicit PES in mangroves, however, remains in an embryonic state. Few examples have been communicated (Fig. 4.2), though efforts are beginning in Kenya (Huxham et al. 2015), Madagascar, Vietnam (Hawkins et al. 2010) and Thailand. This leads to the question: "why is mangrove forest PES lagging so far behind other forest PES initiatives?" This question is particularly pertinent because of the broad range of ecosystem services that can be valorised within a PES scheme: mangrove PES has been proposed primarily to conserve carbon stocks (through "blue carbon" initiatives). Other ecosystem services related to recreation, hydrodynamic flow and wave attenuation for the purposes of disaster risk reduction (DRR) have not yet been an explicit focus of PES discussions, but may also be relevant in the mangrove context.

The aim of this chapter is to identify the key challenges and solutions to implementing PES for mangrove forest ecosystem services, with a particular focus on services related to DRR. Firstly, we discuss the broad range of ecosystem services provided by mangrove forests. Then we highlight three key challenges to the implementation of mangrove PES; (i) how to quantify DRR ecosystem services in a robust manner for PES transactions; (ii) how to ensure long-term permanence of DRR ecosystem services in the dynamic coastal zone; and (iii) how to identify and pair key actors in PES, especially ecosystem service providers and users. A critical and honest discussion of the issues will allow us to identify solutions to

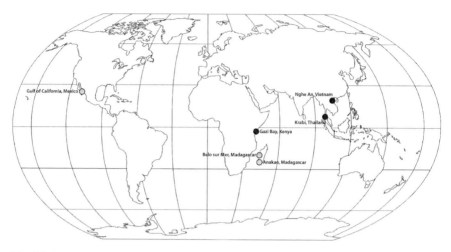

Fig. 4.2 Proposed (*grey*) and pilot (*black*) PES schemes based in mangrove ecosystems across the tropics. Currently, none of these schemes are designed to promote DRR ecosystem services

overcome these challenges and realise the benefits of ecosystem services for DRR for coastal populations that rely on mangroves throughout the tropics.

4.1.1 Mangrove Forests Provide a Multitude of Ecosystem Services

Researchers have described and quantified ecosystem services for decades, although the ecosystem service concept gained wide prominence with the publication of the Millennium Ecosystem Assessment (MA) in 2005. An international effort involving 1300 contributors from 95 countries, the MA (2005) categorised ecosystem services into four major categories:

  (i) Provisioning services – products obtained from an ecosystem;
 (ii) Regulating services – benefits obtained from the regulation of ecosystem processes;
(iii) Supporting services – processes necessary for the production of all other ecosystem services;
 (iv) Cultural services – primarily non-material benefits people obtain from ecosystems through spiritual enhancement, cognitive development, reflection, recreation and aesthetic experiences.

A large literature has now formed around research on the broad range of ecosystem services that mangrove forests in particular provide to coastal populations (Fig. 4.3). Below, we describe particular ecosystem services that are of most relevance to mangrove PES for DRR. Supporting ecosystem services may

Provisioning services	Regulating services
• Raw materials for building construction and fishing equipment • Seafood caught or gleaned in mangroves (e.g. fish, crustaceans, bivalves) • Tannins and waxes acquired from trees • Forest food products such as honey, seeds	• Climate regulation through carbon storage and sequestration • Coastal protection through wave attenuation • Erosion control through sediment stabilization and soil retention • Water purification through pollutant assimilation and nutrient filtering
Supporting services	**Cultural services**
• Maintenance of fisheries as safe nursery grounds and reproductive habitat • Soil formation • Photosynthesis and nutrient production	• Recreation and leisure • Educational opportunities • Aesthetic contribution • Cultural heritage (e.g. community traditions and folklore) • Spritual and religious contributions

Fig. 4.3 A summary of the various ecosystem services provided by mangroves, as classified by the MA 2005 (Based on Barbier et al. 2011; Lacambra et al. 2013; Lau 2013)

either not be of direct relevance to DRR, or are not currently considered for PES, so are not described here.

4.1.1.1 Hydrodynamic Attenuation (Regulating Service)

Mangroves are now well known to interact with and ameliorate incoming hydrodynamic forces such as waves and currents. Hydrodynamic attenuation is equal to the proportion of wave height/current flow reduction per meter of land traversed (Mazda et al. 2006) in a non-linear relationship, and is caused by flow resistance, drag forces, friction and turbulence caused by above-ground vegetative structures. The importance of vegetation in hydrodynamic attenuation means that the magnitude of energy absorption strongly depends on tree density, stem and root diameter, forest width, presence of offshore habitats (e.g. reefs), shore slope, bathymetry, and tidal stage upon entering the forest (Alongi 2008; Koch et al. 2009).

The wave attenuation service of mangroves may be considered the most important in a DRR context, and has been highlighted by recent natural hazards. The role of mangroves in DRR gained the most prominence in response to the 2004 Southeast Asian tsunami. Preliminary surveys after this event suggested that villages behind mangroves suffered less damage and loss of life compared to exposed villages on the coast (e.g. Danielsen et al. 2005; Kathiresan and Rajendran 2005). Mangrove coastal defence services have been calculated at US\$ 672/ha/year in the Philippines (Samonte-Tan et al. 2007) and US\$ 1879/ha/year in Thailand (Barbier et al. 2008). That said, such findings may have been due to statistical correlation and inference rather than hydrodynamic processes, and the mechanisms contributing to tsunami hazard mitigation by mangroves still need more research (Kerr et al. 2006).

Regardless, the perceived importance of mangroves for hazard mitigation has resulted in huge interest in mangrove restoration and their incorporation into coastal defence design throughout the tropics (see Bayani and Barthélemy, Chap. 10). Academic and decision-making contexts are now awash with terms such as "ecological engineering", "building with nature", "nature-based solutions" and "blue/green infrastructure" (see van Wesenbeeck et al., Chap. 8), which all to varying degrees refer to the incorporation of mangroves into coastal engineering design.

4.1.1.2 Shoreline Erosion Control and Adaptation to Sea Level Rise (Regulating Service)

Mangroves have the capacity to reduce shoreline erosion and adapt to sea level rise through two mechanisms. Firstly, mangroves trap and consolidate sediments through their roots, as attenuated water flows encourage sediment to settle out of suspension (Krauss et al. 2003). Roots also contribute to binding the soil and increase soil shear strength. However, the ability of mangroves to encourage deposition, bind sediments and control shoreline erosion may also be species-specific, and mangrove coastlines themselves can erode once species-specific hydrodynamic thresholds are surpassed (Friess et al. 2012).

Secondly, mangroves have the ability to adapt to changing sea levels, if surrounding geomorphological and sedimentological conditions are suitable. Mangroves can increase their surface elevations to potentially keep pace with sea level rise through multiple processes such as sediment trapping and consolidation, and belowground organic matter production (Krauss et al. 2014). Thus, in comparison to traditional hard engineering structures that are fixed at a static elevation, mangroves and other coastal ecosystems may provide an adaptable and flexible coastal defence in some conditions under uncertain sea level rise scenarios (see Whelchel and Beck, Chap. 6). In minerogenic coastal settings,[2] this is reliant on the continued input of sediment into the coastal zone.

4.1.1.3 Carbon Storage (Regulating Service)

The important role of mangroves in carbon production, transport and storage has been known for decades, with mangroves in the United States a particular focus of research. Early research focused on particular processes in the carbon cycle, such as litterfall dynamics (Twilley 1985), aboveground biomass dynamics (Day et al. 1987) and tidal carbon fluxes (Twilley et al. 1986). However, early studies also put carbon into a broader global carbon cycle and climate change perspective (de la Cruz 1982, 1986; Twilley et al. 1992).

---

[2]Made up of mineral materials as opposed to biogenic, i.e. organic, material.

Fig. 4.4 Conducting a
standardized carbon stock
assessment for a mangrove
in northeast Singapore
(Photo by DM Taylor,
reproduced with
permission)

These early research contributions focusing on the role of mangrove forests in climate change mitigation are in some cases forgotten, though are mirrored by similar recent studies that have explored the contribution of mangroves to regional and global carbon budgets (e.g. Bouillon et al. 2008; Donato et al. 2011; Siikamäki et al. 2012). Such studies, bolstered by clear carbon quantification and accounting protocols (e.g. Kauffman and Donato 2012; Fig. 4.4) and new international initiatives (e.g. The International Blue Carbon Initiative) have driven a recent surge in mangrove carbon stock assessments across the tropics (e.g. Donato et al. 2011; Adame et al. 2013; Jones et al. 2014; Thompson et al. 2014). Recent studies are now beginning to extend carbon stock assessments to also incorporate coastal ecosystems adjoining mangroves, providing us with an understanding of where mangrove carbon stocks sit within the broader coastal landscape (Phang et al. 2015). Carbon storage and sequestration is not directly relevant to DRR, but it is a popular mangrove ecosystem service that is the focus of several ongoing mangrove PES proposals. Thus, carbon could be stacked alongside other ecosystem services – such as DRR – to make a potential PES scheme more economically viable (Thompson et al. 2014).

4.1.1.4   Forest Products and Fisheries (Provisioning Services)

Provisioning services relate to products that can be extracted from the mangrove ecosystem. Many products are derived from the vegetation, including timber,

Fig. 4.5 A local fisherman
catching mud crabs to sell to
a local 5 star hotel, Ouvea
atoll, New Caledonia
(Photo by DA Friess)

fuelwood, charcoal, and non-timber forest products such as honey and waxes. Several correlative analyses also suggest that mangroves play an important role in the provision of fisheries (Manson et al. 2005). Although dependent on factors such as geomorphological location and vegetation density/type, mangroves can play a role as a nursery for juvenile fish, or may provide nutrients that are exported to offshore fisheries. Provisioning services can be economically important at multiple scales. Locally, provisioning services can provide subsistence for local coastal communities, or can be sold to local businesses to make small profits and improve local livelihoods (Fig. 4.5). The selling of fish and fuelwood extracted from the mangrove can account for as much as 30 % of a household's income in villages along the east coast of India (Hussain and Badola 2010). Across larger scales and extraction intensities, the value of provisioning services can be considerable; for example, the value of timber extraction, fisheries and other provisioning services across the Sundarbans may reach US$744,000 per year (Uddin et al. 2013).

Unlike hydrodynamic attenuation and shoreline erosion control, provisioning services are not directly related to DRR. However, provisioning services can contribute to a coastal community's adaptive capacity, which may increase its resilience to natural hazards and climate change impacts. Factors such as wealth, health and education are key contributors to adaptive capacity, and a recent global analysis suggests that in general coastal communities have higher levels of all of these factors, compared to communities living inland with less access to the coast and the provisioning services it provides (Fisher et al. 2015). Several reasons account for this. Firstly, coastal fishing as a form of livelihood presents a relatively low cost barrier compared to inland forms of agriculture (Daw et al. 2012). Secondly, coastal communities may (but not always) have easier or closer access to ports and markets for trade (Fisher et al. 2015). Thus, DRR activities (whether or not they are related to PES) would benefit greatly from the incorporation of management interventions that also improve provisioning service usage.

4.1.1.5 Spiritual, Cultural and Heritage Values (Cultural Services)

Cultural values encompass a broad range of ecosystem services that vary greatly in their tangibility and ability to be quantified (Pleasant et al. 2014). Cultural ecosystem services can include clearly tangible and quantifiable recreational and educational values. Spiritual, aesthetic and heritage values are substantially more abstract and intangible, but could have significant value for local coastal populations (Thiagarajah et al. 2015).

At first glance, cultural ecosystem services may not seem directly relevant to DRR. However, local or indigenous knowledge can make a valuable contribution to DRR, although it is often missing from DRR planning, or marginalized in favour of expert scientific knowledge (e.g. Mercer et al. 2010). Marginalization of local knowledge arises due to a perception from some stakeholders of the "unrigorous nature" of local knowledge, or due to unequal power relations between local communities and scientists and decision makers, which come to the fore when knowledge is produced and used (Bohensky and Maru 2011). Cultural value can decrease vulnerability to hazards through inter-generational learning related to warning signs of hazards and how to respond to them (e.g. Furuta and Seino, Chap. 13). For example, indigenous communities in coastal Southeast Asia, such as the Moken sea communities in Thailand, were aware of the warning signs of an impending tsunami, so during the 2004 tsunami they were able to evacuate more quickly than foreign tourists and migrant workers (Mercer et al. 2012). In this example, local knowledge can be regarded as an important, but potentially under-appreciated source of resilience. While substantial challenges may be faced when integrating local and scientific knowledges into decision making, steps in this direction will improve our response to complex socio-ecological challenges (Bohensky and Maru 2011) such as ecosystem-based DRR. The incorporation of cultural ecosystem services into DRR generally is a key research area to pursue in the future.

4.2 Challenge 1 – How to Quantify 'Invisible' DRR Ecosystem Services?

While mangroves provide a variety of ecosystem services, only shoreline protection (wave attenuation and dissipation functions) can be considered directly linked to DRR. Many of the other services described above may contribute to DRR by increasing the adaptive capacity of mangrove-dependent coastal communities, but these may not be suitable for mangrove PES with a focus on DRR, since PES requires a direct ecosystem service that can be explicitly commodified and traded. Tradeable assets require clear quantification and clear ownership rights – traits that are not always possessed by certain ecosystem services. Unlike carbon storage and sequestration, for which clear quantification protocols and market prices exist, it is relatively difficult to commodify shoreline protection services, since (1) very few economic studies have estimated values for them (Barbier 2015); (2) shoreline protection (e.g. wave attenuation) is site-specific and dependent on the local ecological and geomorphological setting; and (3) the amount of attenuation is event-specific, e.g., the amount of hydrodynamic input energy to be attenuated. Yet, despite the difficulties outlined above, several quantification and valuation methods do exist for shoreline protection services, which are discussed in this section.

The first stage in any ecosystem service assessment is to quantify the PES-relevant service in order to establish a baseline against which future performance measures can be compared. Determining wave attenuation involves measuring the current velocity and water level along a cross-shore profile – typically at the open tidal flat, the mangrove fringe, and at systematic points within the mangrove vegetation (Quartel et al. 2007). These hydrodynamic measurements can be taken using pressure sensors and electromagnetic flow devices (such as Acoustic Doppler Velocimeters) that can be mounted on tripods, while bed level height and gradient can be measured at each sample station using a levelling instrument. Wave attenuation is calculated by the difference in initial and final wave height over a specified difference (Mazda et al. 2006). This hydrodynamic data can be combined with biophysical parameters (e.g. stem density, bed roughness, bed gradient), spatial data (shoreline profile, settlement proximity), and data on past events, in order to conduct scenario modelling of hazards such as storm surges (Lau 2013). The outputs are spatial predictions of flood occurrences and hazard levels for each scenario. Such assessments would indicate the current level of shoreline protection services that a particular mangrove provides, where and to whom that service is provided, and how the provision of that service would change with increased or decreased mangrove coverage. This collection of quantitative data and model output can then be subjected to valuation techniques in order to valorise the shoreline protection service – usually an essential step in PES scheme design.

Two cost-based methods can be used to value the shoreline protection service of mangroves: damage costs avoided and replacement costs (Table 4.1) (see Emerton et al., Chap. 2). The former usually requires geographic outputs from scenario models. The method involves estimating the costs of repairing the damages that would be incurred following a reduction in mangrove area, which is used as a proxy for shoreline protection value (Turpie et al. 2010). Damage costs include damage to physical capital such as property, fishing gears, infrastructure (oftentimes the water supply is salinized), and aquaculture/agriculture (e.g. loss of standing crops, fish stocks, or livestock). In addition, more nuanced human capital metrics could be incorporated into the damage cost analysis such as medical expenses or lost household income as a result of injury. The cost of repairing damage sustained

Table 4.1 Advantages and disadvantages of the damage costs avoided and replacement costs methods for valuing the coastal protection service of mangroves

Method	Advantages	Disadvantages
Avoided damage costs	Quantifying wave attenuation can be conducted accurately	Quantifying wave attenuation requires expensive equipment and technical expertise
	Valuation is based on actual market prices	Valuation is based on costs, not benefits
	Not overly data/resource intensive	Very difficult to predict the levels of damage sustained under a particular scenario since values are strongly influenced by the geographic and social (land/property value) context
	An option for locations that are challenging to value by other means	Data on past events is required
	Generally viewed as a better option to replacement costs and contingent valuation	Technical skills (e.g. environmental modelling) is required
		Intra-settlement damage levels and costs could vary greatly
		Land values can change quickly over time as regions gain prosperity or industries go bust
		Difficult to relate damage levels to ecosystem quality and area since there are many other factors
Replacement costs	Quantifying wave attenuation can be conducted accurately	Quantifying wave attenuation requires expensive equipment and technical expertise
	Valuation is based on actual market prices	The valuation is based on costs, not benefits
	Not overly data/resource intensive	Few ecosystems have commensurate artificial substitutes
	An option for locations that are challenging to value by other means	Tends to overestimate actual value of the individual service
		Tends to underestimate actual value of the entire ecosystem since other services that would not be replaced by a manmade alternative are not valued
		Limited application since few environmental actions are based only on cost-benefit comparisons
		Requires strong evidence that the public would demand a manmade alternative if the ecosystem was lost

Based on Pagiola et al. (2004), Turpie et al. (2010), Lau (2013) and Waite et al. (2014)

during past disasters could be used if available (see Bayani and Barthélemy, Chap. 10). Alternatively, if such data were unavailable and modelling was not feasible, the damage costs avoided method can be based on the financial invest- ments that landowners have made in order to protect their assets from possible flood damage (e.g. insurance purchases or spending on anti-flood modifications to their property); this may work better in developed rather than developing countries. The damage costs avoided approach is strongly linked to the geographic (intensity of disaster) and social context (land value, land-use, building type) (Turpie et al. 2010). These values vary greatly both between and within different locales; for example, within the same Bangladeshi village, Hossain (2015) found that poorer residents owned property made out of bamboo with thatched roofs, while high- income earners owned houses made out of bricks with corrugated iron roofing. In this case, both the likelihood of destruction and the rebuild costs of individual buildings will vary greatly. Measurement uncertainty depends partly upon the availability of data on past disasters, but is generally high because it is difficult to model scenarios accurately, and the trajectories, frequencies, and severity of future storms are difficult to predict (Marois and Mitsch 2015). Regardless, this method is generally preferred over the replacement costs method (Lau 2013; Barbier 2015) (see Senhoury et al., Chap. 19).

The replacement costs method estimates the cost of replacing an ecosystem service with an artificial substitute (Pagiola et al. 2004); in the case of mangroves this could mean a groyne or seawall. In order for the valuation to be valid, the man-made alternative must (a) provide a commensurate level of storm protection service, (b) be the cheapest option capable of performing the same role, and (c) society must be willing to incur the cost rather than forgo the service (Pagiola et al. 2004; Waite et al. 2014). Market data are typically available for this method (e.g. an engineer could quote a price for the alternative). However, it has been argued that the replacement cost method overestimates the value of the storm protection services for individual sites, because the approach involves estimating the service benefit primarily by using the costs of constructing groyne or seawalls. Moreover, the artificial substitute is rarely the most cost-effective means of pro- viding the service (Barbier 2007, 2015). In a mangrove storm protection study in Thailand, Barbier (2015) calculated annual welfare losses of US\$ 4,869,720 when using the replacement cost method, which were over an order of magnitude higher than the US\$645,769 calculated when using the avoided damage costs method.

More broadly, however, approaches to quantify and value DRR-related ecosys- tem services (such as storm protection) risk undervaluing the ecosystem as a whole. Artificial substitutes such as sea walls will typically only replace one service (e.g. storm protection), while all other benefits provided by the natural ecosystem will remain lost (Thampapillai and Sinden 2013). For example, in a study conducted by Gunawardena and Rowan (2005) in Sri Lanka, coastal defence was calculated to contribute just 27.6 % of the purported 'total economic value' which also included benefits to the fishery and wood used for building materials.

Recently, choice experiments have been used to value the multiple coastal ecosystem services provided by marine protected areas (Christie et al. 2015).

Coastal protection was one of six services that were used in the experiment, which presented respondents with different combinations of improved, current, and reduced service provision; these service packages were coupled with a hypothetical tax payment that gauged their willingness to pay and allowed the value of each service to be determined (Christie et al. 2015). Similar contingent valuation methods could be suitably applied to mangrove ecosystems, using hypothetical scenarios of declining or increasing service provision. However, contingent valuation can be expensive to implement, requires careful survey design, and is vulnerable to many sources of bias; meanwhile, choice experiments are considered to be technically difficult to implement (Waite et al. 2014).

4.3 Challenge 2 – How to Ensure DRR Service Provision and Permanence During a Disaster?

Once a DRR ecosystem service has been quantified, payments for such a service are dependent upon an agreed level of ecosystem service provision over a specified timescale. The maintenance of ecosystem service provision is related in a non-linear fashion to ecosystem quality, the maintenance of higher trophic levels and species richness (Duarte 2000; Dobson et al. 2006). However, a multitude of anthropogenic and natural stressors can reduce habitat quality and extent, thus impairing sustained ecosystem service provision. Such stressors on mangrove ecosystems may include agricultural land cover change (Webb et al. 2014), land reclamation (Wang et al. 2010), typhoons (Aung et al. 2013) and sea level rise (Krauss et al. 2010), and can have varied impacts from direct habitat conversion and destruction to cryptic declines in habitat quality, while the areal extent of habitat remains the same. In theory, many types of PES should require the reduction or cessation of direct anthropogenic stressors, such as harvesting or land cover conversion. However, many stressors in mangrove ecosystems either cannot be meaningfully reduced due to their process, magnitude and scale (e.g. tropical cyclones), and/or because they originate from a location external to the PES site (e.g. sea level rise), and are thus outside the control of a PES site manager.

External stressors such as tropical cyclones and sea level rise are important in a DRR context as we may promote mangroves to protect coastal communities against their impacts, although these external stressors themselves may have an impact on the mangrove system. An increasing literature exists on the impacts of tropical cyclones and storms on mangrove structure and functioning, especially in the wake of hurricanes, such as Hurricanes Andrew and Mitch, in the Neotropics. Some research has also been conducted in Asia after events such as Cyclone Nargis (Myanmar) and Typhoon Haiyan (Philippines). This body of research has described a number of tropical cyclone and storm impacts on mangroves, which may be immediate or delayed:

- *Defoliation:* species-specific defoliation is a common impact of high winds associated with large storm events, with mangroves in the eye of Hurricane Andrew consistently experiencing 100 % defoliation (Doyle et al. 1995).
- *Tree and branch damage:* strong winds can lead to branch and trunk damage, although damage may be species-specific: in one case study, *Rhizophora mangle* mostly suffered less than 50 % crown damage, while *Laguncularia racemosa* trees suffered 75–100 % crown loss (Sherman et al. 2001).
- *Tree mortality:* tree damage can be so great that mass tree mortality occurs. Mortality can be spatially variable due to species composition, geomorphology, elevation and storm track; in a study in the Dominican Republic after Hurricane Georges, mortality reached 100 % in some plots, with an average of 47.7 % (Sherman et al. 2001).
- *Peat collapse:* tree mortality leads to root death and the cessation of below-ground organic matter production. The peat soil may oxidise and collapse until such time when/if surviving trees and newly recruited individuals begin to produce below-ground organic matter to replace what was lost (Cahoon et al. 2003).
- *Sediment burial:* sediment eroded during a typhoon can be deposited within the mangrove. Such deposits can equal as much as 17 times the annual accretion rate experienced in the mangrove (Castaneda-Moya et al. 2010), which may suffocate the aerial roots of some species.

Sea level rise can also impact upon mangrove habitat quality and extent, with knock-on impacts for ecosystem service provision. Mangrove species distribution is controlled to a large extent by surface elevation and relative tidal inundation (e.g. Watson 1928), which can distribute species according to their tolerance to tidal flooding. Sea level rise – if not matched by similar increases in mangrove surface elevation (Krauss et al. 2014) – can increase tidal inundation beyond species-specific thresholds of tolerance, leading over time to a conversion to more tolerant pioneer mangrove species, and eventually to bare mudflat (Friess et al. 2012).

Thus, tropical cyclones, storms and sea level rise present a particularly interesting quandary: almost by definition, the locations most in need of ecosystem-based solutions for DRR are those that are heavily exposed to hazards. Thus, PES for DRR would provide funding for mangrove conservation to protect populations against short term events such as storms and long term events such as sea level rise, although these very same events can substantially damage the ecosystem in question and impact the provision of the required ecosystem service.

While the presence of external stressors may reduce ecosystem service provision and the effectiveness of PES, this does not mean that PES is untenable in such situations. Friess et al. (2015) describe a number of approaches to deal with external stressors in a PES context. While they vary in design and process, all of these approaches focus on siting a PES scheme in the most suitable biophysical location or reducing the risk of external stressors to financial assets. A three step, hierarchical strategy is proposed (Friess et al. 2015):

(i) *Stressor evaluation.* Ecosystem service provision models (e.g. Villa et al. 2014) must be combined with external stressors models in order to evaluate the risk they pose to a PES scheme. Environmental Impact Assessments on developments surrounding the PES scheme could also be mandated. In theory, these steps will ensure that a PES scheme is located in the most suitable location, for example away from neighbouring human developments, or along a sheltered coastline that is at less risk of storm damage (though this suggests that there could be less need for DRR measures in these areas). However, locating PES schemes in the most suitable locations from an ecosystems service and stressor point of view may not always be feasible, as it neglects political and social imperatives for PES scheme location.

(ii) *Stressor mitigation.* Once a PES scheme is located in an area that gives it the best chance of success, attempts can be made to mitigate the negative impacts of remaining identified external stressors. For anthropogenic external stressors this may require landscape planning and cross-sectoral cooperation. However, it is difficult to mitigate external stressors linked to natural hazards and climate change. For example, tropical cyclones and sea level rise are processes that operate on large scales that cannot be meaningfully mitigated by management interventions.

(iii) *Stressor accommodation.* Under the assumption that natural hazards and climate change stressors cannot be meaningfully mitigated, PES schemes must instead incorporate measures that allow the accommodation and management of risk. Such measures revolve around concepts of Financial Risk Management, particularly reducing uncertainty and investing in insurance measures. These may include third party ecosystem service insurance to pay for unexpected reductions in DRR ecosystem service yield. Bell and Lovelock (2013) propose insurance for mangroves damaged in storms, so that they can be restored and continue to provide DRR ecosystem services. Credit buffers and precautionary savings have also been used in some terrestrial PES sites (e.g. Phelps et al. 2011); more credits are created than are sold, so that there is a buffer to refund credits if the expected ecosystem service provision is not reached.

In summary, when planning a PES scheme to deliver DRR ecosystem services in a location heavily threatened by natural hazards and climate change impacts, scheme locations should ideally be determined through the use of ecosystem service and external stressor evaluation models. This will allow schemes to be situated in locations that maximize ecosystem service provision, while minimizing service impermanence. Once a PES scheme is located correctly, PES scheme planning must incorporate Financial Risk Management measures from the very beginning in order to reduce uncertainty and risk to ecosystem service investors, as natural hazards and climate change-related external stressors may never be fully mitigated.

4.4   Challenge 3 – Ecosystem Service Providers and Users Overlap

PES requires a quantified ecosystem service to be traded. PES involves a transaction between at least one service provider (seller) and service user (beneficiary/buyer) (Wunder 2015). Arguably the most important PES precondition is for a ready user to exist. Potential users include insurance companies, government agencies, NGOs, and local communities (Table 4.2). Providers will likely be local landowners/managers or the local community that implement new management approaches (e.g. mangrove restoration or preservation) in exchange for payment from the ecosystem service user. Thus in the case of local communities, there is the potential for the provider and user to be the same group or stakeholder, which invalidates PES. Additionally, the suitability and structure of mangrove PES for DRR, the types and suitability of users and providers, and their degree of overlap will likely differ between developed and developing economic settings.

Table 4.2 The suitability of different potential PES buyers in developed and developing country settings

Potential buyer	Developed	Developing
Insurance company	Coastal residents likely have property insurance; insurers will need to be convinced that more mangroves means less damage and ultimately less pay-outs (saving them money)	Coastal residents seldom have any insurance cover due to either financial constraints or a deficit of insurers
Government agency	Government may have financial capacity to pay	Government may not have the financial capacity to pay
	Would have to identify situations in which PES would be favoured over command-control regulation and public spending on artificial coastal defences	PES may be more cost-effective than investing in expensive artificial coastal defences
		May be an alternative approach to command-control regulation if compliance is a problem
NGO	Would likely prefer to give financial aid which does not require a return on investment	Would likely prefer to give financial aid which does not require a return on investment
Local community	Potentially could afford payments	May be unable to afford payments
	Potentially overlapping as service users and providers	
Private landowner	Possible that the landowner and land manager may be separate entities. If so, the owner could pay the manager to implement better mangrove restoration/preservation to safeguard the asset being managed	Will likely be unable to afford payments

Insurance companies may have a vested interest in DRR since better-protected coasts will mean less damage and lower pay-outs following a disaster (Forest Trends and The Katoomba Group 2008; Dunn 2011; Lau 2013). The feasibility of insurance companies as users is greater in developed countries in which an established array of insurers and insurance policies exist for property owners to choose from. In developing countries however, coastal residents seldom have insurance cover. This is particularly true for poorer households that will typically own property constructed out of weaker materials (Kolinjivadi et al. 2015), which will therefore be more prone to damage. An insurance company is unlikely to seek improved coastal protection services for coastal settlements that it is not insuring, so in this regard poorer communities may be excluded. Insurance is typically provided on an individual basis and therefore equity issues could arise (in both developed and developing settings) since poorer residents may be priced out.

Government agencies and municipalities responsible for disaster management have also been suggested as potential coastal protection service users (Forest Trends and The Katoomba Group 2008; Lau 2013). In developed countries with ample public spending budgets, it is difficult to see how PES would be a more rational option compared to command-control coastal regulations (that are generally effectively enforced in the developed world) and direct public spending on artificial coastal defences. However, some developing countries will likely have lower public spending budgets, and also more pressing problems to solve – i.e. investing in basic needs such as infrastructure, health, and education. Hence, in such settings, the restoration/conservation of natural barriers may be considered by governments to be more cost-effective than constructing artificial substitutes, which often come at huge installation and maintenance costs. It is feasible that governments may utilise a PES approach to pay local communities that live adjacent to mangroves to reduce mangrove cutting or engage in restoration activities, which can reduce disaster risk in their jurisdiction. This is based on the notion that governments have a duty to ensure the safety of their people.

NGOs can be buyers of ecosystem service credits, particularly to try and nurture carbon-markets. However, in the context of coastal DRR, where a future return on investment is highly unlikely (i.e. climate change exacerbating extreme weather events and sea level rise, thus reducing service provision e.g. *Challenge 2*), it is difficult to see how PES would be favoured over direct aid for which no justification is required other than philanthropy. This is true for both developed and developing settings.

Local communities and private landowners have also been cited as potential users (Lau 2013). This is probably more suited to developed, rather than developing nations. Expecting local communities in developing countries to finance PES seems unfeasible and unjust, because local communities will likely be unable to afford such payments, similar to the equity issues surrounding insurance cover. However, the very notion of local communities (if, due to land tenure issues, they even have control of the ecosystem service in the first place) and private land owners using or buying ecosystem services is controversial, since in almost all foreseeable cases, local communities will also likely be the most suitable service providers (i.e. sellers), as they will be responsible for managing the coastline on which they live. PES requires a transaction to take place between two separate parties, and as such, thinking of local communities as ecosystem service users creates a contradiction, since these beneficiaries would be buying a service that they (in most cases) would also provide.

4.5   Alternatives to PES for DRR

This chapter has described three important issues facing coastal PES as a means of funding ecosystem-DRR activities. All of these issues challenge the fundamental tenets of PES: how can we sufficiently and accurately quantify DRR service provision?; how do we ensure permanence and long-term provision of DRR services in a dynamic coastal environment?; and how do we identify suitable services users and providers, and make sure they are distinct and do not overlap? Ultimately, due to the nascent state of PES for DRR, in many circumstances existing financial mechanisms may be deemed more suitable for DRR and associated disaster relief in mangrove systems, compared to PES. Other financial mechanisms for DRR do exist, although these also tend to vary between developed and developing countries. Three types of mechanisms are primarily relevant to developed countries: compensation, subsidized insurance of assets, and ecosystem service insurance. Within the developing world, financial support for DRR generally comes from a fourth mechanism, donor aid.

Compensation Disaster compensation is a response predominantly confined to the developed world, but is also used increasingly in developing and emerging economies. In considering how socially just such compensation schemes really are, Cooper and McKenna (2008) note that while coastal property owners face a direct financial loss from coastal disasters, compensating them creates accompanying costs to society since the state will use taxpayer's money. It is argued that public interventions are more justifiable at local and short-term scales, but less justifiable at larger geographical and longer time scales since the societal costs to non-coastal tax-paying residents increase due to larger payouts (Cooper and McKenna 2008).

Subsidized Property Insurance Subsidized private insurance offers an alternative to public compensation schemes, especially since private markets are showing an increasing reluctance to underwrite catastrophic risks such as floods (Jaffee and Russell 2006). For example, the US Federal Flood Insurance program subsidizes private insurance premiums to make coastal development more affordable to property owners, and the risks more acceptable to insurance. Similar to compensation, however, this financial benefit for a small group of coastal property owners comes at significant cost to the taxpayer. This approach also perversely encourages development in higher-risk areas (Bagstad et al. 2007). The perverse incentives of subsidized insurance has prompted some economists to question whether governments should be involved in catastrophic risk insurance at all, and have called for private markets to be more robust and take a longer term view of risk and capital (Jaffee and Russell 2006).

State-subsidized insurance is a predominantly developed-nation approach to disaster relief, and potentially for funding DRR activities. However, some have argued for insurance and other public-private programmes to plug the gap between donor pledges and disaster losses. Insurance mechanisms suggested for developing nations include catastrophe insurance pools, catastrophe bonds and risk transfer instruments and derivatives (Linnerooth-Bayer et al. 2005; Linnerooth-Bayer and Mechler 2007).

Ecosystem Service Insurance Payouts from ecosystem service insurance contribute to ecosystem restoration in the event that the ecosystem itself is degraded from an external event (e.g., *Challenge 2*). Bell and Lovelock (2013) proposed a mangrove DRR insurance product focused on protecting coastal land from the impacts of storms. Uptake of such a scheme would rely on property owners understanding that a mangrove forest provides coastal defence for their property. The idea stems from forest carbon credit insurance, wherein buyers can take out insurance as a form of protection for their valuable investment in the event that, for example, the forest is destroyed (Phelps et al. 2011). Premiums could be incorporated into existing property insurance. In designing an ecosystem insurance policy for the DRR services of mangroves, Bell and Lovelock (2013) note the need for: clear specifications of what insured events are covered and excluded; estimates of how much it would cost to rehabilitate the DRR value of mangroves; a prediction of the likely frequency and severity of weather events in the region which will assist with setting the insurance premium; and a protocol for actions the insurer will perform if an insured event occurs. Due to the payments and financial networks required, this is ultimately another financial mechanism most suited for developed countries.

Donor Support State intervention (i.e. compensation) in the aftermath of a disaster is often insufficient in developing nations. Hence, these countries often rely on donor aid for disaster relief, which may be a small percentage of total disaster losses (Linnerooth-Bayer et al. 2005). This is not without major equity issues. For example, international donors contributed US$662.9 million of aid within 3 months after Typhoon Hainan (Philippines), but international assistance still had not reached some affected areas (Lum and Margesson 2014). Much of the aid went to the devastated city of Tacloban which received the most media attention, and assistance was substantially slower to reach rural and small island areas throughout the rest of the archipelago.

4.6 Conclusions

Both coastal populations and mangrove forests continue to face an uncertain future in a coastal zone that is undergoing huge development pressures, exacerbated by the coastal impacts of climate change. PES may be a novel and important instrument to conserve mangroves for their benefits to DRR, but only if current challenges can be overcome. Our ability to quantify DRR ecosystem service provision is lagging behind our knowledge of other ecosystem services such as carbon storage, although several direct and indirect methods of quantification and valuation do exist. Future efforts could focus on how to valorise direct measurements of hydrodynamic attenuation, or how to combine direct measurements with indirect measures of coastal protection such as replacement costs and avoided damage valuation. Ensuring long-term ecosystem service provision is also a challenge in coastal ecosystems that are affected by a host of external stressors that differ markedly in their process, origin, magnitude and scale. These challenges are in no way insurmountable; a series of tools exist to quantify some DRR services, and external biophysical stressors may be mitigated or accommodated in some circumstances.

In addition, some situations may best be tackled through donor support since there are no expectations of a return on investment, which may be unlikely in a dynamic coastal environment. However, at this embryonic stage, we need to take a critical look at PES as an instrument for DRR in mangrove systems. In particular, the mechanics of PES schemes for DRR are uncertain – particularly with regard to the buyer-context and whether these entities overlap or are distinct – as outlined in challenge three. Understanding the three challenges posed in this chapter will ensure that PES is the right funding model to pursue, and will allow us to be more strategic in selecting sites where mangrove ecosystem service delivery, governance and funding arrangements will be most effective over the long term.

Acknowledgements We thank members of the Mangrove Lab (National University of Singapore, www.themangrovelab.com) and J Phelps (Lancaster University) for discussions that have contributed to several of the themes highlighted in this chapter. This research was supported by the Ministry of Education, Government of Singapore (R-109-000-166-112).

References

Adame MF, Kauffman JB, Medina I et al (2013) Carbon stocks of tropical coastal wetlands within the karstic landscape of the Mexican Caribbean. PLoS One 8:1–13

Alongi DM (2008) Mangrove forests: resilience, protection from tsunamis, and responses to global climate change. Estuar Coast Shelf Sci 76:1–13

Aung TT, Mochida Y, Than MM (2013) Prediction of recovery pathways of cyclone-disturbed mangroves in the mega delta of Myanmar. For Ecol Manag 293:103–113

Bagstad KJ, Stapleton K, D'Agostino JR (2007) Taxes, subsidies, and insurance as drivers of United States coastal development. Ecol Econ 63(2–3):285–298

Barbier EB (2007) Valuing ecosystems as productive inputs. Econ Policy 22:177–229

Barbier EB (2015) Valuing the storm protection service of estuarine and coastal ecosystems. Ecosyst Serv 11:32–38

Barbier EB, Koch EW, Silliman BR et al (2008) Coastal ecosystem-based management with nonlinear ecological functions and values. Science 319:321–323

Barbier EB, Hacker SD, Kennedy C et al (2011) The value of estuarine and coastal ecosystem services. Ecol Monogr 81:169–193

Bell J, Lovelock CE (2013) Insuring mangrove forests for their role in mitigating coastal erosion and storm-surge: an Australian case study. Wetlands 33:279–289

Bohensky EL, Maru Y (2011) Indigenous knowledge, science and resilience: what have we learned from a decade of international literature on "integration"? Ecol Soc 16:6

Bouillon S, Borges AV, Castaneda-Moya E et al (2008) Mangrove production and carbon sinks: a revision of global budget estimates. Glob Biogeochem Cycles 22:GB2013

Brander LM, Wagtendonk AJ, Hussain SS (2012) Ecosystem service values for mangroves in South-east Asia: a meta-analysis and value transfer application. Ecosyst Serv 1:62–69

Cahoon DR, Hensel P, Rybczyk J et al (2003) Mass tree mortality leads to mangrove peat collapse at Bay Islands, Honduras after Hurricane Mitch. J Ecol 91:1093–1105

Castaneda-Moya E, Twilley RR, Rivera-Monroy VH et al (2010) Sediment and nutrient deposition associated with Hurricane Wilma in mangroves of the Florida coastal everglades. Estuar Coasts 33:45–58

Chaudhary S, McGregor A, Houston D, Chettri N (2015) The evolution of ecosystem services: a time series and discourse-centered analysis. Environ Sci Pol 54:25–34

Christie M, Remoundou K, Siwicka E, Wainwright W (2015) Valuing marine and coastal ecosystem service benefits: case study of St Vincent and the Grenadines' proposed marine protected areas. Ecosyst Serv 11:115–127

Cooper JAG, McKenna J (2008) Social justice in coastal erosion management: the temporal and spatial dimensions. Geoforum 39:294–306

Danielsen F, Sørensen MK, Olwig MF et al (2005) The Asian tsunami: a protective role for coastal vegetation. Science 310:643

Daw TM, Cinner JE, McClanahan TR et al (2012) To fish or not to fish: factors at multiple scales affecting artisanal fishers' readiness to exit a declining fishery. PLoS One 7, e31460

Day JW, Conner WH, Ley-Lou F et al (1987) The productivity and composition of mangrove forests, Laguna de Terminos, Mexico. Aquat Bot 27:267–284

De la Cruz AA (1982) Wetland uses in the tropics and their implications on the world carbon cycle. Wetlands 2:1–20

De la Cruz AA (1986) Tropical wetlands as a carbon source. Aquat Bot 25:109–115

Dobson A, Lodge D, Alder J et al (2006) Habitat loss, trophic collapse, and the decline of ecosystem services. Ecology 87:1915–1924

Donato DC, Kauffman JB, Murdiyarso D et al (2011) Mangroves among the most carbon-rich forests in the tropics. Nat Geosci 4:293–297

Doyle TW, Smith TJ, Robblee MB (1995) Wind damage effects of Hurricane Andrew on mangrove communities along the southwest coast of Florida, USA. J Coast Res (SI21):159–168

Duarte CM (2000) Marine biodiversity and ecosystem services: an elusive link. J Exp Mar Biol Ecol 250:117–131

Dunn H (2011) Payments for ecosystem services. Department for Environment, Food and Rural Affairs (DEFRA) Evidence and Analysis Series, Paper 4. DEFRA, London, pp 67

EM-DAT (2011) The OFRA/CRED International disaster database. Centre for research on the epidemiology of disasters. Available via www.emdat.be

Fisher B, Ellis AM, Adams DK et al (2015) Health, wealth and education: the socioeconomic backdrop for marine conservation in the developing world. Mar Ecol Prog Ser 530:233–242

Forest Trends, The Katoomba Group (2008) Payments for ecosystem services getting started: a primer. UNEP, Nairobi, p 74

Friess DA, Krauss KW, Horstman EM et al (2012) Are all intertidal wetlands naturally created equal? Bottlenecks, thresholds and knowledge gaps to mangrove and saltmarsh ecosystems. Biol Rev 87:346–366

Friess DA, Phelps J, Garmendia E, Gómez-Baggethun E (2015) Payments for Ecosystem Services (PES) in the face of external biophysical stressors. Glob Environ Chang 30:31–42

Gunawardena M, Rowan JS (2005) Economic valuation of a mangrove ecosystem threatened by shrimp aquaculture in Sri Lanka. Environ Manag 36:535–550

Hawkins S, To PX, Phuong PX et al (2010) Roots in the water: legal frameworks for mangrove PES in Vietnam, Katoomba Group's Legal Initiative Country Study Series. Forest Trends, Washington, DC

Hossain MN (2015) Analysis of human vulnerability to cyclones and storm surges based on influencing physical and socioeconomic factors: evidences from coastal Bangladesh. Int J Disaster Risk Reduct 13:66–75

Hunter JR, Church JA, White NJ, Zhang X (2013) Towards a global regionally varying allowance for sea-level rise. Ocean Eng 71:17–27

Hussain SA, Badola R (2010) Valuing mangrove benefits: contribution of mangrove forest to local livelihoods in Bhitarkanika Conservation Area, east coast of India. Wetl Ecol Manag 18:321–331

Huxham M, Emerton L, Kairo J et al (2015) Applying climate compatible development and economic valuation to coastal management: a case study of Kenya's mangrove forests. J Environ Manag 157:168–181

Jaffee DM, Russell T (2006) Should governments provide catastrophe insurance? Econ Voice 3:1553–3832

Jones T, Ratsimba H, Ravaoarinorotsihoarana L et al (2014) Ecological variability and carbon stock estimates of mangrove ecosystems in northwestern Madagascar. Forests 5:177–205

Jongman B, Ward PJ, Aerts JC (2012) Global exposure to river and coastal flooding: long term trends and changes. Glob Environ Chang 22:823–835

Kathiresan K, Rajendran N (2005) Coastal mangrove forests mitigated tsunami. Estuar Coast Shelf Sci 65:601–606

Kauffman JB, Donato DC (2012) Protocols for the measurement, monitoring and reporting of structure, biomass and carbon stocks in mangrove forests. Working paper 86. CIFOR, Bogor

Kerr AM, Baird AH, Campbell SJ (2006) Comments on "Coastal mangrove forests mitigated tsunami" by K. Kathiresan and N. Rajendran [Estuar. Coast. Shelf Sci. 65 (2005) 601–606]. Estuar Coast Shelf Sci 67:539–541

Koch EW, Barbier EB, Silliman BR et al (2009) Non-linearity in ecosystem services: temporal and spatial variability in coastal protection. Front Ecol Environ 7:29–37

Kolinjivadi V, Grant A, Adamowski J, Kosoy N (2015) Juggling multiple dimensions in a complex socio-ecosystem: the issue of targeting in payments for ecosystem services. Geoforum 58:1–13

Krauss KW, Allen JA, Cahoon DR (2003) Differential rates of vertical accretion and surface elevation change among aerial root types in Micronesian mangrove forests. Estuar Coast Shelf Sci 56:251–259

Krauss KW, Cahoon DR, Allen JA et al (2010) Surface elevation change and susceptibility of different mangrove zones to sea level rise on Pacific high islands of Micronesia. Ecosystems 13:129–143

Krauss KW, McKee KL, Lovelock CE et al (2014) How mangrove forests adjust to rising sea level. New Phytol 202:19–34

Lacambra C, Friess DA, Spencer T, Moller I (2013) Bioshields: mangrove ecosystems as resilient natural coastal defences. In: Renaud FG, Sudmeier-Rieux K, Estrella M (eds) The role of ecosystems in disaster risk reduction. United Nations University Press, Tokyo, pp 82–108

Lau WWY (2013) Beyond carbon: conceptualizing payments for ecosystem services in blue forests on carbon and other marine and coastal ecosystem services. Ocean Coast Manag 83:5–14

Linnerooth-Bayer J, Mechler R (2007) Disaster safety nets for developing countries: extending public-private partnerships. Environ Hazards 7:54–61

Linnerooth-Bayer J, Mechler R, Pflug G (2005) Refocusing disaster aid. Science 309:1044–1046

Lum T, Margesson R (2014) Typhoon Haiyan (Yolanda): U.S. and international response to Philippines disaster. In: Congressional Research Service. pp 1–23

MA (2005) Ecosystems and human well-being: synthesis. Millennium ecosystem assessment. World Resources Institute, Washington, DC

Manson FJ, Loneragan NR, Skilleter GA, Phinn SR (2005) An evaluation of the evidence for linkages between mangroves and fisheries: a synthesis of the literature and identification of research directions. Oceanogr Mar Biol Annu Rev 43:483–513

Marois DE, Mitsch WJ (2015) Coastal protection from tsunamis and cyclones provided by mangrove wetlands – a review. Int J Biodiver Sci Ecosyst Serv Manag 11:71–83

Mazda Y, Magi M, Ikeda Y et al (2006) Wave reduction in a mangrove forest dominated by Sonneratia sp. Wetl Ecol Manag 14:365–378

Mercer J, Kelman I, Taranis L, Suchet-Pearson S (2010) Framework for integrating indigenous and scientific knowledge for disaster risk reduction. Disasters 34:214–239

Mercer J, Gaillard JC, Crowley K et al (2012) Culture and disaster risk reduction: lessons and opportunities. Environ Hazards 11:74–95

Pagiola S, von Ritter K, Bishop J (2004) How much is an ecosystem worth? Assessing the economic value of conservation. World Bank, Washington, DC

Pascual U, Phelps J, Garmendia E et al (2014) Social equity matters in Payments for Ecosystem Services. Bioscience 64:1027–1037

Peduzzi P, Chatenoux B, Dao H et al (2012) Global trends in tropical cyclone risk. Nat Clim Chang 2:289–294

Phang VXH, Chou LM, Friess DA (2015) Ecosystem carbon stocks across a tropical intertidal habitat mosaic of mangrove forest, seagrass meadow, mudflat and sandbar. Earth Surf Process Landf 40:1387–1400

Phelps J, Agrawal A, Webb EL (2010) Does REDD+ threaten to recentralize forest governance? Science 328:312–313

Phelps J, Webb EL, Koh LP (2011) Risky business: an uncertain future for biodiversity conservation finance through REDD+. Conserv Lett 4:88–94

Pleasant MM, Gray SA, Lepczyk C et al (2014) Managing cultural ecosystem services. Ecosyst Serv 8:141–147

Polidoro BA, Carpenter KE, Collins L et al (2010) The loss of species: mangrove extinction risk and geographic areas of global concern. PLoS One 5, e10095. doi:10.1371/journal.pone. 0010095

Quartel S, Kroon A, Augustinus AG et al (2007) Wave attenuation in coastal mangroves in the Red River Delta, Vietnam. J Asian Earth Sci 29:576–584

Samonte-Tan G, White A, Tercero M et al (2007) Economic valuation of coastal and marine resources: Bohol Marine Triangle, Philippines. Coast Manag 35:319–338

Sherman RE, Fahey TJ, Martinez P (2001) Hurricane impacts on a mangrove forest in the Dominican Republic: damage patterns and early recovery. Biotropica 33:393–408

Siikamäki J, Sanchirico JN, Jardine SL (2012) Global economic potential for reducing carbon dioxide emissions from mangrove loss. Proc Natl Acad Sci 109:14369–14374

Thampapillai DJ, Sinden JA (2013) Environmental economics: concepts, methods and policies. Oxford University Press, Melbourne

Thiagarajah J, Wong S, Richards DR, Friess DA (2015) Historical and contemporary cultural ecosystem service values in the rapidly urbanizing city state of Singapore. Ambio 44:666–677

Thompson BS, Clubbe CP, Primavera JH et al (2014) Locally assessing the economic viability of blue carbon: a case study from Panay Island, the Philippines. Ecosyst Serv 8:128–140

Turpie J, Lannas K, Scovronick N, Louw A (2010) Wetland Valuation Volume 1: Wetland Ecosystem Services and Their Valuation: A Review of current Understanding and Practice, Water Research Commission (WRC) report no. TT 440/09. WRC, Cape Town, p 132

Twilley RR (1985) The exchange of organic carbon in basin mangrove forests in a southwest Florida estuary. Estuar Coast Shelf Sci 20:543–557

Twilley RR, Lugo AE, Patterson-Zucca C (1986) Litter production and turnover in basin mangrove forests in southwest Florida. Ecology 67:670–683

Twilley RR, Chen RH, Hargis T (1992) Carbon sinks in mangroves and their implications to carbon budget of tropical coastal ecosystems. Water Air Soil Pollut 64:265–288

Uddin MS, de Ruyter van Steveninck E, Stuip M, Shah MAR (2013) Economic valuation of provisioning and cultural services of a protected mangrove ecosystem: a case study on Sundarbans Reserve Forest, Bangladesh. Ecosyst Serv 5:88–93

UNEP (2014) The importance of mangroves to people: a call to action. In: van Bochove J, Sullivan E, Nakamura T (eds) United Nations Environment Programme World Conservation Monitoring Centre, Cambridge, UK, pp 128

Villa F, Bagstad KJ, Voigt B et al (2014) A methodology for adaptable and robust ecosystem services assessment. PLoS One 9:e91001

Waite R, Burke L, Gray E (2014) Coastal capital: ecosystem valuation for decision making in the Caribbean. World Resources Institute, Washington, DC, pp 88

Wang X, Chen W, Zhang L et al (2010) Estimating the ecosystem service losses from proposed land reclamation projects: a case study in Xiamen. Ecol Econ 69:2549–2556

Watson JG (1928) Mangrove forests of the Malay Peninsula. Malayan Forest Rec 6:1–275

Webb EL, Jachowski NR, Phelps J et al (2014) Deforestation in the Ayeyarwady Delta and the conservation implications of an internationally-engaged Myanmar. Glob Environ Chang 24:321–333

Wunder S (2015) Revisiting the concept of payments for environmental services. Ecol Econ 117:234–243

Wunder S, Wertz-Kanounnikoff S (2009) Payments for ecosystem services: a new way of conserving biodiversity in forests. J Sustain For 28:576–596

Chapter 5
Ecosystem-Based Adaptation and Disaster Risk Reduction: Costs and Benefits of Participatory Ecosystem Services Scenarios for Šumava National Park, Czech Republic

Zuzana V. Harmáčková, Eliška Krkoška Lorencová, and David Vačkář

Abstract The aim of the study was to analyse economic costs and benefits of stakeholder-defined adaptation scenarios for the Šumava National Park, the Czech Republic, and to evaluate their impact on the provision of ecosystem services, primarily focusing on ecosystem-based adaptation options which support disaster risk reduction in a broader region. The study utilised an array of approaches, including participatory scenario building, GIS modelling and economic evaluation. Based on a participatory input by local stakeholders, four adaptation scenarios were created, formulating various possibilities of future development in the area as well as potential vulnerabilities and adaptation needs. The scenarios subsequently served as the basis for biophysical modelling of the impacts of adaptation and disaster risk reduction measures on the provision of ecosystem services with the InVEST modelling suite, focusing on climate regulation, water quality and hydropower production. Finally, a cost-benefit analysis was conducted, quantifying management and investment costs of each adaptation scenario, and benefits originating from the provision of previously modelled regulating ecosystem services, together with a supplementary selection of provisioning services. This study serves as an example of combining stakeholder views, biophysical modelling and economic valuation in the cost-benefit analysis of ecosystem-based adaptation and disaster risk reduction, which provides the opportunity to find shared solutions for the adaptation of social-ecological systems to global change.

Z.V. Harmáčková
Department of Human Dimensions of Global Change, Global Change Research Centre, Academy of Sciences of the Czech Republic, Belidla 986/4a, 60300, Brno, Czech Republic

Faculty of Humanities, Charles University, Prague, Czech Republic

E.K. Lorencová • D. Vačkář (✉)
Department of Human Dimensions of Global Change, Global Change Research Centre, Academy of Sciences of the Czech Republic, Belidla 986/4a, 60300, Brno, Czech Republic
e-mail: vackar.d@czechglobe.cz

© Springer International Publishing Switzerland 2016
F.G. Renaud et al. (eds.), *Ecosystem-Based Disaster Risk Reduction and Adaptation in Practice*, Advances in Natural and Technological Hazards Research 42,
DOI 10.1007/978-3-319-43633-3_5

Keywords Climate change • Ecosystem-based adaptation • Participatory scenarios • Ecosystem services • Cost-benefit analysis

5.1   Introduction

Ecosystem management approaches aiming to enhance ecosystem resilience are considered to provide cost-effective and multifunctional alternatives to climate change adaptation (CCA) (Renaud et al. 2013). In this respect, ecosystem services (in most cases divided into provisioning, regulating and cultural services) have the potential to serve as a framework for assessing the effects of sustainable ecosystem management related to various types of adaptation measures (MA 2005; TEEB 2010). Especially regulating ecosystem services, defined as the benefits obtained from the regulation of ecosystem processes, have been recognised as critically important for climate change mitigation and disaster risk reduction (DRR) (Munang et al. 2013a). As novel tools for quantification, valuation, mapping and modelling of ecosystem services are available (Kareiva et al. 2011), ecosystem services trade-offs under different scenarios of climate change and climate change adaptation can be analysed.

Ecosystem-based Adaptation (EbA) has been defined as "the use of biodiversity and ecosystem services to help people adapt to the adverse effects of climate change and harness opportunities arising from change" (CBD 2009). Current adaptation strategies are predominantly based on technical, structural, social and economic developments; yet, ecosystems and biodiversity can play a significant role in societal adaptation to climate change. Ecosystems and biodiversity will be negatively affected by climate change according to multiple scenarios (MA 2005). However, they provide an ecological infrastructure for adaptation, as they deliver a broad spectrum of mitigation options for addressing climate change impacts, e.g. flood and disaster protection, carbon storage or the prevention of soil erosion (Campbell et al. 2009).

Ecosystem-based approaches to adaptation and DRR have recently been promoted as an alternative approach to buffering the impacts of climate change while sustaining ecosystems and biodiversity, and enhancing the resilience, resistance and performance of ecosystems (Mooney et al. 2009; Jones et al. 2012). EbA has been proposed as a 'natural' solution to adaptation to climate change and is supposed to enhance the adaptation capacity of human society through the sustainable management and restoration of ecosystem services, while providing multiple benefits to human society. EbA surpasses other adaptation approaches by delivering multiple co-benefits and avoiding maladaptation (Munang et al. 2013b). EbA approaches are also closely related to Ecosystem-based Disaster Risk Reduction (Eco-DRR) (Renaud et al. 2013). Examples of EbA measures with related benefits are given in Table 5.1. Other examples of EbA actions include alien species management and enhancing genetic diversity (USGCRP 2008; Naumann et al. 2011).

Cost-benefit analysis presents a framework for comparing costs and benefits of different projects or investments (Hanley and Barbier 2009). Generally, the costs of

Table 5.1 Benefits resulting from EbA

Ecosystem-based Adaptation	Benefits
Restoring fragmented or degraded natural areas	Enhances critical ecosystem services, such as water flow or food and fisheries provision
Protected groundwater recharge zones or restoration of floodplains	Secures water resources so that entire communities can cope with drought and flooding
Connecting expanses of forests, grasslands, reefs or other habitats	Enables people and biodiversity to move better to more viable habitats as the climate changes
Protecting or restoring natural infrastructure such as barrier beaches, mangroves, coral reefs and forests	Buffers human communities from natural hazards, erosion and flooding

implementing EbA and Eco-DRR can be divided into two classes comprising financial and opportunity costs. While financial costs represent the value of resources deployed in the development of EbA components, including the costs of labour, materials, energy, etc., opportunity costs are defined as the value of economic opportunities foregone as a result of EbA, e.g. foregone development, restrictions in resource use and loss of economically utilizable land. Within the financial costs, one-off costs and recurrent costs undertaken to implement certain EbA measures can be distinguished. The one-off costs are necessary e.g. to establish management bodies, conduct surveys and research, or purchase land intended for ecosystem restoration. The recurrent costs are required to run the administrative bodies, maintain and restore ecosystems or monitor the changes of ecosystems. The benefit side of EbA projects comprises primary and secondary benefits. Primary benefits usually include environmental enhancements in the form of ecosystem services provision, (e.g. enhanced carbon storage, habitat creation and water purification and regulation). Secondary benefits are perceived as socio-economic, (e.g. effects on employment and tourism opportunities, quality of life and health improvements (Lange et al. 2010)). Moreover, EbA can bring economic, social and environmental co-benefits and ecosystem services that are both marketable (e.g. livestock and fish production) and non-marketable (e.g., cultural preservation, biodiversity maintenance) (Jones et al. 2012; also see Emerton et al, Chap. 2 and Vicarelli et al, Chap. 3).

The aim of this study was to analyse economic costs and benefits of stakeholder-defined adaptation scenarios for the Šumava National Park, the Czech Republic, and to evaluate their impact on the provision of ecosystem services. The study primarily focused on EbA options which support DRR in the broader region of the Šumava Mountains.

5.1.1 Context of the Case Study

The Šumava Mountains (49.0317878 N, 13.4843789E), located in the southern Czech Republic, present one of the most ecologically valuable forested montane ecosystems in Central Europe (Fig. 5.1). Its landscape is mainly characterised by

Fig. 5.1 Natural character of the case study area, Šumava National Park and Biosphere Reserve, representing Central European mountain type vegetation with peatbogs and lakes (Photo: Josef Brůna)

near-natural and semi-natural coniferous forests (59 %), pastures (14 %), marshes and peat mires (2.4 %), and glacial lakes. The most pristine area of the Šumava Mountains has been protected since the 1960s and declared a National Park (NP) in 1991, surrounded by a buffer zone of a Protected Landscape Area (PLA). Both the NP (680 km^2) and the PLA (996 km^2) comprise the Šumava UNESCO Biosphere Reserve (Fig. 5.2).

In order to capture the broader context of the area, this study focused on the Šumava UNESCO Biosphere Reserve (denoted as Šumava in the following text), not solely on the area of the national park. The study area is situated between 467 and 1378 m above sea level (m a.s.l.), with average temperatures varying between 3 and 6 °C, and average annual precipitation and potential evapotranspiration of approximately 1500 and 450 mm, respectively (Tolasz 2007). The study area comprises 32 municipalities, out of which only ten reach over 500 inhabitants. The area struggles with decreasing population and increasing average age of the inhabitants in local municipalities in the long-term (Novotná and Kopp 2010; Perlín and Bičík 2010). An extensive artificial water reservoir (Lipno) is located in the southern part of the study area, providing numerous assets including drinking water, hydropower and recreational opportunities.

Together with neighbouring Bavarian Forest National Park in southeast Germany, the Šumava NP covers one of the largest forested areas in central Europe, providing a wide array of ecosystem services and high biodiversity levels. The area

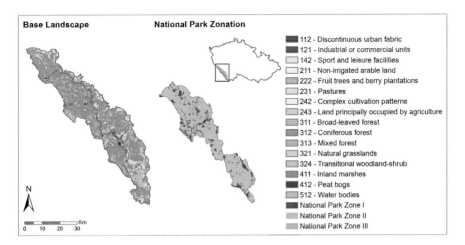

Fig. 5.2 The Šumava National Park and its zonation with surrounding Protected Landscape Area. Both areas comprise the Šumava Biosphere Reserve. Land use and land cover map based on CORINE Land Cover 2006

provides habitats for numerous threatened species such as lynx (*Lynx lynx*) and capercaillie (*Tetrao urogallus*) and contains several sites of pristine Norway spruce (*Picea abies*) forests in higher altitudes. The majority of local habitats are not influenced by human settlements, since most of the former German-speaking inhabitants were expelled after the World War II and the area became a part of the abandoned border zone (Novotná and Kopp 2010). The Šumava NP has been recognised by the International Union for Conservation of Nature (IUCN) (category II – National Park) and reflected in several international conventions, e.g. Ramsar Convention designating the most pristine peat bogs as wetlands of international importance. The Šumava NP is also a part of the Natura 2000 network, a centrepiece of EU biodiversity and conservation policy.

Šumava is covered by the most extensive forest ecosystem in Central Europe; however, the natural composition of the originally mixed beech, pine and spruce forest has been altered, and at present, semi-natural spruce plantations prevail in most of the area. Non-native spruce varieties have been planted in several locations as a result of human demand for wood, especially for glass industry and other commercial demand. Spruce *(Picea abies)* vegetation is not well adapted to local climate and has been susceptible to a range of disturbances such as strong winds and bark beetle (*Ips typographus*) outbreaks (Kindlmann et al. 2012).

Land use and land cover change have been moderate in the past two decades in the study area due to its declaration as a NP. However, intensive tourism and forestry demands have resulted in increasing land use and land cover change, represented mainly by urbanisation and changes in forest management. Both of these changes are limited as the area of the NP is strictly protected. Nevertheless, recent windstorms (in 2007 and 2011) and subsequent bark beetle outbreaks resulted in intensive logging have given rise to strong discussions about the best management approach and the extent of protected areas in the NP (especially non-intervention zones). At the same time, there are numerous development plans intending to build large-scale tourist resorts, which might change the current level of construction in both qualitative and quantitative ways (EIA Servis 2011).

Since the establishment of the NP, the concept of the protected area management has repeatedly changed, which has resulted in several substantial changes in zonation and conservation approaches. The management of the NP is subject to several conflicts, especially between the administration of the NP, environmental groups and non-governmental organizations, scientists and local interest groups, including representatives of municipalities and businesses. The park is split into three zones: Zone I is the most pristine and strictly protected part of the NP, Zone II includes the near-natural ecosystems that were variously influenced by human activities in the past, and Zone III has areas which allow a wide variety of socio-economic activities. Zones I and II present an equivalent to core zones under Czech legislation (Fig. 5.2). At present, Zone I of the NP consists of several small-scale and disconnected patches, scattered around the area of the NP, while some of them are partly non-interventionist. Currently, the legislation designing the NP is being revisited within the process of adjusting the vision of the NP for the future (Křenová and Hruška 2012; Bláha et al. 2012).

5.1.1.1 Climate Change Vulnerabilities

At present, the Šumava NP is threatened by various types of disturbances, including climate change impacts, land use and land cover change. The most pronounced pressures are the growing occurrence of disasters such as extreme weather events and subsequent pest outbreaks, together with intensive tourism and increasing forestry demands. Although localised projections of climate change and its impacts on the ecosystems of the Šumava Mountains have not been conducted, it is possible to derive applicable information from national-wide assessments and local research studies.

Temperatures in the area are expected to rise, with a more pronounced trend in the summer months. The short-term estimate (midpoint in 2030) shows that the average annual air temperature in the Czech Republic will increase, according to the ALADIN-CLIMATE/CZ model, approximately by 1 °C. In the medium-term timeframe (midpoint in 2050), the simulated warming becomes more significant (ME CR 2013). Since the 1960s, average annual and monthly precipitations have not shown a statistically significant trend. However, some changes in the temporal and spatial distribution of precipitation have been observed. Spatially specific heavy rains and droughts are becoming more frequent, related to the overall increase in climate extremity. In Šumava watersheds, the expected decrease in runoff is approximately 20 %, mainly due to increased evapotranspiration, caused by higher temperatures (EEA 2010; Hanel et al. 2012).

Higher winter temperatures are supposed to reduce snowpack and increase evaporation, leading to shifts in annual water outflows. On the other hand, both winter runoff and subsequent risk of spring floods are expected to increase, since water storage in the form of snowpack will be reduced. The period of snow melting is likely to shift from early spring to late winter months (Hanel et al. 2012). A substantial decrease in summer precipitation is projected; however at the same time, intensive precipitation events occurring during summer thunderstorms may result in a greater risk of flash floods (OECD 2013; Hanel et al. 2012). Other expected extreme weather events include windstorms, which are expected to occur more frequently (Beniston et al. 2007).

The near-natural and managed forest ecosystems in the study area have the potential to enhance DRR in a broader region. First, the area has recently suffered from wind storms and subsequent pest outbreaks, which have had serious environmental and socio-economic impacts, including the disintegration of forest ecosystems and substantial conflicts about the management approach. Enhancing the resilience of local ecosystems through EbA has thus the potential to reduce natural hazard impacts (see also McNeely, Chap. 17). Second, the area serves as the source watershed for one of the largest Czech artificial water reservoirs, which provides drinking water and hydropower. Therefore, the potential decrease in water quality due to nutrients and sedimentation may have a detrimental effect on human well-being both on a local and on a national scale. Water-related ecosystem services, provided within the study area, are therefore another important source of DRR potential.

5.2 Methods

5.2.1 Participatory Scenario Building

Since our aim was to base the analysis on participative input, the first step was to elicit local stakeholders' preferences regarding possible future development of the area, the level of nature conservation and economic development In 2014, two participative workshops were organised for various groups of stakeholders, covering all key sectors in the area and representing a broad range of opinions (see Table 5.2). The workshops aimed at (1) creating visions of future development of the study area, by developing a series of storylines describing potential future development of the study area through 2050, or participative scenarios building, and (2) proposing adaptation measures suitable for the study area and matching them to the previously constructed visions. In the first round, we addressed 20 selected stakeholders; however, we had to address another 10 stakeholders in the second round due to a low response rate, eventually gaining 15 attendees.

Since we were aware that the idea of scenario building would be completely new for the stakeholders, the workshops started with introductory presentations explaining the concept of future scenarios (Rounsevell and Metzger 2010) and participative scenario building. Following the introduction, the stakeholders were

Table 5.2 A list of stakeholders involved in the process of participatory scenario building. Specific stakeholders are listed together with the sector of their operation

Sector	Stakeholder institution/agency
Local authorities	Mayors of the municipalities in the South Bohemian Region
Conservation	The administration of the Šumava National Park
Regional development	Regional Development Agency of the Šumava Region
Science/Research	The University of South Bohemia in České Budějovice
Energy	Local energy production agency
Water management authorities	The Vltava Catchment
Agriculture	Local private agricultural enterprises
Tourism/Recreation	Local guides/private tourist enterprises

involved in an array of sub-group discussions and interviews. They were asked to follow a list of key economic sectors and issues characteristic of the area (demographic and economic development, tourism and recreation, agriculture, water management, nature conservation, etc.) and to formulate their preferences and expectations regarding each of the topics. They were also encouraged to add their own insight and issues of interest into the storylines. Due to the substantially different backgrounds of the stakeholders involved, we did not insist on reaching a consensus on a single vision among all stakeholders, but recommended trying to come to an agreement on one vision per stakeholder group. Eventually, the stakeholders identified three storylines denoted as the 'Green scenario', the 'Red scenario' and the 'Shared vision' (see Results). Stakeholder input thus resulted in an array of narratives, which were further translated into land use change scenarios using GIS approaches.

Subsequently, the participants developed a list of adaptation measures suitable for the study area based on their expertise and local experiences (See also Lange et al. Chap. 21). Since the study area is intensively protected, implementing new grey adaptation measures in addition to existing technological facilities (such as existing water treatment plants), e.g. in the form of large-scale construction of technical infrastructure, are not viable. Therefore, the stakeholders focused on green, EbA measures enhancing the resilience of local ecosystems. After each exercise, the sub-groups presented their output to other participants and the workshops were concluded by a follow-up plenary discussion and feedback session.

5.2.2   Land Use Change Scenarios

The storylines created during participative scenario workshops were subsequently translated into an array of land use and land cover change scenarios, using a series of ArcGIS Spatial Analyst tools (ESRI 2013). The land use scenarios also incorporated those adaptation measures that were conveyable in a spatially explicit way (further in this study, we use the term 'land use' to describe both the actual use of

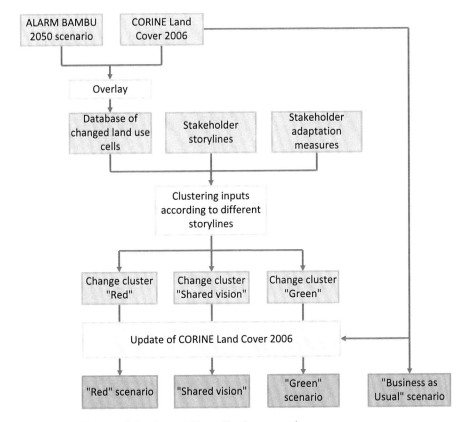

Fig. 5.3 The framework for the modelling of land use scenarios

land and the observable land cover according to the definition by Koomen and Stillwell 2007).

In general, in the process of translating stakeholders' storylines into land use change scenarios, we used the method of combining European-scale dynamic land use scenarios through 2050 with stakeholder inputs, previously developed and tested in the conditions of the Czech Republic (for brief overview see Fig. 5.3 and below, for further methodological details see Harmáčková and Vačkář 2015). As the basis for our analysis, we used dynamic land use scenarios reflecting major European-wide socio-economic and environmental trends, developed within the ALARM project (Rounsevell et al. 2006; Settele et al. 2005) and downscaled to the country-specific level in the Ecochange project (Dendoncker et al. 2006). First, we overlaid the ALARM BAMBU 2050 land use scenario (Spangenberg 2007) with CORINE Land Cover 2006 (CLC2006; EEA 2007) at 100 × 100 m resolution and identified all changed cells. Second, we combined the output with stakeholder storylines and adaptation measures proposed. Third, we grouped these inputs into three clusters corresponding to the storylines. In this step, the land use change trends identified by the stakeholders as highly improbable in the study area were removed; on the contrary, we supplemented each cluster by locally specific trends proposed by the stakeholders. Finally, we used each change cluster as an update to CLC2006, gaining three land use scenarios to 2050 corresponding to stakeholder storylines. Additionally, we introduced a fourth scenario denoted as 'Business as Usual' (BAU), which fully corresponded to Base Landscape (CLC2006, Fig. 5.2) and was used in the cost-benefit analysis as a reference state of landscape assuming no land use and land cover changes.

5.2.3 Assessment of Ecosystem Services

The four land use scenarios developed in the previous step were further used as an input into a biophysical assessment of ecosystem services. Since the extensive forests in the study area may potentially serve as a carbon sink and local water yields contribute to the Lipno reservoir, we selected two relevant regulating and one provisioning ecosystem services for the analyses, specifically climate regulation, water purification (in terms of nitrogen and sediment retention), and hydropower production. The biophysical outputs of the models served two purposes. First, they illustrate the improvement or exacerbation of local ecosystem characteristics related to the provision of selected ecosystem services. Second, they were later used to calculate the economic value of ecosystem services within the cost-benefit analysis.

The provision of ecosystem services was modelled using the InVEST suite of models, developed within the National Capital Project initiative as research and decision-support tools allowing for a spatially explicit evaluation of ecosystem services. InVEST presents a set of models based on land use scenarios as the main inputs, using the approach of ecological production functions, which attribute different levels of ecosystem service provision to specific ecological and socio-economic characteristics of the study location (Kareiva et al. 2011). The models were previously applied in studies encompassing a wide range of geographic and climatic conditions (Nelson et al. 2009; Tallis and Polasky 2009; Goldstein et al. 2012). A detailed description of the modelling assumptions, processes and limitations is provided in Kareiva et al. (2011) and Sharp et al. (2014); therefore, they are not reproduced in this section (See also Whelchel et al, Chap. 6).

Apart from the participatory land use scenarios, which were used in all InVEST analyses, specific data sources and modelling parameters used to model each of the selected ecosystem services are listed in the sections below. We ran the InVEST 3.0.1 models at the resolution of 100 x 100 m (1 ha), which presented a compromise between spatial accuracy and processing requirements. However, only the results for climate regulation are provided in the original resolution; the remaining water-related services were aggregated to the sub-watershed level as recommended by Sharp et al. (2014) (the grid-cell level of the resulting maps is not suitable for direct interpretation). Climate regulation was calculated for the whole study area, while water purification and hydropower production were quantified only for the watersheds contributing to the Lipno reservoir as these are considered to potentially cause excessive sedimentation and eutrophication on the one hand, and change in hydropower production on the other.

5.2.3.1 Climate Regulation

The ecosystem service of climate regulation was modelled in terms of the change in landscape carbon stocks. The input parameters comprised the amount of carbon stored in four carbon pools (aboveground biomass, belowground biomass, soil carbon and dead organic matter) for each land use category. These data were derived from a compilation of studies originating from areas with similar geographic and climatic conditions as the study area (Schumacher and Roscher 2009; IFER 2010; Lindsay 2010; Truus 2011; De Simon et al. 2012; NIR 2012; Schumacher and Roscher 2009). The model sums the amount of carbon stored in each raster cell under the Base Landscape and a future scenario and calculates the difference, which conveys the level of climate regulation reached under a certain scenario.

5.2.3.2 Water-Related Ecosystem Services

Concerning the ecosystem service of water purification, we focused on nitrogen and sediment retention. In accord with the conceptualization of ecosystem services provision by Villamagna et al. (2013) we modelled three aspects of these services using the InVEST model, defined as:

$$\Delta E = D - R \tag{5.1}$$

where D represents the amount of pollutants discharged (nitrogen and eroded soil, respectively), indicating what pressure each scenario imposes on local environmental conditions. R represents the amount of pollutants retained in the landscape as a measure of the capacity to provide an ecosystem service, and ΔE the amount of pollutant exported annually to the stream network. ΔE served as the basis for economic valuation.

Hydropower production was modelled with the corresponding InVEST module based on the annual water yield provided by the study landscape, contributing to the operation of a local hydropower plant at the Lipno reservoir, while taking vegetation, climate and soil parameters into account.

The main data inputs for the analyses were derived from national sources, comprising climate parameters (precipitation and reference evapotranspiration; Tolasz 2007), soil parameters (soil depth, plant available water content, rainfall erosivity and soil erodibility factors; VÚMOP 2014), and watershed and sub-watershed boundaries (TGM WRI 2014). Precipitation and evapotranspiration projections to 2050 were derived from climatic scenarios provided by INGV-

CMCC. Since very limited information on nitrogen export coefficients from various types of land use were available, we used values provided by Reckhow et al. (1980) supplemented by estimates from stakeholders with expertise in local water management. The amount of pollutants discharged and retained, as well as water yield and the amount of hydropower produced were quantified only for the watersheds relevant for the Lipno reservoir.

5.2.4   Cost-Benefit Analysis

As the final step, a cost-benefit analysis of the implementation of different participative scenarios was carried out. Since the primary aim of this study was to assess the impact of scenarios and adaptation measures on the provision of ecosystem services, we focused the cost-benefit analysis solely on the aspects related to ecosystem management within the Šumava NP and PLA and the provision of ecosystem services.

Although the InVEST suite of models provides the option of economic evaluation of ecosystem services provision, we utilised this functionality only in the case of climate regulation, as only this module takes directly into account the process of change between the current state and a future scenario. On the contrary, the remaining InVEST modules solely allow assessing an economic value of an ecosystem service provided by a single landscape at a time, regardless of the current state or a future scenario. Therefore, for the non-carbon regulating ecosystem services incorporated into the cost-benefit analysis, we utilised the biophysical outputs of the InVEST models and used our own calculations (described below) in order to be able to evaluate the gradually changing production of each ecosystem service between the current state and each future scenario.

In all cases, we used the indicator of net present value (NPV), which allows for a long-term accounting for a gradually developing value of regulating ecosystem services, together with fixed revenues from provisioning services, maintenance and investment costs (see also Vicarelli et al. Chap. 3).

Additionally, a sensitivity analysis was conducted focusing on several parameters. First, the NPV of each scenario was calculated at 5 and 1 % discount rates. Second, we used a minimum, mean and a maximum estimate of marginal costs, annual costs and economic revenues, in order to account for potential uncertainty (Table 5.3).

The overall NPV for each scenario in the period 2006–2050 was calculated as

$$NPV = NPV_B - NPV_c \tag{5.2}$$

where NPV_B (NPV_C) stand for the NPV of all benefits (and costs, respectively), linked to the implementation of each scenario.

Table 5.3 Economic parameters used in the cost-benefit analysis

Type of NPV calculated		Variable	Value (in 2010 prices)			Unit	Type of value and source
			Mean	Min	Max		
Nitrogen retention (NPV$_N$)		p$_u$	2.69[a]	–	–	EUR kg N^{-1} year^{-1}	Marginal costs of an additional kilogram of nitrogen removed in artificial wastewater treatment plants (Rybanič et al. 1999).
Sediment dredging (NPV$_S$)		p$_u$	25.64	12.1	52.91	EUR t^{-1} year^{-1}	Marginal costs of an additional ton of sediments dredged. An average of Czech public procurements derived from a database administered by the Ministry of Regional Development of the Czech Republic (MD CR 2014).
Climate regulation (NPV$_{Reg}$)		InVEST parameter	84[a]	–	–	EUR t C^{-1} year^{-1}	Social value of carbon, based on the damage costs associated with releasing an additional ton of carbon (Hönigová et al. 2012)
Management costs and investments (NPV$_M$)	Forest management, operation costs	P$_M$	1,158,642	617,000	1,805,074	EUR year^{-1}	Annual management and investment costs. Annual reports of the Administration of the NP and PLA of Šumava (ANPS 2005, 2006, 2007, 2008).
	Infrastructure		794,721	327,667	1,250,306	EUR year^{-1}	
Hydropower production (NPV$_H$)		p$_u$	0.09	0.07	0.11	EUR kWh^{-1} year^{-1}	An average subsidy for hydropower energy production, specified by the Energy Regulatory Office of the Czech Republic (ERO CR 2014).
Provisioning services (NPV$_R$)	Timber sales	P$_R$	4,379,891	3,849,567	4,893,723	EUR year^{-1}	Annual value of timber and hunting-related services sold. Annual reports and supplementary data provided by the Administration of the NP and PLA of Šumava (ANPS 2005, 2006, 2007, 2008).
	Sales of services related to hunting		35,956	31,129	39,407	EUR year^{-1}	

[a] For carbon sequestration and nitrogen retention, we based our analyses on nationally-specific marginal values (social costs of carbon and nitrogen removal costs); therefore, for these services we did not use a range of values and provide only one estimate per a climate projection

NPV_C included two types of costs:

$$NPV_c = NPV_{CE} + NPV_M \tag{5.3}$$

where NPV_{CE} represents costs potentially influenced by the provision of ecosystem services, and NPV_M costs incurred by the Administration of the NP and PLA of Šumava in order to manage local ecosystems and implement EbA measures.

Specifically, the costs of water treatment (NPV_N) and sediment dredging (NPV_S) at the Lipno reservoir are influenced by the ecosystem service of water purification provided by the adjacent watersheds, and therefore, both were included in NPV_{CE}:

$$NPV_{CE} = NPV_N + NPV_S \tag{5.4}$$

As an input to the calculation of NPV_N and NPV_S, we used the quantification of nitrogen and sediments exported to the stream network from InVEST models. Since the amount of pollutants exported was gradually changing between 2006 (T_0) and 2050 (T) under each scenario, we first calculated the annual amount of a pollutant x (nitrogen or sediments) exported to the stream in each year t and a contributing sub-watershed j (under the assumption of a linear development):

$$n_{jtx} = n_{x0} + d_x t \tag{5.5}$$

$$d_x = \frac{n_{xT} - n_{x0}}{T - T_0} \tag{5.6}$$

where n_{jtx} represents the annual amount of pollutant exported to the stream network, n_{x0} (n_{xT}) the aggregate amount of pollutant exported in year T_0 (T, respectively), and d_x represents the annual increment in the export of the pollutant x. NPV_N (and NPV_S, respectively) were than calculated as:

$$NPV_x = p_{ux} \sum_{t=0}^{T-T_0} \sum_{j=1}^{J} \frac{n_{jtx}}{(1+r)^t} \tag{5.7}$$

where p_{ux} represents the unit costs of nitrogen removal during artificial water treatment (costs of sediment dredging, respectively) and r stands for the discount rate. Apparently, the higher the amount of a pollutant retained in the landscape through the ecosystem service of water purification under each scenario, the lower the costs and the NPV_N (and NPV_S, respectively).

NPV_M, the costs incurred by the administration of the NP and PLA of Šumava for ecosystem management and investments in tourist infrastructure, were calculated based on average annual costs of forest management and operation, restoration of marshes and peat bogs and the construction and maintenance of tourist paths P_M:

$$NPV_M = \sum_{t=0}^{T-T_0} \frac{P_M}{(1+r)^t} \tag{5.8}$$

Similarly, NPV_B included two types of benefits:

$$NPV_B = NPV_{BE} + NPV_R \quad (5.9)$$

where NPV_{BE} represents benefits stemming from the provision of ecosystem services, and NPV_R revenues gained by the administration of the Šumava NP and PLA from the sales of timber and services related to hunting.

The benefits related to ecosystem services (NPV_{BE}) originated from the change in carbon stocks (NPV_{Reg}) and revenues from hydropower production (NPV_H):

$$NPV_{BE} = NPV_{Reg} + NPV_H \quad (5.10)$$

The NPV_{Reg} for each scenario was directly quantified by a functionality of the InVEST Carbon Storage and Sequestration tool, based on the difference between landscape carbon stocks under each scenario and the Base Landscape. NPV_H was calculated according to Eqs. 5.5, 5.6, and 5.7, with n_{jtx} representing the annual contribution of sub-watersheds to hydropower production (an output of InVEST modelling) and p_{ux} representing the legislatively determined price of energy purchased from hydropower plants in the Czech Republic as a proxy (see Table 5.3).

NPV_R was calculated based on average annual revenues from the sales of timber and hunting permissions P_R:

$$NPV_R = \sum_{t=0}^{T-T_0} \frac{P_R}{(1+r)^t} \quad (5.11)$$

The values of unit costs, together with annual costs and revenues entering the equations were based on the review of relevant national sources, summarized in Table 5.3. All the above defined NPVs were calculated for each of the three participatory scenarios (Green, Red and Shared vision) and the reference BAU scenario. Finally, the difference between the NPV of each participatory scenario and the reference BAU scenario was quantified in order to facilitate the ranking of scenarios' economic efficiency.

5.3 Adaptation Scenario Storylines

First, the majority of the stakeholders agreed on two opposite storylines, denoted as the 'Green' storyline prioritizing continued nature conservation and implementation of adaptation measures, and the 'Red' storyline promoting intensive economic development of the area without adaptation. The main reason for this decision was that the current discussion about the future of the Šumava region mainly addresses these two extremes; therefore, these two scenario storylines accurately described

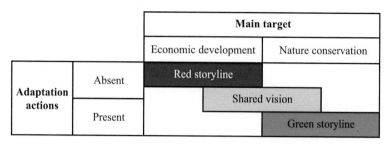

Fig. 5.4 The matrix of final scenario storylines designed for the study area

two contrasting ways of development, which are likely to be pursued in the near future (Fig. 5.4).

Second, the participants created a number of alternative visions during the sub-group exercises. Although the stakeholders were not instructed to try to reach a consensus on a single vision, the storylines resulting from sub-group discussions were very similar and after revising some minor differences in the follow-up discussions, the participants created a 'Shared vision' for the future of the study area.

5.3.1 Definition of Storylines

Red Storyline: Development Without Adaptation   The Red storyline assumed an emphasis on economic development in the study area. The main driving forces in this storyline were population growth, construction of citizen and tourist infrastructure (e.g. tourist centre apartments) and an intensive tourist and recreational use of the area. In this storyline, various development plans such as designation of new ski slopes, ski lifts and paved hiking trails were proposed. Furthermore, the construction of several small-scale artificial water reservoirs was proposed in order to meet the growing water demands. The area of the NPs Zone I was proposed to decrease, while logging would become more intensive in some of the peripheral forested areas of Šumava. Since no part of the study area would be left to a non-intervention regime, this storyline incurred increasing forest management costs. The proportion of urbanized and other intensively used areas increased. This storyline assumed that climate change will not be perceived as a serious threat; therefore, no adaptation measures will be implemented.

Green Storyline: Conservation with Adaptation   The Green storyline assumed that the demographic development in the study area will be stable and the tourism sector will become oriented to long-term sustainability. In comparison with the Red storyline, the investments will enhance the quality of local small-scale accommodation capacities, and will not aim to create new large-scale tourist infrastructure. Therefore, this storyline assumed no growth of urbanised areas outside existing tourist resorts. Zone I of the NP was assumed to be enlarged and united, while all current non-intervention zones will be maintained. In this storyline, substantial emphasis was put on potential impacts of climate change, leading to the implementation of adaptation measures, e.g. restoration of degraded forest areas in the peripheral parts of the NP and complete integration of Zone I. The adaptation measures applied in this storyline were primarily ecosystem-based.

Shared Vision The Shared vision favoured a moderate population growth and opposed implementing incentives to increase local population levels, which would not respect the local social environment and traditional lifestyle, such as large-scale tourist facilities. In terms of tourism development, the vision preferred investments in qualitative, not quantitative improvements, with emphasized low-impact and sustainable forms of tourism, evenly spread throughout the study area. The vision acknowledged the role of the NP in the area and preferred a partial integration of Zone I forest patches and sustainable forestry and agricultural use of peripheral parts. The need for climate change adaptation was recognised within this storyline and the participants preferred EbA measures.

5.3.2 Adaptation Measures in Storylines

The second output of the participatory scenario workshops was a list of adaptation measures suitable for the study area, which could be implemented as a part of the Shared vision and the Green storyline. Since large-scale construction of technological measures is restricted in the area, the participants focused mainly on EbA measures, enhancing the resilience of local ecosystems towards potential impacts of climate change. All proposed adaptation measures complied with differentiated conservation regimes in various zones of the NP and the PLA, assuming less intensive activities in the Zones I and II and, on the contrary, targeting the adaptation measures to the peripheral zones of the study area. In both the Shared Vision and the Green storyline, sustainable forest management and forest conservation were perceived as approaches to EbA. Thus, the costs incurred by sustainable forest management in the study area were considered as adaptation costs in the subsequent cost-benefit analysis.

Green Storyline The adaptation measures proposed for the Green storyline included an enlargement and unification of the NP's Zone I as the primary goal. The unified Zone I would be subject to non-intervention management, leading to an increase in forested area. In the peripheral zones of the study area, revitalization of disturbed ecosystems, sustainable forest management and restoration of forests were proposed as suitable adaptation measures. Specifically, the stakeholders proposed using a variety of genotypes in the forest nursery stock, promoting diverse age classes, species mixes, and a variety of successional stages, and introducing spatially complex and heterogeneous vegetation structure. The Green storyline also proposed large-scale peat-bog and marshlands restoration projects.

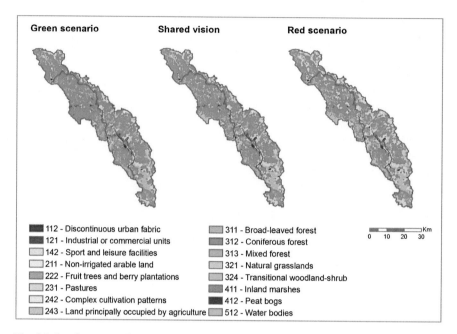

Fig. 5.5 Land use scenarios to 2050, developed on the basis of stakeholder storylines

Shared Vision For the Shared vision, the stakeholders emphasized the threat of water shortages for the future. Therefore, reforestation in the peripheral zones of Šumava together with restoration of peat mires were identified as the most favourable solutions to avoid water shortages. Furthermore, this storyline included implementing soft adaptation measures related to water issues such as reduced water use and construction of more efficient water treatment plants. At the same time, the need for differentiated management and adaptation approaches in the Zone I and the peripheral zones of the NP was stressed.

5.3.3   Land Use Change Scenarios

We visualised scenarios co-defined by stakeholders according to land cover/land use change. Figure 5.5 shows the set of resulting land use change scenarios based on stakeholder input, which were denoted as Green scenario, Red scenario and Shared vision according to the storylines. The BAU scenario (Fig. 5.5), assuming no land use changes between 2006 and 2050 and thus fully corresponding to Base Landscape, was introduced solely for the purpose of scenario comparison in the cost-benefit analysis.

The proportion of different land use classes under each scenario (Table 5.4) indicates that the highest aggregate land use change occurred under the Green

Table 5.4 Proportion of land use and land cover classes in Base Landscape (also used as the BAU scenario in this study) and three scenarios

CLC category	Base Landscape, BAU scenario Area [%]	Green scenario		Shared vision		Red scenario	
		Area [%]	Change 2006–2050 [%]	Area [%]	Change 2006–2050 [%]	Area [%]	Change 2006–2050 [%]
Discontinuous urban fabric	0.49	0.44	−0.04	0.69	0.21	0.70	0.21
Sport and leisure facilities	0.09	0.09	0.00	0.09	0.00	0.21	0.12
Non-irrigated arable land	0.47	0.29	−0.18	0.46	−0.01	0.47	0.00
Pastures	14.31	14.55	0.23	13.75	−0.56	18.93	4.62
Land principally occupied by agriculture	5.66	0.35	−5.31	7.13	1.47	7.17	1.51
Broad-leaved forest	0.60	0.60	0.00	0.59	−0.01	0.55	−0.05
Coniferous forest	58.85	66.28	7.43	60.18	1.34	51.92	−6.93
Mixed forest	5.89	8.26	2.37	5.76	−0.14	5.28	−0.61
Natural grasslands	2.39	1.59	−0.79	2.36	−0.03	2.36	−0.03
Transitional woodland-shrub	6.18	2.14	−4.03	3.51	−2.67	7.95	1.78
Inland marshes	1.55	1.55	0.00	1.54	−0.01	0.71	−0.84
Peat bogs	0.89	0.89	0.00	0.89	0.00	0.70	−0.19
Water bodies	2.46	2.46	0.00	2.46	0.00	2.48	0.02

scenario (10.2 %), followed by the Red scenario (8.5 %) and the Shared vision (3.2 %). However, the quality of land use change under each scenario differed substantially. While the total area of coniferous, mixed and broad-leaved forests increased by 9.8 % under the Green scenario mainly in the non-intervention parts of NP Zone I, replacing earlier successional stages of forest and shrub land. In the Red scenario, forested areas decreased by 7.6 % as a result of transformation to pastures and principally agricultural land in the peripheral parts of Šumava. The slightly increased proportion of area occupied by sport facilities under the Red scenario corresponds to the construction of a ski resort and several small-scale artificial water reservoirs.

5.4 Ecosystem Services Across Scenarios

5.4.1 Climate Change Regulation

Climate regulation represented by carbon storage showed substantial differences between the scenarios (Fig. 5.6a). The spatial pattern of change in carbon storage corresponded to areas where land use category changed between the Base Landscape and each scenario. In the Green scenario, carbon storage increased by 82 t C ha^{-1} on average between 2006 and 2050 as a result of forest growth and the enlargement of forested area. The increase in carbon stocks was less pronounced in the Shared vision (21 t C ha^{-1}). On the contrary, carbon stocks decreased by 87 t C ha^{-1} in the Red scenario, mainly due to logging and an increasing area of agricultural land. The average level of carbon storage under different scenarios varied between 142 and 163 t C ha^{-1} (Table 5.5).

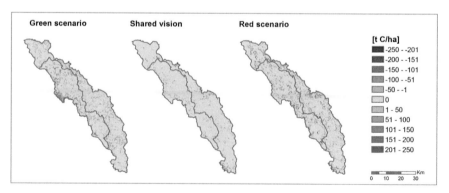

Fig. 5.6a Change in carbon storage for three scenarios in comparison with the Base Landscape

Table 5.5 Aggregate characteristics of carbon storage, water purification and hydropower production for four scenarios

	Total carbon storage [t]	Average carbon storage [t ha^{-1}]	Total nitrogen export to streams [kg year^{-1}]	Average nitrogen export to streams [kg ha^{-1} year^{-1}]
BAU	25,506,000	152	18,100	0.26
Green scenario	27,392,000	163	11,900	0.17
Shared vision	25,576,000	152	18,800	0.27
Red scenario	23,890,000	142	22,000	0.31

	Total sediment export to streams	Average sediment export to streams	Total contribution to hydropower production	Average contribution to hydropower production
BAU	6120	0.09	132,584,000	1890
Green scenario	1450	0.02	131,016,000	1868
Shared vision	6260	0.09	131,859,000	1880
Red scenario	6690	0.10	134,115,000	1912

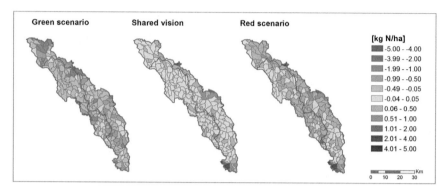

Fig. 5.6b Change in nitrogen discharge for three scenarios in comparison with the Base Landscape

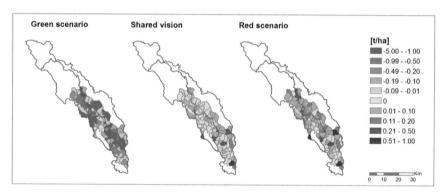

Fig. 5.6c Change in sediment discharge for three scenarios in comparison with the Base Landscape

5.4.2   Water Purification and Sediment Retention

Figure 5.6b indicates that the average level of nitrogen discharge decreased by 0.97 kg ha^{-1} under the Green scenario, resulting from the increase in forested area, since the amounts of nitrogen discharged by forests are generally negligible. In the Red scenario, nitrogen discharge increased by 0.48 kg ha^{-1}, which can be mainly attributed to new agricultural areas in the north-eastern peripheral parts of the study area. The most significant increase in nitrogen discharge occurred in the southern part of the study location under the Red scenario, due to the construction of a new ski resort and related accommodation capacities and infrastructure. The average level of nitrogen eventually exported to local stream network (after subtracting the level of nitrogen retained in the landscape during the modelling process) under different scenarios varied between 0.17 and 0.31 kg N ha^{-1} year^{-1} (Table 5.5).

Trends similar to nitrogen discharge were present in the case of sediment discharge, since both these variables are influenced by similar driving forces (Fig. 5.6c). For the Green scenario, Shared vision and Red scenario, the average

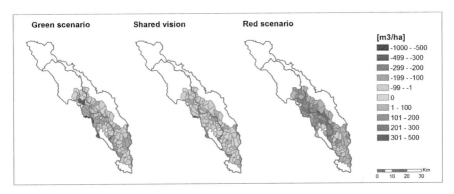

Fig. 5.6d Change in water yield for three scenarios in comparison with the Base Landscape

level of sediment discharge changed by -0.97 t ha^{-1}, 0.04 t ha^{-1} and 0.08 t ha^{-1}, respectively, in comparison with the Base Landscape. The substantial decrease in sediment discharge under the Green scenario resulted from the abandonment of agricultural land and its transformation to forests and pastures. The average amount of sediments exported to the stream network and reaching the Lipno reservoir was 0.02–0.1 t ha^{-1} year^{-1} (Table 5.5).

5.4.3   Hydropower Production

Water yield (as the basic precondition of hydropower production) decreased by 97.4 m^3 ha^{-1} under the Green scenario and by 44.0 m^3 ha^{-1} under the Shared vision (Fig. 5.6d). This trend was caused by the increase in forested area and consequent higher evapotranspiration, which resulted in smaller amounts of water reaching the streams and the Lipno reservoir. Water yield increased by 66.6 m^3 ha^{-1} under the Red scenario, mainly due to an opposite trend in the proportion of forested area. The final contribution of water yield generated by the landscape to hydropower production varied between 1868 and 1912 kWh ha^{-1} year^{-1} under different scenarios (Table 5.5).

5.5   Cost-Benefit Analysis of Adaptation Measures

Figure 5.7 summarizes the results of the cost-benefit analysis for all scenarios, including the comparison of the three participative scenarios and the reference BAU scenario (assuming no land use change) for the time-span of 2006–2050. Detailed cost-benefit analysis results are provided in Table 5.6. In general, benefits exceeded costs in all cases. However, compared to the BAU scenario, the Green

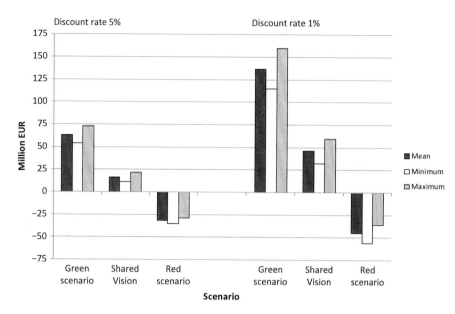

Fig. 5.7 The net present value of three adaptation scenarios compared with the BAU scenario in 2006–2050. The cost-benefit analysis for each scenario was calculated at two levels of discount rate (5 and 1 %) and for the mean, minimum and maximum of marginal costs and annual unit values used as the parameters in the economic evaluation (see Table 5.3)

scenario and the Shared vision proved to have a higher NPV, while the NPV of the Red scenario did not reach the one of the BAU scenario.

The comparison with the reference BAU scenario showed that the most beneficial was the Green scenario, reaching an NPV higher by 63.1 million EUR (results reported at 5 % discount rate and the mean estimate of marginal costs and ecosystem-service unit values). The Green scenario incurred the lowest aggregate costs (22.2 million EUR) as a result of limited forest management in Zones I and II of the NP and lower costs related to water purification. Furthermore, the Green scenario generated the highest benefits (305 million EUR) owing to a substantial increase in carbon storage (contrary to the Red scenario).

The Red scenario was the only one with NPV lower than the BAU (−31.5 milion EUR), mainly due to the negative carbon storage balance. Despite generating the highest profit from hydropower production, timber sales and hunting revenues, the results indicate that the focus on economic development and intensive management of forests under this scenario was outweighed by losses in the provision of ecosystem services.

The Shared vision emphasised sustainability and only moderate landscape changes, and did not introduce substantial development in terms of land use and adaptation measures. Nevertheless, it reached a positive balance in comparison with the BAU scenario (by 16.5 million EUR). In all NPV components, the Shared

Table 5.6 The net present value of costs and benefits related to the implementation of each scenario (in 1000 EUR). The NPV is calculated for the period 2006–2050 with 5% discount rate

				Scenario							
				Discount rate 5%				Discount rate 1%			
				BAU	Green scenario	Shared Vision	Red scenario	BAU	Green scenario	Shared Vision	Red scenario
Costs	Costs related to eco-system services	Sediment dredging	mean	2929	2199	2950	3018	5722	3703	5780	5967
			min	1611	1209	1623	1660	3147	2036	3179	3281
			max	6045	4538	6088	6227	11,809	7641	11,928	12,312
		Nitrogen removal		908	808	922	973	1774	1497	1813	1954
	Annual management costs	Inland marshes and peat bogs restoration	mean	83	121	83	0	164	238	164	0
			min	37	53	37	0	72	105	72	0
			max	338	491	338	0	665	965	665	0
		Forest management, operation costs	mean	21,623	11,313	19,779	24,765	42,239	22,098	38,637	48,376
			min	11,515	6024	10,533	13,188	22,493	11,768	20,575	25,761
			max	33,688	17,624	30,815	38,582	65,805	34,427	60,193	75,366
		Investments in tourist infrastructure	mean	14,832	7760	13,567	16,987	28,972	15,157	26,501	33,182
			min	6115	3199	5594	7004	11,945	6249	10,927	13,681
			max	23,334	12,208	21,344	26,725	45,580	23,846	41,694	52,203
	Sum		**mean**	**40,376**	**22,200**	**37,302**	**45,743**	**78,871**	**42,694**	**72,895**	**89,478**
			min	**20,186**	**11,294**	**18,708**	**22,825**	**39,431**	**21,656**	**36,565**	**44,678**
			max	**64,313**	**35,669**	**59,508**	**72,508**	**125,633**	**68,377**	**116,292**	**141,836**

(continued)

Table 5.6 (continued)

			Scenario							
			Discount rate 5 %				Discount rate 1 %			
			BAU	Green scenario	Shared Vision	Red scenario	BAU	Green scenario	Shared Vision	Red scenario
Benefits	Benefits related to ecosystem services	Hydropower production	mean	222,695	240,161	240,614	241,807	435,007	483,288	484,541	487,837
			min	164,959	177,897	178,233	179,116	322,227	357,991	358,920	361,361
			max	274,932	296,495	297,055	298,527	537,045	596,652	598,199	602,268
		Climate regulation		0	66,758	2484	−57,228	0	128,902	4797	−110,501
	Annual market benefits	Timber sales	mean	81,741	42,764	74,770	93,618	159,671	83,535	146,055	182,871
			min	71,844	37,586	65,717	82,282	140,338	73,421	128,370	160,729
			max	91,330	47,781	83,542	104,601	178,403	93,335	163,189	204,325
		Sales of services related to hunting	mean	671	351	614	769	1311	686	1199	1501
			min	581	304	531	665	1135	594	1038	1300
			max	735	385	673	842	1437	752	1314	1645
	Sum		**mean**	**305,107**	**350,035**	**318,483**	**278,965**	**595,989**	**696,412**	**636,592**	**561,709**
			min	**237,384**	**282,546**	**246,966**	**204,836**	**463,700**	**560,908**	**493,125**	**412,889**
			max	**366,998**	**411,419**	**383,754**	**346,742**	**716,885**	**819,641**	**767,500**	**697,738**
NPV (Benefits – Costs)			mean	264,730	327,834	281,180	233,222	517,117	653,718	563,697	472,231
			min	217,197	271,251	228,258	182,011	424,268	539,252	456,560	368,211
			max	302,684	375,751	324,246	274,234	591,252	751,264	651,207	555,902
Difference in NPV to BAU ($NPV_{scenario} - NPV_{BAU}$)			**mean**	**0**	**63,104**	**16,450**	**−31,509**	**0**	**136,601**	**46,580**	**−44,887**
			min	**0**	**54,054**	**11,060**	**−35,186**	**0**	**114,984**	**32,291**	**−56,057**
			max	**0**	**73,066**	**21,562**	**−28,450**	**0**	**160,012**	**59,955**	**−35,351**

vision ranked between the Green and the Red scenarios, which represented the opposite sides of NPV intervals.

The final ranking of the adaptation scenarios proved robust in the sensitivity analysis, with the Green adaptation scenario performing as the most beneficial, followed by the Shared vision.

5.6 Discussion

In this study, we quantified the costs and benefits of an array of adaptation scenarios, focusing on EbA measures. We aimed to assess a representative sample of ecosystem services provided by forested areas. These were evaluated (Harrison et al. 2010) by focusing on regulating ecosystem services related to climate regulation and water quality, and provisioning ecosystem services. The latter were incorporated in the analysis in two ways; first, through the assessment of hydropower production; and second, through quantifying the revenues from timber and hunting sales.

Cultural services, namely recreation, were not assessed in this study, since only a limited number of visitor use surveys were conducted in the area and detailed tourist use statistics are lacking. The application of economic evaluation methods including travel costs quantification was thus hampered and it was not feasible to include recreation-related services into the cost-benefit analysis. Furthermore, assessing ecosystem services related to landscape aesthetics (Grêt-Regamey et al. 2007), which would require the involvement of stakeholders in preference surveys and contingent valuation exercises, was beyond the scope of this study, as the stakeholders were primarily motivated to take part in future adaptation planning and scenario building for the study area.

However, it can be expected that the provision of both cultural services and other potentially analysed forest ecosystem services, such as pollination or non-timber forest products, would be enhanced by a larger extent of forests and their increased resilience in the Green scenario. Therefore we argue that supplementing the cost-benefit analysis with other types of ecosystem services would not influence the final ranking of scenarios.

The study illustrates that local knowledge can serve as a productive input into the process of CCA linked to DRR (See also Lange et al, Chap. 21). The involvement of local stakeholders brings added value to scenario development and subsequent GIS modelling and economic evaluation by selecting alternatives informative and relevant for local decision-making (Reed et al. 2013). Through the opportunity to gain access to survey results, the stakeholders were motivated to take active part in scenario building and to become familiar with the concepts of EbA and Eco-DRR. However, at the same time, the study highlighted differences between locally-based knowledge and scientific findings. For instance, local stakeholders were convinced that an increase in forested area will enhance water retention in the landscape and safeguard water supplies, which they subsequently suggested as an adaptation measure. However, once the provision of ecosystem services resulting from this adaptation measure was modelled within the Green scenario, the modelling results showed that larger forested area resulted in higher evapotranspiration, which in turn led to smaller water yields and a decrease in hydropower generated. In this case, intuitive locally-based knowledge was in contradiction with modelling outcomes, and further collaboration with stakeholders is vital to communicate potential impacts of various adaptation scenarios based on ecosystem services modelling.

Important question for further research is whether cost and benefit ratios of EbA are higher than for technical solutions (e.g. the construction of traditional flood protection systems). The evidence based on several European case studies shows that the costs of EbA are not necessarily higher than in the case of traditional approaches. Moreover, additional multiple ecological and socio-economic benefits (such as recreation or contribution to regulation ecosystem services) are prevailing the benefits of traditional adaptation measures (Naumann et al. 2011). However, more evidence is required as the literature on valuation of ecosystem services and biodiversity is not always directly applicable to CCA and DRR (Krupnick and Mclaughlin 2012). Our study quantifies only selected benefits and costs associated with the change of ecosystem services provisionin the area of the Šumava National Park. This can serve as a basis for comparisons of costs and benefits associated with EbA and Eco-DRR management in protected areas.

5.7 Conclusions

Climate change and CCA impose novel problems for the management of protected areas, especially in combination with contrasting interests of involved stakeholders. The results of the study suggest that the concept of EbA linked to Eco-DRR provides a new perspective for stakeholders, which allows them to gain different insights and an understanding of local problems, and to suggest new solutions to local issues. While including stakeholders into the process of scenario storyline building ensures that locally relevant phenomena are included and assessed, subsequent cost-benefit analysis of ecosystem services scenarios and different adaptation options have the potential to facilitate the prioritisation of different development options, and to provide assistance to local landscape decision-making and nature conservation planning.

Acknowledgements The research leading to these results has received funding from the European Community's Seventh Framework Programme under Grant Agreement No. 308337 (Project BASE). The text reflects only the authors' views and the EU is not liable for any use that may be made of the information contained therein.

References

ANPS (2005–2008) Annual reports of the Administration of the NP and PLA of Šumava 2005–2008 [in Czech]. http://www.npsumava.cz/cz/1038/sekce/uredni-deska/. Accessed 3 Sep 2014

Beniston M, Stephenson DB, Christensen OB et al (2007) Future extreme events in European climate: an exploration of regional climate model projections. Clim Chang 81:71–95

Bláha J, Romportl D, Křenová Z (2012) Can Natura 2000 mapping be used to zone the Šumava National Park? Eur J Environ Sci 3:57–64

Campbell A, Kapos V, Scharlemann JPW et al (2009) Review of the literature on the links between biodiversity and climate change: impacts, adaptation and mitigation. Secretariat of the Convention on Biological Diversity, Montreal

CBD (2009) Connecting biodiversity and climate change mitigation and adaptation: report of the second ad hoc Technical Expert Group on Biodiversity and Climate Change. Technical series no. 41, Montreal

De Simon G, Alberti G, Delle Vedove G et al (2012) Carbon stocks and net ecosystem production changes with time in two Italian forest chronosequences. Eur J For Res 131:1297–1311

Dendoncker N, Bogaert P, Rounsevell M (2006) A statistical method to downscale aggregated land use data and scenarios. J Land Use Sci 1:63–82

EEA (2007) CLC2006 technical guidelines. European Environment Agency, Copenhagen

EEA (2010) The European environment – State and Outlook 2010: Czech Republic Country Assessment. http://www.eea.europa.eu/soer/countries/cz. Accessed 3 Sept 2014

EIA Servis (2011) Regulatory plan: Connecting Klápa – Hraničník [in Czech]. Environmental Impact Assessment Documentation. EIA Servis s.r.o., České Budějovice. http://portal.cenia.cz/eiasea/detail/EIA_MZP382. Accessed 10 Oct 2014

ERO CR (2014) Price decision 1/2014, specifying subsidies for selected sources of energy [in Czech]. Energy Regulatory Office of the Czech Republic. http://www.eru.cz/en/. Accessed 16 Nov 2014

ESRI (2013) ArcGIS10.2. Environmental System Research Institute, Redlands, CA

Goldstein JH, Caldarone G, Duarte TK et al (2012) Integrating ecosystem-service tradeoffs into land-use decisions. PNAS 109:7565–70

Grêt-Regamey A, Bishop ID, Bebi P (2007) Predicting the scenic beauty value of mapped landscape changes in a mountainous region through the use of GIS. Environ Plan B Plan Des 34:50–67

Hanel M, Vizina A, Máca P et al (2012) A multi-model assessment of climate change impact on hydrological regime in the Czech Republic. J Hydrol Hydromech 60:152–161

Hanley N, Barbier EB (2009) Pricing nature: cost-benefit analysis and environmental policy. Edward Elgar, Cheltenham/Northampton

Harmáčková ZV, Vačkář D (2015) Modelling regulating ecosystem services trade-offs across landscape scenarios in Třeboňsko Wetlands Biosphere Reserve, Czech Republic. Ecol Model 295:207–215

Harrison PA, Vandewalle M, Sykes MT et al (2010) Identifying and prioritising services in European terrestrial and freshwater ecosystems. Biodivers Conserv 19:2791–2821

Hönigová I, Vačkář D, Lorencová E et al (2012) Survey on grassland ecosystem services. Report to the EEA – European Topic Centre on Biological Diversity. Nature Conservation Agency of the Czech Republic, Prague

IFER (2010) Inventorying of landscape – CzechTerra [in Czech]. Ministry of the Environment of the Czech Republic, Prague

Jones HP, Hole DG, Zavaleta ES (2012) Harnessing nature to help people adapt to climate change. Nat Clim Chang 2:504–509

Kareiva P, Tallis H, Ricketts TH et al (2011) Natural capital: theory and practice of mapping ecosystem services. Oxford University Press, New York

Kindlmann P, Matějka K, Doležal P (2012) The Šumava forests, bark beetla and nature conservation [in Czech]. Karolinum, Prague

Koomen E, Stillwell J (2007) Modelling land-use change: theories and methods. In: Koomen E, Stillwell J, Bakema A et al (eds) Modelling land-use change: progress and applications. Springer, Dordrecht

Křenová Z, Hruška J (2012) Proper zonation – an essential tool for the future conservation of the Šumava National Park. Eur J Environ Sci 2:62–72

Krupnick A, Mclaughlin D (2012) Valuing the impacts of climate change on terrestrial ecosystem services. Clim Chang Econ 3:1250021

Lange GM, Dasgupta S, Thomas T et al (2010) Economics of adaptation to climate change – ecosystem services. The World Bank, Washington, DC

Lindsay R (2010) Peatbogs and carbon: a critical synthesis to inform policy development in oceanic peat bog conservation and restoration in the context of climate change (Technical report). University of East London, Environmental Research Group. http://www.rspb.org.uk/images/peatbogs_and_carbon_tcm9-255200.pdf. Accessed 3 Sept 2014

MA (Millennium Ecosystem Assessment) (2005) Ecosystems and human well-being: synthesis. Island Press, Washington, DC

MD CR (2014) The database of public procurement [in Czech]. Ministry of Regional Development of the Czech Republic, Prague, http://www.vestnikverejnychzakazek.cz/. Accessed 15 Oct 2014

ME CR (2013) Sixth national communication of the Czech Republic under the United Nations Framework Convention on climate change. Ministry of Environment of the Czech Republic, Prague

Mooney H, Larigauderie A, Cesario M et al (2009) Biodiversity, climate change, and ecosystem services. Curr Opin Environ Sustain 1:46–54

Munang R, Thiaw I, Alverson K et al (2013a) The role of ecosystem services in climate change adaptation and disaster risk reduction. Curr Opin Environ Sustain 5:47–52

Munang R, Thiaw I, Alverson K et al (2013b) Climate change and ecosystem-based adaptation: a new pragmatic approach to buffering climate change impacts. Curr Opin Environ Sustain 5:67–71

Naumann S, Anzaldua G, Berry P, Burch S, McKenna D, Frelih-Larsen A, Gerdes H, Sanders M (2011) Assessment of the potential of ecosystem-based approaches to climate change adaptation and mitigation in Europe. Final report to the European Commission, DG Environment, Contract no. 070307/2010/580412/SER/B2. Ecologic institute and Environmental Change Institute, Oxford University Centre for the Environment

Nelson E, Mendoza G, Regetz J et al (2009) Modeling multiple ecosystem services, biodiversity conservation, commodity production, and tradeoffs at landscape scales. Front Ecol Environ 7:4–11

NIR (2012) National greenhouse gas inventory report of the Czech Republic. Prague – Ministry of the Environment of the Czech Republic. http://unfccc.int/files/national_reports/annex_i_ghg_inventories/national_inventories_submissions/application/zip/cze-2012-nir-19oct.zip. Accessed 3 Sept 2014

Novotná M, Kopp J (2010) Migration trends in the Bohemian Forest region after 1990 [in Czech]. Silva Gabreta 16:187–206

O'Halloran LR, Borer ET, Seabloom EW et al (2013) Regional contingencies in the relationship between aboveground biomass and litter in the world's grasslands. PLoS ONE 8:e54988

OECD (2013) Czech Republic Country Profile. In: Water and climate change adaptation – policies to navigate uncharted waters. OECD Publishing, Paris

Perlín R, Bičík I (2010) Local development in the Šumava region: Final report to the project analysis of the development of the Šumava National Park in last 15 years [in Czech]. Administration of the NP and PLA of Šumava

Reckhow KH, Beaulac MN, Simpson JT (1980) Modeling phosphorus loading and lake response under uncertainty: a manual and compilation of export coefficients. US Environmental Protection Agency, Washington, DC

Reed MS, Kenter J, Bonn A (2013) Participatory scenario development for environmental management: a methodological framework illustrated with experience from the UK uplands. J Environ Manag 128:345–362

Renaud FG, Sudmeier-Rieux K, Estrella M (eds) (2013) The role of ecosystems in disaster risk reduction. United Nations University Press, New York/Tokyo/Paris

Rounsevell MDA, Metzger MJ (2010) Developing qualitative scenario storylines for environmental change assessment. Wiley Interdiscip Rev Clim Chang 1:606–619

Rounsevell MDA, Reginster I, Araújo MB (2006) A coherent set of future land use change scenarios for Europe. Agric Ecosyst Environ 114:57–68

Rybanič R, Šeffer J, Čierna M (1999) Economic valuation of benefits from conservation and restoration of floodplain meadows. In: Šeffer J, Stanová V (eds) Morava River floodplain meadows – importance, restoration and management. DAPHNE – Centre for Applied Ecology, Bratislava

Schumacher J, Roscher C (2009) Differential effects of functional traits on aboveground biomass in semi-natural grasslands. Oikos 118:1659–1668

Settele J, Hammen V, Hulme P et al (2005) ALARM: assessing Large-scale environmental risks for biodiversity with tested methods. Gaia 14:69–72

Sharp R, Tallis HT, Ricketts T et al (2014) InVEST user's guide. The Natural Capital Project, Stanford

Spangenberg J (2007) Integrated scenarios for assessing biodiversity risks. Sustain Dev 15:343–357

Tallis H, Polasky S (2009) Mapping and valuing ecosystem services as an approach for conservation and natural-resource management. Ann N Y Acad Sci 1162:265–83

TEEB (2010) The economics of ecosystems and biodiversity: ecological and economic foundations. Kumar P (ed). Earthscan, Oxford

TGM WRI (2014) DIBAVOD hydrological database. T. G. Masaryk Water Research Institute, v.v.i. http://www.dibavod.cz/27/struktura-dibavod.html. Accessed 3 Sept 2014.

Tolasz R (ed) (2007) Climate atlas of the Czech Republic [in Czech]. Czech Hydrometeorological Institute, Palacky University, Praha

Truus L (2011) Estimation of above-ground biomass of wetlands. In: Atazadeh I (ed) Biomass and remote sensing of biomass., InTech, http://www.intechopen.com/books/biomass-and-remote-sensing-of-biomass/estimation-of-above-groundbiomass-of-wetlands. Accessed 3 Sept 2014

USGCRP (2008) Preliminary review of adaptation options for climate-sensitive ecosystems and resources. Julius SH, West JM (eds) US Environmental Protection Agency, Washington, DC

Villamagna AM, Angermeier PL, Bennett EM (2013) Capacity, pressure, demand, and flow: a conceptual framework for analyzing ecosystem service provision and delivery. Ecol Complex 15:114–121

VÚMOP (2014) SOWAC-GIS Geoportal. Research Institute for soil and water conservation, v.v.i. http://geoportal.vumop.cz/index.php?projekt=vodni&s=mapa. Accessed 3 Sept 2014

Part II
Decision-Making Tools for Eco-DRR/CCA

Chapter 6
Decision Tools and Approaches to Advance Ecosystem-Based Disaster Risk Reduction and Climate Change Adaptation in the Twenty-First Century

Adam W. Whelchel and Michael W. Beck

Abstract Organisations and governments around the globe are developing methodologies to cope with increasing numbers of disasters and climate change as well as implementing risk reducing measures across diverse socio-economic and environmental sectors and scales. What is often overlooked and certainly required for comprehensive planning and programming are better tools and approaches that include ecosystems in the equations. Collectively, these mechanisms can help to enhance societies' abilities to capture the protective benefits of ecosystems for communities facing disaster and climate risks. As illustrated within this chapter, decision support tools and approaches are clearly improving rapidly. Despite these advancements, factors such as resistance to change, the cautious approach by development agencies, governance structure and overlapping jurisdictions, funding, and limited community engagement remain, in many cases, pre-requisites to successful implementation of ecosystem-based solutions. Herein we provide case studies, lessons learned and recommendations from applications of decision support tools and approaches that advance better risk assessments and implementation of ecosystem-based solutions. The case studies featured in this chapter illustrate opportunities that have been enhanced with cutting edge tools, social media and crowdsourcing, cost/benefit comparisons, and scenario planning mechanisms. Undoubtedly, due to the large areas and extent of exposure to natural hazards, ecosystems will increasingly become a critical part of societies' overall responses to equitably solve issues of disaster risk reduction and climate change adaptation.

Keywords Ecosystem-based solutions • Community resilience building • Risk matrix • Floodplain by design • Water funds • Connecticut • Resilience planning to action framework

A.W. Whelchel (✉) • M.W. Beck
The Nature Conservancy, 55 Church Street, Floor 3, New Haven, CT, USA
e-mail: awhelchel@tnc.org

© Springer International Publishing Switzerland 2016
F.G. Renaud et al. (eds.), *Ecosystem-Based Disaster Risk Reduction and Adaptation in Practice*, Advances in Natural and Technological Hazards Research 42,
DOI 10.1007/978-3-319-43633-3_6

6.1   Ecosystem-Based Risk Reduction and Adaptation

International consortia, national to local governments, academic institutions, and non-governmental organizations are developing methods to cope with an escalating number of disasters and climate change impacts as well as implementing risk reducing measures across diverse socio-economic and environmental sectors and scales. The urgency expressed by recent publications such as the World Risk Report (2012), the Global Assessment Report on Disaster Risk Reduction (2013, 2015), the Intergovernmental Panel on Climate Change 5th Assessment Report (2014) and the United States National Climate Assessment (2014) are serving to accelerate this dialogue across diverse governance structures. What is often overlooked and certainly required for comprehensive planning and programming are better tools and approaches, which explicitly include ecosystems in disaster risk reduction and climate change adaptation. This is particularly true of our collective ability to capture the additional benefits of ecosystems for communities subjected to disaster and climate risks. Fortunately, ecosystems are indeed being increasingly viewed as a critical asset in helping achieve resilience to disasters and climate change (Jones et al. 2012; Renaud et al. 2013; Temmerman et al. 2013; Spalding et al. 2014).

Ecosystems provide protective services among other functions that, if recognized, can be integrated into comprehensive risk management planning and risk reduction actions (Hale et al. 2009; Spalding et al. 2010; World Bank 2016). Recent science supports the ability of globally distributed coastal habitats such as salt marshes (Sheppard et al. 2011; Moller et al. 2014), mangroves (Spalding et al. 2010), oyster reefs (Beck et al. 2011), and coral reefs (Shepard et al. 2005; Ferrario et al. 2014) to reduce risk from flooding and storm surges. Furthermore, governments and businesses are identifying where coastal habitats can be cost-effective defenses (CCRIF 2010; van den Hoek et al. 2012; Temmerman et al. 2013; NYC Special Initiative for Rebuilding and Resiliency 2014). The benefits of intact, vegetated watersheds, inland wetlands and riparian zones have also been recognized as critical for reducing downstream flood risks (Warner et al. 2013).

What is also clear are the co-benefits provided through the integration of ecosystems into disaster risk reduction and climate change adaptation (Eco-DRR/CCA). In addition to shoreline protection, Eco-DRR/CCA can help sustain local livelihoods (Green et al. 2009) and regulate climate via carbon sequestration (Pritchard 2009). With a vast majority of people on earth depending on freshwater supplied from rivers and lakes (Morris et al. 2003), coupled with escalating degradation and anticipated water shortages for two-thirds of the world's population by 2025 (WWAP 2009), the imperative to relieve risks where feasible through freshwater ecosystems management is paramount.

6.2   Rationale for Eco-DRR/CCA Tools and Approaches

One of the central challenges in ensuring ecosystems are mainstreamed into DRR/CCA is the limited knowledge about the many facilitative tools and approaches, or more importantly, understanding how they can and have been used to support decisions for DRR and CCA (see also Krol et.al., Chap. 7). In the broad sense, there are a growing number of tools and approaches but with fewer examples of how these have actually advanced decisions involving Eco-DRR/CCA. Central to the practitioner's ability to remedy this challenge, therefore, rests on addressing the following critical questions:

1. What tools and/or approaches are used or could potentially be used to design and implement Eco-DRR/CCA?
2. How can these tools and/or approaches help with the implementation of Eco-DRR/CCA?
3. What are the limitations or gaps in existing tools and/or approaches to operationalise Eco-DRR/CCA, either at project or programmatic levels across diverse and interconnected scales?

Clearly, an examination of available and future tools and approaches is required to better understand how Eco-DRR/CCA can be integrated into existing planning (i.e., integrated watershed management, protected area/fire/drought management) as well as identify other pre-requisites. Such pre-conditions include the ability to connect the right expertise with planning efforts that are enabled by financial and policy incentives and supported within governance structures. As discussed below, there is a growing call for integrating ecosystems in immediate and long-term resiliency efforts.

6.2.1   Distinguishing Between "Tools" and "Approaches"

In this chapter, we make a distinction between tools and approaches in the context of Eco-DRR/CCA. Generally, tools consist of software or documented methods used to support decision-making and help a community through various information-gathering endeavors towards a more comprehensive understanding of a particular situation. Many tools with potential for advancing Eco-DRR/CCA implementation focus on the geospatial presentation of environmental and/or socio-economic data guided by planning needs, with some tools allowing for future scenarios runs. Some tools are in the public domain; others must be purchased or licensed, and the degree of technical training required to operate the tools varies considerably. In some data rich parts of the world, more advanced tools provide complex modeling and quantitative analysis of disasters and climate change impacts to natural and/or human systems (e.g., coastal engineering tools such as Delft3D and Mike21). Often a combination or suites of tools are used to provide for a robust planning process. Cutting edge tools are able to illustrate spatially and quantitatively the consequences of risk management decisions. Regardless of a tool's sophistication, community-based efforts often benefit by having tools integrated into collaborative processes that are connected to ongoing or upcoming action plans and management efforts.

Approaches include qualitative, semi-quantitative, and/or quantitative processes; from informal panels of experts to community-driven applications intended to aid Eco-DRR/CCA. Many approaches used for Eco-DRR/CCA planning were not developed specifically for that purpose. Many approaches are drawn from other applications such as land-use planning, environmental monitoring, and fire management, which in many cases already recognize Eco-DRR/CCA as a co-benefit. As with any newly expanding field, the diversity of approaches being put into practice presents a challenge for practitioners in search of transferability, reliability, and consistency.

Comprehensive and effective Eco-DRR/CCA planning and implementation can and is being enhanced with decision-support tools and approaches by addressing several core considerations:

- Knowledge of type, intensity, frequency, spatial distribution and duration of disasters (past, current and/or future events) and relationship with climatic variables (e.g., precipitation, temperature, sea level rise) over time;
- Assessment of disaster and climate vulnerabilities (e.g. assessing ecosystems, infrastructure or populations) and strengths (e.g., healthy/intact natural infrastructure, availability/accessibility of social services) over time;
- Prioritization of adaptive strategies to reduce risk and reinforce resilience;
- Governance structure and stability/diversity of partnerships (i.e., private/public/NGO) coupled with incentives to induce and sustain action.

These core considerations can be integrated into and used to advance a stepwise, planning-to-action framework as presented here:

1. Identify near-term and long-term disaster and climate change impacts;
2. Construct risk profiles and prioritize strengths and vulnerabilities;
3. Develop initial and sequenced adaptation strategies for highest priorities;
4. Link strategies to ongoing decision making;
5. Prepare and implement adaptation plans;
6. Monitor and reassess effectiveness of actions taken;
7. Routinely re-integrate best available disaster and climate change data and tools.

The challenge for practitioners, of course, lies in knowing which tools and approaches are best suited to address these core considerations and planning-to-action framework steps at an appropriate scale (e.g. from multi-national to local community) in order to ensure that Eco-DRR/CCA is integrated and operationalised across disciplines, sectors, and management constructs. Herein resides one of the principal opportunities and constraints for Eco-DRR/CCA. A

Table 6.1 Approaches and tools used to advance the planning-to-action framework steps

Approach	Tool(s)	Steps (see text above)
Community Resilience Building:	Risk Matrix[c]	#2, #3, #4, #5, #6
Community Resilience Building - Connecticut[a]	Coastal Resilience Tool	#1, #7
Watershed Management:	InVEST	#1, #2, #3, #6, #7
Floodplain by Design[a]	Community Engagement	#4, #5
Watershed Management:	RIOS/Financial Incentives[c]	#1, #2, #3, #6, #7
Monterrey Metropolitan Water Fund-Mexico[b]	Community Engagement	#4, #5
Coastal Zone Management: Belize[b]	InVEST/Scenarios	#1, #2, #3, #6, #7
	Community Engagement	#4, #5
Additional Tools Available		
	Climate Wizard	#1, #7
	Coastal Defense[c]	#1, #2, #6, #7
	Crowd Sourcing/Social Media	#1, #2, #6

[a]Focused on Eco-DRR/CCA as outcome
[b]Recognizes Eco-DRR/CCA as a co-benefit
[c]Provides for balance between Eco-DRR/CCA and socio-economic tradeoffs

summary of the approaches and associated tools featured in this chapter, along with their respective connections to the planning-to-action framework steps described above, are provided in Table 6.1, which serves as a guide to the different case study examples presented.

6.3   General Resources and Case Studies

There is a multitude of approaches and tools currently available for many areas of the globe that can deliver actionable information on the core considerations and support the planning-to-action framework steps identified above. In addition to a summary of web-based portals, a series of case studies are provided below to generate lessons learned and recommendations for decision makers and practitioners. The following materials are not meant to be exhaustive nor prescriptive, but simply a window into real-world situations that have employed approaches and tools for Eco-DRR/CCA.

A summary of the more prominent web-based portals providing data, tools, approaches, and case studies applicable to the core considerations and planning-to-action framework steps, as discussed above, are provided in Table 6.2.

Table 6.2 Prominent web-based, freely accessible portals and tool-sheds

Name	Managing entity	Web address
Climate Adaptation Knowledge Exchange	EcoAdapt	http://www.cakex.org/
Climate Change Knowledge Portal	The World Bank Group	http://sdwebx.worldbank.org/climateportal/
Adaption Learning Mechanism	United Nations Development Programme	http://www.adaptationlearning.net/
weAdapt	Stockholm Environmental Institute	https://weadapt.org/
Digital Coast	National Oceanic & Atmospheric Administration	http://coast.noaa.gov/digitalcoast/

6.3.1 Planning-to-Action Framework Steps and Case Studies

Where obstacles such as lack of available resources (i.e., data, expertise, funding, governance, etc.) have been minimized, a proliferation of tools that focus directly or indirectly on Eco-DRR/CCA has emerged. In many situations, these tools can be instrumental at enabling the incorporation of ecosystems into DRR/CCA efforts. Tools can also be used as stand-alone assessment independent of or towards the beginning of a DRR/CCA process; particularly for framework step #1 (identify near-term and long-term impacts), #2 (construct risk profiles), and #7 (routinely re-integrate best available data). To move comprehensively through the planning-to-action framework steps (see 6.2.1), a broader and more collective approach that seeks to integrate available tools is required to successfully advance Eco-DRR/CCA. In particular, step #3 and #4 (development, prioritizations, sequencing and linkage of adaptation strategies) are ideally derived through community-based engagement, adaptation strategy synthesis, and/or consensus building approaches. As is often the case, these approaches naturally lead to implementation of step #5 and #6 (prepare and implement plan; monitor and reassess effectiveness). The following sections provide case studies of approaches (refer to Table 6.1) that integrate tools to enable Eco-DRR/CCA via the planning-to-action framework steps.

6.3.1.1 Community Resilience Building in Connecticut (USA)

Along the eastern seaboard of the United States – particularly in the aftermath of Tropical Strom Irene (August 2011) and Sandy (October 2012) - it has become apparent that the operationalisation of Eco-DRR/CCA requires further investment in certain pre-requisites that focus on process and stakeholder engagement. In essence, tools and applications (apps) are instrumental in identifying near and long-term impacts (step #1) and initial construction of risk profiles (step#2) but are most impactful when integrated within a flexible and adaptive, community-based approach (steps #2 - #6). This critical learning leap resulted in the launch of a Community Resilience Building Workshop (www.CommunityResilienceBuilding. com) process in Connecticut (USA) developed by The Nature Conservancy (TNC) to assist federal and state agencies, regional planning agencies, municipalities, corporations, and other stakeholders (Whelchel 2012). The Workshop process helps to build resilient communities and mainstream Eco-DRR/CCA by providing a way to combine tools within a facilitated community-engagement construct. One such tool is the Coastal Resilience decision-support platform.

The Coastal Resilience (www.coastalresilience.org) decision-support platform was partially initiated due to the recognition that Eco-DRR/CCA was not being fully integrated into disaster and climate planning (Ferdaña et al. 2010; Gilmer and Ferdaña 2012; Beck et al. 2013). From its origins in New York and Connecticut (USA) beginning in 2007, this web-based tool has expanded to include 10 states (USA) and several other nations (Honduras, Guatemala, Belize, Mexico, Grenada, Saint Vincent and the Grenadines). This tool focuses on spatially defining the risk reduction characteristics of ecosystems within disaster (i.e., storm surge, inland flooding) and climate change (i.e., sea level rise, precipitation) applications, alongside socio-economic considerations from local to national scales. The tool is being applied internationally in places such as Grenada in partnership with the Red Cross to assess social and ecological vulnerability as well as by international organizations to develop Coasts at Risk indicators (UNU-EHS 2014) and the World Risk Report's (2012) Index. The tool provides decision makers a much needed suite of map layers and apps via an intuitive, user-friendly interface. For Coastal Resilience, the overarching framework includes: (1) awareness of hazards, (2) risk assessment of strengths and vulnerabilities, (3) development of choices, and (4) evaluating the impact of resilient actions (Beck et al. 2013) (see also discussion by Krol et al., Chap. 7).

At the core of the Community Resilience Building Workshop approach is the focus on obtaining a diverse suite of stakeholders engaged as planning commences, during, and afterwards to ensure the community champions the outcomes. Such a process often requires expanding beyond the disaster response professionals to include among others: elected officials, planners, employers, neighborhood or community representatives, natural resource managers, health care providers, finance professionals, and legal counsel. Essentially, the approach must include those who make decisions, have influence over decisions, or are impacted by the decisions made. Arguably this is one of the most important - yet under-emphasised - foundational requirements to ensuring comprehensive, community-driven support for actions that will incorporate Eco-DRR/CCA projects and policies.

Once assembled, the community representatives are asked to develop 'profiles' for hazards in their communities as well as for ecosystems, infrastructure, and societal sectors (Fig. 6.1). To do this, the Risk Matrix tool, is used along with a facilitated, participatory-mapping exercise. The Risk Matrix allows the participants to collaboratively identify vulnerable sectors and those assets that already support resilience in their community. Identified community assets often

Fig. 6.1 Community-based participatory mapping during Community Resilience Building Workshops in Bridgeport, Connecticut (USA) (Whelchel 2012) (Author's own photo)

include natural resources such as wetlands, beaches and dunes, and floodplains, which reinforce the community's recognition of ecosystems in a risk management context. Participants then utilize base maps to mark vulnerabilities and strengths as well as identify ownership or responsibility for those community elements. This process serves to spatially translate the dialogue and generate an overall profile of ecological, infrastructure, and societal elements along with overlaps/proximity of inter-dependent, complex situations (e.g., routine flooding on a road that is used by an elderly population, who are surrounded by protective salt marshes and floodplains). Participants then identify actions that either reduce the vulnerability or reinforce the strength for each identified community element. Once completed, participants are asked to relatively rank the importance (high, medium, low) and determine the urgency (ongoing, short-term, long-term) of each community-based action. Finally, participants are asked to further prioritize all the high importance, short-term actions through the community's Risk Matrix (Fig. 6.2) and select the three top priority needs across the three 'profiles' for the community to pursue immediately. This helps to ensure that the community is fully embracing Eco-DRR/CCA as a priority in the communities' overall approach to resiliency.

The Workshop process using the Risk Matrix is flexible enough to address all hazards (e.g., extreme heat, drought, storm surge, tornadoes, sea level rise, landslides, tsunamis), in any setting (e.g., inland, coastal, high elevation, deserts, urban), across multiple governance/societal structures (e.g., neighborhood, municipal, multi-municipal, regional, national, multi-national) and at any geographic scale. To date, 24 municipalities in Connecticut (USA) serving over 787,000

Community Resilience Building Workshop Risk Matrix (www.communityresiliencebuilding.com)

H-M-L priority for action over the Short or Long term (and Ongoing)
V = Vulnerability S = Strength

V or S	Vulnerabilities and Assets	Location	Ownership	Coastal Flooding: Sea Level Rise/Storm Surge	Inland Flooding: Rain Events	Ice and Snow	Wind	Priority H-M-L	Time S/L term ongoing
	Infrastructure								
V	Evacuation Routes - Roads/Intersections	Municipal-wide	State/Utility	Implement supportive communication program and install highly visible evacuation route signage.				H	S
V	Electrical Distribution System	Multiple	Utility	Within floodplain area need plan to address protection and long-term relocation of equipment.		Maintain power line protection zone (tree trimming).		H	S-L
V	Dams - (Inland and Coastal)	Multiple	Private		Prevent possibility of castastrophic dam failure; prepare contingency plan(s).			M	L
V	Landfill/Rubbush Facility	Specific	Private	Secure facility against future flooding and rain events.				L	L
S	Commercial/Industrial Facilities (Zoning Regulations)	Multiple	Municipal	Current building codes and zoning overlays limiting development in high risk areas.					Ongoing
S	New Ambulance/Heath Center	Specific	Municipal	Continue to support ambulance services and awareness amongst vulnerable populations.					Ongoing
	Societal								
V	At-Risk Neighborhoods and Populations	Municipal-wide	Private	Education on best practices to reduce risk; initiate "Neighbor helping Neighbor Program".				H	S
V/S	Coastal Homeowners	Coastline	Private	Review building codes and zoning regulations; Continue frequent communication about risks and evacuation procedures.				M	Ongoing
S	Full-time Emergency Managers (Fire and Police)	Municipal-wide	Municipal	Continue support (well equipped and experienced) to further strengthen services.					Ongoing
S	Regional Cooperative Agreements	Multiple	Municipal/Utilities	Maintain regional cooperative service agreement over time (sheltering, food and water supply, etc.)					Ongoing
	Ecosystem								
V	Forest (uniform age structure and diversity)	Municipal-wide	Municipal/State	Seeks management that diversifies the age structure of forests; assess key vulnerabilities from tree fall on eletrical system				H	S
V/S	Coastal Natural Infrastructure (Salt Marsh, Beaches, Dunes)	Multiple	State/Private	Assess risk reduction potential from existing and future wetland advancement zones. Restore/maintain beaches and dunes.				H	L
V	River Systems/Riparian Corridors (Inland Erosion)	Multiple	Municipal/Private			Education/Map impacts/identify solutions to reduce risk to people and property/ green infrastructrue, restoration, abandonment.		M	S
S	Protected Open Space/Public Amenities	Multiple	State/Municipal	Maintain existing open space to help reduce risk to Town/Seek to increase open space with the highest risk reduction charateristics.					Ongoing

Top Priority Hazards (sea level rise, wildfire, flooding, tornado, ice, heat wave, hurricanes, etc.)

Fig. 6.2 Abbreviated community-driven Risk Matrix used during the Community Resilience Building Workshop to generate comprehensive and prioritized resilience action plans across and between sectors (infrastructure, societal, ecosystems) that incorporate Eco-DRR/CCA (Whelchel and Ryan 2014) (Author's own graphic)

residents have completed this workshop process resulting in prioritized action plans to improve resilience that feature Eco-DRR/CCA (Box 6.1).

One key effort undertaken in advance of the Workshops is a full analysis of existing ecosystems, alongside projections of the future distribution of critical habitat such as salt marshes given ongoing increases in sea levels (Hoover et al. 2010; Hoover and Whelchel 2015). For each of the 24 coastal communities in Connecticut, a Salt Marsh Advancement Zone Assessment was conducted that identifies where the future habitat will be at the parcel scale (i.e. finest scale of land ownership and land-use decisions) (Horton et al. 2014; Ryan and Whelchel 2015). This helps to facilitate community dialogue on potential conflicts and opportunities arising from the current built environment and protected natural management areas, respectively (Fig. 6.3). The assessments are critical in shaping risk considerations at the community scale by requiring recognition of ecosystems and their risk avoidance services for people and property; and not just the recognition of the exposure and vulnerability of infrastructure and society to hazards.

Box 6.1 Common Community-Derived Prioritized Actions Via Community Resilience Building Workshops Using the Risk Matrix Tool
Environmental/Ecosystems

- Protection of conserved lands, natural buffers around waterways, and ongoing maintenance of wetlands.
- Resilient Conservation Practices: Anticipate changes in location, size, and distribution of wetlands and waterways under future conditions and prioritize acquisition to reduce development in risk-prone areas.
- Develop and/or strengthen low impact development policies and green infrastructure projects.

Infrastructure/Facilities

- Design and plan for infrastructure (transportation, sanitary, communications, etc.) conversion during redevelopment and prioritized upgrades. Prior to improvements carefully consider the future "design storms" for infrastructure given anticipated changes in precipitation patterns (3 cm/ 24 h. vs. 12 cm/24 h.).
- Prioritize the location of water retention systems, maximize infiltration rates, and increase separation between storm-water runoff and sewer systems. Design to minimize polluted discharges to wetlands, rivers, and other potable water sources.
- Modify existing land use and development policies to reduce the risk to building stock and public amenities over time (i.e., building codes, zoning

(continued)

Box 6.1 (continued)
overlays, voluntary buy-outs followed by ecosystem restoration, increased
density in lower risk areas).

Societal/Community Fabric

• Improve sheltering capacity for and preparedness of citizens.
• Strengthen support for ecosystems as protective features that reduce exposure of people and property within communities to disasters.

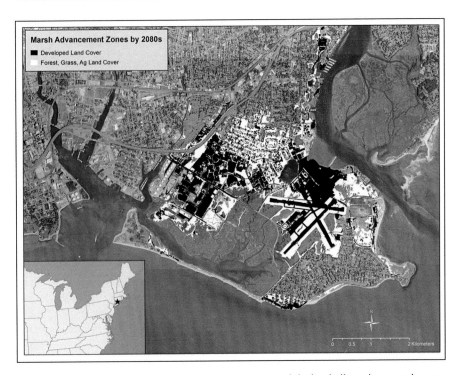

Fig. 6.3 The Salt Marsh Advancement Zone Assessment tool depicts built environment impacts
due to inundation (*developed land cover (black)*) and potential salt marsh advancement zones
(*undeveloped land cover –currently forest, grass, and agriculture (white)*) using downscaled sea
level rise projections (1.32 m by 2080s depicted) in Stratford, Connecticut (USA) (Ryan and
Whelchel 2014) (Author's own graphic)

The Community Resilience Building Workshop approach is currently being
promoted for national deployment in the USA and internationally. This approach
is also being used to build and integrate resilient communities into a larger
regional framework for resilience in the central coast of Connecticut (USA),
including the metropolitan areas of greater Bridgeport and New Haven (30 % of
Connecticut's coast with 591,000 people). Application of the approach highlights

one of the most critical aspects of integrated Eco-DRR/CCA, i.e. broad yet directed engagement and consensus-building with communities around risks, planning, and actions.

In some cases, the recognition of ecosystem importance and their incorporation into resiliency approaches requires a triggering event. The impact of Tropical Storm Irene and Sandy (National Weather Service 2013) on the eastern seaboard of the USA has resulted in the incorporation of Eco-DRR/CCA principles in the recovery plans at the federal (Hurricane Sandy Building Task Force 2013) and state (New York 2100 Commission 2013; Ambrette and Whelchel 2013) level. These two storm events have also facilitated progressive funding for significant, resilience-orientated projects (i.e., Rebuild by Design – Resilient Bridgeport (Connecticut)). Approaches that integrate tools as illustrated by this Coastal Resilience case study have been instrumental in setting the standard for enhanced resiliency amongst coastal and inland communities affected by major disasters and subjected to increasingly intense rainfall in the USA (Horton et al. 2014).

6.3.1.2 Floodplain by Design – Integrating Flood Risk Reduction in Puget Sound (USA)

The state of Washington is currently one of the most flood-prone in the USA. Currently, there are 57,000 flood insurance policies in the state providing insurance coverage for assets totaling $13 billion (USD), with 35 % of those policies outside of the federally designated flood areas (Sumioka et al. 1998; Washington Department of Ecology 2004). Across the Puget Sound watershed (Fig. 6.4), flood management efforts are lagging the pace of population expansion and development resulting in more people and property in flood-prone areas, water quality declines, and loss of fish habitat (Fig. 6.5). While there is an understanding of the short and long term characteristics of flood risk (types, locations, re-occurring costs) in the watershed, the systems for managing the floodplain are recognized as disjointed, uncoordinated, and inadequately resourced. As is often the case in larger, multi-jurisdictional geographies, the impediment to advancing priority strategies is fragmentation or overlap within decision-making/regulatory systems and structures. To adjust that prognosis in the watershed, the Floodplain by Design (FbD) approach is being implemented.

The FbD approach seeks to ensure better management of shared floodplain resource through the integration of flood hazard reduction, habitat protection and restoration, and improved water quality and outdoor recreation. The FbD is a merger between a science-driven framework known as the Active River Area (Smith et al. 2008) that requires consideration of the dynamic connections and interactions of land and water through which a river flows and a modeling application that maps ecosystem service values and trade-offs between conservation and development. The modeling application used is the Natural Capital (NatCap) Project's Marine Integrated Valuation of Environmental Services and Tradeoffs (InVEST) program (Sharp et al. 2014; see also Bayani and Barthélemy,

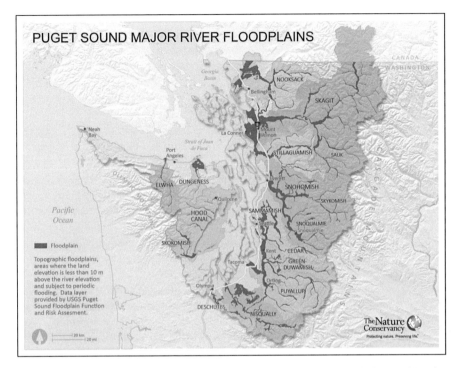

Fig. 6.4 Map of Puget Sound watershed in state of Washington (USA) depicting the 17 major rivers and current distribution of floodplains contributing to Floodplain by Design (Graphic reproduced or used with permission)

Fig. 6.5 Extreme flooding on the Snoqualmie Valley within the Puget Sound watershed in Washington (USA) (Photo reproduced or used with permission)

Chap. 10). The intended outcomes of FbD are to make river dependent or surrounding communities safer, improve the ecological health of the river, and increase the cost-effectiveness of long-term river management and immediate post-disaster recovery of the communities. This approach relies on a tool to satisfy many of the planning-to-action framework steps (see 6.2.1) alongside state/regional partnership and an incentivized community engagement process to link strategies and plan implementation (framework steps #4 and #5).

FbD is originating a new private-public partnership across local, state and federal agencies and organization that could simultaneously achieve floodplain management and ecosystem recovery goals in the most cost-effective manner possible. This innovative and collaborative FbD partnership seeks to reduce impediments to achieving collective actions by linking decision-making to actions through funding incentives, in effect changing the collective paradigm towards better management of the entire watershed. An overarching framework is used across the Puget Sound watershed to advance the FbD approach: (1) Implement integrated floodplain projects across the 17 largest rivers; (2) Craft regional vision and work plan (10-year) for each river; (3) Match funding to needs via vision/work plan by sustaining existing, securing new, and aligning state and federal funding programs with these regional visions (i.e. coordinating investment); and (4) Build technical and permitting assistance capacity to ensure further integration across jurisdictions. This FbD framework is a main driver to advance Eco-DRR/CCA efforts in the entire Puget Sound watershed (Box 6.2).

Box 6.2 Key Eco-DRR/CCA Principles of Floodplain by Design
Step 1: Maximize Natural Infrastructure Use – work with, not against, natural processes such as flooding frequency and extent (annual, 100 year, 100–500 year) by incorporating floodplains, wetlands and open areas in management decisions. Some key tactics to assist with this step may include:

- **Setback Levees**: levees or berms constructed or moved farther from the river and ideally out of the floodplain, thereby allowing rising rivers more room to adjust and flood.
- **Connected Floodplains**: connected or never "cut off" from the river by levees or other structures or "reconnected" by the removal or management of levees.

Step 2: Diversify Portfolio of Flood-Risk Management Techniques – tailor techniques to specific requirements of the watershed. In addition to dams and levees as well as setback levees and connected floodplains, such techniques can include floodways and flood bypasses, which are large-scale floodplain reconnections for storage and conveyance of water.
Step 3: Maximizing Community Benefits – from initial identification of community needs/values, seek to enhance benefits of floodplains and rivers to

---

(continued)

> **Box 6.2** (continued)
>
> local entities by improving access, safety and health of river systems through collaborative consideration of solutions; not only reducing flood risk but also improving habitat for fish and wildlife and water quality impacts at the source.
>
> **Step 4: Plan and Implement Resilient "Whole-River" Practices** - dams, levees, floodways, natural areas, topography, croplands, existing and planned developments, and river uses – such as for recreation, municipal water supply, irrigation, and navigation – are all inter-related and must be managed as such.
>
> **Step 5: Develop Mosaics of Accommodating Land Uses** – a mosaic of diverse land uses that are both resilient to floods and consistent with vibrant communities; tailor land use for the average frequency and duration of floods the area is subjected to.

The principal vehicle to orchestrate this systemic change is a funding program administered by the Washington Department of Ecology. Nine projects using the FbD approach have been funded via a $33 million (USD) investment by the state matched by $80 million (USD) from other sources. For example, an integrated floodplain plan was developed in response to funding opportunities for the Puyallup River (one of the 17 major rivers in Puget Sound watershed) that was designed to reconnect floodplains and estuary habitat, permanently preserve 600 acres of farmland through conservation easements, provide critical habitat to support populations of Chinook salmon, and reduce flood risk to municipalities and shared infrastructure. An early investment in 2014 in the Puyallup River of $4.7 million (USD) has been matched with over $17.5 million (USD) in state, county, and local funding sources, reflecting an investment leverage ratio of 3.7 to 1.

State grant criteria continue to be the principal mechanism to ensure projects like the Puyallup River meet the requirements of FbD. The criteria awarded more points and subsequent higher ranking for projects that demonstrate effectiveness at advancing multiple benefits, such as flood risk reduction, floodplain ecosystem protection and restoration, agricultural viability, water quality and open space access. Additional points are awarded for proposals that avoid ongoing costs including maintenance and emergency response and longer-term changes in hydrology, sedimentation, and water supply due to extreme weather events. State grant criteria also serves to prioritize pilot and design projects that seek creative solutions, fill funding gaps at the local level, and favor underserved communities and social justice issues. Eligible applicants across the watershed have readily accepted the state grant criteria, as evident through the 71 proposals submitted towards a second call for proposals.

Recognition that different governance structures and regulatory mechanisms are needed to realize collective and cumulative gains is not enough to generate the implied transformation. In the Puget Sound watershed, introduction of state grant criteria that favor the integration of multiple objectives has been well-received and will likely over the long-term incentivize a more resilient future at ever increasing scales via locally-driven creative solutions that mainstream Eco-DRR/CCA.

6.3.1.3   Water Funds – Financially Linking Watershed Management with Risk Reduction

In 2000, a catalytic approach to integrated watershed management known as 'water funds' was launched in Quito, Ecuador (Tallis et al. 2008). Since then, this approach has been successfully replicated through over 60 water funds across South America, Australia, Central America, USA, and East Africa (Goldman-Brenner et al. 2012). The approach brings water users (typically large businesses, government agencies, municipalities) together to jointly invest via a financial mechanism that directs funds to top priority ecosystem-based projects within defined watersheds. The joint investments, often private-public partnerships, result in benefits via returns to all the investors. These water fund collaboratives also provide a governance structure to collectively derive and sustain decisions on priority funding needs and water resource management (i.e., conservation, power generation, drinking water supply). The success of the water fund approach is due in large part to flexibility of the financial mechanism or investment vehicle (i.e., endowment, direct incentives to landowners, direct investment towards actions) through which objectives are funded. The pooling and leveraging of funds through an independent fiduciary administrator towards common outcomes is paramount to maintaining existing programmes and attracting other regions to water funds. Water funds typically rely on tools and financial incentives to advance through many of the planning-to-action framework steps, namely facilitating fiduciary and action-orientated partnerships and community engagement (step #4 and #5).

Once established, each water fund defines the core objective(s) of watershed management and goes about identifying and prioritizing opportunities. To ensure that capital derived through water funds is allocated to (1) achieve the greatest return for multiple objectives, (2) quantify improvement through various investment portfolios, and (3) compare these improvements against the ongoing status-quo management, the Resource Investment Optimization System (RIOS) (http://www.naturalcapitalproject.org/RIOS.html) tool was developed for water funds. The tool couples biophysical data (i.e., soils, land use, slope, flood risks) with water consumer demand (i.e., population density and distribution) to geospatially determine the optimal places to maximize returns on conservation investment (ROCI) within a defined watershed. The tool provides a relative ranking of optimal places for conservation investment, informed by the most urgent needs of stakeholders (e.g. tackling floods, drought, groundwater supply) and taking into account constraints (e.g. security risks, policy restrictions). For example, if a water fund manager is looking to reduce downstream flood risk, tools such as RIOS can now help determine the most prudent suite of investments, such as buying farmland along streams, reconnecting floodplains through restoration and/or voluntarily relocating at-risk populations to higher ground. Ecosystem services tradeoffs of various investment portfolios are estimated by RIOS and can be monitored and adapted over time for greater effectiveness on the ground. Tools like RIOS are particularly attractive to decision makers because they generate reliable and comparable estimates on locally relevant ROCI and provide a way to monitor action effectiveness. In addition, the application and outputs from RIOS can effectively establish a regional platform from which Eco-DRR/CCA can be incorporated into a supportive financial and governance construct.

The integration of tools into initial design and scoping of water fund projects is also being expanded in several locations to incorporate forecasts of disasters and climate change. This type of consideration is of particular concern to large water users/providers and governments when assessing flood and drought risks. One foremost example is the Monterrey Metropolitan Water Fund (FAMM) centered in the watersheds of Monterrey, Mexico, which is one of the most important industrial capitals in Latin America and home to over four million people who are routinely subjected to devastating floods and extreme drought (Gonzalez 2011). The FAMM is part of the Latin America Water Fund Partnership established in 2011 by TNC, FEMSA Foundation, The Inter-American Development Bank and the Global Environmental Facility to advance the 14 water funds underway and the 18 under evaluation across Brazil, Colombia, Panama, Venezuela, and Mexico.

With over 40 partners engaged, including various business sectors, academia, conservation groups, civil society organizations, and multiple levels of governments, the FAMM is specifically designed to improve water management through compensating and incentivizing actions that reduce flood risks and increase availability of drinking water during droughts through aquifer recharge. The focus of this water fund is on the Cumbres de Monterrey National Park (Fig. 6.6) upstream from

Fig. 6.6 Cumbres de Monterrey National Park within the San Juan River Watershed above City of Monterrey, Mexico (Photo reproduced or used with permission)

the city of Monterrey, all located within the San Juan River watershed. The Park meets approximately 60 % of the water consumption needs but is also the principal origin of flash flood risks to downstream communities such as Monterrey. Reforestation and soil conservation projects funded through FAMM are intended to significantly reduce the speed and peak volume of downstream runoff. The FAMM is also directing capacity to educating Monterrey residents and consumers on water conservation measures. In this regard, this water fund provides a meaningful example of an approach informed by tools and driven by partnerships and financial mechanisms towards common goals and outcomes with Eco-DRR/CCA priorities.

6.3.1.4   Integrating Coastal Zone Management in Belize

The Government of Belize tasked the Coastal Zone Management Authority and Institute (CZMAI) with the design of the Integrated Coastal Zone Management Plan for the entire coast of Belize. To inform its development, the CZMAI partnered with World Wildlife Fund and NatCap, to focus on three critical ecosystem services: lobster fisheries productivity, recreational activities, and coastal risk reduction. The NatCap developed an integrated database on biodiversity, habitats, and marine and coastal uses. Then, together with local stakeholders, the team formulated three possible future scenarios: (1) a conservation scenario emphasizing sustainable use and investment in coastal habitats; (2) a compromise ('informed management') scenario that advanced development and conservation; and (3) an infrastructure development scenario. These scenarios were analyzed with InVEST (Sharp et al. 2014) to determine the tradeoffs among options, the quantity of services provided, and iterations of other possible scenarios. Similar to the other case studies presented in this chapter, the integrated coastal zone management planning approach in Belize employs a tool and various scenarios to advance through the framework steps and contributes directly to partnerships and community engagement processes (steps #4 and #5; and steps #2 and #3 for scenario generation) (see also Bayani and Barthélemy, Chap. 10).

The importance of coastal risk reduction in the scenarios was made clear. The benefits in terms of disaster damages avoided totaled billions (in Belize Dollars or BZD), whereas other benefits (i.e. tourism and lobster fisheries) totaled in the millions (BZD). However, there were significant tradeoffs with respect to benefits. For example, more development would generate a higher recreation value, but also much higher disaster damages to infrastructure due to the loss of coastal habitat risk reduction services. By categorizing and integrating marine and coastal uses and visualizing them in maps, stakeholders were better informed with potential conflicts arising from different land-use and the opportunities for negotiating between competing interests.

The development of alternative scenarios has proven to be one of the greatest difficulties because stakeholders are often not able to visualize and articulate multiple and inter-dependent future scenarios, particularly at a national level

(Gleason et al. 2010; Halpern et al. 2012). In summary, the CZMAI was tasked with developing a coastal zone management plan (submitted September, 2013) with the help of alternatives assessed with InVEST, and the scenarios developed in stakeholder workshops were useful in presenting land-use tradeoffs to decision-makers. The integration of Eco-DRR/CCA as a key variable at the front end of this effort is instructive and was critical in determining disaster and climate resilient outcomes. This case study highlights a growing trend in the use of scenario planning or 'future visioning' that allows for comparisons (i.e. costs/benefits, effectiveness) between various, individual or sequenced series of risk avoidance actions (Dawson et al. 2011; Mahmoud et al. 2011) and represents a critical next step for tool development that balance ecosystem and socio-economic tradeoffs in a disaster and climate altered future (Shepard et al. 2011).

6.3.2   Additional Tools Available for Select Planning-to-Action Framework Steps

The following provides an additional set of tools that have been proven effective for stand-alone assessments independent of or towards the beginning of a DRR/CCA process and for fulfilling the core considerations and specific framework steps – particularly steps #1, #2, and #7 (see Table 6.1).

6.3.2.1   Climate Wizard – Future Climate Change Projections for Decision Makers

The Climate Wizard tool suite arose in 2009 from the need to provide modelled projections of future climates in a format and at a scale useful for decision makers. TNC along with partners from the University of Washington, Santa Clara University, The University of Southern Mississippi, and Lawrence Livermore National Laboratory worked together to create tools to view and access current climate change information, and visualize observed and expected temperature and precipitation as well as derived climate variables such as moisture deficit, moisture surplus trends and measurements of extreme precipitation and heat events anywhere on earth. Climate Wizard tools offer a straightforward interface for processing and visualizing numerous climate variables for both past climate and future climate models and greenhouse gas emission scenarios (Fig. 6.7). Users can download map images and graphics for three time periods (past 50 years (1951–2006); mid-century (2040–2069); end of century (2070–2099) as well as annual, monthly and seasonal time steps. This tool has provided a valued resource for planners addressing framework steps #1 and #7 independently or as part of a more comprehensive approach.

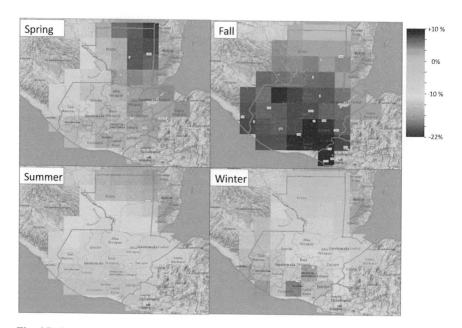

Fig. 6.7 An ensemble analysis from the Climate Wizard tool of 16 General Circulation Models showing the projected changes in precipitation quantity (mm/year) and distribution by 2050s (2040–2069) for the A2 emissions scenario across the Sierra Madre de Chiapas (Mexico, Guatemala, El Salvador) (Graphic reproduced or used with permission)

One of the key abilities of Climate Wizard is to bridge the divide between climate science and practitioners through the production of novel, downscaled, future-climate data sets, thus making climate change information more relevant and useable. Recent advancements through the Climate Wizard Custom framework provide globally, daily downscaled climate projections for a range of future projections which have been adopted by The World Bank via their Climate Change Knowledge Portal (see Table 6.2) (http://climateknowledgeportal.climatewizard. org). This availability of climate projections highlights a pre-requisite to refine and customize tools to inform decisions on climate impacts to water, agriculture and ecosystems. In this case, the tool demonstrates future aridity impacts by modeling the interactions of precipitation and rising temperature patterns. It also provides unprecedented access to future projections globally for various aridity metrics (Aridity Index, Climate Moisture Deficit and Surplus) for nine general circulation models.

A Mandarin version of Climate Wizard with data developed by the Chinese National Climate Center was released in 2014 to support a national future flood risk assessment and investment plan for floodplains (http://www.climatewizard.org.cn. s3-website-us-west-1.amazonaws.com). Applications of the tool along critical waterways like the Yangtze River illustrate the potential to influence flood risk reduction projects throughout China and in countries where Chinese companies invest. Ultimately, tools like Climate Wizard increase the accessibility to locally relevant projections with actionable visualization of climate change, which could then be used to forecast the implications of adaptive actions that incorporate Eco-DRR/CCA.

6.3.2.2   Coastal Defense Application

Coastal Defense Application resides in the Coastal Resilience tool as an open source app that integrates coastal hazards with social, ecological, economic, and coastal engineering to match adaptation with priority needs (framework steps #1, #2, #6, #7). This app helps to advance Eco-DRR/CCA by identifying the coastal protection value of existing reefs (Fig. 6.8) and wetlands and allowing the user to design and tailor implementation of natural infrastructure projects. More specifically, this app helps (1) identify areas that may be at risk of coastal erosion and inundation from wave action and storm surge; (2) interactively examine the role of coastal habitats in attenuating wave height and energy (Fig. 6.9); and (3) in a broader planning context determine appropriate disaster risk and climate adaptation strategies that incorporate green (habitats) and grey (seawalls and other man-made structures) infrastructure trade-offs. To generate these outputs the model InVEST (Sharp et al. 2014) builds in standard engineering techniques to calculate the reduction of wave height and energy in the presence and absence of coastal habitat. The app allows the user to define the value range for model variables within an

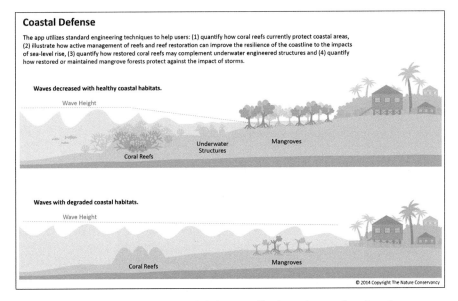

Fig. 6.8 Conceptual diagram of coastal defense application using coral reefs and mangroves protection and restoration to assist with disaster risk reduction and climate adaptation (Graphic reproduced or used with permission)

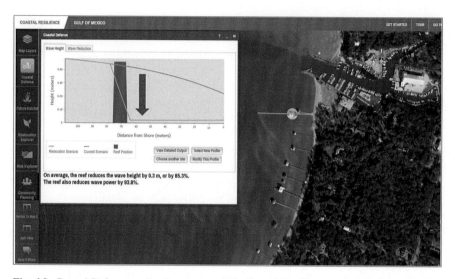

Fig. 6.9 Coastal Defense application output within Coastal Resilience tool depicting the reduction in wave height by oyster reefs designed with specified height characteristic in coastal Alabama (USA) (Graphic reproduced or used with permission)

intuitive and user-friendly interface thus reflecting real world scenarios. For the Coastal Defense app this includes user-specified offshore forcing conditions (wave and surge characteristics), a sea-level rise value, locations of restored or degraded coastal habitats and built infrastructure. In the USA, the app has been deployed in Puget Sound, Washington (tidal marshes), Mobile Bay, Alabama (oyster beds), and the Florida Keys (coral reefs and mangroves), with the potential for replication around the globe. In addition, the app has been used to assist in the identification of appropriate Eco-DRR/CCA projects in the Gulf of Mexico following the Deepwater Horizon oil spill (see also Bayani and Barthélemy, Chap. 10).

6.3.2.3   Crowd Sourcing/Social Media Tools

Emerging technological trends have resulted in a proliferation of decision-support tools that harness social media venues, specifically crowd sourcing. If harnessed appropriately, crowd sourced data can help to inform framework steps #1 and #2, and most importantly, help to monitor in real time during major events the effectiveness of actions taken that incorporate Eco-DRR/CCA. The use of crowd sourcing has expanded in the context of flood risk management (Haklay et al. 2014; Wan et al. 2014) principally because geographic information systems and technology are already an integral part of flood preparation activities. The information derived helps to reinforce the flood reduction services provided by ecosystems through eye-witness accounts and ultimately helps build local acceptance for ongoing and future actions that establish Eco-DRR/CCA solutions.

In a growing number of places like Brazil (Degrossi et al. 2010), the Philippines (Pineda 2012), and Jakarta, Indonesia (Holderness and Turpin 2015), citizen-derived reports sent through electronic messages assist emergency managers and responders by providing immediate, local flooding assessments across large areas. The use of technology in this way can help to direct disaster response efforts to areas of greatest need. Over time, data from multiple events help to drive flood risk reduction actions, such as the voluntary relocation of people followed by floodplain restoration in those self-identified locations. Of concern, however, is the level of accuracy in citizen reports, the ability of emergency management systems to process increased data volumes, and ultimately, the capacity of disaster response structures to incorporate the information and efficiently respond in appropriate timeframes (i.e., crowd sourcing outpacing the adaptive capacity of emergency management).

6.4   Lessons Learned and Recommendations

What is clear from these case studies (see Table 6.1) and many others (see Table 6.2) is that decision-support tools and approaches have improved rapidly in the last decade and continue to demonstrate the importance of Eco-DRR/CCA. A deeper understanding by decision makers, stakeholders, and practitioners of what mechanisms are being used to implement Eco-DRR/CCA, how these mechanisms can be used, and their inherent limitations, remains a critical challenge as illustrated by the case studies above. Despite the advances, external factors such as governance and funding remain pre-requisites to successful implementation. This is keenly evident in the Puget Sound watershed example whose successes thus far are largely driven by publicly-sourced finance commitments and funding processes (see Box 6.2) and by larger-scale collaboration around multi-objectives, including Eco-DRR/CCA.

Further lessons learned from the Coastal Resilience Program in Connecticut (USA) include the need to engage diverse stakeholders through a community-driven workshop approach that integrates tools within the planning-to-action framework steps. The recommendation therefore is to engage a broad suite of stakeholders at the beginning, during, and routinely thereafter, with particular emphasis on elected and appointed officials (i.e. decision makers), as a community works through the framework steps (see Box 6.1). This case study also highlights the importance of a trigger event (e.g. Tropical Storm Irene and Sandy) to advance Eco-DRR/CCA through recovery efforts.

The integrated coastal management efforts in Belize further reinforce this need to activate stakeholders more broadly through proactive engagement processes. The work in Belize, however, also highlights one of the ongoing challenges for decision-support tools and subsequent framework steps: the limited ability of tools to help stakeholders visualize alternative and inter-dependent future scenarios across larger geographies. A recommendation, therefore, is to develop tools that generate comparative outcomes from decisions or scenarios (i.e., cost of 'no-action', delayed action(s), and/or action sequences) that are easily understandable by stakeholders and are coupled with a progression through the framework steps. For example, this need is directly linked to the ability to sequence adaptation strategies (step #3) and assess action effectiveness (step #6). Of course, the critical consideration for 'future visioning' efforts is the ability to display comparisons of costs/benefits and effectiveness of Eco-DRR/CCA policies and projects. Social media that generate crowd sourcing of information in places like Brazil, Philippines, and Indonesia, have shown promise in fostering greater community receptivity towards scenario planning with Eco-DRR/CCA as a desired outcome, as well as in prioritizing voluntary relocation and subsequent ecological restoration to reduce flood risks.

In the case of the Water Funds approach and projects like the Monterrey FAMM, the importance of private-public partnerships in a financial construct can result in the prioritization and implementation of Eco-DRR/CCA projects at a watershed scale. One recommendation to improve the Eco-DRR/CCA linkages is to include in the prioritization process information on the size, configuration, and proximity of various habitats that can optimize benefits to society such as flood prevention. Establishment of a dedicated and sustainable funding source is certainly key to success with Water Funds throughout Central and South America and serves as a core enabling factor for Eco-DRR/CCA implementation (which is also a lesson derived from the Puget Sound watershed example). Another clear recommendation is the need to support efforts that prioritize projects and quantify the true cost-effectiveness of Eco-DRR/CCA over time. This would require standardization in the design and specifications for Eco-DRR/CCA projects in order for engineers to assign comparative costs for implementation and maintenance over the longer term, alongside traditional hard engineering projects.

Undoubtedly, organisations and governments around the globe will continue to develop tools and approaches in response to the mounting ecological, social and economic costs of disasters and climate change. These tools and approaches will continue to collectively enhance societies' ability to capture the additional and protective benefits of ecosystems. Nonetheless, decision makers and practitioners also need to point out the limitations of existing tools and approaches and express urgency for improvements. As illustrated within this chapter, it is clear that Eco-DRR/CCA decision-support mechanisms have improved rapidly in the last decade. Despite these advancements, factors such as resistance to change, the cautious approach by development agencies, governance structure and overlapping jurisdictions, funding, and limited community engagement remain, in many cases, pre-requisites to successful implementation of ecosystem-based solutions. The planning-to-action framework steps outlined in this chapter help guide communities to overcome these challenges and work towards maximizing resilience opportunities. What is certain is that ecosystems will increasingly be a critical part of societies' overall response to equitably solving issues associated with disasters and climate change in the decades and centuries to come.

References

Ambrette B, Whelchel AW (2013) Adapting to the rise: a guide for Connecticut's Coastal Communities. The Nature Conservancy, Coastal Resilience Program. Publication 13–5, New Haven, CT. https://www.conservationgateway.org/ConservationPractices/Marine/crr/library/Documents/TNC%20Adapting%20to%20the%20Rise.pdf. Accessed 15 Mar 2015

Beck MW, Brumbaugh RD, Airoldi L et al (2011) Oyster reefs at risk and recommendations for conservation, restoration, and management. Bioscience 61:107–116

Beck MW, Gilmer B, Whelchel AW et al (2013) Using interactive decision support to integrate coast hazard mitigation and ecosystem services in Long Island Sound, New York and Connecticut USA. In: Renaud F, Sudmeier-Rieux K, Estrella M (eds) Linkages between ecosystems, livelihoods and disaster risk reduction. PEDRR, UNU Press, Bonn, pp 140–163

CCRIF (2010) Enhancing the climate risk and adaptation fact base for the Caribbean: preliminary results of the economics of climate adaptation study, vol 28. Caribbean Catastrophic Risk Insurance Facility, Cayman Islands

Dawson RJ, Ball T, Werritty J et al (2011) Assessing the effectiveness of non-structural flood management measures in the Thames Estuary under conditions of socio-economic and environmental change. Glob Environ Chang 21(2):628–646

Degrossi LC, de Albuquerque JP, Fava MC, Mendiondo EM (2010) Flood citizen observatory: a crowdsourcing-based approach for flood risk management in Brazil. http://www.agora.icmc.usp.br/site/files/papers/degrossi-seke2014.pdf. Accessed 2 Oct 2014

Global Assessment Report on Disaster Risk Reduction (2013) http://www.preventionweb.net/english/hyogo/gar/2013/en/home/GAR_2013/GAR_2013_2.html. Accessed 20 Jan 2015

Global Assessment Report on Disaster Risk Reduction (2015) http://www.preventionweb.net/english/hyogo/gar/2015/en/gar-pdf/GAR2015_EN.pdf. Accessed 28 Aug 2015

Ferdaña Z, Newkirk S, Whelchel AW et al (2010) Building interactive decision support to meet management objectives for coastal conservation and hazard mitigation on Long Island, New York, USA In: Andrade Pérez A, Herrera Fernandez B, Cazzolla Gatti R (eds) Building resilience to climate change, Gland, IUCN, Switzerland, pp 72–87

Ferrario F, Beck MB, Storlazzi C et al (2014) The effectiveness of coral reefs for coastal hazard risk reduction. Nat Commun 5(3794):1–9

Gilmer B, Ferdaña Z (2012) Developing a framework for assessing coastal vulnerability to sea level rise in Southern New England, USA. In: Otto-Zimmermann K (ed) Resilient cities 2: cities and adaptation to climate change proceedings of the global forum, Bonn, August 2011. Springer Science and Business Media BV

Gleason M, Mccreary S, Miller-Henson M et al (2010) Science-based and stakeholder-driven marine protected area network planning: a successful case study from north central California. Ocean Coast Manag 53(2):52–68

Goldman-Brenner RL, Benitez S, Boucher T et al (2012) Water funds and payments for ecosystem services: practice learns from theory and theory can learn from practice. Oryx 46(1):55–63

Gonzalez CA (2011) Severe flooding from extreme meteorological events in Monterrey, Mexico. In: Valentine EM, Apelt CJ, Ball J, Chanson H et al (eds) Proceedings of the 34th World Congress of the International Association for Hydro- Environment Research and Engineering, Australia, pp 615–622

Green A, Smith SE, Lipsett-Moore G et al (2009) Designing a resilient network of marine protected areas for Kimbe Bay, Papua New Guinea. Oryx 93(4):488–498

Haklay ME, Antoniou V, Basiouka S et al (2014) Crowdsourced geographic information use in government. Global Facility for Disaster Reduction & Recovery, World Bank, London

Hale LZ, Meliane I, Davidson S et al (2009) Ecosystem-based adaptation in marine and coastal systems. Renewable Resour J 25(4):21–28

Halpern BS, Diamond J, Gaines S et al (2012) Near-term priorities for the science, policy and practice of coastal and marine spatial planning. Mar Policy 36:198–205

Holderness T, Turpin (2015) Assessing the role of social media for civic co-management duing monsoon floooding in Jakarta, Indonesia. http://petajakarta.org/banjir/en/research/. Accessed 28 Aug 2015

Hoover M, Whelchel AW (2015) Tidal marsh classification approaches and future marsh migration mapping methods for Long Island Sound, Connecticut, and New York. In: Tiner RW, Lang MW, Klemas VV (eds) Remote sensing of wetlands: applications and advances. CRC Press, Boca Raton

Hoover M, Civco DL, Whelchel AW (2010) The development of a salt marsh migration tool and its applications in Long Island Sound. ASPRS, San Diego, http://clear.uconn.edu/publications/research/tech_papers/Hoover_et_al_ASPRS2010.pdf. Accessed 20 Oct 2014

Horton R, Yohe G, Easterling W et al (2014) Chapter 16: Northeast. Climate change impacts in the United States. In: Melillo JM, Richmond TC, Yohe GW (eds) The third national climate assessment. U.S. Global Change Research Program, pp 371–395. doi:10.7930/J0SF2T3P

Hurricane Sandy Rebuilding Task Force (2013) Hurricane sandy rebuilding strategy: stronger communities, a resilient region., U.S. Department of Housing and Urban Development, http://portal.hud.gov/hudportal/documents/huddoc?id=HSRebuildingStrategy.pdf. Accessed 16 Oct 2014

Intergovernmental Panel on Climate Change 5th Assessment Report (2014) http://www.ipcc.ch/report/ar5/. Acessed 16 Oct 2014

Jones HP, Hole DG, Zavaleta ES (2012) Harnessing nature to help people adapt to climate change. Nat Clim Chang 2:504–509

Mahmoud MI, Gupta HV, Rajagopal S (2011) Scenario development for water resources planning and watershed management: Methodology and semi-arid region case study. Environ Model Softw 26(7):873–885

Moller I, Kudella M, Rupprecht F et al (2014) Wave attenutation over coastal salt marshes under storm surge conditions. Nat Geosci Lett 7:727–731

Morris BL, Lawrence ARL, Chilton PJC et al (2003) Groundwater and its susceptibility to degradation: a global assessment of the problem and options for management. Early warning and assessment report series, RS 03–3. United Nations Environment Programme, Nairobi, Kenya

National Weather Service (2013) Hurricane/Post-Tropical cyclone sandy (October 22–29, 2012). Service Assessment. Department of Commerce, United States of America http://www.nws.noaa.gov/os/assessments/pdfs/Sandy13.pdf. Accessed 3 Nov 2014

New York 2100 Commission (2013) Recommendations to improve the strength and resilience of the empire state's infrastructure. http://www.governor.ny.gov/assets/documents/NYS2100.pdf. Accessed 21 Oct 2014

NYC Special Initiative For Rebuilding And Resiliency (2014) A stronger, more resilient New York. City of New York. http://www.nyc.gov/html/sirr/html/report/report.shtml. Accessed 05 Jan 2015

Pineda MVG (2012) Exploring the potentials of a community-based disaster risk management system (CBDRMS), the Philippine experience. Int J Innov Manag Technol 3(6):708–712

Pritchard D (2009) Reducing emissions from deforestation and forest degradation in developing countries (REDD) – the link with wetlands. Foundation for International Environmental Law and Development, London

Renaud FG, Sudmeier-Rieux K, Estrella M (eds) (2013) The role of ecosystems in disaster risk reduction. United Nations University Press, Tokyo

Ryan A, Whelchel AW (2014) A Salt Marsh Advancement Zone Assessment of Stratford, Connecticut. The Nature Conservancy, Coastal Resilience Program. Publication Series #1-F, New Haven, Connecticut. https://www.conservationgateway.org/ConservationPractices/Marine/crr/library/Documents/Stratford%20Salt%20Marsh%20Advancement%20Zone%20Assessment%20September%202014.pdf. Accessed 10 Oct 2014

Ryan A, Whelchel AW (2015) A Salt Marsh Advancement Zone Assessment of Connecticut. The Nature Conservancy, Coastal Resilience Program. Publication Series #1 Final (A-W), New

Haven, Connecticut. http://coastalresilience.org/project-areas/connecticut-solutions/. Accessed 10 February 2015

Sharp R, Chaplin-Kramer R, Wood S et al (eds) (2014) InVEST user's guide: integrated valuation of environmental services and tradeoffs. The Natural Capital Project, Stanford

Shepard C, Dixon DJ, Gourlay M et al (2005) Coral mortality increases wave energy reaching shores protected by reef flats in the Seychelles. Estuar Coast Shelf Sci 64:223–234

Shepard C, Crain C, Beck MW (2011) The protective role of coastal marshes: a systematic review and metaanalysis. PLoS One 6(11):272–283

Sheppard SRJ, Shaw A, Flanders D et al (2011) Future visioning of local climate change: a framework for community engagement and planning with scenarios and visualisation. Futures 43(4):400–412

Smith MP, Schiff R, Olivera A, MacBroom JG (2008) Active river area: a conservation framework for protecting rivers and streams. The Nature Conservancy, Boston, http://www.floods.org/PDF/ASFPM_TNC_Active_River_%20Area.pdf. Accessed 17 Sept 2015

Spalding M, Kainuma M, Collins L (2010) World atlas of mangroves. Earthscan, London

Spalding MD, McIvor A, Beck MW et al (2014) Coastal ecosystems: a critical element of risk reduction. Conserv Lett 7:293–301

Sumioka SS, Kresch DL, Kasnick KD (1998) Magnitude and frequency of floods in Washington. U.S. Geological Survey Water-Resources Investigations Report: 97–4277

Tallis H, Kareiva P, Marvier M, Chang A (2008) An ecosystem service framework to support both practical conservation and economic development. Proc Natl Acad Sci U S A 105 (28):9457–9464

Temmerman S, Meire P, Bouma PJ et al (2013) Ecosystem-based coastal defence in the face of global change. Nature 504(7478):79–83

United States National Climate Assessment (2014) http://nca2014.globalchange.gov/. Accessed 2 Oct 2014

UNU-EHS (2014) Coasts at risk: an assessment of coastal risks and the role of environmental solutions. http://www.crc.uri.edu/download/SUC09_CoastsatRisk.pdf. Accessed 24 Oct 2014

van den Hoek R, Brugnach M, Hoekstra A (2012) Shifting to ecological engineering in flood management: introducing new uncertainties in the development of a Building with Nature pilot project. Environ Sci Pol 22:85–99

Wan Z, Hong Y, Khan S et al (2014) A cloud-based global flood disaster community cyber-infrastructure: development and demonstration. NASA Publications. Paper 146. http://digitalcommons.unl.edu/nasapub/146. Accessed 8 Oct 2014

Warner JF, van Buuren A, Edelenbos J (eds) (2013) Making space for the river: governance experiences with multifunctional river flood management in the US and Europe. IWA Publishing, London

Washington Department of Ecology (2004) Map moderization business plan: Washington State. http://www.floods.org/PDF/SBP_WA_04.pdf. Accessed 4 Nov 2014

Whelchel AW (2012) City of Bridgeport, Connecticut community resilience building workshop summary of findings. The Nature Conservancy, New Haven, https://www.conservationgateway.org/ConservationPractices/Marine/crr/library/Documents/Bridgeport%20Climate%20Preparedness%20Workshops%20Report%20August%202012.pdf. Accessed 11 Nov 2014

Whelchel AW, Ryan A (2014) Town of Madison, Connecticut community resilience building workshop summary of findings. The Nature Conservancy Coastal Resilience Program – Publication #14-1, New Haven https://www.conservationgateway.org/ConservationPractices/Marine/crr/library/Documents/Madison%20Hazards%20and%20Community%20Resilience%20Workshop%20Summary%20of%20Findings%20Final.pdf. Assessed 20 June 2015.

World Risk Report (2012) http://www.worldriskreport.com/uploads/media/WRR_2012_en_online.pdf. Accessed 2 Oct 2014

World Bank (2016) Managing coasts with natural solutions: guidelines for measuring and valuing the coastal protection services of mangroves and coral reefs. In: Beck MW, Lange G-M (eds) Wealth Accounting and the Valuation of Ecosystem Services Partnership (WAVES). World Bank, Washington, DC, http://documents.worldbank.org/curated/en/2016/02/25930035/man aging-coasts-natural-solutions-guidelines-measuring-valuing-coastal-protection-services-man groves-coral-reefs. Accessed 15 Jun 2016

WWAP (2009) The United Nations world water development report 3: water in a changing world. UNESCO/Earthscan, Paris/London, http://webworld.unesco.org/water/wwap/wwdr/wwdr3/ pdf/WWDR3_Water_in_a_Changing_World.pdf. Accessed 10 Oct 2014

Chapter 7
The Use of Geo-information in Eco-DRR: From Mapping to Decision Support

Bart Krol, Luc Boerboom, Joan Looijen, and Cees van Westen

Abstract Ecosystem services can play an important role as measures for disaster risk reduction. At the same time it is important to find out where and how ecosystem-based disaster risk reduction really can make a difference. If we want to find out what will be the effect of alternative risk reduction measures, how ecosystem services can play a role in this context, and how they compare with other types of interventions, then there is a clear role for geo-information. Geographical information, such as obtained from spatial-temporal simulation modelling and spatial multi-criteria evaluation, is used for analyzing and monitoring what could be the effect of alternative development scenarios on the exposure to natural hazards, or of different combinations of engineered, ecosystem-based and other non-structural risk reduction measures. This helps to set management priorities and propose actions for risk reduction and risk-informed spatial planning. With the help of a spatial decision support system, the effect of risk reduction alternatives and their effect on risk reduction – now and in the future – can be analyzed and compared. This can support the selection of 'best' alternatives. The recently developed RiskChanges is presented, which is a web-based, open-source spatial decision support tool for the analysis of changing risk to natural hazards. It is envisaged that the use of the RiskChanges will support the provision of relevant geo-information about risk and changes in risk, and thus provides input for structured risk reduction-, disaster response-, and spatial development-planning.

Keywords GIS • Planning • Spatial Decision Support System (SDSS) • Simulation modelling • Uncertainty

B. Krol (✉) • L. Boerboom • J. Looijen • C. van Westen
Faculty of Geo-Information Science and Earth Observation (ITC), University of Twente, Enschede, The Netherlands
e-mail: bart.krol@utwente.nl

© Springer International Publishing Switzerland 2016
F.G. Renaud et al. (eds.), *Ecosystem-Based Disaster Risk Reduction and Adaptation in Practice*, Advances in Natural and Technological Hazards Research 42, DOI 10.1007/978-3-319-43633-3_7

161

7.1   Introduction

Ecosystem services can play an important role as measures for disaster risk reduction. At the same time it is important to find out where and how ecosystem-based disaster risk reduction (Eco-DRR) really can make a difference. For example, insight is needed about what will be the risk-reducing effect of ecological interventions, and how ecosystem services compare with other risk reduction alternatives. This requires access to geo-information and structural development of capacity to both generate and use this geo-information.

Geo-information helps planners and decision makers to be informed about which areas are exposed to hazards, where this exposure increases and where greater risk may develop, or where the occurrence of multiple-hazards will further increase vulnerability of local communities and corresponding disaster risk. Altan et al. (2010) provide a useful demonstration of the possibilities of using geo-information technology in disaster risk management, targeted at decision-makers and disaster management practitioners. Using geo-information (e.g. spatial-temporal simulation models), it is possible to analyze and monitor what could be the effect of alternative development scenarios on the exposure to natural hazards (see for example Sliuzas et al. 2013a). With the help of geo-information, different combinations of engineered, ecosystem-based and other non-structural risk reduction measures can be compared. This helps risk managers to set management priorities and propose actions for risk reduction. Especially in areas where land is scarce, it is important to have adequate spatial and temporal information available, to support the analysis of costs and benefits of ecosystem services as measures for disaster risk reduction.

The parties (stakeholders) involved in planning, design and decision-making in disaster risk reduction typically have different views and priorities (Peters Guarin et al. 2012). Modelers and planners, for example, may have different perspectives on the uncertainty of hazards and risk and judge risk reduction alternatives and trade-offs differently. An effective Decision Support System (DSS) facilitates collaborative decision-making by different groups of stakeholders, dealing with the different perspectives they may have. An example is the Planning Kit DSS that was developed to support the design process of the 'Room for the River programme' in The Netherlands (Kors 2004; de Bruijn 2007). A Spatial Decision Support System (SDSS), in addition, facilitates the use of geographical data and models that use these data; it also includes models for the structuring of spatial decision making processes and methods for decision support, such as spatial multi-criteria evaluation (see for example Sugumaran and Degroote 2010).

In this chapter, we present the use of geo-information in Eco-DRR to analyze how and where ecosystem functions can be beneficial for risk reduction and how these may change over time; to find out what are the trade-offs of different ecosystem services; to carry out risk assessments and to share risk information with stakeholders; to compare the risk reduction effect of different intervention alternatives using simulation modelling; and to facilitate collaborative decision-making. This chapter also introduces RiskChanges: a web-based SDSS for the analysis of changing risk to natural hazards. Its aim is to support the evaluation of the effect of different risk reduction alternatives (involving both structural and non-structural, including ecosystem-based measures) on reducing disaster risk, both now and in the future (see also Whelchel et al., Chap. 6; Bayani and Barthélemy, Chap. 10).

7.2   Geo-information and Ecosystem Services

Depending on their biophysical properties, ecosystems have the potential to supply services. Healthy and well-managed ecosystems help communities to cope with the impacts of more frequent and extreme hazard events and therefore adapt to climate change (Renaud et al. 2013). Ecosystem services that aim to reduce disaster risk are mainly regulating services. All over the world, particularly ecosystems' regulating services are declining, often due to an increase in the use of provisioning ecosystem services, to produce more food, fuel and other products (Millennium Ecosystem Assessment 2005). Increase in agricultural production systems, for example, often comes at the cost of biodiversity and/or other regulating services (for instance those that help control erosion). This causes an imbalance in available ecosystem services that will only increase human exposure to extreme events.

Both the supply and demand for ecosystem services are spatially explicit and may differ from place to place. The production of ecosystem services, for example, is often expressed as a function of land use, climate and environmental variation (Maes et al. 2011). The analysis of ecosystem services and their benefits for different users involves their valuation to reflect human attitudes and preferences. For the assessment of trade-offs between different ecosystem services, proper spatial indicators are required for ecosystem functions and services (Crossman et al. 2013; de Groot et al. 2010). Such an assessment requires the development of geographical information in maps and models: to quantify the benefits received from ecosystem services, to estimate where they are produced, to quantify changes in ecosystems and the services they (can) provide over time, and also to describe the production of ecosystem services as a function of land use, climate and environmental variation. For example, to reduce the risk of flooding, proxies to estimate water retention capacities are calculated as a function of vegetation cover and soil type. A model-based approach of mapping ecosystem services will result in a better exploration of risk reducing scenarios and policy alternatives. Different value maps of ecosystem services can be produced and combined using weighted overlaying techniques, depending on the priorities of the planners and stakeholders involved.

7.2.1 Example from the Netherlands

A good and by now well-known example is the national Room for the River programme in the Netherlands (see at: www.ruimtevoorderivier.nl). This integrated flood risk management programme represents a governmental response to coping with higher water levels in the Dutch rivers without simply raising and strengthening river dikes. An approach of 'working with nature' (see also Meyer 2009; De Vriend and Van Konigsveld 2012) instead of fighting against it has resulted in 34 different flood risk reduction projects spread over the Netherlands, most of which have been finalized in 2015. Two of these projects are introduced in Box 7.1 and Box 7.2. Selected ecosystem-based flood risk reducing measures, such as the restoration of floodplains and wetlands, have a double function in many of these projects: they also enhance the re-establishment of natural values (e.g. the presence of given plant- and animal-species, scenic beauty) and promote the development of recreational activities.

A relevant decision support tool in the Room for the River programme is the Box of Blocks software. This is a combined hydraulic model and scenario planning tool that calculates the hydraulic effects of combinations of structural (e.g. river channel widening) and non-structural (e.g. wetland development) measures for flood risk reduction and thus supports the design and selection of measures (Schut et al. 2010; Dutch Ministry of Water Management, Transport and Public Works 2013). This Box of Blocks tool includes 600 different measures with potential for water level reduction. It was made available to the stakeholders involved in the different projects, who have used it to evaluate and visualize the effectiveness and interdependencies of their proposed measures to reduce water levels. This tool also displays the costs of each measure and the effects on agriculture production and natural values, amongst others. It has also facilitated the dialogue and cooperation between policymakers from different regions, by demonstrating the interdependencies of river management at the national level (Schut et al. 2010).

Box 7.1 River Dike Re-location and Construction of a Flood By-pass at Nijmegen, The Netherlands
Source: www.ruimtevoordewaal.nl; www.infranea.eu

Room for the Waal River at Nijmegen is one of the projects in the Dutch Room for the River programme. Its aim is to protect the city of Nijmegen and its surroundings from future floods and at the same time increase the spatial quality of the urban environment in the project area. The Waal River forms a bottleneck for water discharge in a sharp river bend near Nijmegen. This has recently resulted in high water levels, and caused severe flooding in 1993 and 1995. To protect the inhabitants of the city against floodwater, an existing dike is re-located 350 m inland. In addition, an ancillary river channel is constructed in the river's flood plain, also including the construction of three

---

(continued)

Box 7.1 (continued)
bridges and a new quay. This will create an island in the Waal River and a unique urban river park with many additional possibilities for recreation, cultural activities, and re-establishing of natural values.

For the planning, coordination and modelling of this project a so-called Building Information Modelling (BIM) system is used. This BIM provides three-dimensional (3D) representations of the physical and functional aspects of the planned infrastructural designs considered in the project. These are combined with ecological and water management information available in a Geographical Information System (GIS). In this way, the possible effect of proposed interventions can be modelled and potential conflicts – between design components but also between stakeholder interests – can be identified and discussed. This approach of geographical information sharing and collaborative decision-making supports the different parties involved in designing and managing this complex project (Fig. 7.1).

Fig. 7.1 3D-impression (downstream view) of the expected results of the Room for the River project at Nijmegen. River bend in Waal river (*left side*) and construction of an ancillary river channel (*right sight*) create a new island for recreation and re-establishing natural values (Image: Room for the Waal Nijmegen, (www.ruimtevoordewaal.nl), used with permission).

Within the boundary conditions for lowering of river water levels and connected flood risk reduction set by the national Room for the River programme, it is left to the regional and local stakeholders in the respective projects to negotiate and decide for a mix of structural and non-structural – including ecosystem-based – flood protection measures. This decentralized approach also holds for the selection of additional tools and techniques to support the design, planning and management of projects.

Box 7.2 Flood By-Pass Development Near the Dutch Town Kampen
Source: www.ruimtevoorderivierijsseldelta.nl

A combination of increased water discharge (rainfall-induced) by the IJssel River and expected sea level rise make the Dutch towns of Kampen and Zwolle and their hinterland increasingly more vulnerable to the effects of flooding. To increase the resilience to climate change and at the same time improve the spatial quality of the area, a new flood channel, the Reeve Deep by-pass, will be constructed in the IJssel river delta. Apart from flood protection measures, there are several other spatial issues to be considered in the development of an integrated flood protection plan, including: attention to nature management (the development of a new wetland area, in particular), interests of the agricultural sector, options for recreation, the development of new housing areas, and the presence of a railway and several highways. For the spatial design of the flood by-pass a Digital Elevation Model (DEM) was used. Both the average and expected extreme water levels are projected on this model. This helps to obtain a better geographical understanding of the delta landscape and the potential wetland areas. Taking into account the hydraulic requirements set by the national Room for the River programme, this has ultimately led to the development of an integrated spatial plan for the IJssel delta (Fig. 7.2).

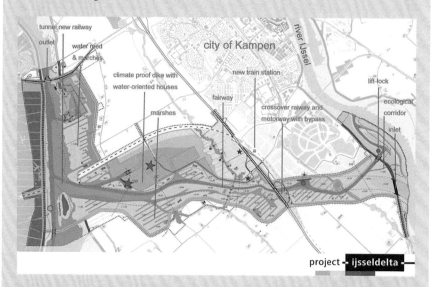

Fig. 7.2 Overview of the spatial development plan for the IJssel delta near Kampen (IJssel river, new flood channel and other water bodies in *blue colours*; wetlands and other vegetation cover in *green colours*) (Image, courtesy of A. Otten, Province of Overijssel)

7.3 Geo-information and Risk Assessment

Disaster risk can be defined as the probability for harmful consequences or losses, in a given area and over a period of time (Birkmann et al. 2013). This makes risk a geographical problem, with both spatial and temporal aspects playing a role. The assessment of risk requires a geographical analysis, because its different components – i.e. the assessment of natural hazards, of elements at risk and their vulnerability – both differ and vary in space and time (van Westen 2010). This dynamic character of the risk concept makes the collection of geographical data – of past and present hazard events, of elements at risk and their vulnerability – and their spatial-temporal analysis often a complex task. This is even more so if multiple hazards are considered, for example hazards sharing the same triggering event or occurring as a cascade of hazard events (van Westen 2013).

At the same time geographical data, GIS and remote sensing technology are to date widely applied for the analysis of natural hazards and disaster risk. In the form of GIS-based risk maps, risk related information is supplied in many countries to mandated agencies and authorities. Increasingly, also the general public is informed about risk and changes in risk in their living environment (Basta et al. 2007). An example is the systematic delivery of geographical risk information in The Netherlands using the on-line risk information portal: 'risicokaart' (see at www.risicokaart.nl). A number of relevant examples of the application of geo-information in a disaster risk context are presented elsewhere in this book.

7.3.1 Qualitative and Quantitative Approaches

There is an increasing need for quantitative forms of risk assessment that express risk as probability of a given level of loss together with the associated uncertainties (see for example Corominas et al. 2014; Crozier and Glade 2005). Quantitative methods are expected to allow for an objective and reproducible way of risk assessment also in a multi-hazard risk context (Kappes et al. 2012). At the same time, however, quantitative risk assessment methods mostly focus on physical vulnerability aspects, whereas qualitative risk assessment approaches tend to also incorporate other (i.e. economic, social, ecological, institutional, cultural) vulnerability aspects (van Westen 2013).

Qualitative and semi-quantitative approaches to risk assessment are often considered when the availability of (numerical) geographical data is limited. This kind of risk assessment is also considered as an initial screening process to identify natural hazards and risk (van Westen 2013). An international example is the annual World Risk Report (2014) that uses a risk index approach to rank countries worldwide based on their potential disaster risk. In a multi-criteria type of analysis using 28 different indicators influencing risk, a so-called World Risk Index value was computed for each country considered. A national level example applying the risk index approach is the development of a landslide risk index map for Cuba (Castellanos Abella and van Westen 2007). The adopted approach involves the use of multiple spatial indicators as input for a Spatial Multi-Criteria Evaluation (SMCE). In the absence of reliable landslide inventory data, hazard indicator maps are used representing both conditional factors (e.g. slope, geology, land cover) and triggering factors (e.g. earthquakes, rainfall). Geographical data about population distribution, transportation and housing, amongst others are used to represent physical aspects of vulnerability. In fact, Castellanos Abella and Van Westen (2007) label their approach as semi-quantitative because of the use of weighing certain indicators to allow for better representation of the spatial variability present in the available data. The resulting national risk index map of Cuba provides geographical information that supports decision makers in prioritizing resources for further risk assessments at provincial, municipal and local levels.

A risk assessment using SMCE can also be carried out at the sub-national level, for example for a province, district or municipality. As a qualitative approach, SMCE can be labelled as subjective and mainly useful if data are lacking for a more quantitative risk analysis. But it can offer more than just that. Applying SMCE, it is possible to use expert knowledge – from engineers, economists, authorities, local communities, amongst others – and to include 'soft' information like perception and preferences in a risk assessment (Alkema and Boerboom 2012). The active involvement of these multiple stakeholders – frequently with initially conflicting views and perceptions – in an SMCE procedure facilitates collaborative decision-making processes (Alkema and Boerboom 2012).

Geographical information about risk and also about expected changes in risk over time can be used to evaluate and compare the expected effects of different strategies for risk reduction. With the help of geographical data, modelling techniques and GIS-software tools, so-called 'what if' type of analyses can be carried out and alternative future scenarios can be generated and compared to support decision-making processes (Longley et al. 2005).

7.3.2   Risk Assessment Tools

Two well-known examples of a combined methodology and open-source software tool for quantitative, probabilistic multi-hazard risk assessment are HAZUS-MH and CAPRA. HAZUS-MH (www.fema.gov/hazus) was developed by the Federal Emergency Management Agency in the USA. CAPRA, the Central American Probabilistic Risk Assessment Program (www.ecapra.org) was initiated by the Center for Coordination of Natural Disaster Prevention in Central America (CEPREDENAC), the United Nations International Strategy for Disaster Reduction (UNISDR) and the World Bank.

To support the building of capacity in disaster risk management in national and local governments, The World Bank's Global Facility for Disaster Reduction and Recovery (GFDRR) has recently reviewed 31 open-source and open-access software packages for the quantitative analysis of natural hazards and risks (GFDRR 2014). Increasingly, free and open-source GIS software tools are also extended with new functionalities that are specifically relevant in a hazard and risk analysis context. An example is the functionality for SMCE in the ILWIS GIS software package (http://52north.org/communities/ilwis/ilwis-open). Another example is the QGIS software package (www.qgis.org) with its INASAFE (http://insafe.org) plugin that is used to generate hazard impact scenarios in support of disaster preparedness and response planning. A new initiative is the development of RiskChanges, a web-based, open-source SDSS for analyzing changing hydro-meteorological risk (van Westen et al. 2014). RiskChanges is described in more detail later in this chapter.

7.3.3   Spatial-Temporal Simulation Modelling

Predictive modelling is increasingly used for analyzing and monitoring what could be the effect of alternative development scenarios on the exposure to natural hazards, or of different combinations of engineered, ecosystem-based and other non-structural risk reduction measures in space and time. In this manner, possible trends or future situations can be considered, together with alternative policy options and interventions for risk reduction. In Box 7.3 an example of flood simulation modelling in Kampala, Uganda is presented, where the development and application of a scenario-based urbanization and flood modelling approach has created an information environment that facilitates the development of an integrated flood management strategy (Sliuzas et al. 2013a).

Unfortunately, in practice the link between the modelling and prediction of (hazardous) natural processes and corresponding risks on the one hand and their management and governance on the other hand is still rather weak (Greiving et al. 2014). Scientific developments in hazard and risk assessment and the needs and demands of decision-makers and end-users of risk information are still not well connected (van Westen 2013). Additional challenges are posed by the often-existing uncertainty in space and time about the possible roles and effects of urban growth processes, land use trends, climate change, and other future scenarios. A decision support mechanism can bring different stakeholders (representing different disciplines, sectors, etc.) together more easily in the assessment of risk and the search for effective risk management strategies. This interaction between stakeholders is, for example, an integral part of joint planning of flood risk reduction projects in the Room for the River programme in The Netherlands (see for example: Roth and Winnubst 2014).

Box 7.3 Scenario-Based Modelling of Current and Future Flood Risk in Kampala, Uganda
Source: Sliuzas et al. (2013a, b).

Accelerated urban growth and increasing rainfall-induced flood problems have motivated Kampala – the capital city of Uganda – to join UN-HABITAT's Cities and Climate Change Initiative (CCCI). As part of CCCI's Integrated Flood Management Project in Kampala, researchers and students from the University of Twente, Makerere University and a German consultant have analyzed the current and possible future flood risk situation in Lubigi catchment inside Kampala. In this catchment (approx. area = 28 km^2) a system of lined channels connects populated hills to a system of central drains in the catchment's main valley – increasingly populated as well – that subsequently drain into a natural wetland system further downstream. Residents and business owners have developed a number of mechanisms to cope with the effect of flooding, but the frequent rainfall-induced floods are a nuisance and also pose a risk with significant costs, both economic as well as health related.

The open-source spatial-temporal modelling environment OpenLISEM (http://blogs.itc.nl/lisem) was used to simulate a 10-year rainstorm event of 1000 mm in a day, considering a series of possible future scenarios, including:

- Maintaining the current situation of unimproved drainage and unregulated urban development, i.e. a scenario of 'no change';
- Physical improvement of the drainage system with structural interventions in the main drain and culverts in secondary channels, i.e. a 'hard engineering' scenario;
- A 'green engineering' scenario involving a number of so called Sustainable Drainage System (SuDS, see also Woods-Ballard et al. 2007) options for improving the functioning of drainage channels, using a mix of widening and deepening of drains, creation of grassed waterways, identification of areas for temporary water storage;
- A 'planning only' scenario consisting of urban development control, including identification of flood hazard zones and restriction of housing.

These scenarios were further refined considering an urban growth projection for 2020, using annual growth rates of 4.2% ('trend') and 6.5% ('high').

Based on the predictive flood modelling using OpenLISEM the 'no change' scenario shows severe flooding (up to 2 m and for more than 24 h) in areas along the primary drainage channel; in large areas flood water stays up to 24 h until the water level decreases to manageable levels. The 'hard engineering' scenario is expected to reduce the extent and duration of flooding but will not eliminate it. Using the modelling results, the researchers

(continued)

Box 7.3 (continued)

also observed that improved culverts in the secondary channels reduced local flooding, but at the same time cause water to be delivered more rapidly to areas downstream and thus potentially increase flood problems elsewhere. Given the modelling results, it is expected that the increased water infiltration in the 'green engineering' scenario will also contribute to a reduction of the flood problem. The planning scenario shows the importance of controlling and regulating urban development for dealing with flood problems in the future. The scenario-based flood modelling has also resulted in the identification of a number of areas that face chronic flooding: flooding hot spots where urban development control and dedicated planning measures are especially important.

The results obtained by the research team show that for this Kampala case, urban growth and disregard of planning will have a stronger effect on flooding and flood related problems than any possible future climate change. In Lubigi catchment the best flood reduction effect is expected from a mix of 'hard engineering' measures in the central valley, 'green engineering' on the hills slopes, together with improved urban planning strategies and housing regulations (Fig. 7.3).

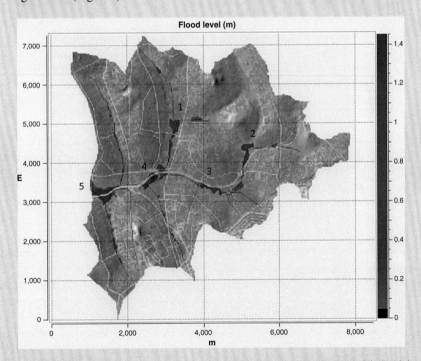

Fig. 7.3 Flooding hotspots [*blue colors*: recurrent flood water depth in meters] in Lubigi catchment based on multiple scenario analyses (Sliuzas et al. 2013a; used with permission)

7.4 Spatial Decision Support for Eco-DRR

An ideal DSS for Eco-DRR would allow for exploring different options and arriving at a decision, for example for a particular intervention measure. The essence of a decision was very well captured by Von Foerster (1992 p.14) when he stated that, "Only those questions that are in principle undecidable, we can decide." In other words: only if we feel that there is a trade-off between our options, because no single option is the best, are we making a decision. Since this sense of trade-off will remain, we remain undecided. Therefore, a DSS should not just describe our physical or societal environment in tables and maps in the way that databases and information systems do. Nor should it describe the behavior of our environment the way models – such as ecosystem models, rainfall runoff models or landslide risk assessment models – do. A DSS should capture the 'undecidable', i.e. the trade-offs (Ackoff 1981) of often nested, chained, and poorly structured decisions. DSSs – as a class of software tools – can support decision makers both when judgment about trade-offs is important in the decision making process and when the human information processing capacity limits the decision making process. When such DSS address spatial decision problems using geographical data we speak of Spatial Decision Support Systems (SDSS), which help us decide between spatial alternatives (Rauscher 1995).

7.4.1 Dealing with Uncertainty: Modelers' and Decision Makers' Perspectives

In the context of disaster risk reduction, DSSs not only support the judgment about trade-offs, but also about the uncertainty related to hazards and risk. Even if uncertainty is minimized by the quantification of risk – and hence becomes, by metaphor, a "controllable island in the sea of uncertainty" (Nowotny et al. 2001 p.14) – DSSs still need to support decision making in a sea of uncertainty. This uncertainty can be further distilled to (i) uncertainty due to variability, i.e. stochastic or ontological uncertainty, and to (ii) uncertainty due to limited knowledge, i.e. epistemic uncertainty (van Asselt and Rotmans 2002). However, this is a modeler's perspective on uncertainty. Ambiguity is an additional source of uncertainty (Brugnach et al. 2008), which is defined here as the "existence of two or more equally plausible interpretation possibilities" (Dewulf et al. 2005 p. 115), as is often resulting from clearly different (stakeholder) perceptions about what is at stake (Dewulf et al. 2005). These three concepts of uncertainty and risk are illustrated in Box 7.4 using a recent study by Petr (2014) about changes in the provision of forest ecosystem services in British national forest estates, under the influence of climate change-induced drought effects on stands of tree species.

Box 7.4 Scenarios for Uncertain Climate Change and Yield Decline in British Forests
Source: Petr (2014)

In 2009 a climate change projection for the UK was released (Murphy et al. 2009). It was the first probabilistic projection for the UK, considering two spatial resolutions (25 and 5 km) and temporal resolutions over 30 year periods, starting from the 2020s (2010–2039) until the 2080s (2070–2099).

For the calculation of total probable risk of tree yield change, spatial and temporal resolutions, both probabilistic data of moisture deficit and drought vulnerability response curves for forest stand yields of three tree species (i.e. *Sitka spruce*, *Scots pine* and *Pedunculate oak*) were used. Total probable risk is expressed as the sum of all probable yield changes of a tree species in each of the spatial and temporal ranges (Petr et al. 2014).

Given the uncertainty of future climate change scenarios, three scenarios by the Intergovernmental Panel on Climate Change (IPCC 2000) were used to prepare risk maps of Britain for each decade. The resulting tree yield changes were translated to predict the loss of ecosystem service provisions until the end of the 21st century (especially production and carbon sequestration) and with predicted losses of up to 50% for some species in certain localities (Petr et al. 2014). If ecosystem service functions continue to decline during the 21st century, existing adaptation options – such as the tree species currently selected or area expansion of tree species – will reach their expiry date. Beyond this point of expiry, policymakers will need to shift to new policy options to achieve a required adaptation. The described approach follows a method of 'dynamic adaptation policy pathways', introduced by Haasnoot et al. (2013) for decision making in a context of uncertain changes.

When forest planners in Scotland were exposed to the policy pathways options and the possibility to assess expiry dates of certain species choices, their framing of adaptation was observed to diverge (Petr et al. 2015). For instance, forest planners in two districts decided as a group that expiry dates for keeping spruce, which is the dominant tree species in all districts, occurred much later, which varied from their individual decisions. This gives reason to suspect that individual planners frame the role of climate change in species choice differently in terms of urgency, and for some reason seem to ignore this ambiguity in a collective decision, i.e. when they decide together.

7.4.2   An SDSS That Addresses Risk Uncertainty

Using a spatial decision support system for Eco-DRR, we first of all expect to be able to assess disaster risk. But we also expect the availability of tools to make judgments about trade-offs between different spatial and temporal criteria regarding alternative ecosystem services and other possible interventions that can reduce disaster risk. Finally, we expect to be able to assess uncertainty, both from a modeler's and decision maker's point of view. Stochastic uncertainty is typically addressed using methods for sensitivity analysis. Epistemic uncertainty and ambiguity can be addressed using different models or different decision problem formulations. Scenario development plays a crucial role, both for varying exogenous variables that could affect intervention options (Engelen 2000) in different ways (epistemic uncertainty) and to express ambiguity.

These characteristics have also been considered in the development of the RiskChanges SDSS that is presented in more detail in the next section. RiskChanges allows for the assessment of risk while assuming multiple scenarios (e.g. population growth, development policies) that can affect different intervention alternatives at different moments in time, while SMCE is applied for the assessment of trade-offs of different interventions and scenarios. Stochastic uncertainty is addressed through the probabilistic nature of the hazards. Since RiskChanges is an open system that can be used for any disaster risk assessment, the definition of alternatives and their indicators allows for dealing with epistemic uncertainty and to some extent variation in the framing of problems or risks and alternative solutions.

7.5 RiskChanges: A Web-Based SDSS for Analyzing Changing Hydro-Meteorological Risk

RiskChanges is a new SDSS that enables the geographical analysis of the effect of risk reduction planning alternatives on the reduction of current and future risk. It supports decision makers in selecting 'best alternatives' for intervention. The RiskChanges SDSS is developed in the context of two EU-funded research projects: the INCREO project (www.increo-fp7.eu) and the CHANGES project (www.changes-itn.eu). This overview of RiskChanges is drawn from the presentation and description of the system by van Westen et al. (2014).

RiskChanges is targeted at three main groups of stakeholders involved in risk assessments. The envisaged end-users of RiskChanges include agencies involved in planning of risk reduction measures, and that also have the capacity to analyze and visualize geographical data at the municipal level. Examples are civil protection organizations that develop plans for disaster response; expert organizations involved in the technical design of structural measures (e.g. dams, dikes) and/or the development of non-structural and ecosystem-based risk reduction measures; organizations with a development planning mandate. A second group of stakeholders involves information providing organizations that are responsible for the production, the provision and monitoring of hazard-related information (e.g. flood scenario maps). A third main stakeholder group involves organizations that typically provide information (e.g. cadastral, transportation) about elements at risk.

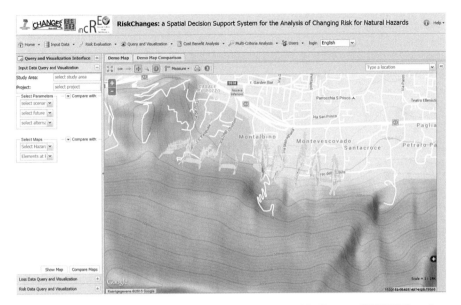

Fig. 7.4 Opening screen of the web-based RiskChanges SDSS (Source: CHANGES project website)

RiskChanges is a web-based system, designed based on open-source software and following open standards. RiskChanges is available online in the CHANGES project website. Its opening screen is shown in Fig. 7.4. It is possible to use RiskChanges for multi-hazard risk assessment at different spatial-temporal resolutions, in different countries and within different legal settings.

7.5.1 Different Risk Assessment Workflows

RiskChanges can be used for four different types of risk assessment workflows:

1. *Analyzing the current level of risk.* Using geographical data about natural hazards, elements at risk and their vulnerability, it is possible to perform an evaluation of current (multi-) hazard risk level.
2. *Analyzing 'best' alternatives for risk reduction.* In this workflow, stakeholders first identify a number of risk reduction alternatives – structural, non-structural, ecosystem-based – and request expert organizations to provide them with updated hazard maps and information about elements at risk and their vulnerability reflecting the consequences of these alternatives. The new risk level is analyzed and compared with the current level of risk in order to estimate levels of risk reduction. A subsequent evaluation of costs and benefits (in financial terms and/or in terms of other constraints) per alternative helps to make a

selection of a 'best' risk reduction alternative. Note that this workflow can also be used in a case of 'best' disaster response planning, or as the basis for early warning system design.

3. *Evaluating the possible consequences on risk of different future scenarios.* In this workflow, the effect of possible future risk scenarios of population growth, land use change, climate change or other trends that cannot be controlled by the (local) planning organizations involved in the risk assessment are analyzed.

4. *Evaluating how different risk reduction alternatives can lead to risk reduction under different future scenarios.* This is the workflow in which current risk, the potential effect of risk reduction alternatives, and the different future scenarios come together.

Central to RiskChanges is risk assessment. The RiskChanges system itself does not include facilities to generate natural hazard maps and maps of elements at risk; relevant information is produced outside the system. Hazard maps and information about elements at risk can be uploaded using the system's *Data input module.* After data preparation, they are fed as input data into the system's *Risk Evaluation module.* A spatial risk assessment can be carried out ranging from simple exposure analysis to quantitative analysis resulting into risk curves. After a loss calculation, users can opt for different types of risk assessments, for example hazard-specific or specific elements at risk, concentrating on economic risk or population risk, for identified risk reduction alternatives and future scenarios. In a *Cost-benefit analysis module* users can analyze the costs of identified risk reduction alternatives, also taking into account how costs and benefits may change in time (for example depending on future scenarios). A *Multi-Criteria Analysis module* supports the users in determining the most optimal risk reduction alternative using a spatial multi-criteria evaluation approach. Thus, the pros and cons of different engineering-oriented, ecosystem-based and other non-structural risk reduction alternatives can be critically evaluated and contrasted. The results of risk assessment are presented using RiskChanges' *Visualization module,* as maps but also in the form of risk curves, tables and graphs. This also includes tools for the visualization of temporal changes.

7.6   A Role for Geo-information in Eco-DRR

Over the years the use of geo-information in disaster risk reduction has moved from a mere focus on the generation of hazard and risk maps by specialists for specialists to the use of geo-information in processes of collaborative decision-making and planning of risk reduction strategies. As is also shown in the examples used in this chapter, in practice often a mix of both structural and non-structural measures, of engineering and ecosystem-based interventions, are considered as part of strategies to cope with expected future risk scenarios.

If we want to find out what will be the effect of alternative risk reduction measures, how ecosystem services can play a role in this context, and how they compare to other types of interventions, then there is a clear role for geo-information in the field of Eco-DRR. Moreover, the use of geo-information also facilitates the communication between different stakeholder groups, including hazard and risk specialists, land users, development planners, decision makers, local communities and the public in general.

Using SDSSs, it is not only possible to assess disaster risk, but also to make judgments about the trade-offs of different ecosystem services and other possible risk reducing interventions. It is envisaged that the use of the RiskChanges SDSS will support the provision of relevant geographical information about risk and changes in risk, and thus provide input for structured risk reduction planning, disaster response planning, and spatial development planning.

Of course a number of challenges to SDSS implementation in the risk reduction context remain. They concern, for example, data availability, the proper linkage of different components of an SDSS, user guidance and the presentation of outputs of a decision making process. In addition, an important implementation-related challenge is about participatory development: how to engage users in the development of a decision support mechanism. If these challenges can be properly addressed, RiskChanges can play an important role in supporting the selection of 'best' alternatives for multi-hazard risk reduction, under different future scenarios, and including Eco-DRR options.

Acknowledgement The authors would like to thank the two anonymous reviewers for their constructive criticism and comments, from which this chapter has benefited.

References

Ackoff RL (1981) The art and science of mess management. Interfaces 11(1):20–26

Alkema D, Boerboom L (coordinators) (2012) Development and testing of spatial multi-criteria evaluation for selected case sites. Deliverable 5.6. SafeLand Project – 7th Framework Programme Cooperation Theme 6 Environment (including climate change) Sub-Activity 6.1.3 Natural Hazards

Altan O, Backhaus R, Boccardo P, Zlatanova S (eds) (2010) Geoinformation for disaster risk management – examples and best practices. JBGIS and UNOOSA/UN-SPIDER, Copenhagen, Available via: http://www.un-spider.org/about/portfolio/publications/jbgis-unoosa-booklet

Basta C, Neuvel JMM, Zlatanova S, Ale B (2007) Risk-maps informing land-use planning processes; a survey on the Netherlands and the United Kingdom recent developments. J Hazard Mater 145:241–249

Birkmann J, Cardona OD, Carreño ML et al (2013) Framing vulnerability, risk and societal response: the MOVE framework. Nat Hazards 67:193–211

Brugnach M, Dewulf A, Pahl-Wostl C, Taillieu T (2008) Toward a relational concept of uncertainty: about knowing too little, knowing too differently, and accepting not to know. Ecol Soc 13(2):30

Castellanos Abella EA, Van Westen CJ (2007) Generation of a landslide risk index map for Cuba using spatial multi-criteria evaluation. Landslides 4(4):311–325

Corominas J, van Westen CJ, Frattini P et al (2014) Recommendations for the quantitative analysis of landslide risk: open access. Bull Eng Geol Environ IAEG 73(2):209–263

Crossman ND, Burkhard B, Nedkov S et al (2013) A blueprint for mapping and modelling ecosystem services. Ecosyst Serv 4:4–14

Crozier MJ, Glade T (2005) Landslide hazard and risk: issues, concepts, approaches. In: Glade T, Anderson MA, Crozier MJ (eds) Landslide hazard and risk. Wiley, Chichester

de Bruijn K (2007) Survey of existing DSS Tools in the Netherlands. Deltares. Final version of report for the FLOODsite project (project website: www.floodsite.net)

de Groot RS, Alkemade R, Braat L et al (2010) Challenges in integrating the concept of ecosystem services and values in landscape planning, management and decision making. Ecol Complex 7 (3):260–272

De Vriend HJ, Van Konigsveld M (2012) Building with nature: thinking, acting and interacting differently. EcoShape, Building with Nature, Dordrecht, The Netherlands

Dewulf A, Craps M, Bouwen R et al (2005) Integrated management of natural resources: dealing with ambiguous issues, multiple actors and diverging frames. Water Sci Technol 52 (6):115–124

Dutch Ministry of Water Management, Transport and Public Works (2013) Tailor made collaboration. A clever combination of process and content. Ando bv, The Hague, pp 60

Engelen G (2000) The WadBos policy support system: information technology to bridge knowledge and choice. Technical paper prepared for the National Institute for coastal and Marine Management/RIKZ. The Hague, The Netherlands

GFDRR (2014) Understanding risk – review of Open Source and Open Access Software Packages. Available to Quantify Risk from Natural Hazards. Washington, Global Facility for Disaster Risk Reduction, World Bank Group, 67 p

Greiving S, van Westen CJ, Corominas J et al (2014) Introduction: the components of risk governance. In: van Asch T et al. (eds) Mountain risks: from prediction to management and governance, advances in Natural and Technological Hazards Research 34, Springer Science + Business Media, Dordrecht 2014, pp 1–27

Haasnoot M, Kwakkel JH, Walker WE, ter Maat J (2013) Dynamic adaptive policy pathways: a method for crafting robust decisions for a deeply uncertain world. Glob Environ Chang 23 (2):485–498

IPCC (2000) Special report on emissions scenarios. Cambridge University Press, Cambridge, 599 pp

Kappes MS, Keiler M, von Elverfelft K, Glade T (2012) Challenges of analyzing multi-hazard risk: a review. Nat Hazards 64:1925–1958

Kors A (2004) The DSS 'Planning Kit' and its application in the Spankracht study. Lowland Technol Int 6(2):67–73

Longley PA, Goodchild MF, Maguire DJ, Rhind DW (2005) Geographical information systems: principles, techniques, management and applications. Wiley, Hoboken

Maes J, Paracchini ML, Zulian G (2011) A European assessment of the provision of ecosystem services: towards an atlas of ecosystem services. European Commission, Joint Research Centre. Publications Office of the European Union, pp 82

Meyer HN (2009) Reinventing the Dutch delta: complexity and conflicts. Built Environ 35 (4):432–451

Millennium Ecosystem Assessment (2005) Ecosystems and human well-being: synthesis report. World Resources Institute, Island Press, Washington, DC

Murphy J, Sexton D, Jenkins G et al (2009) UK climate projections science report: climate change projections. Met Office Hadley Centre, Exeter

Nowotny H, Scott P, Gibbons M (2001) Re-thinking science: knowledge and the public in an age of uncertainty. Polity Press, Cambridge

Peters Guarin G, McCall MK, van Westen C (2012) Coping strategies and risk manageability: using participatory geographical information systems to represent local knowledge. Disasters 36(1):1–27

Petr M (2014) Climate change, uncertainty, and consequent risks: opportunities for forest management adaptation in Britain. Ph.D. dissertation. Enschede, University of Twente Faculty of
Geo-Information and Earth Observation (ITC), pp 171

Petr M, Boerboom LGJ, van der Veen A, Ray D (2014) Spatial and temporal drought risk
assessment of three major tree species in Britain using probabilistic climate change projections: open access. Clim Chang 124(4):791–803

Petr M, Boerboom LGJ, Ray D, van der Veen A (2015) New climate change information modifies
frames and decisions of decisions makers: an exploratory study in forest planning. Regional
Environmental Change (2015) in press (open access), 10 p

Rauscher HM (1995) Natural resource decision support: theory and practice. AI Appl 9(3):1–2

Renaud FG, Sudmeier-Rieux K, Estrella M (2013) Opportunities, challenges and future perspec-
tives for ecosystem based disaster risk reduction. In: Renaud FG, Sudmeier-Rieux K, Estrella
M (eds) The role of ecosystems in disaster risk reduction. United Nations University Press,
Tokyo

Roth D, Winnubst M (2014) Moving out or living in a mound? Jointly planning a Dutch flood
adaptation project. Land Use Policy 41:233–245

Schut M, Leeuwis C, van Paassen A (2010) Room for the river: room for research? The case of
depoldering the Noordwaard, the Netherlands. Sci Public Policy 37(8):611–627

Sliuzas R, Flacke J, Jetten V (2013a) Modelling urbanization and flooding in Kampala, Uganda.
In: Proceedings of the 14th N-AERUS/GISDECO conference, 12–14 September 2013,
Enschede, Netherlands, 16 p

Sliuzas RV, Lwasa S, Jetten VG et al (2013b) Searching for flood risk management strategies in
Kampala. In: Planning for resilient cities and regions: proceedings of AESOP-ACSP joint
congress, 15–19 July 2013, Dublin, Ireland, 10 p

Sugumaran R, Degroote J (2010) Spatial decision support systems: principles and practices. CRC
Press, Boca Raton

van Asselt MBA, Rotmans J (2002) Uncertainty in integrated assessment modelling – from
positivism to Pluralism. Clim Chang 54(1-2):75–105

van Westen CJ (2010) GIS for the assessment of risk from geomorphological hazards. In:
Alcantara-Ayala I, Goudie A (eds) Geomorphological hazards and disaster prevention. Cambridge University Press, Cambridge, pp 205–219

van Westen CJ (2013) Remote Sensing and GIS for Natural Hazards Assessment and Disaster Risk
Management. In: Shroder JK, Bishop MP (eds) Treatise on geomorphology, pp 259–298.
Academic Press, San Diego (Remote Sensing and GIScience in Geomorphology vol 3)

van Westen CJ, Bakker WH, Andrejchenko V et al (2014) RiskChanges: a spatial decision support
system for analysing changing hydro-meteorological risk: extended abstract. Presented at:
analysis and management of changing risks for natural hazards: international conference,
18–19 November 2014, Padua, Italy, 13 p

von Foerster H (1992) Ethiks and second-order cybernetics. Cybern Hum Knowing 1:9–19

Woods-Ballard B, Kellagher R, Martin P et al (2007) The SUDS manual, CIRIA C697, RP697

World Risk Report (2014) Published by Buendnis Entwicklung Hilft (Alliance Development
Works) and UNU-EHS, p. 74, Available via www.worldriskreport.org

Chapter 8
Nature-Based Approaches in Coastal Flood Risk Management: Physical Restrictions and Engineering Challenges

Bregje K. van Wesenbeeck, Myra D. van der Meulen, Carla Pesch, Huib de Vriend, and Mindert B. de Vries

Abstract Ecosystem destruction not only incurs large costs for restoration but also increases hydraulic forces on existing flood defence infrastructure. This realisation has made the inclusion of ecosystems and their services into flood defence schemes a rapidly growing field. However, these new solutions require different design, construction and management methods. A close collaboration between engineers, ecologists and experts in public administration is essential for adequate designs. In addition, a mutual understanding of the basic principles of each other's field of expertise is paramount. This chapter presents some simple approaches for the integration of ecosystem-based measures into coastal engineering projects, which may be of use to experts from a range of fields. Further, it stresses the importance of ecological processes which determine the persistence and health of coastal ecosystems, a point which is rarely emphasised in coastal engineering. The main aim of this chapter is to highlight the role of ecosystem properties for flood defence to stimulate the coastal engineering community in adopting an ecosystem view. In the near future the hope is that greater awareness of ecosystem processes will lead to more sustainable and climate-robust designs. For this, engineers, ecologists and social scientists involved in coastal defence projects need to develop a common language, share the same design concepts and be willing to share the responsibility for these innovative designs.

Keywords Nature-based coastal defence • Flood risk mitigation • Ecosystem services • Coastal engineering • Coastal management • Design • Ecosystem-based management

B.K. van Wesenbeeck (✉)
Unit for Marine and Coastal Systems, Deltares, P.O. Box 177, 2600 MH Delft, The Netherlands

Faculty of Civil Engineering and Geosciences, Delft University of Technology, P.O. Box 5048, 2600 GA Delft, The Netherlands
e-mail: bregje.vanwesenbeeck@deltares.nl

© Springer International Publishing Switzerland 2016
F.G. Renaud et al. (eds.), *Ecosystem-Based Disaster Risk Reduction and Adaptation in Practice*, Advances in Natural and Technological Hazards Research 42,
DOI 10.1007/978-3-319-43633-3_8

181

8.1 Introduction

Uncertain future projections of sea level rise, river runoff and storminess in combination with the increasing call for sustainable development have given rise to a whole suite of concepts that attempt to embed ecosystem-based approaches into water management (Barbier et al. 2008; Borsje et al. 2011; Gedan et al. 2011a). The reasoning behind this is that nature can help in providing adaptive and cost-efficient no/low-regret flood risk management solutions that will be especially suitable in light of the uncertain climate change scenarios (Cheong et al. 2013). Nevertheless, putting these ideas into practice has proven challenging. First, there are several gaps in knowledge that are not yet properly addressed (Bouma et al. 2014; Spalding et al. 2014; Renaud et al., Chap. 1), such as the role that ecosystems play during extreme events (Möller et al. 2014) and long-term stability of ecosystems. Second, new coastal defence design principles, coastal risk management routines that include ecosystem considerations and tailor-made methods to assess safety levels of flood defence structures are required to standardize these approaches. Although all this is technically possible and in the end may well be more cost-effective than traditional construction and management practices, deviating from standard procedures involves additional efforts and project risks, which may be an impeding factor for large-scale application of ecosystem-based flood risk mitigation.

Currently, there are multiple names for concepts that aim to integrate ecosystems into infrastructural developments (Box 8.1; Renaud et al., Chap. 1). These concepts all include an integrated approach which takes into account multiple interests, combining ecological, technical and socio-economic needs. Terms differ from very broad concepts, such as eco-engineering and eco-technology that are applicable across systems and for a variety of functions, to more focused concepts, such as natural coastal defence. Several concepts, such as green adaptation and ecosystem-based adaptation, focus specifically on adaptation to the consequences of climate change, as climate change seems to be an important driver for the application of green concepts (Cheong et al. 2013; Temmerman et al. 2013). Green infrastructure is used more in an urban context. A recurring theme in all these concepts is making use of natural processes and ecosystem services for functional purposes, often in relation to water management. Although all concepts bridge between engineering and ecological approaches, the originally strict separation between these two disciplines has not yet fully disappeared (Cheong et al. 2013). A new form of engineering can be defined that starts from a system perspective and co-creates

M.D. van der Meulen • M.B. de Vries
Unit for Marine and Coastal Systems, Deltares, P.O. Box 177, 2600 MH Delft,
The Netherlands

C. Pesch
DA – Applied Research Group, HZ University of Applied Sciences, Research Group Building
with Nature, P.O. Box 364, 4380 AJ Vlissingen, The Netherlands

H. de Vriend
Delft University of Technology, Delft, The Netherlands

ecosystem-based solutions with experts from both disciplines. This does not conflict with the main principles of engineering but coincides with the main principles in flood risk management.

Box 8.1 Glossary of Ecosystem-Based Approaches to Engineering

Eco-DRR
Ecosystem-based disaster risk reduction (Eco-DRR) is the sustainable management, conservation and restoration of ecosystems to reduce disaster risk, with the aim of achieving sustainable and resilient development (Estrella and Saalismaa 2013: 30)

Eco-engineering/ecological engineering
The design of sustainable ecosystems that integrate human society with its natural environment to stimulate both (Mitsch and Jørgensen 2003).

Eco-technology
Advancing technology beneficial for humans while minimizing ecological impact and adopting ecology as a fundamental basis with a holistic problem view (https://en.wikipedia.org/wiki/Ecotechnology).

Ecosystem-based adaptation
Helping people adapt to climate change by making use of ecosystem services and biodiversity. This includes the sustainable management, conservation and restoration of specific ecosystems that provide key services and increase resilience of communities to climate change effects (Colls et al. 2009).

Building with Nature
Building with Nature is a new design philosophy in hydraulic engineering. Natural elements such as wind, currents, flora and fauna are utilized in designing a hydraulic engineering solution, thereby creating additional benefits for nature, recreation and the local economy (www.ecoshape.nl).

Building for Nature
Optimizing ecological functions of grey infrastructure (www.ecoshape.nl).

Natural infrastructure
Natural infrastructure (sometimes called green or sustainable infrastructure) is the interconnected network of natural and undeveloped areas needed to maintain and support ecosystems (http://www.epa.gov/region3/green/infra structure.html). The term is also used for improving the natural values of grey infrastructure.

Green infrastructure
Green infrastructure is the strategically planned network of high quality green spaces   (http://ec.europa.eu/environment/nature/ecosystems/index_en.htm). The EPA uses the term green infrastructure for its approach to use vegetation and soil for managing storm water runoff on the spot in local communities (http://water.epa.gov/infrastructure/greeninfrastructure/index.cfm).

Some differences between ecosystems as a coastal defence component and traditional designs need to be taken into account. First, a dyke or levee is a structure that is built in a short time-span and the flood defence properties of the structure will start from the moment it is in place. To a certain extent this also holds for ecosystem-based designs, however in some cases ecosystems need to build up biomass and trap sediment in order to become effective in mitigating the hydrodynamic and soil mechanical loads. If these properties of ecosystems are used in combination with dikes or levees this may yield an optimal combination as levees on soft soil show subsidence over time and therefore need additional maintenance or upgrading. If levees are designed in combination with ecosystems in front of the levee, this ecosystem has several years to fully develop and can then compensate for levee degradation or deficiency by building up soil and increasing stability and by increasingly attenuating waves. This way, safety levels may stay the same and expensive levee improvements might be avoided or postponed. A full life-cycle analysis, which is becoming more popular for engineering projects, shows that in case of changes in risk, for instance because of sea level rise, a naturally accreting ecosystem in front of the dike can provide compensation by trapping sediment or can be easily adapted to worsening external conditions by extending the vegetated area or by sediment nourishment. For predicting ecosystem stability in the near future, existing data on actual ecosystem dynamics and services can be used. Of course this is not possible at an infinite time-scale, as uncertainty of these predictions will increase on longer time-spans, but it can be done at similar timescales as for traditional designs (decades).

Dykes are built to last several decades and regular monitoring and management is organized to ensure that safety levels are met through the entire lifespan of a levee. The same holds for ecosystems; once a certain stage is reached, the ecosystem is likely to stay in place and deliver its flood defence services for several decades and this should be ensured through a management and monitoring process. Monitoring should mainly focus on the ecosystem's health (e.g. growth, absence of disease), since this determines its life span and flood defence properties. If monitoring results show poor ecosystem health, or a change in the species distribution of the ecosystem, management interventions should be considered, such as replanting vegetation or excluding grazers.

This chapter attempts to highlight basic ecological principles relevant for flood risk management and the project phases that follow during implementation of measures. The chapter starts by placing ecosystems as a central element in integrated flood risk management and by a solid problem analysis that needs to distinguish between flood type and primary causes of flooding. This chapter presents a framework for project implementation that draws parallels between designs including ecosystem components and traditional designs. Then, it presents simple rules for ecosystem selection based on environmental conditions, followed by a description of coastal ecosystems and their flood risk mitigation functions which are described and ranked for their effectiveness. Finally, the chapter discusses biotic conditions that influence long-term ecosystem functioning and management.

The main focus in this chapter lies on coastal and estuarine ecosystems, but similar principles can be applied to freshwater environments.

8.2 Infrastructure or Ecosystem?

In general grey measures are defined as infrastructural measures that do not integrate ecosystem functions or presence. Examples of grey measures are structural measures that are man-made, such as levees and dams. Although green measures can be considered structural measures too, in that they do intervene on the hazard intensity directly, they consist of ecosystems that are naturally present in the area or that can be restored or recreated if they are degraded or have disappeared. Hybrid measures are a combination of green and grey strategies and constitute, for example, a mangrove forest that reduces wave impact, but also has an earthen levee in the back that blocks surges. A systematic approach for implementation of measures for flood risk reduction is illustrated by the flood risk management cycle (Sayers et al. 2013). Ideally, integrated flood risk management includes the evaluation of natural and socio-economic systems both in identifying of root causes of flooding and in defining a preferred strategy and accompanying measures. Figure 8.1 illustrates how the flood risk management cycle is related to project phases and to ecosystem inclusion.

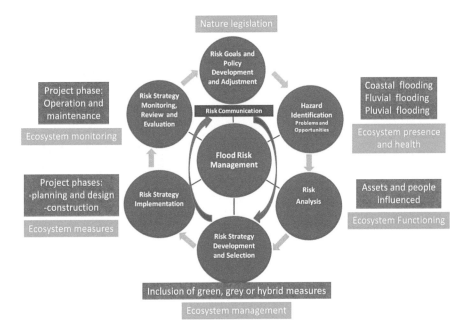

Fig. 8.1 Flood risk management cycle with in grey more information on project phases and in green ecosystem parameters to be assessed. Modified by authors from Sayers et al. (2013)

The cycle in Fig. 8.1 starts by identification of hazards, which is basically a solid problem detection and description. Three types of flood hazards are distinguished; coastal flooding, fluvial flooding and pluvial flooding. Groundwater flooding is not considered here as green measures are not so appropriate to mitigate the effects of groundwater flooding. As part of the identification of the hazard a more detailed analysis of hazard frequency and hazard intensity is required in order to obtain a clear overview of problems and opportunities that these hazards pose. As a part of this process the current ecosystem presence and health status should be evaluated as ecosystems influence the intensity of the hazard. The next question is what consequences the hazards will have? This question is answered in the risk analysis that evaluates the impacts a hazard has on people and assets. This impact is related to the functioning of present ecosystems and their role in mitigating hazard impact through wave attenuation and reduction of winds and currents. As a next step, a risk reduction strategy should be developed. This strategy should include advice for managing ecosystems to improve their health or effectiveness. The risk analysis will also contain advice on potential measures, including the potential restoration of ecosystems as a part of green or hybrid measures.

As in any engineering project, risk strategy implementation of project phases, apply to green, grey and hybrid measures (Fig. 8.2). Monitoring of the implementation strategy falls under the project phase that we define as 'operation and maintenance'. For strategies that include ecosystem-based measures it is advisable to follow an adaptive monitoring and management strategy (van Wesenbeeck et al. 2014). We will go into more detail on the three project phases and how these translate to ecosystem-based measures in the rest of this chapter.

Fig. 8.2 Simplified project phases for engineering projects (own figure)

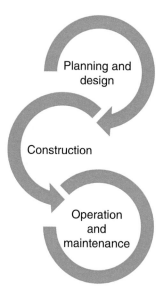

Planning and design

Construction

Operation and maintenance

8.2.1 Risk Strategy Development and Implementation

All flood risk assessments start with a detailed problem analysis. This entails a proper understanding of the appropriate measures. This problem description also sheds more light on the required functionality of measures and interventions. In the problem analysis phase, one has to think of the main functionalities that a design should have and what the requirements of the design are in terms of sustainability, costs, ecological value, but also in terms of the technical requirements. Currently there is a movement to shift from mono-functional designs to benefit- oriented designs that are optimized not only from an engineering viewpoint, but also for yielding maximum co-benefits (Vriend et al. 2015). Implementation of measures has a strong focus on- the-ground project perspective and is therefore mostly implemented in the project phases 'planning and design','construction' and 'operation and maintenance' (Fig. 8.2). These project phases can aid in identifying knowledge gaps and requirements for adopting ecosystem-based approaches. A similar approach is taken in the large Dutch Building with Nature program (www. ecoshape.nl). In the following section, we go through these project phases and outline what ecosystem knowledge can be merged into these and what the main caveats are.

8.2.2 Design

Ecosystem effectiveness in performing a certain function needs to be taken into account when intentionally integrating ecosystems into flood management schemes. There is considerable quantitative evidence in the literature from field and modelling studies that illustrates the capacity of ecosystems and vegetation to reduce currents and waves. This caused by the fact that there is a structure in place, which is supported by other functions, such as the capturing of sediments to decrease erosion. Ecosystems, in contrast to grey structures, are constrained by environmental conditions and develop naturally if environmental conditions are suitable, implying that not all ecosystems can establish in a specific environment. In this respect, ecosystem restoration is very different from the construction of man-made objects such as buildings or engineering structures. However, there are several parallels between development of ecosystem-based solutions and development of traditional infrastructure. Firstly, a construction process always aims to achieve certain functionality. In terms of an ecosystem this functionality can be translated into a service that an ecosystem can provide, such as attenuating waves or reducing erosion (Fig. 8.3). A design that provides this functionality can be based on building blocks. For example, if the desired function is accommodation for living, the building blocks can consist of a house, an apartment complex, a tent, or an igloo. Which building blocks are chosen depends on the external conditions, such as climatic conditions and available construction materials. Similarly, an ecosystem can be considered a building block for an ecosystem-based flood defence

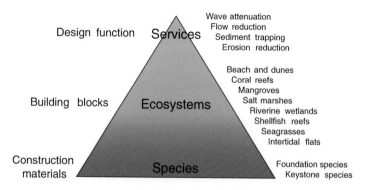

Fig. 8.3 Parallels between the traditional design and construction process and ecosystem-based alternatives for coastal ecosystems (own figure)

system. Which ecosystem is chosen depends on the environmental conditions and the availability of building materials, in this case available species and seed sources (Fig. 8.3).

The building block 'ecosystem' in turn consists of smaller elements, hence, plant and animal communities and species. For the construction or restoration of an ecosystem the species level is the most important, as species, like bricks, form the basis of the entire ecosystem. Two types of species are particularly important: foundation species and keystone species. Foundation species are dominant in terms of effect or abundance. They are able to set the formation of an entire ecosystem into motion. Examples are oysters that construct oyster reefs, or certain grass species that initiate marsh formation. Foundation species create habitat and thereby attract many other species. Keystone species, on the other hand, are often crucial to ecosystem persistence, health and structure. These species are mostly grazers or predators that maintain the subtle balance between several other species. Without the presence of the keystone species, often a single other species tends to dominate the ecosystem and eradicate other species. This development is considered undesirable as it is likely to influence specific characteristics of the ecosystem. For example, local extinction of sea otters, a keystone species in kelp forest, has been shown to cause disappearance of those forests (Estes and Palmisano 1974). Sea otters eat urchins and urchins eat kelp. If sea otters disappear, urchins thrive and overgraze the kelp, which may disappear completely (Estes and Palmisano 1974). These kind of changes in species composition might change the functionality of the system (e.g. wave attenuation), or it may influence the resilience and stability of the system.

8.2.2.1 Ecosystem Effectiveness

Coastal ecosystems can play a role in flood risk mitigation by attenuating waves, by reducing or deflecting currents and by forming a physical barrier between land and water (Borsje et al. 2011; Gedan et al. 2011b; Shepard et al. 2011). One of the most useful and best quantified factors is attenuation of waves to mitigate coastal flooding. There are several predefined factors that contribute to the process of wave attenuation. From an engineering perspective these factors influence the most relevant parameters in wave attenuation models, such as the water depth, the bottom roughness and the length that the wave travels over the feature compared to the wave characteristics. From an ecological perspective the most important variables are constituted by the living and dead material built up by ecosystems. In case of coral reefs this is determined by the height of the reef crest, the coral biomass on top of the reef and the width of the reef (Ferrario et al. 2014). In the case of salt marshes and mangroves, soil elevation and vegetation biomass are essential in achieving wave attenuation. In most wave attenuation models that include vegetation, vegetation presence is represented as bottom roughness or in the parameters vegetation height, stem density, stem diameter and a bulk drag coefficient (Mendez and Losada 2004). The product of these parameters constitutes the so-called 'vegetation factor' (Mendez and Losada 2004). Vegetation flexibility is accounted for in the bulk drag coefficient that often is regarded as a calibration factor. However, it is more often found that this factor varies with wave conditions (Möller et al. 2014) and this is not yet incorporated well into numerical models.

The level to which ecosystems actually contribute to flood risk mitigation is dependent on the type of ecosystem present. This efficiency also depends on the underlying mechanisms. For example, coral reefs will mainly cause wave breaking and are in that way considered very effective as the wave height behind the reef will be considerably reduced (Ferrario et al. 2014). Salt marshes and sea grasses will mainly attenuate waves rather than breaking them (Möller and Spencer 2002; Ondiviela et al. 2014). Several ecosystems that play a role in flood risk mitigation are ordered by the level of protection against flooding they offer (see Table 8.1; Koch et al. 2009; Gedan et al. 2011b; Bouma et al. 2014). It should be noted that quantitative information on flood defence properties of ecosystems is not available for all systems. Furthermore, these values are often based on a limited number of studies.

Morphological systems such as beaches and dunes are known to be very effective flood defence systems as they can be used without any additional hard defence measures, provided that they have a sufficient erosion buffer. They mainly protect the hinterland against storms and flooding, by dissipating wave energy and providing a physical barrier against high water levels (Defeo et al. 2009). Sand dunes have also shown to be effective during extreme events, such as tropical storms and tsunamis. For example, they were reported to block surges up to 3.7 m in India during a tsunami (Mascarenhas and Jayakumar 2008) and to break waves, thereby reducing wave energy up to 97 % (Ferrario et al. 2014). Full-grown coral reefs that extend to mean sea level are very effective in breaking wind waves. Although they do not protect the land against flooding, a reduction of wave heights by 20–50 % (Harborne et al. 2006) and a reduction of tidal current speeds by 30 % (Harborne et al. 2006) have been measured. An advantage of coral reefs is that they form a hard physical barrier. Therefore, the coastal defence function of coral reef

Table 8.1 Quantitative overview of flood defensive properties of different ecosystems. It has to be noted that these values were mainly measured in the first tens of meters of the ecosystems, as it is usually this area that contributes the most to coastal defence

Ecosystem	Coastal defence property	Value	Reference
Beaches & dunes	Block waves	Waves up to 3.7 m	(Mascarenhas and Jayakumar 2008)
Coral reefs	Reduce waves, reduce tidal current speed	20–97 % reduction in wave energy, 30 % reduction current speeds	(Harborne et al. 2006; Ferrario et al. 2014)
Mangroves	Wave attenuation	20–60 %	(Mazda et al. 1997; Gedan et al. 2011b)
Salt marshes	Wave attenuation, foreshore stabilization	1.1–2.1 % per m of marsh	(Möller and Spencer 2002)
Shellfish reefs	Wave breaking	40 % with low water levels and wave heights (for the oyster *Crassostrea sp.*)	(Borsje et al. 2011)
Sea grass	Wave attenuation	40 %; 7,3 mm of wave attenuation per m of sea grass	(Fonseca and Cahalan 1992; Bouma et al. 2005)

systems, like that of beach and dune systems, can be considered relatively robust as the physical structure will not deteriorate immediately in case of mortality of the living components. This allows for recovery of systems and implies that there is no immediate loss of coastal defence function.

Salt marshes, mangroves and shellfish reefs, such as oysters and mussels, have a clear coastal defence potential. Both build rather robust structures and attenuate waves (Möller and Spencer 2002; Shepard et al. 2011). An important drawback of these ecosystems is that they are most effective at low water depths (Feagin 2008). Yet, their general effects on wave reduction should not be underestimated. The reduction in depth by building extensive shallow platforms, such as intertidal flats and salt marshes, reduces wind fetch, hence wave growth, and limits the maximum wave height, as wave height is a function of water depth. Moreover, the presence of shallow areas in front of the dike is thought to stabilize the dike, allowing for a less costly dike design. Mangrove forests mostly contribute to attenuation of waves and reduction of storm surges through their structure of stems and leaves (McIvor et al. 2012a, b). If mangroves are healthy and present for areas that exceed over a kilometre they can even have a positive effect on reducing large waves, such as tsunamis (Marois and Mitsch 2015). Finally, the effects of sea grasses on wave attenuation are moderate. Most of the quantitative studies were conducted in the laboratory, and resulted in values of 40 % wave height attenuation for low water levels and low waves (Fonseca and Cahalan 1992). Bouma et al. (2005) found 7.3 mm of wave attenuation per m of sea grass in a flume experiment with a plant density of 13,400 plants per m^2. However, there is a lack of large-scale measurements that enable predicting the effects of sea grasses on a landscape scale. Hence it is currently impossible to effectively include effects of sea grass into coastal defence schemes.

8.2.2.2 Ecosystem Limitations

The challenge of integration of ecosystem functions into coastal defence schemes is to translate complex ecosystem behaviour into simple generic rules that can be related to engineering. Therefore, we tried to depict abiotic constraints in a decision diagram as used more often in decision-making for coastal engineering design (CIRIA et al. 2013). Limiting environmental conditions provides simple guidance on which ecosystems could possibly establish at a specific location. This has to be done at different scales, as different drivers determine critical conditions for ecosystem occurrence on a global, regional and local scale. On a global scale, climate, hence latitude is often leading. For example, many species are limited by the occurrence of temperatures below zero degrees Celsius. In the case of ecosystems for coastal defence purposes we therefore distinguish temperate and tropical climates, as coastal ecosystems in either of these climates are quite distinct (Fig. 8.4).

After identifying large-scale climate conditions, a regional-scale parameter that determines the possible ecosystem type is the salinity of the water. In Fig. 8.4 a distinction is made between fresh and salt water. Here, we will elaborate on saline ecosystems and their function in flood risk mitigation. On a local scale, an important factor determining ecosystem occurrence is exposure to impact of waves which is reflected mostly by the fetch (e.g. distance that wind can blow without blocking to generate wave set-up). Some systems, such as beaches and dunes, are suited for highly dynamic wind and wave conditions, whereas others need a more sheltered environment (e.g. salt marshes and mangroves). This factor is linked with the sediment type, as fine muddy sediments are usually found in sheltered environments and coarser sediments in more dynamic environments. Sediment size and composition are also related to the amount of nutrients present in the soil. Muddy sediments generally contain more nutrients, which is critical to the occurrence of

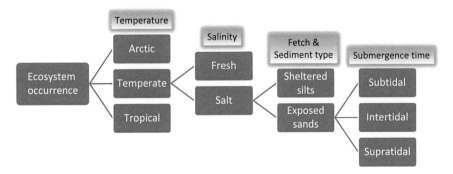

Fig. 8.4 Environmental conditions that influence ecosystem occurrence (own figure)

certain vegetation. Finally, the selection of a suitable ecosystem in a given area is also determined by submergence time. For example, coral reefs need to be inundated permanently, whereas marshes and mangroves only tolerate limited periods of inundation, and therefore need to be situated around or above mean high water level.

8.2.3   Construction of Measures

Implementation of ecosystems into flood risk reduction schemes implies that in the construction phase an ecosystem either needs to be conserved if it is already present, or that it needs to be created if it is not already there, or if it is in a degraded state and does not perform desired functions to a maximum extent. Therefore, this project phase has strong parallels with restoration ecology. Restoration ecology strives to restore physical, chemical and ecological conditions for ecosystem recovery and it pays attention to ecosystem structure, such as species diversity, as well as to ecosystem functioning (Bradshaw 1996). Restoration knowledge differs between ecosystems and requires different methods for each system. For example, restoration and conservation of sea grasses has not been very successful (van Katwijk et al. 2009).

In other cases there may be ample experience with restoration, but methods may be questionable. For example, there is a lot of experience with planting of mangroves, but there is also on-going debate on the efficiency of planting (Lewis 2005; van Wesenbeeck et al. 2015). Basically, planting does not focus on restoration of abiotic and chemical conditions for ecosystem recovery but only on ecological conditions. Even from an ecological point of view planting is not always desirable as it is often done with a single species and therefore it pays no attention to structural complexity of ecosystems. However, general knowledge of preferred abiotic conditions of mangroves is present and there is massive experience with restoration of mangrove forests for nature conservation purposes. Especially after the Indian Ocean tsunami in 2004, mangrove restoration became a major issue. Although many of the projects focused on replanting and therefore did not cover very large areas, they generated knowledge on how to improve these efforts. Instead of planting mangroves, restoration of abiotic conditions that allow for natural recruitment is a preferable method (Schmitt et al. 2013; Winterwerp et al. 2013), likely to have longer-term and larger-scale results.

To integrate ecological principles and structural complexity of ecosystems during the operation and construction phase, attention should at least be paid to main biological indicators that will put biotic constraints on ecosystem recovery, such as grazing. Table 8.2 summarizes available knowledge on foundation species and keystone species for specific ecosystems. As stated before, both types of species are crucial for ecosystem formation and persistence. Furthermore, interactions above and below ground have to be taken into account. Experiments in grasslands have shown that restoration success increases substantially once

Table 8.2 Key biotic controlling factors for ecosystem health, functioning and restoration

Ecosystem	Foundation species	Keystone species	Engineering capacities	Specifics	References
Beaches and coastal sand dunes	Marram grass (*Amophilla arenaria*), or other dune grass species	–	Effective trapper of wind-blown sand, builds high sand dunes	Can be harmed by belowground mycorrhiza	(Eppinga et al. 2006)
Coral reefs	Certain coral species, such as *Acropora sp.* that is known to set off formation of the bicarbonate structure	Urchins or grazing fishes, such as parrot fish, that graze on macro-algae that otherwise overgrow corals	Built bicarbonate reef structure	Sensitive and threatened ecosystem; causes of decline not always clearly understood	(McClanahan et al. 2002; Mumby et al. 2006)
Mangroves	No evidence for pioneer species being a foundation species, but likely that *Avicennia sp.* fulfils this role	Sesarmid and fiddler crabs that oxygenate soil by bioturbation	Attenuation of waves by roots and shoots of trees	Efficient carbon sequestration in these ecosystems; many planting efforts fail due to poor ecosystem knowledge	(Mazda et al. 1997; Slim et al. 1997)
Salt marshes	Several cordgrass species and other grass species	There is evidence pointing at blue crabs for US salt marshes	Trap sediment and create elevated platform	Facilitation between marsh plants is important structuring mechanism	(Altieri et al. 2012)
Shellfish reefs	Mussels, oysters		Reef builder that reduces erosion	Often threatened by overharvesting, restoration can be done effectively using several low tech methods. Presence of hard substrate needed for settlement	(Carranza et al. 2009)
Seagrass beds	Several seagrass species	In some cases urchins that graze on algae	Entrapment of small silty particles	Link with below ground community	(van der Heide et al. 2012a; b)

subterranean communities are transplanted to the restoration sites (Kardol et al. 2009). Very little is known on the role of above and below ground interactions in coastal landscapes. Evidence is emerging, however, that these interactions might be equally important there as in the terrestrial environment (van der Heide et al. 2012b) (Table 8.2). Although ecosystems facilitate survival and persistence of other ecosystems through physical protection, trophic relations or chemical processes, implications of these interactions for their coastal defence functions is not yet fully comprehended. However, it should be kept in mind that as our ecosystem knowledge advances, new insights will challenge earlier assumptions.

8.2.4   Operation and Maintenance

In addition to including ecosystems into the design process of coastal infrastructure there is a need for testing and evaluation methods and for management guidance. In the Netherlands, flood safety is regulated by law and therefore dikes are regularly monitored to assess safety level. Ecosystem properties are, however, not yet integrated in this monitoring methodology, making it difficult to take these into account. For assessment of ecosystems as part of flood risk mitigation measures, a range of properties should be monitored. In the Dutch case where a willow forest was planted in front of a levee to attenuate waves (Borsje et al. 2011) a monitoring protocol was established for the levee manager. This protocol included assessment of the amount of vegetation and the length of the vegetation field (especially in relation to the angle of incidence the waves). The minimum values for these properties, as with a traditional design, depend on the requirements that were set for the particular design.

Standardization of monitoring of ecosystem properties can be established through generic metrics such as the amount of biovolume. Biovolume is a measure for the volume of vegetation per unit of height (m^3/m). Here, the amount of branches per square meter should be known, as well as the diameter of the branches. It is unclear however, how generally this standardization method can be applied, since the testing methodology was tailor-made for a specific case. More practical implementations should be constructed in order to increase the body of knowledge on this vegetation property. A second way to assess ecosystem properties in relation to safety is to look at the failure mechanisms of the ecosystem in a similar probabilistic way as is done for failure mechanisms of levees. For the Noordwaard case, the standard mechanisms such as piping and instability were not assessed; however, other mechanisms that could affect the integrity of the willow forest were taken into account. Examples are disease and consumption by grazers, erosion, fires and (moving) ice. Indirectly these are also taken into account when looking at the biovolume, since these mechanisms may affect the density, height and width of the vegetation field. The extent and manner in which these failure mechanisms can be reversed however, could require a different management regime than those of biovolume.

For most ecosystems little is known on management actions for flood risk reduction purposes specifically. However, as ecosystems are dynamic by nature it is likely that adaptive management methods are best suited for ecosystem-based measures (van Wesenbeeck et al. 2014). Adaptive management methods are designed for constant change and include a constant cycle of monitoring, data analyses, validation of assumptions and adapting designs or management actions based on monitoring results. There is a long tradition in adaptive management of sandy coasts for flood defence in the Netherlands (Van Koningsveld and Mulder 2004; Mulder et al. 2011), which implies that this can be implemented if sufficient financial resources and institutional arrangements are in place. However, it remains doubtful how this will translate to other countries with fewer funds and less complex governance settings. Moreover, including other ecosystems into flood risk mitigation schemes requires similar knowledge on management of these ecosystems. For mangroves we observe that although there is ample evidence for good restoration practices these rarely translate to the ground (Lewis 2005; van Wesenbeeck et al. 2015). In addition, there is little knowledge on ecosystem management at larger scales. For example, large-scale restoration of fully functioning coral reefs has been proven extremely complicated (Rinkevich 2005; Young et al. 2012). This implies that it might not be possible to manage most ecosystems from a coastal defence perspective.

8.3   Conclusions

Using services of coastal and fluvial ecosystems in flood defence systems is a new and promising development. This can be undertaken in addition to traditional structural measures such as levees and dikes. In light of climate change, the uncertainty of future climate scenarios and the unpredictable development of safety norms and design conditions, including ecosystems in flood risk mitigation provide us with a new set of adaptive and few regrets solutions. In the near future, these combinations especially may provide safe, robust, adaptive and cost-effective alternatives to current approaches (Cheong et al. 2013).

Application of flood defence designs that make use of ecosystem services is not yet common practice. Integrating ecosystems into flood defences requires: (1) thorough knowledge of natural processes, ecological restoration and ecosystem behaviour, (2) engineers to acquire some basic ecosystem understanding, (3) ecologists to obtain basic comprehension of engineering, and (4) simple guidelines and design rules on how to implement an ecosystem in a flood risk mitigation scheme. The latter should be taken further than presented here and incorporate systems in engineering protocols and standards. For mainstreaming ecosystem-based interventions for flood risk mitigation, coastal engineers should adopt a system view, based on insight into hydrodynamics, sediment transport, morphology and ecosystem functioning. Also, ecologists should not refrain from carrying similar responsibility as engineers for flood defence designs that integrate ecosystems. This may even require adaptation of standard mono-disciplinary educational systems. Additionally, it implies that organisations now involved in management of natural values of these landscapes are instrumental in the maintenance of healthy and therefore functional ecosystem-based solutions for flood safety. This may initiate ecosystem restoration for flood risk mitigation and, more importantly, provide an incentive to conserve existing coastal and floodplain ecosystems, to avoid costly engineering measures that would arise if they were destroyed.

References

Altieri AH, Bertness MD, Coverdale TC et al (2012) A trophic cascade triggers collapse of a salt-marsh ecosystem with intensive recreational fishing. Ecology 93:1402–1410

Barbier EB, Koch EW, Silliman BR et al (2008) Vegetation's role in coastal protection – response. Science 320:177–1773

Borsje BW, van Wesenbeeck BK, Dekker F et al (2011) How ecological engineering can serve in coastal protection. Ecol Eng 37:113–122

Bouma TJ, De Vries MB, Low E et al (2005) Trade-offs related to ecosystem engineering: a case study on stiffness of emerging macrophytes. Ecology 86:2187–2199

Bouma TJ, van Belzen J, Balke T et al (2014) Identifying knowledge gaps hampering application of intertidal habitats in coastal protection: opportunities & steps to take. Coast Eng 87:147–157

Bradshaw AD (1996) Underlying principles of restoration. Can J Fish Aquat Sci 53:3–9

Carranza A, Defeo O, Beck M (2009) Diversity, conservation status and threats to native oysters (Ostreidae) around the Atlantic and caribbean coasts of south america. Aquat Conserv Mar Freshwat Ecosyst 19:344–353

Cheong S-M, Silliman B, Wong PP et al (2013) Coastal adaptation with ecological engineering. Nat Clim Chang 3:787–791

CIRIA, Ecology M.o., USACE (2013) The international levee handbook (C731) CIRIA

Colls A, Ash N, Ikkala N (2009) Ecosystem-based Adaptation: a natural response to climate change. IUCN Gland, Switzerland

Defeo O, McLachlan A, Schoeman DS et al (2009) Threats to sandy beach ecosystems: a review. Estuar Coast Shelf Sci 81:1–12

Eppinga MB, Rietkerk M, Dekker SC et al (2006) Accumulation of local pathogens: a new hypothesis to explain exotic plant invasions. Oikos 114:168–176

Estes JA, Palmisano JF (1974) Sea otters: their role in structuring nearshore communities. Science 185:1058–1060

Estrella M, Saalismaa N (2013) Ecosystem-based disaster risk reduction (Eco-DRR): an overview. In: Renaud FG, Sudmieier-Rieux K, Estrella M (eds) The role of ecosystems in disaster risk reduction. United Nations University Press, New York

Feagin RA (2008) Vegetation's role in coastal protection. Science 320:176–177

Ferrario F, Beck MW, Storlazzi CD et al (2014) The effectiveness of coral reefs for coastal hazard risk reduction and adaptation. Nat Commun 5

Fonseca MS, Cahalan JA (1992) A preliminary evaluation of wave attenuation by four species of seagrass. Estuar Coast Shelf Sci 35:565–576

Gedan KB, Altieri AH, Bertness MD (2011a) Uncertain future of New England salt marshes. Mar Ecol Prog Ser 434:229–237

Gedan KB, Kirwan ML, Wolanski E et al (2011b) The present and future role of coastal wetland vegetation in protecting shorelines: answering recent challenges to the paradigm. Clim Change 106:7–29

Harborne AR, Mumby PJ, Micheli F et al (2006) The functional value of Caribbean coral reef, seagrass and mangrove habitats to ecosystem processes. Adv Mar Biol 50:57–189

Kardol P, Bezemer TM, Van Der Putten WH (2009) Soil organism and plant introductions in restoration of species-rich grassland communities. Restor Ecol 17:258–269

Koch EW, Barbier EB, Silliman BR et al (2009) Non-linearity in ecosystem services: temporal and spatial variability in coastal protection. Front Ecol Environ 7:29–37

Lewis RR III (2005) Ecological engineering for successful management and restoration of mangrove forests. Ecol Eng 24:403–418

Marois DE, Mitsch WJ (2015) Coastal protection from tsunamis and cyclones provided by mangrove wetlands–a review. Int J Biodivers Sci Ecosyst Serv Manag 11:71–83

Mascarenhas A, Jayakumar S (2008) An environmental perspective of the post-tsunami scenario along the coast of Tamil Nadu, India: role of sand dunes and forests. J Environ Manag 89:24–34

Mazda Y, Magi M, Kogo M, Phan Nguyen H (1997) Mangroves as a coastal protection from waves in the Tong King Delta, Vietnam. Mangrove Salt Marshes 1:127–135

McClanahan T, Polunin N, Done T (2002) Ecological states and the resilience of coral reefs. Conserv Ecol 6

McIvor A, Möller I, Spencer T, Spalding M (2012a) Reduction of wind and swell waves by mangroves. The Nature Conservancy and Wetlands International

McIvor A, Spencer T, Möller I, Spalding M (2012b) Storm surge reduction by mangroves. The Nature Conservancy and Wetlands International

Mendez FJ, Losada IJ (2004) An empirical model to estimate the propagation of random breaking and nonbreaking waves over vegetation fields. Coast Eng 51:103–118

Mitsch WJ, Jørgensen SE (2003) Ecological engineering: a field whose time has come. Ecol Eng 20:363–377

Möller I, Spencer T (2002) Wave dissipation over macro-tidal saltmarshes: effects of marsh edge typology and vegetation change. J Coast Res 36:506–521

Möller I, Kudella M, Rupprecht F et al (2014) Wave attenuation over coastal salt marshes under storm surge conditions. Nat Geosci 7:727–731

Mulder JPM, Hommes S, Horstman EM (2011) Implementation of coastal erosion management in the Netherlands. Ocean Coast Manag 54:888–897

Mumby PJ, Dahlgren CP, Harborne AR et al (2006) Fishing, trophic cascades, and the process of grazing on coral reefs. Science 311:98–101

Ondiviela B, Losada IJ, Lara JL et al (2014) The role of seagrasses in coastal protection in a changing climate. Coast Eng 87:158–168

Rinkevich B (2005) Conservation of coral reefs through active restoration measures: recent approaches and last decade progress. Environ Sci Technol 39:4333–4342

Sayers P, Yuanyuan L, Galloway G et al (2013) Flood risk management: a strategic approach. UNESCO, Paris

Schmitt K, Albers T, Pham TT, Dinh SC (2013) Site-specific and integrated adaptation to climate change in the coastal mangrove zone of Soc Trang Province, Viet Nam. J Coast Conserv 17:545–558

Shepard CC, Crain CM, Beck MW (2011) The protective role of coastal marshes: a systematic review and meta-analysis. PLoS One 6, e27374

Slim FJ, Hemminga MA, Ochieng C et al (1997) Leaf litter removal by the snail Terebralia palustris (Linnaeus) and sesarmid crabs in an East African mangrove forest (Gazi Bay, Kenya). J Exp Mar Biol Ecol 215:35–48

Spalding MD, McIvor AL, Beck MW et al (2014) Coastal ecosystems: a critical element of risk reduction. Conserv Lett 7:293–301

Temmerman S, Meire P, Bouma TJH et al (2013) Ecosystem-based coastal defence in the face of global change. Nature 504:79–83

van der Heide T, Eklöf JS, van Nes EH et al (2012a) Ecosystem engineering by seagrasses interacts with grazing to shape an intertidal landscape. PLoS One 7

van der Heide T, Govers LL, De Fouw J et al (2012b) A three-stage symbiosis forms the foundation of seagrass ecosystems. Science 336:1432–1434

van Katwijk MM, Bos AR, de Jonge VN et al (2009) Guidelines for seagrass restoration: importance of habitat selection and donor population, spreading of risks, and ecosystem engineering effects. Mar Pollut Bull 58:179–188

Van Koningsveld M, Mulder JPM (2004) Sustainable coastal policy developments in the Netherlands. A systematic approach revealed. J Coast Res 20:375–385

van Wesenbeeck BK, Mulder JPM, Marchand M et al (2014) Damming deltas: a practice of the past? Towards nature-based flood defenses. Estuar Coast Shelf Sci 140:1–6

van Wesenbeeck B, Balke T, van Eijk P et al (2015) Aquaculture induced erosion of tropical coastlines throws coastal communities back into poverty. Ocean Coast Manag 116:466–469

Vriend dH, Koningsveld vM, Aarninkhof S, Baptist M (2015) Sustainable hydraulic engineering through building with nature. J Hydro Environ Res 9(20):159–171

Winterwerp JC, Erftemeijer PLA, Suryadiputra N et al (2013) Defining eco-morphodynamic requirements for rehabilitating eroding mangrove-mud coasts. Wetlands 33:515–526

Young CN, Schopmeyer SA, Lirman D (2012) A review of reef restoration and Coral propagation using the threatened genus Acropora in the Caribbean and western Atlantic. Bull Mar Sci 88:1075–1098

Chapter 9
Overview of Ecosystem-Based Approaches to Drought Risk Reduction Targeting Small-Scale Farmers in Sub-Saharan Africa

Julia Kloos and Fabrice G. Renaud

Abstract Rain-fed agriculture in Sub-Saharan Africa (SSA) provides major but highly climate-dependent sources of livelihoods. Recurrent dry spells and droughts can impact SSA's agro-ecosystems in multiple ways, negatively affecting local social-ecological systems (SES). Droughts not only destroy crops and livestock and degrade natural resources but also impact a large variety of ecosystem services. However, ecosystems can also frequently be powerful agents for drought mitigation and resilient livelihoods. Ecosystem-based approaches mitigate drought impacts while providing multiple co-benefits which contribute to poverty alleviation and sustainable development, food security, biodiversity conservation, carbon sequestration and livelihood resilience. In drought risk management, ecosystem-based solutions have always been important, even if not explicitly acknowledged as such. Based on available literature, this chapter provides an overview of approaches for drought risk reduction in SSA in the context of ecosystem-based disaster risk reduction (Eco-DRR) and ecosystem-based adaptation (EbA). Using selected criteria, the review found many types of approaches, which strengthen functionality of the ecosystem and offer substantial environmental and socio-economic benefits, and thus help to mitigate drought impacts. More information on the limits of these approaches is needed in order to integrate them effectively into Eco-DRR and EbA programmes and complement them with more traditional disaster risk reduction strategies.

Keywords Drylands • Agro-ecosystems • Ecosystem services • Ecosystem-based disaster risk reduction (Eco-DRR) • Ecosystem-based adaptation (EbA)

J. Kloos (✉) • F.G. Renaud
United Nations University Institute for Environment and Human Security (UNU-EHS),
Platz der Vereinten Nationen 1, 53113 Bonn, Germany
e-mail: julia.kloos@gmx.de

© Springer International Publishing Switzerland 2016
F.G. Renaud et al. (eds.), *Ecosystem-Based Disaster Risk Reduction and Adaptation in Practice*, Advances in Natural and Technological Hazards Research 42,
DOI 10.1007/978-3-319-43633-3_9

199

9.1   Introduction

A high degree of climatic and seasonal variability and recurrent extreme events such as droughts and floods are typical in Sub-Saharan African (SSA) drylands. Climatic hazards and dependency on rain-fed agriculture, together with socio-economic and environmental specificities of the region, make SSA's drylands highly vulnerable to the impacts of climate change (Niang et al. 2014). SSA's limited infrastructural development, low levels of per capita income, mostly subsidence-based rural population, and partial reliance on international food aid and disaster relief weaken the coping and adaptive capacities of social-ecological systems (SES)[1] (Benson and Clay 1998; Shiferaw and Okello 2011). In addition, land in SSA is often characterized by low inherent soil fertility, a poor capacity of most soils to retain moisture, and widespread soil degradation (Lahmar et al. 2012). This predisposition, together with population growth, high poverty rates, and a lack of capacity to invest in more sustainable agricultural practices are important factors that contribute to increasing land degradation (Holden and Binswanger 1998; Shiferaw and Okello 2011; Shiferaw et al. 2014). As a result, small-scale farmers find themselves confronted with the twin problems of drought and desertification, which are intrinsically linked (Falkenmark and Rockström 2008). In this context, there is an urgent need to mitigate drought impacts through adaptation processes which go hand in hand with economic development programs, improved food security, poverty reduction initiatives and sustainable environmental management.

The role of ecosystems in climate change adaptation (CCA) and disaster risk reduction (DRR) is increasingly acknowledged (e.g., Colls et al. 2009; Sudmeier-Rieux 2010; Estrella and Saalismaa 2013; Niang et al. 2014) and a growing body of literature and practical applications exist for numerous hazard contexts and under diverse socio-economic conditions. While ecosystem-based approaches for e.g. coastal hazards, river floods or landslides are well established, drought as a slow onset hazard is still under-represented in the discourse around Eco-DRR. This is starting to change as, for example, the theme of the 2014 World Day to Combat Desertification focused on ecosystem-based adaptation, emphasizing the importance of mainstreaming climate change adaptation into sustainable land management.

It is important to note that in the context of droughts, most mitigation[2] strategies, particularly those developed traditionally by small-scale farmers, are ecosystem-

---

[1] A system that includes societal (human) and ecological (biophysical) subsystems in mutual interaction (Gallopin 2006:294).

[2] Drought mitigation in the disaster risk reduction community is usually understood as a set of programs, measures and actions, which are undertaken in advance of a drought event in order to reduce the expected impacts of a drought and facilitate recovery. Mitigation includes proactive elements of drought preparedness. The term "drought mitigation" therefore corresponds to the term "adaptation" in the climate change community. Drought mitigation does not address the reduction of greenhouse gas emissions as usually associated with the term "mitigation" in a climate change context (Wilhite et al. 2014; WMO and GWP 2014).

based (Estrella et al. 2013). Locally-adapted, sustainable agricultural practices and strategies that strengthen ecosystem functioning exist in order to address challenges such as land degradation, food insecurity and a lack of access to agricultural inputs (Liniger et al. 2011; Jones et al. 2012; Munang et al. 2014). These have the potential to help reduce the susceptibility of the agro-ecosystem (croplands, rangelands, agro-forests, etc.) to droughts, increase preparedness and spread drought risks through diversification of agricultural production. Thus, they can contribute to healthier, more resilient ecosystems which produce a wide variety of ecosystem services. In such SES, the capacity of nature is used to buffer farmers and communities against the impacts of climate change and natural hazards. Additionally, the provision of a wider variety of ecosystem services results in many social, economic and cultural co-benefits which help to increase the resilience of SES facing climatic variability (Doswald and Estrella 2015).

This chapter aims to provide an overview of ecosystem-based approaches used mainly for dryland agriculture in SSA and discusses their suitability to support CCA and DRR objectives. To undertake this overview of ecosystem-based approaches to drought mitigation, we followed the definitions of Eco-DRR and EbA as outlined in Chap. 1, and adapted them to the drought context. This resulted in some key criteria for identifying suitable ecosystem-based approaches.

In order to collect the relevant literature on ecosystem-based approaches we draw on the review of Doswald et al. (2014) on EbA.[3] From the list of publications these authors used for the review, we selected all papers dealing with drought or rainfall variability focusing on SSA drylands and agricultural management and added more recent publications (2012–2014) through Scopus and Google Scholar searches. This allowed us to include a large number of papers, but because there are numerous concepts and approaches – that sometimes overlap – the overview of approaches is not fully exhaustive. The review, however, provides insights into the main and more common ecosystem-based approaches and agricultural techniques that can be used as part of EbA/Eco-DRR in the context of droughts.

The chapter starts by linking the concepts of Eco-DRR and EbA to the characteristics and impacts of droughts. From the definitions and conceptualizations around EbA and Eco-DRR (Chap. 1) we developed criteria described in Box 9.1 to identify suitable approaches and agricultural techniques. These approaches and techniques can be considered to be ecosystem-based, while at the same time they reduce drought risks and facilitate CCA. The key environmental, social and economic benefits, which contribute to greater livelihood resilience, are summarized. Furthermore, important drawbacks that may hinder application, as observed in the scientific and applied literature, are also highlighted. We also discuss to what extent the approaches help to solve multiple goals, operate at multiple scales and are locally adapted. The chapter concludes with a summary of the advantages of applying an ecosystem-based approach to drought risk reduction and boosting adaptation in SSA, but also points to current knowledge gaps.

---

[3]The procedure for identifying ecosystem-based adaptation measures for the review is described in Munroe et al. (2012).

Box 9.1 Criteria for Identifying Suitable Approaches and Agricultural Techniques for Ecosystem-Based Approaches Addressing Droughts
Approaches addressing droughts were characterised as suitable when they:

- Strengthen functionality of the ecosystem and use natural processes to provide multiple services;
- Provide drought mitigation (strengthening of regulating services of the ecosystem);
- Generate social, economic and cultural co-benefits. Through the improved functionality of ecosystems, ecosystem services and biodiversity are improved/maintained leading to multiple co-benefits- (e.g. improved yields, empowerment of marginalized groups, cultural value of diverse and healthy agricultural landscapes, etc.);
- Address multiple goals, e.g. minimize trade-offs and maximize benefits with development objectives (Andrade et al. 2011);
- Are applicable at multiple scales;
- Combine different sources of knowledge to generate locally adapted and well-negotiated approaches.

9.2 Linking Drought Risk Reduction to the Principles of Ecosystem-Based Approaches

In order to identify ecosystem-based approaches to reduce drought risks, this section first sheds light on the specific characteristics of droughts as slow-onset hazards and their major impacts on the ecosystem services provided by agro-ecosystems. Based on this background, principles of ecosystem-based approaches in a drought context are derived which are then used for the identification of approaches suitable to reduce drought risks.

9.2.1 Droughts and Their Impacts on Agro-Ecosystems

A drought is broadly defined as *"sustained, extended deficiency in precipitation"* (WMO 1986:2), or more specifically when *"precipitation has been significantly below normal recorded levels, causing serious hydrological imbalances that adversely affect land resource production systems"* (UNCCD 2012a:1).[4] Generally, droughts are divided into four classes with increasing focus on the impact on SES (Wilhite and Glantz 1985):

---

[4]While insufficient rainfall is the primary cause of drought, this often goes together with increased potential evapotranspiration (IPCC 2012).

- **Meteorological drought**: when precipitation is lower than the long-term normal for a prolonged period.
- **Agricultural drought:** when there is insufficient soil moisture to meet the needs of a particular crop at a particular time.[5]
- **Hydrological drought**: when deficiencies occur in surface and subsurface water supplies.
- **Socio-economic drought**: when human activities are affected by reduced precipitation and related water availability. This form of drought associates human activities with elements of meteorological, agricultural, and hydrological drought and becomes evident when drought affects health, the well-being and quality of life of the population.

As these categories only broadly identify the respective drought types in agro-ecosystems, useful guidance can be taken from Rockström (2003), who refers to a meteorological drought as: insufficient rainfall to generate a harvest and seasonal rainfall which differs from long-term seasonal average.

SSA is characterized by seasonal rainfall patterns and rain-fed agriculture, thus timing of agricultural activities with respect to rainfall is critical. Dry spells are different from droughts, as they describe rainfall deficits over a period of several weeks during the agricultural production period (Rockström 2003) and are forecasted to intensify in East and southern Africa in the future (Niang et al. 2014:1206). West Africa shows increased drought and flood risks towards the late 21 century (Sylla et al. 2015). This review therefore targets both agricultural droughts and dry spells in the general context of climate variability.

The impact of any type of drought or dry spell is very much dependent on its length, timing and frequency, in addition to its severity, intensity, magnitude and areal extent (see e.g., Kallis 2008), and simultaneously on the vulnerability of exposed systems. Hence droughts/dry spells can impact a range of ecosystem services and reduce the capacity of agro-ecosystems to provide the benefits on which people depend. They can negatively impact provisioning services, resulting in reduced productivity of agro-ecosystems and reduced availability of fresh water (quantity and quality). They affect services that regulate the quality and quantity of water and soil, habitat services, as well as cultural services such as spiritual and religious values or cultural heritage. They can also negatively impact biodiversity and increase the likelihood of other, potentially hazardous events, such as wildfires.[6]

For agricultural droughts, the distribution of rainfall in relation to crop requirements matters more than total seasonal rainfall. The impact of agricultural droughts and dry spells depends very much on critical plant growth stages. Plants which have already been impacted by previous water shortages show a reduced capacity to take

---

[5]Also called "*soil moisture drought*" to refer to the fact that soil moisture deficits have wider effects than only those on the agro-ecosystem (IPCC 2012).

[6]Assessments of the full range of ecosystem services that are impacted by droughts and of interactions with the social system are rare (see e.g. Banerjee et al. (2013) for an example).

up water in the root zone (Falkenmark and Rockström 2008). Soil conditions, such as water holding capacity and water infiltration, have an impact on the manifestation of an agricultural drought, as they directly affect soil moisture content. Plant conditions, in particular water uptake capacity, further determine the degree of plant water stress and eventually yield reductions.

In order to mitigate the impacts of agricultural droughts and dry spells, the most direct entry point is an integrated agricultural management of water, soils and crops. Falkenmark and Rockström (2008) state that agricultural droughts/dry spells can be strongly influenced by existing management practices related to water, soils and crops. Building resilience to droughts therefore depends on increasing the ability to implement these management practices optimally for drought risk reduction. This is a crucial angle for ecosystem-based approaches to address. Approaches that strengthen resilience to droughts need to reduce the susceptibility of the SES to drought impacts, for instance through improving the water holding capacity of soils, or through crop diversification to balance crop water requirements. Such measures can simultaneously improve the resilience and sustainability of rural livelihoods, for example through increased yields, income, and food stocks, and thus reduce the need for migration.

9.2.2 Principles of Ecosystem-Based Approaches for Drought Risk Reduction

Social and ecological systems are not just linked but are interconnected and co-evolving across spatial and temporal scales (Stockholm Resilience Center 2007). Accordingly, Eco-DRR and EbA approaches should be implemented as integrated, holistic and interdisciplinary approaches recognizing these interconnectivities between and within systems (Sudmeier-Rieux 2010; Jones et al. 2012; Munang et al. 2014). EbA builds on the links between climate change, biodiversity, ecosystem services and sustainable resource management in order to increase the resilience[7] of livelihoods climate change impacts. Eco-DRR similarly aims for sustainable development and disaster resilience, based on managing, restoring and conserving ecosystems. Both emphasize the use of biodiversity and ecosystem services in a sustainable way (see Chap. 1 and the discussion on ecosystem-based DRR and CCA) and refer to the restoration of degraded/transformed agro-systems. The articulation of ecosystem services can be useful in implementing an ecosystem-based approach because it allows all the flows of services from the ecosystem to be captured. This provides the basis for them to be managed in such a way that they provide the greatest benefit to

---

[7]Resilience: *"The capacity of social, economic, and environmental systems to cope with a hazardous event or trend or disturbance, responding or reorganizing in ways that maintain their essential function, identity, and structure, while also maintaining the capacity for adaptation, learning, and transformation"* (IPCC 2014:5).

humans whilst still ensuring ecosystem function. Importantly, both Eco-DRR and EbA stress the generation of economic, social and environmental co-benefits when compared to other, more conventional DRR and CCA measures (e.g., dams and dikes). The creation of co-benefits contributes to improving the resilience of the SES (see Box 9.1).[8]

There are a number of approaches that try to explicitly link multiple objectives in an integrated way in order to tackle climatic, environmental, social and economic challenges (UNDP 2011). Review papers on EbA and Eco-DRR stress that many such approaches are not completely new ideas, but already exist in traditional natural resources management and ecosystem restoration efforts, or can be part of DRR or CCA measures (Munroe et al. 2012; Estrella and Saalismaa 2013).

As mentioned in Box 9.1, when considering ecosystem-based approaches for DRR and CCA, taking a landscape perspective is critical. Typically, decisions on land use changes need to be informed by the flow of ecosystem services at the landscape scale (Vignola et al. 2009). The multifunctional landscapes approach supports the idea that ecosystem service flows need to be managed at multiple scales and integrate ecological principles at the field, farm and landscape scales (McGranahan 2014). The ecosystem-based approaches described in Sect. 9.3 should be seen in the context of *"people-centered landscape approaches to environmental management"* as defined by Sayer et al. (2013:8349), particularly when several approaches are being implemented in a landscape, addressing both livelihoods, economic development and conservation goals. Ecosystem-based approaches are not only about the measures themselves, but also how to best combine them at the field, farm and landscape levels to maximise DRR, CCA and development objectives. Some of the approaches discussed in the next section target the landscape level, while others are farm or field level, but can be part of landscape level approaches.

9.3 Ecosystem-Based Approaches to Drought Risk Reduction in Sub-Saharan Africa

As the impacts of droughts are manifold, so are the coping and adaptation strategies of small-scale farmers in the drylands of SSA. A recent overview by Shiferaw et al. (2014) summarises response strategies of rural households in SSA and discusses key technological, institutional and policy strategies for drought mitigation and adaptation, such as improved crop varieties, improved soil fertility and water management, and index-based insurances. Here, we focus only on ecosystem-based approaches which have a direct link to agricultural activities, but of course these need to be complemented by integrated technological, institutional and political measures and include governance and management aspects (see

---

[8]For a closer comparison of EbA and Eco-DRR see Doswald and Estrella (2015).

e.g. Andrade et al. 2011, for the latter). Among them, there are several scientific and applied literature approaches which address how to increase water efficiency and agricultural productivity while reducing land degradation. These techniques are all explicitly or implictly suitable for drought mitigation.

In a drought context, existing publications on EbA and Eco-DRR refer to the sustainable management of grasslands and rangelands (UNCCD 2012c); agricultural practices that maintain vegetation cover, conserve soils and restore natural vegetation; shelter belts and green belts (Estrella and Saalismaa 2013); water harvesting and conservation farming (Colls et al. 2009); protected area management (e.g., Dudley et al. 2015) and others.

To extend this list and make it more specific as to how many of these approaches contribute to the principles of Eco-DRR/EbA, we compiled additional information through the literature review. The following section provides an overview of (existing) approaches which were identified based on our set criteria (Box 9.1). But it cannot provide a comprehensive list of all existing approaches and techniques.

We begin by presenting the broader classes of approaches that go beyond addressing purely agricultural goals. These approaches entail multiple strategies and agricultural techniques, of which some or all can be considered as ecosystem-based and address multiple goals tackling the links between water, land, and biota. Aiming for sustainability and resilience, some approaches explicitly refer to being holistic, by calling for collaboration, flexible management, local knowledge and participation at multiple geographical scales and hence overlap with our set criteria (Box 9.1).

After discussing the broader classes of approaches, more narrow or targeted approaches and agricultural techniques are described. The length of description for each approach reflects the amount of available literature per approach. Table 9.1 gives a systematic overview of goals, environmental and social benefits, scale issues and highlights drawbacks that may hinder the application of these more targeted tools and practices. However, this overview can only be general, as the specific impacts depend on the local context.

9.3.1 Broader Classes of Approaches

Resource-conserving agriculture and **sustainable intensification**[9] have many commonalities with the aim to make best use of natural resources and ecosystem services in a sustainable manner and to simultaneously promote social, environmental and health objectives and while increasing productivity (Pretty et al. 2006; Bennett and Franzel 2013). Bossio et al. (2010:5) consider a wide range of measures that belong to resource-conserving agriculture, such as eco-agriculture,

---

[9]For a discussion of the differences and commonalities between the concepts of sustainable intensification and ecological intensification see Tittonell (2014).

Table 9.1 Overview of identified approaches and agricultural techniques for an ecosystem-based approach to drought risk reduction

Agricultural practice	Multiple goals	Environmental benefits, Ecosystem Services	Socio-economic benefits	Scale	Relevance of locally adapted techniques using diverse knowledge sources	Drawbacks
Organic agriculture	Agro-ecosystem health and biodiversity conservation, sustainable and healthy livelihoods, food security, climate change adaptation (e.g., FAO/WHO 2006; Müller et al. 2013)	Water holding capacity, soil fertility, biodiversity all increased, reduced pest pressure, carbon sequestration (Müller 2009)	Net income increases, reduced health risks, reduced economic risks, improved food security, reduced migration (Panneerselvam et al. 2013)	Field → catchment	Use of locally available natural inputs, complementing organic agricultural practices with local knowledge (Barron 2006)	Labour and knowledge intense, availability and transport of organic matter, lower yields compared to intensive, conventional systems with inorganic fertilizer application, certification costs, risk of drop in demand (Müller et al. 2013; Panneerselvam et al. 2013)
Conservation Agriculture	Sustainable, resource-efficient agricultural production, environmental conservation, improved livelihoods, strengthening resilience and adaptation (Niang et al. 2014; FAO 2015d)	Fertility increase, water holding capacity improvement through build-up of soil organic matter, higher rainfall infiltration, soil moisture, a gradual increase of soil carbon, biodiversity increase, (Thierfelder	Increase economic gross margins, increased and stable crop yields (Thierfelder et al. 2013)	Field → catchment	Building on traditional technologies practiced by local farmers for example for intercropping agricultural production systems (Lahmar et al. 2012)	Low biomass productivity and multiple land uses hence high opportunity costs of crop residues, labour intense during aggradative phase, and due to weeding, but later labour requirements decrease (Lahmar et al. 2012;

(continued)

Table 9.1 (continued)

Agricultural practice	Multiple goals	Environmental benefits, Ecosystem Services	Socio-economic benefits	Scale	Relevance of locally adapted techniques using diverse knowledge sources	Drawbacks
		et al. 2013), gradual soil quality improvement (Corbeels et al. 2014)				Corbeels et al. 2014), complex interactions between soil types, seasonal effects and tillage (Baudron et al. 2012)
Agro-forestry	Income security, food security, strengthening resilience and adaptation, environmental protection (Niang et al. 2014; van Noordwijk et al. 2014)	Microclimate improvement, soil protection, soil fertility increase, carbon storage increase, biodiversity increase, better water and nutrient cycling, ground water recharge, reduced run-off, erosion and flooding control (Bayala et al. 2014)	Availability of easily accessible forest products at critical times, diversity of products and improved diet (Bossio et al. 2010; Bayala et al. 2014)	Field → catchment	Farmer managed natural regeneration (Reij et al. 2009), role of local knowledge emphasized for successful agroforestry (Chazdon 2008)	A dense tree canopy can also reduce crop yields, and increases diseases, competition between the plants for nutrients and water (Bayala et al. 2014)
Water harvesting	Drought mitigation, adaptation, food security, strengthening social and ecological resilience	Water productivity increase, reduced erosion, soil fertility increase, improved soil moisture, runoff and groundwater	Yield increase, reduced agricultural expansion, improved water availability, reduced flood risk	Field → Catchment	*Zaï* and half-moon techniques are build on traditional knowledge (Reij et al. 2009)	Ex-situ: evaporation and seepage losses, siltation, capital intensive, community institutions needed, knowledge

	(Biazin et al. 2012; Dile et al. 2013)	recharge, reduced risk of erosion (Yosef and Asmamaw 2015), increase in biodiversity, local groundwater recharge, land rehabilitation, reduced stream flow pollution, increased biodiversity and carbon sequestration	(Yosef and Asmamaw 2015)			on the downstream impacts (Lasage and Verburg 2015), risk of vector borne diseases (Yosef and Asmamaw 2015) In situ: water logging, labor intense, no protection against poor rainfall distribution, water logging, labor intense (Biazin et al. 2012)
Evergreen agriculture	Improve livelihoods, sustain resource base, sustainable food security, environmental resilience (Garrity et al. 2010)	Soil fertility increase, soil carbon increase, water infiltration and water holding capacity of soils increase (Garrity et al. 2010)	Yield increase, higher benefit to cost ratio (Ajayi et al. 2009), income diversification, increased food security, fodder production	Field → catchment	Farmer managed natural regeneration and further locally adapted strategies exist (Reij et al. 2009; Garrity et al. 2010)	Slow initial growth phase of fertilizer tree Faidherbia, knowledge intensive, initially very labor intensive (Garrity et al. 2010)
Ecological intensification	Making use of the regulating functions of nature at landscape level (Bommarco et al. 2013; Tittonell 2014)	Better provision of climate change related ecosystem services, such as carbon sequestration, energy use efficiency, soil water holding capacity, resilience to drought and excessive	Integrated pest and disease management, enhanced productivity (Ochola et al. 2013)	Landscape level	Techniques to increase soil organic matter content are an important part of many traditional farming systems, complemented with scientific research	Knowledge and labor intense (Ochola et al. 2013), an understanding of relations of land management and ecosystem services at different scales is needed, research/

(continued)

Table 9.1 (continued)

Agricultural practice	Multiple goals	Environmental benefits, Ecosystem Services	Socio-economic benefits	Scale	Relevance of locally adapted techniques using diverse knowledge sources	Drawbacks
		rainfall (Rossing et al. 2013), in field and off-field diversity (Crowder et al. 2010; Bommarco et al. 2013), soil fertility improvement and ecological adaptability (Ochola et al. 2013)				knowledge gaps (Bommarco et al. 2013), transition period until benefits occur (Tittonell 2014)

conservation agriculture, water harvesting, organic agriculture, integrated pest management and others. These approaches draw on the capabilities of smallholders to be innovative and manage and conserve land through participatory methods of decision-making, implementation and capacity building. Integrated pest and nutrient management, conservation tillage, agroforestry, cover crops, aquaculture, water harvesting and livestock integration are usually referred to as sustainable intensification approaches (Pretty et al. 2011). Pretty et al. (2011) compiled evidence from African farmers, applying a wide range of approaches for sustainable intensification and found evidence of reduced soil erosion; increased resilience to climate-related shocks such as droughts; increased soil carbon content; improved water productivity; reduced debt and production costs; livelihood diversification; and improved household-level food security and income. These approaches, therefore, meet many of the criteria as detailed in Box 9.1.

Sustainable land management (SLM) is "*land managed in such a way as to maintain or improve ecosystem services for human well-being, as negotiated by all stakeholders*" (Winslow et al. 2009:63). It includes other approaches such as soil and water conservation, natural resources management and integrated ecosystem management and aims to achieve productive and healthy ecosystems by integrating social, economic, physical and biological needs and values in a holistic manner (Liniger et al. 2011). Among others, Thomas (2008) emphasizes the role SLM could play in simultaneously addressing problems of land degradation, climate change adaptation and mitigation and biodiversity conservation. Liniger et al. (2011) stress the ability to prevent, mitigate and rehabilitate land degradation and address water scarcity, low soil fertility, lack of organic matter and reduced biodiversity. With the primary objectives of enhancing food production, addressing land degradation and providing sustainable and resilient livelihoods (UNCCD 2012b), these techniques are also associated with substantial economic, social and environmental co-benefits such as timber and fuel wood production, non-timber forest production, cultural preservation, biodiversity maintenance, recreation and tourism – all provided at usually low costs (Jones et al. 2012; Munang et al. 2014; Davies et al. 2015). UNCCD (2012d) report growing evidence that SLM can reduce poverty and lead to sustainable economic growth. All these benefits simultaneously strengthen the resilience of farmers and make their agricultural production less susceptible to droughts.

Climate smart agriculture is built on three main pillars, namely sustainably increasing agricultural productivity, increasing resilience to climate change and reducing greenhouse gases emissions (FAO 2010, 2013c). Context-specific and locally adapted techniques that address prevailing risks and livelihood situations are favored (Zougmoré et al. 2014). Climate-smart agriculture embraces all strategies that integrate land and water management, contribute to the build-up of soil organic matter and use varieties well adapted to changing climatic conditions. It therefore directly addresses the principles of EbA/Eco-DRR and supports CCA and DRR.

Ecological intensification is the use of all resources, such as land, water, biodiversity and nutrients, in an efficient, regenerative manner, while minimizing negative impacts. Therefore, it is considered by FAO as "*a knowledge-intensive* *process that requires optimal management of nature's ecological functions and biodiversity to improve agricultural system performance, efficiency and farmers' livelihoods"* (FAO 2015a:1). It is a context-specific, ecosystem-based, *"smart use of the natural functionalities of the ecosystem (support, regulation) to produce food, fiber, energy and ecological services in a sustainable way"* (Tittonell 2014:58), recognizing the role of local and indigenous knowledge. Ecological intensification fosters the management of regulating and supporting services, while enhancing the productivity of agricultural systems and reducing anthropogenic inputs (Bommarco et al. 2013). It includes approaches based on agro-ecology, organic agriculture, some diversified farming systems, nature mimicry, some forms of conservation agriculture, agro-forestry and evergreen agriculture (Tittonell 2014). Through the direct management of regulating services, entry points for drought mitigation are inherent characteristics of this approach.

Eco-agriculture is an approach operating at the landscape level with the goal of maintaining biodiversity and ecosystem services and of managing agricultural production in a sustainable way, in order to improve rural livelihoods. The approach stresses links between different ecosystems and ecosystem functions at the landscape level (Scherr and McNeely 2008; Bossio et al. 2010). It provides opportunities to include ecosystem services that contribute to drought risk mitigation and drought resilience in multifunctional landscape planning. Eco-agriculture aims to advance multiple goals in the same landscape and hence provides room for explicitly targeting drought risk reduction and adaptation objectives, in addition to other objectives.

9.3.2 More Specific Approaches and Agricultural Techniques

Organic agriculture *"is a holistic production management system which promotes and enhances agroecosystem health, including biodiversity, biological cycles, and soil biological activity. It emphasizes the use of management practices in preference to the use of off-farm inputs, taking into account that regional conditions require locally adapted systems. This is accomplished by using, where possible, cultural, biological and mechanical methods, as opposed to using synthetic materials, to fulfil any specific function within the system. (...)"* (FAO/WHO 2006:2).

Organic agriculture is often recognized as an approach to sustainable livelihoods in the context of sustainable development and vulnerability reduction (Milestad and Darnhofer 2003; Borron 2006; Bennett and Franzel 2013; Müller et al. 2013). Key strategies for organic agriculture are crop diversification and increasing soil organic matter. Crop diversification contributes to a more efficient use of nutrients and water, with multiple sowing dates for different crops, which could decrease the risk of crop failures due to dry spells and increase livelihood resilience to such threats by providing different crops at different points in time and fostering biodiversity.

Increasing soil organic matter enhances nutrient levels in the soil and thus maintains/increases soil productivity. Soil organic matter also improves the soil's water holding capacity so that available water in the plant root zone is increased (Reganold et al. 1987; Emerson 1995; Pimentel et al. 2005). Therefore, organic agriculture is less susceptible to dry spells (see e.g., Kloos and Renaud 2014 for a study in northern Benin) and other extreme weather conditions such as drought, flood and waterlogging, as well as reduced wind and water erosion (IPCC 2007). Increased soil fertility and soil moisture have been found to increase yield (in low yield environments) and net income increases were observed, together with additional benefits such as improved food security, investment in improved housing conditions, school attendance of children and reduced migration (Panneerselvam et al. 2013).

Water harvesting is the collection of runoff for productive purposes. Water harvesting techniques can be classified into macro-catchment systems, micro-catchment systems and in-situ systems (Dile et al. 2013).[10] Ex-situ water harvesting systems have been shown to mitigate intra-seasonal dry spells and increase water productivity, which leads to yield improvements (Mwenge Kahinda et al. 2007). In-situ techniques prevent soil erosion, increase deposition of nutrients and organic matter and thereby improve soil fertility. They increase the soil water content in the root zone and therefore help bridge dry spells. Water harvesting systems have been found to contribute to yield improvements and to sustain ecosystems in agricultural landscapes. Through these mechanisms, social and ecological resilience to natural hazards are strengthened, and climate change and food insecurity are addressed. (Agro-)ecosystems can be stabilized, while additional benefits to people are provided (Biazin et al. 2012; Dile et al. 2013).

Among water harvesting techniques, there are many, sometimes traditional practices, which can be considered as Eco-DRR/EbA according to the criteria in Box 9.1, and which are multi-functional.[11] *Zaï* and half-moon techniques[12] are examples of traditional practices whereby run-off water and organic matter are concentrated in small pits, thereby conserving water and soil (Barry et al. 2008). These traditional methods have been extensively promoted and were well adopted by farmers in Burkina Faso (Kaboré and Reij 2004; Reij et al. 2009). Planting pits concentrate water and nutrients directly where needed by the crops, restore soil fertility, increase water holding capacity and directly collect water. This helps crops to survive long dry spells or dry spells during critical stages of crop growth (Reij et al. 2009). Due to the application of organic matter concentrated in the planting pits, trees and shrubs also germinate and are often found to be protected by farmers

---

[10]Dile et al. (2013) provide an overview of different types of water harvesting systems (ex-situ and in-situ), their biophysical and ecological functions, mechanisms, social implications and drawbacks.

[11]There are many mixed approaches. Applying organic matter or fostering the growth of nitrogen fixing trees is simultaneously a measure of nutrient management or agro-forestry.

[12]Small pits are called *"Zaï"* or *"tassa"* and larger, half-moon shaped holes *"Demi-lunes"* (half-moons).

in order to establish agro-forestry systems using *Zaï* for reforestation (Reij et al. 2009). However, problems of waterlogging may occur in very wet years (Lee and Visscher 1990). As land preparation is carried out during the dry season, labor for other crops is available during the start of the rainy season.

Overall, *Zaï* has proven very beneficial for badly degraded areas, such as in the Central Plateau region of Burkina Faso, where between 200,000 and 300,000 ha of land have been rehabilitated using the technique alone or in combination with stone bunds and/or agroforestry systems (Reij et al. 2009). This has been shown to improve food security by reducing the number of months without food deficits, to enable vegetation regrowth with additional benefits and to reduce migration rates due to improved livelihoods. The example from Burkina Faso is referred to as a successful EbA (Reij et al. 2009; Munang et al. 2014). In addition, there are other water harvesting approaches that combine traditional knowledge with scientific knowledge:

- Soil and water conservation structures such as terracing systems in steep zones or stone lines are known throughout Africa. Contour stone/rock bunds or vegetative barriers slow down and filter run-off, which facilitates infiltration and the capture of sediments, thereby increasing soil water and reducing erosion. '*Fanya juu*' terraces common in East Africa are built together with a ditch, along the contour of a sloping terrain.[13]
- Rainwater harvesting and catching runoff in small dams or waterholes is practiced in wide areas, as well as specific small-scale irrigation systems such as "*Ndiva*" (Enfors and Gordon 2008). Among the different water harvesting techniques, traditional micro-catchment approaches have been shown to attract the greatest uptake in the Sahelian zone of West Africa (Barry et al. 2008).

Evergreen agriculture involves integrating trees into cropping systems (Garrity et al. 2010) and is a combination of conservation agriculture and agro-forestry (see below) practices within the same location. Through the inter-cropped trees, a green vegetation cover is maintained throughout the year. Nitrogen-fixing trees increase the nutrient supply and trees generate organic matter, with positive impacts on water infiltration and the water holding capacity of soils, thus supporting drought resistance.

Garrity et al. (2010) describe case studies from Zambia, Malawi, Niger and Burkina Faso that present a variety of locally adapted strategies combined under the umbrella of evergreen agriculture. Many of these approaches show a reduction of climatic risks under evergreen agriculture.

Conservation agriculture (CA) is based on three principles: minimal soil disturbance; permanent soil cover; and crop rotations, in order to achieve

---

[13]Stone lines on low slopes are mainly found in West Africa (Burkina Faso, Mali, Niger); Earth bunds/ridges mainly in East Africa (Ethiopia, Kenya) and Southern Africa (Malawi, Zambia, Zimbabwe, etc.); *Fanya juu* mainly in East Africa (Kenya; also Ethiopia, Tanzania, Uganda); vegetative strips throughout Africa especially in the more humid parts (Liniger et al. 2011).

sustainable and profitable agriculture and improved livelihoods (FAO 2015d). It is increasingly promoted as a measure to address land degradation, mitigate droughts and increase economic gross margins. The soil organic matter content is often low in SSA; hence, the permanent organic soil cover in CA benefits the water balance and biological activity and contributes to the in-situ build-up of soil organic matter in the soil. However, manure and other organic matter are often scarce in African subsistence agriculture as it is used for multiple purposes (e.g. for fodder, building activities etc.), and this can hinder the success of CA in SSA (Lahmar et al. 2012). CA can help to conserve soil moisture because soil is covered by crop residues, which makes it an effective technology for mitigating the negative effects of erratic rainfall or dry spells (Corbeels et al. 2014). The recent IPCC chapter of WG II on Africa recognizes with high confidence that conservation agriculture, including approaches such as agro-forestry and farmer-managed natural tree regeneration, conservation tillage, contouring, terracing and mulching, *"provides a viable means for strengthening resilience in agro-ecosystems and livelihoods that also advance adaptation goals"* (Niang et al. 2014:1203).[14]

Agro-forestry means that trees are managed together with crops and/or animal production systems in agricultural settings (FAO 2013a). In general, agro-forestry systems are classified into agrosilvicultural ("trees with crops"), silvopastoral ("Trees with livestock") and agrosilvopastoral ("Trees with both crops and livestock"). Agro-forestry enables farmers to better withstand drought and climate change, enhances biodiversity, reduces erosion and contributes to water and nutrient cycling (Bayala et al. 2014). Forest resources, such as non-timber forest products, can provide safety nets in case of shocks as they are available when other resources may be affected by droughts. Trees are recognized for the multifunctional value they provide (Bossio et al. 2010). Agro-forestry parklands are traditionally used among Sahelian farmers (Boffa 1999; Bayala et al. 2014). Additional provisioning services such as food, fuel, fodder, medicine, wood and building materials become available for farmers and the local population. Furthermore, regulating services, such as micro-climate regulation and ground water recharge, and supporting services, which are needed to maintain other services, are provided. In particular, soil carbon sequestration, nutrient cycling and reduced greenhouse gas emissions are supplied (Bayala et al. 2014), and these services are also recognized to contribute to soil fertility improvements and water conservation.

The loss of traditional agro-forestry systems could be addressed by assisted natural vegetation, so-called Farmer Managed Natural Regeneration techniques (FMNR), which have resulted in a significant increase of tree cover in the Zinder and Maradi regions in Niger, compared to a few decades before. Reij et al. (2009) found that these techniques have reduced the villages' risks of food shortages[15]

---

[14]For more details on drivers and constraints for adaption of conservation agriculture in SSA see Corbeels et al. (2014).

[15]Agroforestry in Western Kenya has increased food security during drought and flooding by 25 % due to increased income and improved livelihoods (Thorlakson and Neufeldt 2012).

caused by droughts or other factors. Trees reduce wind speed, evaporation and the need to re-sow (Larwanou et al. 2006), freeing labor for other activities (Garrity et al. 2010). Furthermore, reduced migration, lowered infant mortality and empowerment of women were observed in villages with FMNR (Reij et al. 2009).

Agro-forestry and reforestation have been promoted and used in many SSA countries as a way to mitigate climatic variability, deal with droughts and reduce desertification (Fisher et al. 2010). Restoring forests seems to be more successful when approaches include local knowledge of tree characteristics, use diverse species with particular economic or ecological benefits and integrate forest rehabilitation into general development strategies (Chazdon 2008). However, it is important to be aware of potential trade-offs. While in parkland environments, for instance, trees have been shown to yield multiple benefits in terms of microclimate and soil fertility improvements, some questions remain about their effects on crop production through competition for resources, depending on the crop-tree combinations (Bayala et al. 2014).

Crop-livestock integration is very common in rain-fed farming systems in SSA (Powell et al. 2004). In these mixed systems, productivity and management of croplands, rangelands and livestock are closely linked via nutrient cycling (e.g. grazing, fodder, manure), income (availability, investment and storage), and labor (e.g. animal power).

An integrated farming system[16] is based on a coordinated framework in which the waste or by-products of one component are used as an input for other components (IFAD 2010). Due to the cyclic nature of production, management decisions related to one component may affect the others. Approaches to maintain and enhance the productivity of mixed crop-livestock systems require a thorough understanding of the socio-economic and biophysical components and interactions (Powell et al. 2004). In terms of DRR and CCA, a balanced approach is needed that incorporates minimization of risks associated with rainfall variability and/or dry spells and droughts through e.g. improved water productivity (see e.g. Herrero et al. 2010; Amede et al. 2011 for more information). However, currently, there is still a gap in the literature on how best to use the interactions in mixed crop-livestock systems to buffer farmers against climate change and droughts (Thornton and Herrero 2015).

In order to be fully effective, these approaches and agricultural techniques, including strategies to improve water and land management, need to be linked to

---

[16]Diversified systems consist of components such as crops and livestock that coexist independently from each other. In this case, integrating crops and livestock serves primarily to minimize risk and not to recycle resources. In an integrated system, crops and livestock interact to create a synergy, with recycling allowing the maximum use of available resources (FAO 2001 as cited in IFAD 2010).

complementary approaches such as integrated plant nutrient management,[17] integrated pest management,[18] timing of activities and diversification, as well as integrating livestock management, post-harvest management and marketing and institutional aspects (Rockström 2003). Integrated approaches that combine well-adapted, diverse crops and crop varieties, agricultural management practices that conserve soil and water and increase the resilience of the agro-ecosystem to droughts, and institutional and policy options of drought risk management (e.g. forecasting and early warning systems, input/output market development and insurance systems) can strengthen the resilience of the social-ecological systems at multiple scales (Shiferaw et al. 2014).

9.4 Discussion

This overview of different agricultural approaches and techniques is not exhaustive,[19] but shows that there are multiple ways of managing agro-ecosystems in order to provide benefits which strengthen ecosystem functions (in particular the directly drought-related variables of water-holding capacity and infiltration rates, which are both linked in part to the organic carbon content of the soils) and increase the ecological buffer capacity. Soil protection and better water and nutrient cycling are additional aspects that reduce the susceptibility of the environmental system to drought, but also improve soil fertility, which has direct positive implications for the social system. Furthermore, a suitable micro-climate and the maintenance of diverse species, generating multifunctional agro-ecosystems, may increase the response diversity and hence the capacity of ecosystems to buffer against droughts (Liniger et al. 2011). Additionally, carbon sequestration is one environmental benefit that contributes to climate change mitigation goals. Table 9.1 summarizes some important socio-economic co-benefits that are linked to the approaches. These benefits, such as increases in income, improved food and water security, improved health and reduced economic risks, all contribute to poverty reduction and sustainable development goals linked to EbA/Eco-DRR approaches (as described in

---

[17]Integrated Plant Nutrient Management *"aims to optimize the condition of the soil, with regard to its physical, chemical, biological and hydrological properties, for the purpose of enhancing farm productivity, whilst minimizing land degradation"* (FAO 2015c).

[18]Integrated pest management is an ecosystem-based approach to crop production and protection that aims to ensure the growth of healthy crops with the least possible disruption to agro-ecosystems and encourages natural pest control mechanisms (FAO 2015b). Using an ecosystem approach to control pests, a *"coordinated integration of multiple complementary methods to suppress pests in a safe, cost-effective, and environmentally friendly manner"* is needed (Parsa et al. 2014:3889). Prevention of pests is addressed through developing ecosystem resilience and diversity for pest, disease, and weed control. Pesticides are only used when other options are ineffective (Pretty et al. 2011).

[19]We did not include livestock-related agricultural practices, nor specific forestry management systems or fish production.

Sect. 9.2, Box 9.1). Overall, these strategies ensure the flow of a wider range of ecosystem services, even when faced with climatic shocks, such as droughts.

While many of the reviewed approaches have particular links to EbA and Eco-DRR, there are also potential pitfalls. Increasing agricultural productivity is still an important aspect of many of the described approaches. When applying these approaches in an Eco-DRR or EbA context, the focus shifts towards ensuring productivity in drought-prone years and under difficult rainfall conditions, rather than aiming to maximize yields during years where rainfall patterns correspond well to crop water needs (Davies et al. 2015).

It is important for risk management to differentiate between manageable droughts, where improved management and livelihood resilience can help mitigate impacts at the farm or watershed level, and unmanageable droughts where the preparedness and coping capacities at the small-scale farmers' level are overwhelmed and mechanisms outside the watershed are required (Rockström 2003). However, we found that this issue is not very much discussed in the reviewed approaches and techniques. The degree of risk reduction that can be provided by these approaches, for individual farmers but also at larger landscape scales, seems to be less well researched.

Many of the approaches can increase the capacity of an agro-ecosystem to maintain its functionality in case of droughts and dry spells, but often, information is missing about the duration of dry spells that a particular approach is able to tackle. This hinders, for instance, the useful combination of ecosystem-based approaches with other DRR or CCA strategies, such as structural measures or disaster preparedness, in order to more efficiently reduce risks.

For example, Barbier et al. (2009) show in a case study in Northern Burkina Faso that micro-level water harvesting techniques are beneficial, but have their limits and are insufficient in order to substantially reduce vulnerability and poverty. Garrity et al. (2010) observe that some quantified impacts, at least for the FMNR-approaches at larger scales, are available, which could provide useful information for disaster risk managers. Another example is a study by Ajayi et al. (2009), which quantifies the impact of evergreen agriculture on food security by estimating the number of additional food secure days in a household. It is one example of how to quantify drought risk reduction impacts that goes beyond improved yields. Such an estimate could then provide useful information for a comprehensive risk reduction strategy (see Garrity et al. 2010).

Increasingly, there is a call for multifunctionality at the landscape level, as such a perspective can help to meet multiple objectives of food production, biodiversity conservation, land rehabilitation and also drought mitigation (Minang et al. 2015). Many processes have impacts off-site that require management at a broader scale in a systemic manner. As different ecosystem services require management strategies at different temporal or spatial scales, a landscape perspective is needed to ensure that broader scale ecosystem services are not negatively affected. For example, pollination services or biological weed control (e.g. within IPM) are affected by farm-level activities, but are also strongly influenced by the spatial configuration and diversity of the surrounding landscape (Bommarco et al. 2013). While soil-related services may be best managed at the farm-scale, *Zaï* and half-moon techniques, for instance, can increase water level and tree cover if applied at the watershed scale (Bayala et al. 2014). Some of the existing approaches, such as ecological intensification, explicitly operate at the landscape level; other site-level or farm-level approaches could also be applied at a landscape level. In the case of rainwater harvesting approaches, Karpouzoglou and Barron (2014) argue that successful generalization of the adoption of such technologies requires an understanding of the processes in play at various spatial scales which influence adoption. This includes the shifting ideology associated with food production systems (from purely productivist systems to factoring in equity and sustainability concepts in food production), integrating agro-ecological approaches into agricultural research and development, as well as putting more emphasis on traditional and local knowledge.

EbA and Eco-DRR span different spatial scales, from the local/community level to the subnational, national and sometimes international level. Because they are multisectoral and multidisciplinary, EbA and Eco-DRR require communication and consensus-building among all stakeholders.[20] Stakeholder communication, negotiation and participation are integral parts of sustainable land management. As local conditions need to be considered in Eco-DRR/EbA, specific, well-adapted and negotiated approaches are required.

A suitable Eco-DRR or EbA approach for drought risk reduction would therefore consist of multiple complementary tools, for instance drawing from the various agro-ecological approaches as well as other approaches. But before replications of successful approaches can be considered at larger scales, scientific evidence of the effects of the approach on the environment and livelihoods is required, so that it can be adapted to a given site. Many of the approaches presented above are extremely knowledge intensive due to the complexity of ecological processes, particularly when operating at different spatial scales. Local knowledge and traditional techniques play a key role in many approaches described above. Some of the examples of water harvesting techniques or agro-forestry are traditional practices (e.g. *Zaï*, *Fanya juu*), which have been revived and further developed and are well supported through institutions and policies.

To plan EbA and Eco-DRR at the landscape level, easy access to information about the approaches is required (such as ecosystem services provision, but also including governance and institutional aspects). One useful development in that

---

[20]There are a few examples in the region of successful large scale implementation of ecosystem-based measures to reduce the impacts of climatic droughts. In terms of large scale implementation, the Great Green Wall (GGW) initiative, which is an African partnership to tackle desertification in the Sahel and Sahara, is perhaps the most contemporary one. This initiative encompasses 13 countries and addresses the desertification problem through a variety of interventions (i.e. not limited to planting a tree barrier). It also aims to support the efforts of local communities in the sustainable management of their resources. By doing so, the initiative contributes to climate change mitigation and adaptation and to the improvement of the livelihoods of the communities in the region (FAO 2013b).

direction is the World Overview of Conservation Approaches and Technologies (WOCAT) platform, which aims to unite the efforts in knowledge management and decision support for up-scaling SLM among all stakeholders, including national governmental and non-governmental institutions and international and regional organisations and programmes. It provides a wealth of knowledge on sustainable land management, including global online databases (WOCAT 2015). A systematic assessment based on the concept of ecosystem services, and including long term impacts, could be helpful in comparing and selecting between complementary approaches and tools for an Eco-DRR/EbA strategy. A continuous dialogue between scientists and local farmers is needed in order to generate knowledge and exchange experiences (Tittonell 2014).

Governance is an important aspect of Eco-DRR and EbA and is explicitly referred to in many of the above described approaches and techniques. Resource-conserving agriculture, for instance, emphasises the use of participatory processes for decision-making, implementation and capacity building. Governance aspects of Eco-DRR and EbA could not be tackled in depth in this review chapter. However, it is important to stress that many successful approaches are built on customary governance schemes. These approaches strengthen the role of local practices and existing resource-governing institutions. Particularly from an EbA perspective, the existing governance and institutional systems need to be capable of supporting flexible and adaptive management,[21] given the prevailing uncertainties, non-linear effects, cross-scale effects and thresholds of social-ecological systems under climate change. Such systems should incorporate mechanisms for experimentation, innovation and learning, and management approaches at all levels need to be kept flexible and adaptive (Liniger et al. 2011).

9.5   Conclusions and Outlook

There are multiple approaches that apply ecological principles to ensure agriculturally productive farms and landscapes and the continuous flow of ecosystem services, even when hazards such as droughts occur. As drought prevention and exposure reduction options are very limited in drylands, strengthening the resilience of agro-ecosystems and reducing their susceptibility to drought impacts are necessary, while enhancing their capacities to cope and recover.

The literature reviewed still focuses very much on provisioning services, in particular, yield potentials. Increasingly though, studies are including additional ecosystem services and, in particular, longer term impacts on livelihoods, food and water security or off-site effects. However, assessments that quantify a wide range

---

[21] Instead of aiming to minimize disturbances and uncertainties, adaptive management strives to strengthen resilience by providing space for experimentation, learning and understanding of ecological processes (Darnhofer et al. 2010).

of ecosystem services and how they are impacted by droughts or how they can help provide resilient livelihoods have not yet been comprehensively researched. The lack of coherence in the existing assessments makes it difficult to compare studies in terms of impacts. More systematical assessments would be important in order to be able to combine approaches and agricultural techniques for Eco-DRR/EbA in a complementary way.

While many studies highlighted some socio-economic benefits from the approaches that can contribute overall to drought risk reduction and resilience, more research is needed specifically from an Eco-DRR perspective. How much do these co-benefits support disaster risk reduction when hazards of different intensities strike?

Despite all the positive aspects of Eco-DRR and EbA, an honest discussion of what Eco-DRR/EbA can and cannot provide is important (see e.g., Cook et al. 2015 for a similar discussion on ecological intensification). While many of the described approaches are considered to be win-win situations (Liniger et al. 2011), this is not always the case, and some require trade-offs in terms of ecosystem service deliveries (e.g. among different provisioning services, vis a vis regulating services). Missing information on the limits of Eco-DRR and EbA poses a challenge to their effective integration into DRR and CCA planning. This chapter has shown that many of the existing drought risk reduction approaches are Eco-DRR/EbA in nature, but that they are not the sole answer to mitigate drought risks in SSA's drylands.

References

Ajayi O, Akinnifesi F, Sileshi G, Kanjipite W (2009) Labour inputs and financial profitability of conventional and agroforestry- based soil fertility management practices in Zambia. Agrekon 48(3):276–292

Amede T, Tarawal S, Peden D (2011) Improving water productivity in crop-livestock systems of drought-prone regions. Editorial comment. Exp Agric 47(S1):1–5. doi:10.1017/S0014479710001031

Andrade A, Cordoba R, Dave R et al (2011) Draft principles and guidelines for integrating ecosystem-based approaches to adaptation in projects and policy design: a discussion document. Serie Técnica. Boletin Técnico no. 46. IUCN-CEM, CATIE, Turriabla

Banerjee O, Bark R, Connor J, Crossman ND (2013) An ecosystem services approach to estimating economic losses associated with droughts. Ecol Econ 91:19–27

Barbier B, Yacouba H, Karambiri H et al (2009) Human vulnerability to climate variability in the sahel: farmers' adaptation strategies in Northern Burkina Faso. Environ Manag 43:790–803

Barron S (2006) Building resilience for an unpredictable future: how organic agriculture can help farmers adapt to climate change. Food and Agriculture Organization of the United Nations, Sustainable Development Department

Barry B, Olaleye AO, Zougmoré R, Fatondji D (2008) Rainwater harvesting technologies in the Sahelian zone of West Africa and the potential for outscaling. IWMI Working Paper 126. International Water Management Institute, Colombo

Baudron F, Tittonell P, Corbeels M et al (2012) Comparative performance of conservation agriculture and current smallholder farmingpractices in semi-arid Zimbabwe. Field Crop Res 132:117–128

Bayala J, Sanou J, Teklehaimanot Z et al (2014) Parklands for buffering climate risk and sustaining agricultural production in the Sahel of West Africa. Curr Opin Environ Sustain 6:28–34

Bennett M, Franzel S (2013) Can organic and resource-conserving agriculture improve livelihoods? A synthesis. Int J Agric Sustain 11(3):193–215. doi:10.1080/14735903.2012.724925

Benson C, Clay E (1998) The impact of drought on Sub-Saharan African economies: a preliminary examination. Technical Paper No. 401. World Bank, Washington, DC

Biazin B, Sterk G, Temesgen M et al (2012) Rainwater harvesting and management in rainfed agricultural systems in Sub-Saharan Africa – a review. Phys Chem Earth Parts A/B/C 47–48:139–151

Boffa JM (1999) Agroforestry Parklands in Sub-Saharan Africa, FAO conservation guide. FAO, Rome

Bommarco R, Kleijn D, Potts S (2013) Ecological intensification: harnessing ecosystem services for food security. Trends Ecol Evol 28(4):230–238

Borron S (2006) Building resilience for an unpredictable future: how organic agriculture can help farmers adapt to climate change. Food and Agriculture Organization of the United Nations, Rome

Bossio D, Geheb K, Critchley W (2010) Managing water by managing land: addressing land degradation to improve water productivity and rural livelihoods. Agric Water Manag 97:536–542

Chazdon RL (2008) Ecosystem services on degraded lands. Science 320:1458–1460. doi:10.1126/science.1155365

Colls A, Ash N, Ikkala N (2009) Ecosystem-based adaptation: a natural response to climate change. IUCN, Gland

Cook S, Silici L, Adolph B (2015) Sustainable intensification revisited. Briefing. IIED

Corbeels M, De Graaff J, Ndah H et al (2014) Understanding the impact and adoption of conservation agriculture in Africa: a multi-scale analysis. Agric Ecosyst Environ 187:155–170

Crowder DW, Northfield TD, Strand MR, Snyder WE (2010) Organic agriculture promotes evenness and natural pest control. Nature 466:109–112

Darnhofer I, Bellon S, Dedieu B, Milestad R (2010) Adaptiveness to enhance the sustainability of farming systems: a review. Agron Sustain Dev 30(3):545–555

Davies J, Ogali C, Laban P, Metternicht G (2015) Homing in on the range: enabling investments for sustainable land management. Technical Brief 29/01/2015. IUCN/CEM, Nairobi

Dile YT, Kalrberg L, Temesgen M, Rockström J (2013) The role of water harvesting to achieve sustainable agricultural intensification and resilience against water related shocks in Sub-Saharan Africa. Agric Ecosyst Environ 181:69–79

Doswald N, Estrella M (2015) Promoting ecosystems for disaster risk reduction and climate change adaptation. Discussion paper UNEP

Doswald N, Munroe R, Roe D et al (2014) Effectiveness of ecosystem-based approaches for adaptation: review of the evidence-base. Clim Dev 6(2):185–201

Dudley N, Buyck C, Furuta N et al (2015) Protected areas as tools for disaster risk reduction. A handbook for practitioners. MOEJ/IUCN, Tokyo/Gland

Emerson WW (1995) Water retention, organic carbon and soil texture. Aust J Soil Res 33:241–251. doi:10.1071/SR9950241

Enfors EI, Gordon LJ (2008) Dealing with drought: the challenge of using water system technologies to break dryland poverty traps. Glob Environ Chang 18(4):607–616

Estrella M, Saalismaa N (2013) Ecosystem-based disaster risk reduction (Eco-DRR): an overview. In: Renaud FG, Sudmeier-Rieux K, Estrella M (eds) The role of ecosystems in disaster risk reduction. United Nations University Press, Tokyo/New York/Paris

Estrella M, Renaud FG, Sudmeier-Rieux K (2013) Opportunities, challenges and future perspectives. In: Renaud FG, Sudmeier-Rieux K, Estrella M (eds) The role of ecosystems in disaster risk reduction. United Nations University Press, Tokyo/New York/Paris

Falkenmark M, Rockström J (2008) Building resilience to drought in desertification-prone savannas in Sub-Saharan Africa: the water perspective. Nat Res Forum A United Nations Sust Dev J 32(2):93–102

FAO (2001) Stratégie Régionale harmonisée de mise en oeuvre de l'initiative « Grande muraille verte pour le Sahara et le Sahel ». Food and Agriculture Organization of the United Nations

FAO (2010) "Climate-Smart" Agriculture. Policies, Practices and Financing for Food Security, Adaptation and Mitigation. Food and Agriculture Organization of the United Nations, Rome

FAO (2013a) Advancing agroforestry on the policy agenda: a guide for decision-makers, by Buttoud G in collaboration with Ajayi O, Detlefsen G, Place F, Torquebiau E. Agroforestry Working Paper no. 1. Food and Agriculture Organization of the United Nations, Rome

FAO (2013b) Africa's great green wall reaches out to new partners. Food and Agriculture Organization of the United Nations. http://www.fao.org/news/story/en/item/210852/icode/. Accessed 25 Mar 2015

FAO (2013c) Climate-smart agriculture sourcebook. Food and Agriculture Organization of the United Nations, Rome

FAO (2015a) AGP – ecological intensification. Changing paradigms of agriculture. http://www.fao.org/agriculture/crops/thematic-sitemap/theme/biodiversity/ecological-intensification/en/. Accessed 25 Mar 2015

FAO (2015b) AGP – integrated pest mangement. Food and Agriculture Organization of the United Nations. http://www.fao.org/agriculture/crops/thematic-sitemap/theme/pests/ipm/en/. Accessed 25 Mar 2015

FAO (2015c) AGP – what is integrated plant nutrient management? Food and Agriculture Organization of the United Nations. http://www.fao.org/agriculture/crops/thematic-sitemap/theme/spi/scpi-home/managing-ecosystems/integrated-plant-nutrient-management/ipnm-what/en/. Accessed 25 Mar 2015

FAO (2015d) Agriculture and consumer protection department. Conservation agriculture. http://www.fao.org/ag/ca/. Accessed 25 Mar 2015

FAO/WHO (2006) Codex Alimentarius: organically produced foods.

Fisher M, Chaudhury M, McCusker B (2010) Do forests help rural households adapt to climate variability? Evidence from Southern Malawi. World Dev 38(9):1241–1250

Gallopin GC (2006) Linkages between vulnerability, resilience, and adaptive capacity. Glob Environ Chang 16:293–303

Garrity DP, Akinnifesi FK, Ajayi OC et al (2010) Evergreen agriculture: a robust approach to sustainable food security in Africa. Food Sec 2:197–214

Herrero M, Thornton PK, Notenbaert AM et al (2010) Smart investments in sustainable food production: revisiting mixed crop-livestock systems. Science 327:822–825

Holden ST, Binswanger HP (1998) Small farmers, market imperfections, and natural resource management. In: Lutz E, Binswanger HP, Hazell P, McCalla A (eds) Agriculture and the environment. Perspectives on sustainable rural development. The World Bank, Washington, DC

IFAD (2010) Integrated crop-livestock farming systems. Livestock thematic papers tools for project design. International Fund for Agricultural Development, Rome

IPCC (2007) Climate change 2007: impacts, adaptation and vulnerability. Contribution of working group II to the fourth assessment report of the IPCC. Cambridge University Press, Cambridge, UK

IPCC (2012) Managing the risks of extreme events and disasters to advance climate change adaptation: special report of the Intergovernmental Panel on Climate Change. Cambridge University Press, Cambridge, UK

IPCC (2014) Summary for policy makers working group II contribution to the fifth assessment report of the Intergovernmental Panel on Climate Change IPCC

Jones HP, Hole DG, Zavaleta ES (2012) Harnessing nature to help people adapt to climate change. Nat Clim Chang 2:504–509

Kaboré D, Reij C (2004) The emergence and spreading of an improved traditional soil and water conservation technique in Burkina Faso. EPTD Discussion paper no. 14. IFPRI – Environment and Production Division

Kallis G (2008) Droughts. Annu Rev Environ Resour 33:85–118

Karpouzoglou T, Barron J (2014) A global and regional perspective of rainwater harvesting in sub-Saharan Africa's rainfed farming systems. Phys Chem Earth 72–75:43–53

Kloos J, Renaud FG (2014) Organic cotton production as an adaptation option in north-west Benin. Outlook Agric 43(2):91–100

Lahmar R, Bationo BA, Lamso ND et al (2012) Tailoring conservation agriculture technologies to West Africa semi-arid zones: building on traditional local practices for soil restoration. Field Crop Res 132:158–167

Larwanou M, Abdoulaye M, Reij C (2006) Etude de la régénération naturelle assistée dans la Région de Zinder (Niger): Une première exploration d'un phénomène spectaculaire. International Resources Group for the. U.S. Agency for International Development, Washington, DC

Lasage R, Verburg PH (2015) Evaluation of small scale water harvesting techniques for semi-arid environments. J Arid Environ 118:48e57

Lee MD, Visscher TJ (1990) Water harvesting in Five African Countries. IRC Occasional Paper No. 14

Liniger HP, Mekdaschi Studer R, Hauert C, Gurtner M (2011) Sustainable land management in practice, Guidelines and best practices for Sub-Saharan Africa TerrAfrica. World Overview of Conservation Approaches and Technologies (WOCAT)/Food and Agriculture Organization of the United Nations (FAO), Rome

McGranahan DA (2014) Ecologies of scale: multifunctionality connects conservation and agriculture across fields, farms and landscapes. Land 3:739–769

Milestad R, Darnhofer I (2003) Building farm resilience: the prospects and challenges of organic farming. J Sustain Agric 22(3):81–97. doi:10.1300/J064v22n03_09

Minang PA, van Noordwijk M, Freeman OE et al (eds) (2015) Climate-smart landscapes: multifunctionality in practice. World Agroforestry Centre (ICRAF), Nairobi

Müller A (2009) Benefits of organic agriculture as a climate change adaptation and mitigation strategy in developing countries. Environment for Development Discussion Paper Series EfD DP 09-09

Müller A, Osman-Elasha B, Andreasen L (2013) The potential of organic agriculture for contributing to climate change adaptation. In: Halberg N, Müller A (eds) Organic agriculture for sustainable livelihoods. Earthscan food and agriculture. Routledge, London, pp 101–125

Munang R, Andrews J, Alverson K, Mebratu D (2014) Harnessing ecosystem-based adaptation to address the social dimension of climate change. Environ Sci Pol Sust Dev 56(1):18–24

Munroe R, Roe D, Doswald N et al (2012) Review of the evidence base for ecosystem-based approaches for adaptation to climate change. Environ Evid 1:13

Mwenge Kahinda J, Rockstrom J et al (2007) Rainwater harvesting to enhance water productivity of rainfed agriculture in the semiarid Zimbabwe. Phys Chem Earth 32:1068–1073

Niang I, Ruppel OC, Abdrabo MA et al (2014) Africa. In: Barros VR, Field CB, Dokken DJ et al (eds) Climate change 2014: impacts, adaptation, and vulnerability, Part B: Regional aspects. Contribution of working group II to the fifth assessment report of the Intergovernmental Panel on Climate Change. Cambridge University Press, Cambridge/New York, pp 1199–1265

Ochola D, Jogo W, Ocimati W et al (2013) Farmers' awareness and perceived benefits of agroecological intensification practices in banana systems in Uganda. Afr J Biotechnol 12 (29):4603–4613

Panneerselvam P, Halberg N, Lockie S (2013) Consequences of organic agricutlure for smallholder farmers' livelihood and food security. In: Halberg N, Müller A (eds) Organic agriculture for sustainable livelihoods. Routledge, London, pp 21–44

Parsa S, Morse S, Bonifacio A et al (2014) Obstacles to integrated pest management adoption in developing countries. Proc Nat Acad Sci USA 111:3889–3894

Pimentel D, Hepperly P, Hanson J et al (2005) Environmental, energetic, and economic comparisons of organic and conventional farming systems. Bioscience 55(7):573–582. doi:10.1641/0006-3568(2005)055[0573:EEAECO]2.0.CO;2

Powell JM, Pearson RA, Hiernaux PH (2004) Review and interpretation. Crop-livestock interactions in the West African Drylands. Agron J 96:469–483

Pretty JN, Noble AD, Bossio D et al (2006) Resource-conserving agriculture increases yields in developing countries. Environ Sci Technol 40(4):1114–1119

Pretty J, Toulmin C, Williams S (2011) Sustainable intensification in African agriculture. Int J Agric Sustain 9:5–24

Reganold JP, Elliott LF, Unger YL (1987) Long-term effects of organic and conventional farming on soil erosion. Nature 330(6146):6370–6372. doi:10.1038/330370a0

Reij C, Tappan G, Smale M (2009) Agroenvironmental transformation in the Sahel: another kind of "Green Revolution". IFPRI Discussion Paper 00914

Rockström J (2003) Resilience building and water demand management for drought mitigation. Phys Chem Earth 28:869–877

Rossing WAH, Modernel P, Tittonell P (2013) Diversity in organic and agro-ecological farming systems for mitigation of climate change impact, with examples from Latin America. In: Fuhrer J, Gregory P (eds) Climate change impact and adaptation in agricultural systems. CAB International, Wallingford, pp 69–87

Sayer J, Sunderland T, Ghazoul J et al (2013) Ten principles for a landscape approach to reconciling agriculture, conservation, and other competing land uses. Proc Nat Acad Sci USA 110(21):8349–8356

Scherr SJ, McNeely JA (2008) Biodiversity conservation and agricultural sustainability: towards a new paradigm of 'ecoagriculture' landscapes. Philos Trans R Soc B 363:477–494

Shiferaw B, Okello JJ (2011) Stimulating smallholder investment in sustainable land management: overcoming market, policy and institutional challenges. In: Kohlin RBG (ed) Agricultural investments, livelihoods and sustainability in East African agriculture. Resources for the Future, Washington, DC

Shiferaw B, Tesfaye K, Kassie M et al (2014) Managing vulnerability to drought and enhancing livelihood resilience in sub-Saharan Africa: technological, institutional and policy options. Weather Clim Ext 3:67–79

Stockholm Resilience Center (2007) http://www.stockholmresilience.org. Accessed 03 Feb 2015

Sudmeier-Rieux K (2010) Ecosystem approach to disaster risk reduction. Basic concepts and recommendations to governments, with a special focus on Europe. European and Mediterranean Major Hazards Agreement (EUR-OPA)

Sylla M, Giorgi F, Pal J, et al. (2015) Projected changes in the annual cycle of high intensity precipitation events over West Africa for the late 21st century. J Climate, in press. doi:10.1175/JCLI-D-14-00854.1

Thierfelder C, Mwila M, Rusinamhodzi L (2013) Conservation agriculture in eastern and southern provinces of Zambia: long-term effects on soil quality and maize productivity. Soil Tillage Res 126:246–258

Thomas RJ (2008) 10th anniversary review: addressing land degradation and climate change in dryland agroecosystems through sustainable land management. J Environ Monit 10:595–603

Thorlakson T, Neufeldt H (2012) Reducing subsistence farmers' vulnerability to climate change: evaluating the potential contributions of agroforestry in western Kenya. Agric Food Sec 1(15):1–13

Thornton PK, Herrero M (2015) Adapting to climate change in the mixed crop and livestock farming systems in Sub-Saharan Africa. Nat Clim Chang 5:830–836

Tittonell P (2014) Ecological intensification of agriculture – sustainable by nature. Curr Opin Environ Sustain 8:53–61

UNCCD (2012a) About the convention. Part 1, Article 1. http://www.unccd.int/en/about-the-convention/Pages/Text-Part-I.aspx. Accessed Nov 2014

UNCCD (2012b) Sustainable land management for food security. http://www.unccd.int/en/programmes/Thematic-Priorities/Food-Sec/Pages/FS-SLM.aspx. Accessed Mar 2015

UNCCD (2012c) What is ecosystem-based adaptation? http://www.unccd.int/en/programmes/Event-and-campaigns/WDCD/Pages/What-is-Ecosystem-Based-Adaptation.aspx. Accessed Mar 2015

UNCCD (2012d) Zero Net land degradation. A sustainable development goal for Rio + 20. Policy brief. Bonn

UNDP (2011) Mainstreaming drought risk management – a primer. http://frameweb.org/adl/en-US/7123/file/962/Mainstreaming%20DRM-English.pdf. Accessed Mar 2015

van Noordwijk M, Bizard V, Wangpakapattanawong P et al (2014) Tree cover transitions and food security in Southeast Asia. Glob Food Sec 3(3–4):200–208

Vignola R, Locatelli B, Martinez C, Imbach P (2009) Ecosystem-based adaptation to climate change: what role for policy-makers, society and scientists? Mitig Adapt Strateg Glob Chang 14(8):691–696

Wilhite DA, Glantz MH (1985) Understanding: the drought phenomenon: the role of definitions. Water Int 10(3):111–120

Wilhite DA, Sivakumar MVK, Pulwarty R (2014) Managing drought risk in a changing climate: the role of national drought policy. Weather and Climate Extremes 3:4–13

Winslow M, Sommer S, Bigas H et al (2009) Understanding desertification and land degradation trends. In: Proceedings of the UNCCD first scientific conference, 22–24 September 2009, during the UNCCD ninth conference of parties, Buenos Aires

WMO (1986) Report on drought and countries affected by drought during 1974–1985. World Meteorological Organization, Geneva

WMO, GWP (2014) National drought management policy guidelines: a template for action, Integrated Drought Management Programme (IDMP) tools and guidelines series 1. World Meteorological Organization/Global Water Partnership, Geneva/Stockholm

WOCAT (2015) Knowledge base. World Overview of Conservation Approaches and Technologies. https://www.wocat.net/en/knowledge-base.html. Accessed 25 Mar 2015

Yosef BA, Asmamaw DK (2015) Rainwater harvesting: an option for dry land agriculture in arid and semi-arid Ethiopia. Int J Water Res Environ Eng 7(2):17–28

Zougmoré R, Jalloh A, Tioro A (2014) Climate-smart soil water and nutrient management options in semiarid West Africa: a review of evidence and analysis of stone bunds and zaï techniques. Agric Food Sec 3:16. doi:10.1186/2048-7010-3-16

Chapter 10
Integrating Ecosystems in Risk Assessments: Lessons from Applying InVEST Models in Data-Deficient Countries

Niloufar Bayani and Yves Barthélemy

Abstract The linkages between ecosystem conditions and disaster risk reduction have gained increasing international attention. Despite this growing awareness, national and local decision makers often lack the tools to visualize disaster risk under different ecosystem conditions. As a result, the importance of ecosystems continues to be under-appreciated in decision-making processes related to disaster risk reduction and climate change adaptation. While spatial models have commonly been applied in both ecological assessments and disaster management, there have been relatively few studies that merge these two applications. This chapter demonstrates applications of the InVEST (Integrated Valuation of Ecosystem Services and Tradeoffs) tool in data-deficient countries where the United Nations Environment Programme (UNEP) is currently implementing ecosystem-based field interventions to reduce disaster risk. InVEST software (developed by the Natural Capital Project) provides spatial tools for assessing ecosystems and disaster risk even when limited data is available. The first study presented in this chapter takes into consideration the role of coastal and marine ecosystems in reducing exposure to coastal hazards in a small municipality in the south of Haiti. It provides an example of a qualitative assessment of exposure to storm surges and coastal flooding under different ecosystem management scenarios. The second study examines realistic land use change scenarios such as reforestation and urbanization and their impacts on soil erosion and sedimentation in a river basin in the Democratic Republic of the Congo. Through detailed examination of the two case studies, this chapter aims to demonstrate how integrated models such as InVEST could function as decision-support tools for considering ecosystem-based solutions for disaster risk reduction and

N. Bayani (✉)
United Nations Environment Programme, 11-15 Chemin des Anémones, Châtelaine,
CH-1219 Geneva, Switzerland
e-mail: Niloufarbayani@gmail.com

Y. Barthélemy
SUZA, P.O.Box: 146, Zanzibar, Tanzania
e-mail: yves.barthelemy@obscom.eu

© Springer International Publishing Switzerland 2016
F.G. Renaud et al. (eds.), *Ecosystem-Based Disaster Risk Reduction and Adaptation in Practice*, Advances in Natural and Technological Hazards Research 42,
DOI 10.1007/978-3-319-43633-3_10

227

climate change adaptation. The limitations, challenges and areas for improvement of each model application, as well as implications for local decision-making and awareness-raising, are discussed.

Keywords Coastal exposure • Disaster risk reduction • Eco-DRR • Ecosystem management • Erosion • GIS tool • ICZM • IWRM • Natural infrastructure • Scenario • Sedimentation

10.1   Introduction

The linkages between ecosystem conditions and disaster risk reduction have gained increasing international attention. The Sendai Framework for Disaster Risk Reduction (SFDRR), which serves as the global framework to guide disaster risk reduction (DRR) efforts from 2015 to 2030, clearly identifies ecosystem-based disaster risk reduction as a key solution to building resilience and reducing risk (UNISDR 2015). This is a major advancement since the Hyogo Framework for Action (UNISDR 2005). For the first time, sustainable management of ecosystems is recognized as a way to build disaster resilience in several priorities within the SFDRR. Another key achievement is that ecosystems will need to be taken into account in undertaking risk assessments under SFDRR Priority 1. This would enable decision makers to gain important insight into the linkages between specific ecosystem conditions and disaster risk (Estrella et al. 2013). Despite this growing awareness, national and local decision makers still lack the tools to visualize disaster risk under different ecosystem management or land use change scenarios (Beck et al. 2013). Integrated spatial models[1] are promising tools for this purpose.

The demand for risk modeling tools in the disaster risk management community is growing (GFDRR 2014). As a result, much effort has focused on developing risk models and building capacity among development professionals in applying such tools (GFDRR 2014). Many open access (for example, Delft-3D suite, HEC Suite, CAPRA Flood, InaSAFE Flood, TOMAWAC Wave, OsGEO Tsunami) and commercially licensed (for example, RiskScape Storm; RiskScape Wind) tools have been developed to assess the risk posed by various natural hazards to people and property.

Spatial models have also been widely used in assessing ecosystem conditions (see review by DeFries and Pagiola 2005) and ecosystem management (for example Yang et al. 2011). The Millennium Ecosystem Assessment (2005) was critical in stimulating greater interest in tools and models for assessing ecosystem services and their inclusion in development planning (Mooney 2011; Guerry et al. 2012). Various countries have since applied spatial models to assess ecosystem services

---

[1]Spatial models are analytical procedures applied in geographic information systems (GIS) that simulate real-world conditions using spatial data.

(for example UK NEA 2011; EME 2011; Japan Satoyama Satoumi Assessment 2010). However, relatively few spatial models merge the assessment of risk and ecosystem services (Arkema et al. 2013; Peduzzi et al. 2013).

In recognition of this gap, a number of initiatives have been recently developed that build on existing risk assessment methodology to integrate ecosystems as natural infrastructure against hazards. RiVAMP is a particularly innovative method of coastal risk assessment that integrates ecosystems and climate change factors (UNEP 2010). It quantifies the protection offered by coastal ecosystems against beach erosion and flooding (Peduzzi et al. 2013). However, the intensive data requirements of this methodology limit its applicability in data poor countries (Doswald and Estrella 2015). More accessible online tools are therefore needed. This chapter demonstrates applications of the InVEST modeling tools in data-deficient countries where the United Nations Environment Programme (UNEP) is implementing ecosystem-based disaster risk reduction field interventions with support from the European Commission.

Designed by The Natural Capital Project, InVEST (Integrated Valuation of Environmental Services and Tradeoffs) is an open source toolset of spatial models with relatively low data requirements (Tallis and Polasky 2011).[2] The set of models are scenario driven and can provide science-based information about how changes in ecosystems and land use influence the flow of natural capital to people, called "ecosystem services" (Mooney 2011; Guerry et al. 2012). While the software was not developed exclusively for risk assessment, it includes models to evaluate regulating services that are linked to disaster risk (e.g. coastal protection, sediment retention) and is therefore a suitable tool for ecosystem-based disaster risk reduction and climate change adaptation. InVEST has a tiered design which can accommodate a range of data availability and modeling expertise (Daily et al. 2011).[3] We selected qualitative models that could be easily applied in data-poor countries; however other InVEST models exist that are more data intensive and can produce quantitative outputs using ecological production functions.

In the first study presented, we applied InVEST Coastal Vulnerability model to a small municipality in the south of Haiti. Our aim was to qualitatively assess the exposure of people and infrastructure to storm surges and coastal flooding under different ecosystem management scenarios, while taking into consideration the role of coastal and marine ecosystems in reducing exposure. In the second study, we examined the impact of land use change scenarios on soil erosion and sedimentation

---

[2]InVEST is available for download at www.naturalcapitalproject.org

[3]Lower tiers are qualitative and have minimal data requirements, while higher tiers can quantify ecosystem services through complex calculations and more intensive data requirements (Guerry et al. 2012). Tier 0/1 models such as Coastal Vulnerability and Sediment Retention are most suitable for identifying patterns of ecosystem services and supporting planning or priority setting exercises with limited data (Daily et al. 2011). If more data is available, more complex InVEST models can be applied, for example to quantify the protective value of ecosystems based on a particular hurricane (Guerry et al. 2012).

in the Lukaya River Basin in the Democratic Republic of the Congo. Through detailed examination of the two case studies, this chapter aims to demonstrate how integrated spatial models such as InVEST enable progress in ecosystem-based solutions to disaster risk reduction and climate change adaptation. We discuss the limitations, challenges and areas for improvement of each model application, as well as implications for local decision-making and awareness-raising.

10.2 Coastal Exposure in Port Salut, Haiti

The InVEST Coastal Vulnerability (CV) model was applied in Port Salut, Haiti, to identify areas of the coastline that are more exposed to coastal flooding and storm surges and areas where habitats have the greatest potential to defend coastal communities against these hazards.[4] The main output is a qualitative and relative assessment of exposure. InVEST CV model was selected because of its two main advantages: (1) it takes into account both the geophysical and ecological characteristics of the area in measuring coastal exposure, and (2) it has relatively low data requirements and is therefore suitable for data-poor countries.

10.2.1 Study Area: Port Salut, Haiti

Port Salut is a small municipality in the South Department of Haiti with a population of 18,000 inhabitants (Fig. 10.1). In 2013 the Government of Haiti declared the coastal zone of Port Salut as one of the first marine protected areas in the country (Government of the Republic of Haiti 2013). The white sandy beach attracts many visitors each year, and the tourism industry is growing fast. However, like other areas of Haiti, Port Salut is frequently hit by tropical storms (CSI 2012b). Storm surges cause flooding of the coastal zone, while excess rainwater runoff results in overflow of rivers. People of Port Salut and their assets are caught between floods originating from upstream watersheds and the sea, and frequently incur losses. In addition, climate change will likely exacerbate the impacts of coastal hazards by increasing the maximum wind speed of tropical cyclones in the Caribbean (Handmer et al. 2012; Climate and Development Knowledge Network 2012) and causing sea level rise, which will intensify tropical storm surge impacts.

Coastal habitats in Port Salut are highly degraded (Reef Check Foundation 2013; Société Audubon Haiti 2013). Overfishing, deforestation, and inadequate agricultural practices leading to high sedimentation are some of the major causes of coastal

---

[4]It should be noted that while InVEST CV model calculates exposure to coastal flooding and storm surges, it does not measure inland flooding from storm water.

Fig. 10.1 Location map of the Municipality of Port Salut, Haiti (Source: GoogleEarth)

degradation. Sand extraction and construction on sand dunes have further resulted in severe beach erosion (CSI 2012a). Due to the combined effects of unsustainable land use, natural hazards and climate change, the protective role of ecosystems against hazards is diminishing, and the economic and touristic potential of Port Salut is at risk.

10.2.2 Methodology: InVEST Coastal Vulnerability (CV) Model

10.2.2.1 Design of ecosystem management Scenarios

Assumptions Coastal ecosystems are considered as effective buffers against inundation and erosion (Peduzzi et al. 2013). We ranked ecosystems' protective role under the common assumption that degraded ecosystems are less effective in protecting coastal communities than healthy ecosystems (European Union 2014). For example low-density agroforestry near the shore was considered less protective than dense vegetation. We also assumed that fixed and stiff habitats that penetrate the water column (e.g. coral reefs, mangroves) provide more protection than flexible and seasonal habitats (e.g. seagrass) (Guannel et al. 2015). The distance

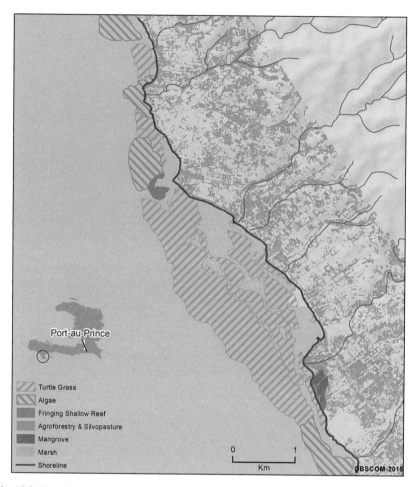

Fig. 10.2 Port Salut ecosystem map was developed based on remote sensing and ground-truthed through marine and terrestrial field surveys. Transient ecosystems such as seagrass beds were considered less effective in attenuating wave energy than fixed and stiff ecosystems such as coral reefs and coastal vegetation (Sources: Terrestrial and Marine Habitat from the classification of a WorldView 2 satellite image by UNEP/OBSCOM/Reefcheck/Société Audubon Haiti)

between ecosystems and the shore was also taken into consideration in ranking the protective role of each ecosystem type (Natural Capital Project 2014).

Mapping of Ecosystems An ecosystem map was developed through remote sensing on a World View 2 satellite image (0.5 m panchromatic/4 m multispectral) from 2011. The initial map was ground-truthed (Fig. 10.2) by comparing land use and land cover types with results of field surveys (Reef Check Foundation 2013; Société Audubon Haiti 2013).

Scenarios To evaluate the role of coastal ecosystems in reducing exposure to storms, we developed three ecosystem management scenarios:

1. "Current ecosystem conditions" assumes current degradation level of ecosystems;
2. "Without ecosystems" assumes ecosystems are completely degraded, therefore do not provide protection to the coastal zone;
3. "Restored ecosystems" assumes all ecosystems are restored to their pristine state and fulfill their full potential in protecting the coastal zone.

"Without ecosystems" and "restored ecosystems" scenarios are not intended to be plausible reflections of the future. Instead, the objective was to evaluate where and to what extent ecosystems play a significant role in protecting the Port Salut community.

10.2.2.2 Calculating Coastal Exposure

To estimate the relative exposure of the segments of the coastline of Port Salut (8 km stretch of shoreline) under different ecosystem management scenarios, we calculated an index of coastal erosion and inundation using the CV model of InVEST (Version 3.0.0). The model follows previous approaches (Hammar-Klose and Thieler 2001; Gornitz 1990; Thieler and Hammar-Klose 1999) by incorporating the role of ecosystems in calculating exposure. An important assumption of the model is that "more ecosystem" means "more protection from storm impacts" and therefore "less exposure to storm surge and inundation" (Natural Capital Project 2014).

Coastal exposure was compared under different ecosystem management scenarios and the role of ecosystems in protecting the coastline was identified for each segment. The results were in terms of relative exposure to coastal hazards of each 50×50 m segment compared to all other segments of the coastline. The model also provides exposure outputs in terms of the contribution from individual variables (wind, wave and storm surge potential).

Exposure Index The tool produces a qualitative index, which ranks segments of the coastline based on relative exposure. It calculates the effects of storms on exposure by incorporating observed data on a number of variables: waves, wind, shoreline type and relief are mandatory inputs, while ecosystem type, surge potential and sea level rise are optional inputs. Multiple tide gauges that collect sea level change data over a long period of time are needed to measure variations in relative sea level change at the local scale of this study. As such data were lacking, we ignored sea level change in our calculations. Exposure is calculated as the geometric mean of all the variable ranks, ranging from low (rank = 1) to high (rank = 5), based on a combination of user- and model-defined criteria (Arkema et al. 2013):

$$\text{Exposure Index} = \left(R_{\text{Ecosystem}} R_{\text{ShorelineType}} R_{\text{Relief}} R_{\text{Wind}} R_{\text{Waves}} R_{\text{SurgePotential}} \right)^{1/6}$$

$$(\text{xx}.1)$$

where R represents the ranking of the bio-geographical variable (i.e., ecosystem, relief, etc.). All variables are weighted equally. Surge potential depends on the amount of time wind blows over relatively shallow water and is calculated as a function of the length of the continental shelf fronting the shoreline (Natural Capital Project 2014).

Input Data Topographic relief was developed by combining the Digital Elevation Model (DEM) from SRTM V4 and bathymetry based on GEBCO data and sonar measurements from a boat (Reef Check Foundation 2013). Google Earth images, aerial Ortho-Photos and field verification were used to classify segments of the shoreline into rocky shore, sandy beach, river mouth and built structure. Wind and wave exposure were calculated based on eight years (2006–2013) of National Oceanographic and Atmospheric Administration WAVEWATCH III model hindcast re-analysis (Tolman 2009).

10.2.3   Results

The InVEST Coastal Vulnerability model produces a qualitative estimate of exposure to coastal erosion and flooding induced by storms. Exposure index values can range from 1 to 5 and are relative to other pixels. We classified exposure into low, medium, and high categories based on the distribution of index values across all segments (161 in total) and three habitat scenarios on a 50 x 50 m scale for the coastline of Port Salut. We classified the distribution of results (ranging from 1.02 to 3.55) into quartiles that define areas of highest ($>2.59 =$ upper 25 %), intermediate ($2.1–2.59 =$ central 50 %) and lowest hazard ($<1.77 =$ lower 25 %) (After Arkema et al. 2013).

Exposure with Current Ecosystem Conditions Under the "current" ecosystem scenario, 57 % of the segments of the coastline are highly exposed, while 29 % are at medium and 15 % are in the low exposure categories (Fig. 10.3). Exposure to coastal hazards and the importance of ecosystems vary across the shoreline of Port Salut (Fig. 10.4b). The sandy beaches are the most exposed, while the rocky shores in the north appear to be the least exposed.

Exposure Without Ecosystems We compared exposure under "current" and "without ecosystems" scenarios to identify where ecosystems reduce the exposure of the community to hazards. Without ecosystems, the entire coastline experiences higher exposure compared to the current scenario; 81 % will be exposed to the highest hazard, and only 1 % will be exposed to low hazard (Figs. 10.3 and 10.4). The center of town falls into the high exposure category.

Exposure with Restored Ecosystems Restoration of ecosystems to a healthy, functioning state will provide additional protection. We compared the restoration

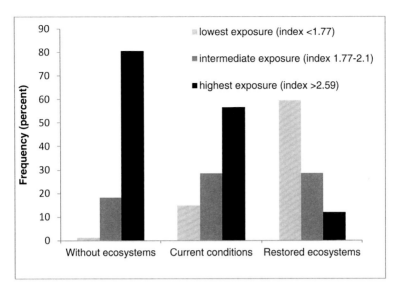

Fig. 10.3 Exposure of the segments of the shoreline under different ecosystem management scenarios

scenario with current conditions and found that the entire coastline would experience lower hazard exposure. The area of lowest exposure increases to 59 %, while the area of highest exposure drops to 12 % (Fig. 10.3). Reduced exposure is especially significant at the popular sandy beaches of Port Salut (Fig. 10.4c).

We found that ecosystems of particular importance in Port Salut were mangroves, the marshland near Port Salut Market and the shallow fringing coral reef fronting the Point Sable sandy beach (see Fig. 10.1 for the location of reference points). The presence of these ecosystems alone reduces exposure index by >0.65.

InVEST CV results was overlaid with population density and locations of key infrastructure (e.g. bridges and roads) and economic centers (e.g. Port Salut Market) to identify the population and infrastructure exposed. Figure 10.5 shows that the densely populated areas, the market and important bridges are located in some of the most highly exposed parts of Port Salut.

10.2.4 Policy Implications

Our objective of applying the CV model to Port Salut was to experiment with a relatively simple and freely accessible tool that incorporates ecosystems in exposure assessment and which could support decision-making. Our analysis demonstrates where ecosystems play an important role in protecting the community of Port Salut from coastal hazards. Comparison of ecosystem management scenarios

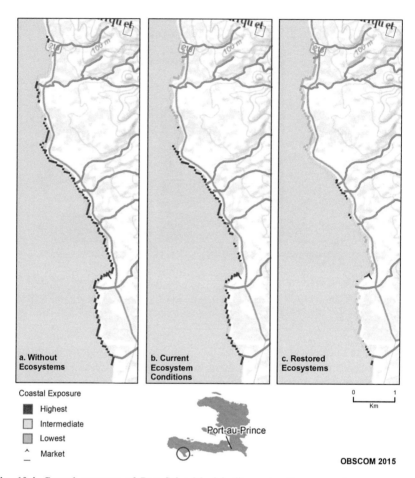

Fig. 10.4 Coastal exposure of Port Salut Municipality under three ecosystem management scenarios (Source: Background Map from ESRI)

shows that loss of ecosystems would result in more severe storm impacts across the municipality, while their restoration would provide additional protection.

Coral reefs, mangroves and other ecosystems in Port Salut are highly threatened from unsustainable use. The declaration of Port Salut as a marine protected area provides an opportunity for protecting and restoring ecosystems for multiple benefits to the local community and the growing tourism industry. The results of InVEST modelling can inform the development of a sustainable, resilient land use plan and coastal zoning as well as raise awareness about the role of ecosystems in protecting people and livelihoods from hazards.

Benefits provided by coastal ecosystems (coral reefs, mangroves, wetlands and coastal forests) to property development and tourism are well documented (McKenzie et al. 2011; Turner et al. 1998) but not always publicly understood.

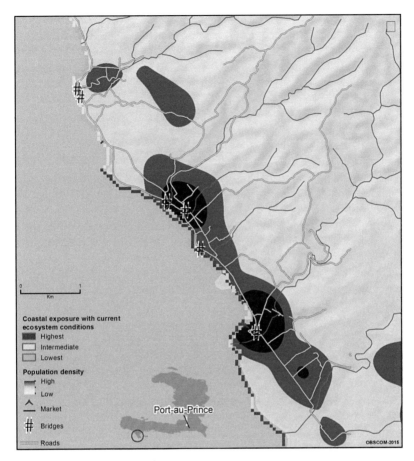

Fig. 10.5 Population and infrastructure exposed (Population density is based on interpolation from a house counting produced by UNDP on a 2010 aerial image)

InVEST CV model results can be used to identify areas where investments might be lost if the existing ecosystems in Port Salut are further degraded. Exposure maps that integrate geophysical and ecological information can also guide future investments in tourism and property development in Haiti to safer locations, while promoting ecosystem conservation that attracts more visitors. Success stories, showing how bad investments can be avoided and ecological potential maximized, could serve as triggers for the hope needed to rehabilitate degraded ecosystems in other areas of Haiti.

UNEP is working with the Government of Haiti to implement an ecosystem-based disaster risk reduction (Eco-DRR) project in Port Salut, through ecosystem conservation and improved coastal zone management. Field activities include re-vegetation, mangrove restoration and reforestation along shorelines, riparian forests and streams that are exposed to flooding and coastal storm surge impacts, as well as supporting fisheries with a low impact on near-shore ecosystems.[5] InVEST results show where ecosystems effectively mitigate hazards and where the community benefits the most from the protection provided by ecosystems (Arkema et al. 2013; Langridge et al. 2014), and can therefore be used to identify the best locations to target for Eco-DRR interventions.

10.2.5 Challenges and Areas for Improvement

Our results highlight the hazard mitigation service that ecosystems provide and the consequences of their loss. However, we do not aim to suggest that ecosystems alone are sufficient to protect populations and infrastructure. A more detailed assessment of hazard mitigation measures, including engineered or hybrid measures, is required to identify the best solutions. Other ecosystem services in Port Salut that can provide sustainable sources of livelihoods and local resilience to disasters should also be considered in selecting the most suitable DRR measures. For example shifting fishing pressure from the shallow continental shelf to offshore stocks may provide a more sustainable source of food, while enabling the natural rehabilitation of ecosystems that mitigate hazards.

The InVEST CV model can be applied at different spatial scales, depending on the accuracy and resolution of the available datasets. Our application of InVEST CV demonstrates that it is possible to apply qualitative tools even in data poor countries such as Haiti. While application of the model to a small area in Port Salut required resource-intensive data acquisition and local fieldwork, global and national datasets may be sufficient to run InVEST CV model at a national scale. Such studies may then be used to highlight the importance of Haiti's ecosystems for enhancing the country's resilience to disasters and help prioritize ecosystem conservation in critical areas. Further studies could explore InVEST Wave Attenuation and Erosion Reduction, a more data intensive model that quantifies the role of habitats in reducing erosion. They can also include scenarios of sea level rise projections and assess exposure under different climate change scenarios.

We caution that outputs of our assessment should not be generalized to other areas of Haiti, nor should decision-making on DRR rely solely on the results of this model. It is also possible that sea level rise may overwhelm the mitigation capacity of certain ecosystems, especially given their current degradation level. Additional studies are required to determine the extent of protection provided by ecosystems to future hazards and climate change impacts. Climate change is expected to

---

[5]For more information see: http://www.unep.org/disastersandconflicts/Introduction/ DisasterRiskReduction/Countryactivities/tabid/104431/Default.aspx

aggravate flooding and storm surge impacts in Haiti (Handmer et al. 2012). It is therefore important to take a multi-hazard approach in assessing future exposure.

10.3   Sediment Retention in the Lukaya River Basin, Democratic Republic of the Congo

10.3.1   Introduction

Natural vegetation protects watercourses, regulates the timing and amount of water flows and maintains water quality by reducing sedimentation (Mendoza et al. 2011). In the Democratic Republic of the Congo (DRC), UNEP has taken an integrated approach to water resource management that takes into account disaster risks such as severe soil erosion and flooding, while improving river water quality. The pilot demonstration project is in the Lukaya River Basin near the capital, Kinshasa, where deforestation and unsustainable land use have resulted in excessive soil erosion, extremely high sedimentation rates in the river and subsequently high risk of flooding. We applied the InVEST Sediment Retention Model to the Lukaya River Basin to provide a technical basis for decision-making on land use management that reduces disaster risk and maximizes the provision of ecosystem services such as natural filtration of sediment by vegetation.

10.3.2   Study Area: The Lukaya River Basin

Located in the provinces of Kinshasa and Kongo Central, the Lukaya River Basin is a small watershed of approximately 355 km^2 in the southern edge of Kinshasa (Fig. 10.6). The Basin is a sub-catchment of the N'Djili River Basin, which flows into the Congo River. From its source in Kongo Central, the Lukaya River traverses nearly 55 km to its confluence with the N'Djili River. The catchment may be divided into two parts: an upstream section centred on the town of Kasangulu, which is a quintessentially rural region, and a downstream zone around the densely urbanized Kinshasa neighbourhood of Kimwenza where a large part of the anthropogenic pressure is concentrated. Quarrying, slash and burn agriculture, charcoal production and horticulture along the river have resulted in rampant deforestation, excessive sedimentation rates, and high incidence of flooding.

The population of Kinshasa and surroundings rely upon major watersheds such as the Lukaya River Basin to meet their basic water needs. The water treatment plant in Kimwenza is one of the four sources of drinking water to Kinshasa, and supplies water to an estimated 40,000 people (N.Z. Yanga, personal communication). The plant is operated by REGIDESO, the public utility company in charge of production and supply of water in the country. Excessive sediment load in the river

Fig. 10.6 Location of the
Lukaya River Basin,
southwest of the capital,
Kinshasa, in the Democratic
Republic of the Congo.
The REGIDESO water
treatment plant is located
downstream near the
Lukaya River

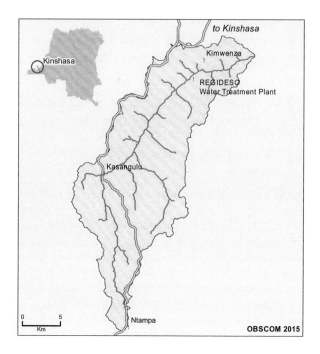

has reduced water quality and raised water production costs. Water supply is
regularly interrupted to manually remove excess sediment from the treatment
plant (N.Z. Yanga, personal communication).

10.3.3   Methodology: InVEST Sediment Retention Model

There is scientific evidence that natural vegetation holds soil in place and captures
sediment moving overland (Conte et al. 2011). The InVEST Sediment Retention
model (version 3.0.0) was applied to the Lukaya River Basin to identify key areas in
the basin for sediment retention and transport. The model estimates the capacity of
a land parcel to retain sediment by using information on geomorphology, climate,
vegetation cover and management practices. It can answer questions such as
"Where would reforestation achieve the greatest downstream water quality benefits?". Estimated soil loss and sediment transport can be used to determine the
avoided sedimentation service provided by the ecosystem. The major assumption of
the model is that "more vegetation" means "less sedimentation". The model may
also be used to value the landscape in terms of water quality maintenance or
avoided sedimentation, and to determine how land use changes may impact the
cost of sediment removal (Natural Capital Project 2014).

10.3.3.1 Calculating Sedimentation

InVEST Sediment Retention model focuses on sheetwash erosion processes and is based on the Universal Soil Loss Equation (USLE) (Wischmeier and Smith 1978). This InVEST model then builds on USLE to also include the influence of vegetation in retaining soil loss from upstream areas. It predicts erosion as a function of the energetic ability of rainfall to move soil, the erodibility of soil types, slope, the erosion protection results of management practices and the presence of vegetation. The model routes the sediment originating on each parcel of land or pixel along its flow path. Vegetated pixels retain some of the sediment depending on their efficiency. The remaining sediment is exported to the next pixel along the path.[6] Outputs are in terms of the total sediment load exported to streams from the watershed as an annual average (tonnes/year). While calculations are at the pixel scale, interpretations of the model should be limited to the observed patterns of soil loss at the sub-watershed scale, rather than quantitative measures at the pixel scale.

Input Data Model inputs include elevation, land use and land cover (LULC), rainfall erosivity (i.e. duration and intensity of precipitation), and soil erodability (i.e. soil types). Other inputs include the stream network, sub-watersheds and features of the water treatment plant (or reservoir). Data was collected through a combination of remote sensing and field measurements. An LULC map of the Basin was produced based on a LANDSAT image, filtered to minimize noise and down-scaled to 90 m per pixel (Mfumu 2013). This LULC reflects conditions in 2013. A digital elevation model (DEM) was developed from SRTM images at 30 m resolution. Soil types were assessed through the analyses of soil samples collected in the field (Makanzu 2013). Precipitation data (1961–2013) provided by the National Agency of Meteorology and Remote Sensing of DRC (METTELSAT) were used to calculate rainfall erosivity. Biophysical factors – i.e., effects of vegetation cover and management (C factor), effects of specific support practices such as terracing (P factor) and sediment retention efficiency of land use types – were selected through literature review (see Tombus et al. 2012; FAO 2013; EPA 2007; Baja et al. 2014; USDA 1997).

10.3.3.2 Design of Land Use Change Scenarios

In the Lukaya Basin, anarchical urbanization is occurring along the national road. The towns of Kimwenza and Kasangulu in particular are experiencing high rates of urbanization. In addition, houses are being built on deforested slopes. Different scenarios were developed to examine the effects of land use change on soil erosion potential in the basin. Our simple scenarios reflect observed trends of urbanization

---

[6]We used version 3.0.0 of InVEST Sediment Retention model, which has D-infinity flow direction instead of D-8 and therefore offers a more accurate calculation of flow paths than previous versions of the model.

and potential for reforestation activities in the watershed. For the sake of brevity, we will only present scenarios of land use change in and around the town of Kasangulu (17.57 km^2).

The InVEST Sediment Retention model was run under three land use change scenarios in Kasangulu:

1. "Current" land use conditions
2. "Urbanization" assumes urban land replaces savanna and bare soil
3. "Reforestation" assumes forest replaces agriculture and savanna

As in the case of the previous example in Haiti, scenarios are not intended to be precise simulations of the future. Instead, the objective was to evaluate where and to what extent vegetation plays a role in reducing sedimentation in the Lukaya Basin.

10.3.4 Results

10.3.4.1 Key Findings Under Current Conditions

The InVEST Sediment Retention model was applied to identify key areas in the basin for sediment retention and transport. Under current land use conditions, total potential soil erosion in the Lukaya River Basin is estimated at 53,154 (K tonnes/ year) (Table 10.1). Vegetation cover retains 49,564 (K tonnes/year) of sediment while the remaining 3,418 (K tonnes/year) is exported to the Lukaya River.

Figure 10.7 left shows potential soil erosion at the scale of individual sub-watersheds. It suggests that the steep slopes in the northwest of the basin have the highest potential for soil erosion, likely due to extensive de-vegetation. Figure 10.7 middle shows estimated sediment retention rates per sub-watershed and suggests that the northern sub-watersheds also play a particularly important role in retaining excessive sediment. In these sub-watersheds, agricultural land and small patches of secondary forest can be found downstream of denuded slopes and likely retain much of the sediment on its flow path. Figure 10.7 right shows sediment export rates per sub-watershed; those that export the highest quantity of sediment to the River are displayed in *red*. They can be considered as priority areas for ecosystem restoration as they lack downstream vegetation that can retain detached sediment before reaching the river, and/or are located adjacent to the river.

10.3.4.2 Kasangulu Land Use Change Scenarios

Urbanization. Expansion of the town of Kasangulu will increase potential soil erosion from the basin by 227 K tonnes/year (i.e. 0.43 %; Table 10.1). Due to higher erosion rates, the remaining vegetation cover in the area will retain an additional 17 K tonnes/year (i.e. 0.03 %) of sediment. However, the retention capacity does

Table 10.1 Comparison of total potential soil loss, sediment export and sediment retention potential from the watershed under different land use change scenarios in Kasangulu

Scenario	Current	Urbanization of Kasangulu (Compared to current)	Reforestation of Kasangulu (Compared to current)
Potential soil loss i.e., USLE (K tonnes/yr)	53,154	53,381 (+0.43 %)	51,754 (−2.63 %)
Total sediment retained (K tonnes/yr)	49,564	49,581 (+0.03 %)	48,437 (−2.27 %)
Total sediment export (K tonnes/yr)	3,418	3,621 (+5.61 %)	3,151 (−7.88 %)

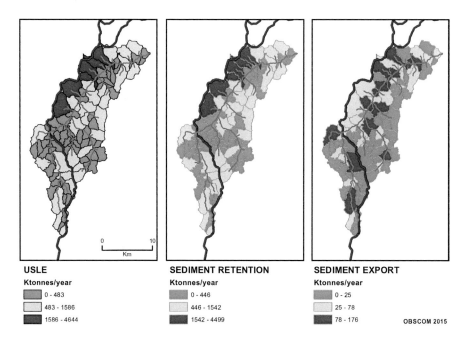

USLE
Ktonnes/year
- 0 - 483
- 483 - 1586
- 1586 - 4644

SEDIMENT RETENTION
Ktonnes/year
- 0 - 446
- 446 - 1542
- 1542 - 4499

SEDIMENT EXPORT
Ktonnes/year
- 0 - 25
- 25 - 78
- 78 - 176

OBSCOM 2015

Fig. 10.7 Outputs of InVEST Sediment Retention model under current land use conditions in the Lukaya River Basin: *left* potential soil erosion (USLE); *center* sediment retention; *right* sediment export

not match the higher erosion rate. As a result, an additional 203 K tonnes/year (i.e. 5.61 %) of sediment are exported to the river. Figure 10.8 *left* shows that with the exception of one sub-watershed, urbanization will increase soil erosion potential in all sub-watersheds of Kasangulu. Similarly, Fig. 10.9 *left* shows that all sub-watersheds in Kasangulu will export more sediment to the river under the urbanization scenario.

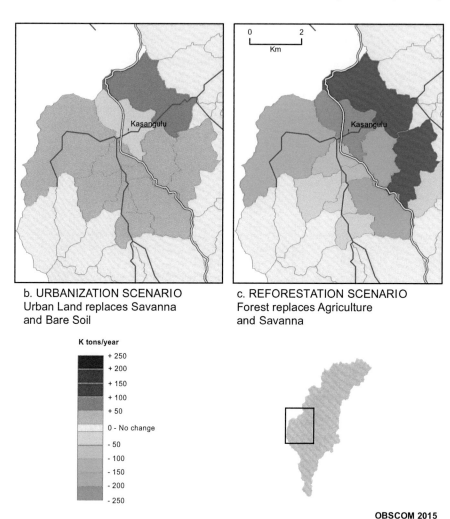

Fig. 10.8 Change in potential soil loss (USLE) under Kasangulu land use change scenarios: *left* urbanization; *right* reforestation. Change is expressed in K tonnes/year as the difference between USLE in each sub-watershed under the different scenarios and USLE under current land-use. A positive value means an increase, while a negative value means a decrease in soil erosion potential

Reforestation. Reforestation of the same area in Kasangulu reduces potential soil erosion (USLE) by 1,400 K tonnes/year (i.e., –2.63 %). As a result, the quantity of sediment exported to the river will be reduced by 267 K tonnes/year (i.e., –7.88 %). Lower potential soil erosion also means that the amount of sediment retained by vegetation is reduced to 1,127 K tonnes/year (i.e., –2.27 %). Figure 10.8 right shows that reforestation of Kasangulu will reduce soil erosion potential in all sub-watersheds, while Fig. 10.9 right shows sediment export from all sub-watersheds will also decrease.

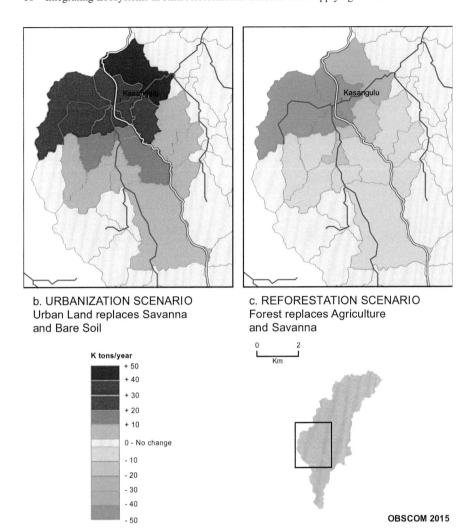

Fig. 10.9 Change in sediment export under Kasangulu land use change scenarios: *left* urbanization; *right* reforestation. Change is expressed in K tonnes/year as the difference between sediment export from each sub-watershed under the different scenarios and sediment export under current land-use. A positive value means an increase, while a negative value means a decrease in sediment export

10.3.5   Policy Implications

Our application of the InVEST Sediment Retention Model provides an informative picture of the expected outcomes of land use change in the Lukaya River Basin. The model was used to simulate erosion and sedimentation dynamics under different land use change scenarios. The outputs show that: (a) currently, vegetation cover is providing an important ecosystem service by retaining sediment and reducing erosion potential; (b) areas that currently export the highest amount of soil to the Lukaya River are already degraded land and/or close to the river.

An examination of land use change scenarios in Kasangulu reveals some of the environmental costs of unplanned urbanization (increased sedimentation; deteriorating river water quality), and the potential benefits of reforestation (reduced sedimentation; improved river water quality). They also highlight forest parcels that currently or potentially offer the greatest sediment retention benefits. This information can inform land use planning to maximize returns on reforestation and conservation investments in Kansangulu.

To reduce the damages and costs associated with sedimentation, decision makers require information regarding the extent to which different parts of a landscape contribute to sediment retention, and how land use changes may affect this retention. The InVEST Sediment Retention Model is a powerful tool, which can help the government, community-based organizations, land use planners, and local NGOs to define priority areas for nature conservation, scale up reforestation activities or restrict urbanization. Our application of this model has enabled us to identify the most sensitive areas in the basin in terms of sediment production, which can guide land use decisions that decrease flood risk to local communities in the Lukaya River Basin.

Reforestation and restoration of existing forested land in the watershed can significantly improve the provision of ecosystem services to Kinshasa. UNEP is working with the Lukaya Water Users Association (AUBR-L) to implement agroforestry activities where they would be most effective. Re-vegetation activities are also being implemented in the most sensitive sub-watersheds to reduce sediment export to the river. The intervention sites are selected based on a combination of field and modelled assessments as well as community and government consultations.

10.3.6 Challenges and Areas for Improvement

The USLE methodology predicts erosion from sheet wash alone and therefore ignores other types of erosion (e.g. gullies, stream-bank) (Natural Capital Project 2014). While USLE is considered as a standard method to calculate soil loss (de la Rosa et al. 1998; Sivertun and Prange 2003; Devatha et al. 2015), there is little consensus about parameter values in non-agricultural lands (Karpilo and Toy 2004), which posed a challenge as we selected parameters based on literature review. However, in order to reduce uncertainties in the future, data from the field should be used to calibrate model parameters and compare findings with field- measurements of soil erosion and sedimentation.

InVEST Sediment Retention model includes a step that estimates the monetary value of the ecosystem service (i.e., sediment retention) provided by each land parcel in terms of avoided water production cost. UNEP and partners are taking steps to collect the necessary economic data to take advantage of this feature of InVEST; these include collaboration with the management of the REGIDESO water treatment plant in Kimwenza to follow sediment load in the plant, estimate the cost of sediment removal over time and the costs associated with inconsistent water supply to Kinshasa due to excessive sediment load (see the location of the plant in Fig. 10.6).

10.4 Lessons Learnt

There is ample evidence of the role that ecosystems can play in disaster risk reduction (DRR) and climate change adaptation (CCA) (for example Costanza et al. 2008; Campbell et al. 2009; Bebi et al. 2009; Dolidon et al. 2009; Hettiarchi et al. 2013; Peduzzi et al. 2013; Lacambra et al. 2013). However, decision makers and investors still have doubts about the benefits of ecosystems compared to engineered infrastructure (Estrella et al. 2013). InVEST can be used to promote ecosystem-based DRR/CCA as a decision-support tool as well as for advocacy and awareness-raising.

10.4.1 InVEST as an Eco-DRR/CCA Decision Support Tool

10.4.1.1 Advantages

Ecosystem-based measures have gained increasing global attention as valuable approaches to local DRR/CCA. However in practice, decision makers at the local level often face numerous conflicting interests and typically have limited capacity to map and integrate the potential impacts of different ecosystem management scenarios in decision making (Tallis and Polasky 2011). There are still relatively few examples of studies that examine risk under ecosystem management scenarios at scales relevant to decisions that affect ecosystems (Beck et al. 2013). Based on our application of two InVEST models, we contend that such integrated models can serve as flexible, scientifically-based and practical tools to support local decision making. They can provide information about where ecosystem services are provided; who is affected and which management decisions affect their provision (also see Whelchel & Beck, Chap. 6 for other tools and approaches).

Rather than prescribing the best decision, such models inform decision-making by highlighting the trade-offs and potential outcomes of different decisions (Guerry et al. 2012) and can therefore be used as a basis for negotiations among stakeholders (Ghazoul 2007; McKenzie et al. 2011). While they are not comprehensive, they can serve as first assessments of exposure (in terms of the area exposed to a certain hazard) and shed light on the mitigation role of ecosystems. The outputs and visuals can be easily interpreted by different stakeholders and used to prioritize areas for conservation of existing ecosystems or for their restoration in order to reduce the risk of disasters or to adapt to clime change impacts (e.g. changes in hydrological cycles). As described in our case studies, the private sector can also benefit from integrated modeling to invest in safer areas, minimize economic loss to hazards, and reduce operation costs.

An important feature of these models for DRR/CCA decision-making at the local level is that they enable comparisons between land use or ecosystem management decisions. Without visualizing scenarios, decision makers tend to ignore ecosystem services (McKenzie et al. 2011). InVEST makes it possible to explore various management options. For instance, it points out differences in the output metrics (i.e., exposure in Haiti; sediment export in DRC), between ecosystem management scenarios, and therefore highlights the protective services provided by ecosystems.

InVEST can also be applied at various scales. For example, while we applied the CV model to a small municipality in Haiti, others have applied this model to inform coastal and marine spatial planning in two counties in California (Langridge et al. 2014), the West Coast of Vancouver Island in Canada (i.e., 10–100 km^2; Guerry et al. 2012), and at national scales in Belize (Arkema et al. 2014) and the United States (Arkema et al. 2013).

Another advantage of InVEST is its accessibility to all users regardless of GIS expertise.[7] InVEST is an open source and stand-alone software. The relatively low data requirements make the models suitable for application in even the most data-poor countries. This is partly because the basic structure of InVEST can be run with the minimum amount of data. Additional functions are embedded as optional steps, which can be applied or skipped depending on data availability.

It is clear that ecosystem benefits go beyond DRR concerns as they provide many more services (e.g. critical sources of livelihood, food, water, and building materials) than what is captured by single models. Future work could explore synergies and tradeoffs across multiple services and scenarios in the same area by applying scenario building tools (WWF 2012) and a combination of spatial models to assess different ecosystem services.

10.4.1.2 Challenges

The main difficulty of applying InVEST models to small areas in Haiti and DRC was to find reliable data with high resolutions. As a result, we were obliged to acquire site-specific information through field surveys and remote-sensing on purchased satellite images. To avoid this challenge, other studies in data-deficient countries could focus on larger areas to reduce reliance on resource-intensive data acquisition.

---

[7] InVEST models can be applied in as little as one month, although depending on the location, scale of work and capacity of the team, it may take up to 24 months to gather data and run models (Eichelberger 2013).

Given time constraints and data availability, our scenarios in both case studies were first-cut, simplified scenarios. They were also snapshots in time and focused on the protection provided by ecosystems once they are fully restored or fully degraded. However, in reality, in between states are more common, while ecological infrastructure also needs time to mature, and therefore demonstration of their multiple benefits over the short- and long-term is needed (Estrella et al. 2013). Communicating uncertainty, such as time lags and future climate changes, in scenario effects to non-scientific stakeholders can be a significant challenge but should be reflected while disseminating the results (Walker et al. 2003; Guerry et al. 2012).

In addition, our scenarios only account for ecosystem-based measures. Clearly, there are limits to the protection provided to people and their assets by ecological infrastructure (Renaud et al. 2013). Hybrid solutions may be more successful and cost-effective than green infrastructure alone (for example Rao et al. 2013; van Wesenbeeck et al., Chap. 8 and David et al., Chap. 20). In the future, scenarios could delve deeper and assess the tradeoffs between different DRR/CCA measures. There exist a variety of resources in order to move beyond simple scenarios to possible futures used to illuminate multiple choices and consequences (See WWF 2012).

The effectiveness of ecosystems in mitigating hazards depends on features such as their health and composition (species, density, size), geology and topography of the area as well as the intensity and type of hazards (Estrella et al. 2013). In InVEST models, the mitigation role of ecosystems is accounted for through a combination of model and user-defined criteria (Natural Capital Project 2014). It is therefore a challenge to the user to define the relative protection provided by different local ecosystems of various sizes, compositions and health levels, especially in poorly studied areas. We recommend that users seek advice from ecologists in categorizing the use-defined criteria and contend that further research is required to better understand the characteristics that lend an ecosystem more or less suitable as a natural buffer and to build scientific consensus.

Another challenge is to balance the information needs of local decision makers with ecologically meaningful spatial scales. Functional units of ecosystems often exceed the boundaries of jurisdictional planning units. Finding the right scale becomes especially troublesome when dealing with marine ecosystems due to the high level of connectivity among populations of marine organisms (McCook et al. 2009).

Finally, as with any model, outputs are only as reliable as input data. Users should keep in mind that interpretation of InVEST results should only go as far as there is certainty and at appropriate scales. For example, detailed land use planning should not rely on a single run of a model. Nor should the models be directly applied to other locations or be used in place of quantitative assessments of risk and vulnerability.[8]

---

[8]Comprehensive risk assessments entail examination of vulnerability, quantitative potential damage over time and at various intensities of hazards, assessment of exposed economic resources and people.

10.4.2   InVEST as a Tool for Eco-DRR/CCA Advocacy and Awareness-Raising

In addition to decision-making support, integrated models such as InVEST are valuable tools for advocacy and raising awareness. Opportunities to use ecosystem-based DRR/CCA approaches appear not only at the local level but also in various policy mechanisms and economic sectors. To take advantage of these opportunities, it is important that all actors and decision makers (e.g. municipal, provincial and national government; NGOs) understand the benefits of ecosystems in protecting people. By allowing the integration of ecosystems within risk assessments, such integrated modeling tools demonstrate the value of Eco-DRR/CCA measures by providing specific examples at scales that resonate with decision makers and can therefore promote ecological thinking in decision-making processes, for example to advocate for banning of environmentally harmful activities or promote ecologically beneficial decisions (McKenzie et al. 2011). Science-based modeling, even if limited to qualitative results, could also attract interest and investment in more substantial, quantitative assessments to inform decision-making in DRR/CCA.

Despite increasing interest in economic valuation of ecosystem services at the global level, local estimates are more useful to decision makers than global estimates (Estrella et al. 2013). InVEST can be used to assess the economic value of ecosystem services related to DRR/CCA in specific locations. Values could be the hazard mitigation services or provisioning services that improve community resilience by providing livelihood sources. As such, the InVEST tool can be used to promote Eco-DRR/CCA more effectively.

While there are close linkages between DRR and CCA, these two approaches have been largely applied in parallel. The international community and national governments have recognized that integration of these approaches presents an opportunity to gain a more holistic understanding of risks (Doswald and Estrella 2015). We find InVEST to be a useful tool for this purpose as it allows scenario building that incorporate both current and future hazards. For example, climate change impacts such as sea level rise can be included in exposure calculations in the CV model. In addition, certain InVEST models are designed specifically to assess climate change impacts such as changes in hydrological cycles (Lawler et al. 2011). These models can also be used as a basis for more advanced CCA analyses, for example to model shifts in the distribution of ecosystems under climate change, which may impact their hazard mitigation role. These changes are not automatically embedded in InVEST models but can be incorporated in the design of scenarios (For example see Guerry et al. 2012).

In summary, based on our applications, we conclude that integrated models such as InVEST are valuable tools for Eco-DRR/CCA because they can assess the impacts of various management scenarios and emphasize the need to maintain healthy and functioning ecosystems. However, modelling is only the first step; incorporating ecosystems in DRR/CCA decisions has many more requirements (political will, community engagement, regulations, etc.) to translate this knowledge into action (McKenzie et al. 2014). Integrated models should therefore be used as convincing tools alongside other resources for promoting Eco-DRR and CCA.

Acknowledgements We would like to acknowledge Antoine Mfumu, Fils Makanzu, Souhail Elmdari, and Isavella Monioudi who helped us with parts of the analyses presented in this chapter. We are also grateful to the following organizations in Haiti and DR Congo that provided data: AUBR-L, METTELSAT, REGIDESO Kimwenza, CNIGS, CIAT and UNDP Haiti. And lastly, thanks are due to Greg Guannel, Marisol Estrella, Maximilien Pardo, Celine Jacmain, Adonis Velegrakis and two anonymous reviewers for their insightful comments.

References

Arkema KK, Guannel G, Verutes G et al (2013) Coastal habitats shield people and property from sea-level rise and storms. Nat Clim Chan. Published online: 14 July 2013. doi: 10.1038/NCLIMATE1944

Arkema KK, Verutes G, Bernhardt JR et al (2014) Assessing habitat risk from human activities to inform coastal and marine spatial planning: a demonstration in Belize. Environ Res Lett 9:114016 (11 pp). doi:10.1088/1748-9326/9/11/114016

Baja S, Nurmiaty U, Arif S (2014) GIS-based soil erosion modelling for assessing and suitability in the urban watershed of Tallo River, South Sulawesi, Indonesia. Mod Appl Sci 8(4):50–60. Canadian Center of Science and Education

Bebi P, Kulakowski D, Christian R (2009) Snow Avalanche disturbances in forest ecosystems – state of research and implications for management. For Ecol Manag 257:1883–1892

Beck WM et al (2013) Increasing the resilience of human and natural communities to coastal hazards: supporting decisions in New York and Connecticut. In: Renaud FG, Sudmeiere-Rieux K, Estrella M (eds) The role of ecosystems in disaster risk reduction. United Nations University Press, Tokyo, pp 140–163

Campbell A, Kapos V, Scharlemann JPW et al (2009) Review of the literature on the links between biodiversity and climate change: impacts, adaptation and mitigation. Secretariat of the Convention on Biological Diversity, Technical series 42

Climate and Development Knowledge Network (2012) Managing climate extremes and disasters in Latin America and the Caribbean: lessons from the SREX report. CDKN. Available via www.cdkn.org/srex

Conte M, Ennaanay D, Mendoza G et al (2011) Retention of nutrients and sediment by vegetation. In: Kareiva P, Tallis H, Ricketts T et al (eds) Natural capital: theory and practice of mapping ecosystem services. Oxford University Press, Oxford. Kindle Edition

Costanza R, Pérez-Maqueo OM, Martínez ML et al (2008) The value of coastal wetlands for hurricane protection. Ambio 37:241–248

CSI [Cote Sud Initiative] (2012a) Extrait du rapport d'evaluation des problemes environnementaux du littoral de la Cote Sud Ouest d'Haiti: La place de Port Salut. August 2012

CSI (2012b) Impactes de l'ouragan Sandy dans le Department du Sud et recommendations pour relèvement immédiat. 4 Novembre 2012

Daily GC, Kareiva PM, Polasky S, Ricketts TH, Tallis H (2011) Mainstreaming natural capital into decisions. In: Kareiva P, Tallis H, Ricketts TH, Daily GC, Polasky S (eds) Natural capital: theory and practice of mapping ecosystem services. Oxford University Press, Oxford. Kindle Edition

DeFries R, Pagiola S (2005) Analytical approaches for assessing ecosystem condition and human well-being. In: Hassan R, Scholes R, Ash N (eds) Ecosystems and human well-being: current state and trends. Island Press, Washington, DC, pp 37–71. Available via http://www.unep.org/maweb/documents/document.271.aspx.pdf

de la Rosa D, Mayol F, Moreno JA, Bonsón T, Lozano S (1998) An expert system/neural network model (ImpelERO) for evaluating agricultural soil erosion in Andalucia region, southern Spain. Agric Ecosyst Environ 73(3):211–226

Devatha CP, Deshpande V, Renukaprasad MS (2015) Estimation of soil loss using USLE model for Kulhan Watershed, Chattisgarh – a case study. International conference on water resources, coastal and ocean engineering. Aquatic Procedia 4:1429–1436

Dolidon N, Hofer T, Jansky L, Sidle R (2009) Watershed and forest management for landslide risk reduction. In: Sassa K, Canuti P (eds) Land- slides: disaster risk reduction. Springer, Berlin, pp 633–646

Doswald N, Estrella M (2015) Promoting ecosystems for disaster risk reduction and climate change adaptation: opportunities for integration. UNEP Discussion Paper. United Nations Environment Programme Post-Conflict and Disaster Management Branch. http://www.unep. org/disastersandconflicts/portals/155/publications/Eco_DRR_discussion_paper.pdf

Eichelberger B (2013) InVEST: a tool for mapping and valuing ecosystem services. Presentation at Ecosystem Knowledge Network. March 2013. http://ecosystemsknowledge. net/sites/default/files/wpcontent/uploads/2013/06/InVEST_Introduction_EKN_BE.pdf. Accessed 4 May 2015

EME [Evaluación de los Ecosistemas del Milenio de España] (2011) Síntesis de resultados. Fundación Biodiversidad. Ministerio de Medio Ambiente, y Medio Rural y Marino. Madrid: Spain.

EPA [Environmental Protection Agency] (2007) Basins/HSPF training appendix E: using the TMDL USLE tool to estimate sediment loads. 23 January 2007. http://water.epa.gov/scitech/ datait/models/basins/upload/2007_01_23_ftp_basins_training_b4appe.pdf. Accessed 27 Apr 2015

Estrella M, Renaud FG, Sudmeier-Rieux K (2013) Opportunities, challenges and future perspectives for ecosystem-based disaster risk reduction. In: Renaud FG, Sudmeier-Rieux K, Estrella M (eds) The role of ecosystems in disaster risk reduction. United Nations University Press, Tokyo

European Union (2014) Mapping and assessment of ecosystems and their services: indicators for ecosystem assessments under Action 5 of the EU Biodiversity Strategy to 2020. 2nd Report. February 2014. http://ec.europa.eu/environment/nature/knowledge/ecosystem_assessment/ pdf/2ndMAESWorkingPaper.pdf. Accessed 28 Apr 2015

FAO [Food and Agriculture Organization of the United Nations] (2013) Wishmeier and Smith's Empirical Soil Loss Model (USLE). FAO Corporate Document Repository. Produced by FAO Natural Resources Management and Environment Department. http://www.fao.org/docrep/ t1765e/t1765e0e.htm. Accessed 15 Apr 2015

GFDRR [Global Facility for Disaster Reduction and Recovery] (2014) Review of open source and open access software packages available to quantify risk from natural hazards. International Bank for Reconstruction and Development, The World Bank, Washington, DC

Ghazoul J (2007) Recognising the complexities of ecosystem management and the ecosystem service concept. GAIA 16(3):215–221

Gornitz V (1990) Vulnerability of the east coast, USA to future sea level rise. J Coast Res 9:201–237

Government of the Republic of Haiti (2013) Arrêté déclarant aire protégée le complexe marin et côtier situé dans le Sud-Ouest de la péninsule Sud sous la dénomination d'Aire protégée de Ressources Naturelles Gérées de Port-Salut/Aquin. Le Moniteur 168 (156). Port-au-Prince, Haiti. http://ciat.gouv.ht/artpublic/bibliotheque/File/Kiliweb/arretes/20130826_ 156PortSalutAquin_.pdf. Accessed 18 Apr 2015

Guannel G, Ruggiero P, Faries J et al (2015) Integrated modeling framework to identify the coastal protection services supplied by vegetation. J Geophys Res Oceans 120(1):324–345

Guerry A, Ruckelhaus MH, Arkema KK et al (2012) Modeling benefits from nature: using ecosystem services to inform coastal and marine spatial planning. Int J Biodivers Sci Ecosyst Serv Manage 8(1-2):107–121

Hammar-Klose ES, Thieler ER (2001) Coastal vulnerability to sea-level rise: a preliminary database for the US Atlantic, Pacific and Gulf of Mexico Coasts. US Geological Survey Digital Data Series, vol 68

Handmer J, Honda Y, Kundzewicz ZW et al (2012) Changes in impacts of climate extremes: human systems and ecosystems. In: Field CB, Barros V, Stocker TF et al (eds) Managing the risks of extreme events and disasters to advance climate change adaptation. A special report of working groups I and II of the Intergovernmental Panel on Climate Change (IPCC). Cambridge University Press, Cambridge, UK/New York, pp 231–290

Hettiarchi SSL, Samarawickrama SP, Fernando HJS et al (2013) Investigating the performance of coastal ecosystems for hazard mitigation. In: Renaud FG, Sudmeier-Rieux K, Estrella M (eds) The role of ecosystems in disaster risk reduction. United Nations University Press, Tokyo

Japan Satoyama Satoumi Assessment (2010) Satoyama-Satoumi ecosystems and human well-being: socio-ecological production landscapes of Japan – summary for decision makers. United Nations University, Tokyo

Karpilo RD, Toy TJ (2004) Non-agricultural C and P values for RUSLE. In: National meeting of the American Society of Mining and Reclamation conference proceedings April 18–24, 2004. Lexington, KY. Downloaded at: http://www.asmr.us/Publications/Conference%20Proceedings/2004/0995Karpilo%20COpdf. Accessed 3 April 2015

Lacambra C, Friess DA, Spencer T, Möller I (2013) Bioshields: mangrove ecosystems as resilient natural coastal defences. In: Renaud FG, Sudmeier-Rieux K, Estrella M (eds) The role of ecosystems in disaster risk reduction. United Nations University Press, Tokyo

Langridge SM, Hartge EH, Clark R et al (2014) Key lessons for incorporating natural infrastructure into regional climate adaptation planning. Ocean Coastal Manag 95:189–197

Lawler JJ, Nelson E, Conte M et al (2011) Modeling the impacts of climate change on ecosystem services. In: Kareiva P, Tallis H, Ricketts TH, Daily GC, Polasky S (eds) Natural capital: theory and practice of mapping ecosystem services. Oxford University Press, Oxford. Kindle Edition

Makanzu IF (2013) Rapport d'activités de la Phase 2. Report produced for UNEP

McCook LJ, Almany GR, Berumen ML et al (2009) Management under uncertainty: guidelines for incorporating connectivity into the protection of coral reefs. Coral Reefs 28:353–366

McKenzie E, Irwin F, Ranganathan J et al (2011) Incorporating ecosystem services in decisions. In: Kareiva P, Tallis H, Ricketts TH, Daily GC, Polasky S (eds) Natural capital: theory and practice of mapping ecosystem services. Oxford University Press, Oxford. Kindle Edition

McKenzie E, Posner S, Tillmann P et al (2014) Understanding the use of ecosystem service knowledge in decision making: lessons from international experiences of spatial planning. Environ Plan C Govern Pol 32(2):320–340

Mendoza G, Ennaanay D, Conte M et al (2011) Water supply as an ecosystem service for hydropower and irrigation. In: Kareiva P, Tallis H, Ricketts TH, Daily GC, Polasky S (eds) Natural capital: theory and practice of mapping ecosystem services. Oxford University Press, Oxford. Kindle Edition

Mfumu KA (2013) Projet pilote de Gestion Intégrée des Ressources en Eau du Bassin Versant de la Lukaya; Rapport de mi-parcours de la consultation. Report produced for UNEP

Millennium Ecosystem Assessment (2005) Ecosystems and human well-being: synthesis. Island Press, Washington, DC. http://www.millenniumassessment.org/documents/document.356.aspx.pdf

Mooney H (2011) Getting there. In: Kareiva P, Tallis H, Ricketts TH, Daily GC, Polasky S (eds) Natural capital: theory and practice of mapping ecosystem services. Oxford University Press, Oxford. Kindle Edition

Natural Capital Project (2014) InVEST coastal vulnerability model 3.0.0 documentation http://www.naturalcapitalproject.org/models/coastal_vulnerability.html

Peduzzi P, Velegrakis A, Estrella M, Chatenoux B (2013) Integrating the role of ecosystems in disaster risk and vulnerability assessments: lessons from the Risk and Vulnerability Assessment Methodology Development Project (RiVAMP). In: Renaud FG, Sudmeier-Rieux K, Estrella M (eds) The role of ecosystems in disaster risk reduction. United Nations University Press, Tokyo

Rao NS, Carruthers TJB, Anderson P et al (2013) An economic analysis of ecosystem-based adaptation and engineering options for climate change adaptation in Lami Town, Republic of the Fiji Islands. A technical report by the Secretariat of the Pacific Regional Environment Programme– Apia, Samoa: SPREP 2013, 62 pp

Reef Check Foundation (2013) Port Salut: marine habitat survey; Recommendations for InVEST vulnerability model for CSI. October 11, 2013

Renaud FG, Sudmeier-Rieux K, Estrella M (2013) In: Renaud FG, Sudmeier-Rieux K, Estrella M (eds) The role of ecosystems in disaster risk reduction. United Nations University Press, Tokyo

Sivertun A, Prange L (2003) Non-point source critical area analysis in the Gisselö watershed using GIS. Environ Model Softw 18(10):887–898

Société Audubon Haiti (2013) Port Salut coastal revegetation plan; findings and recommendations by Société Audubon Haiti to Cote Sud Initiative. 14 October 2013

Tallis H, Polasky S (2011) Assessing multiple ecosystem services: an integrated tool for the real world. In: Kareiva P, Tallis H, Ricketts TH, Daily GC, Polasky S (eds) Natural capital: theory and practice of mapping ecosystem services. Oxford University Press, Oxford. Kindle Edition

Thieler ER, Hammar-Klose ES (1999) National assessment of coastal vulnerability to future sea-level rise: preliminary results for the US Atlantic Coast [Internet]. US Geological Survey. Open-File Report. Available from: http://pubs.usgs.gov/of/of99-593/

Tolman HL (2009) User manual and system documentation of WAVEWATCH III version 3.14. Technical Note. US Department of Commerce, National Oceanographic and Atmospheric Administration, National Weather Service, National Centers for Environmental Predictions.

Tombus FE, Yuksel M, Sahin M, Ozulu IM, Cosar M (2012) Assessment of soil erosion based on the method USLE; Corum Province example. Technical Aspects of Spatial Information II. FIG Working Week 2012. Rome, Italy, 6–10 May 2012. http://www.fig.net/pub/fig2012/papers/ts05e/TS05E_tombus_yuksel_et_al_5848.pdf

Turner RK, Lorenzoni I, Beaumont N et al (1998) Coastal management for sustainable development: analysing environmental and socio-economic changes on the UK coast. Geogr J 164:269–281

UK NEA [National Ecosystem Assessment] (2011) The UK National ecosystem assessment: Technical Report. UNEP-WCMC, Cambridge

UNEP [United Nations Environment Programme] (2010) Risk and Vulnerability Mapping Project (RiVAMP) linking ecosystems to risk and vulnerability reduction: the case of Jamaica.

UNISDR (2015) Sendai framework for disaster risk reduction 2015–2030. United Nations International Strategy for Disaster Reduction, Geneva. 18 March 2015

UNISDR [United Nations International Strategy for Disaster Reduction] (2005) Hyogo framework for action 2005–2015: building the resilience of nations and communities to disasters. United Nations International Strategy for Disaster Reduction, Geneva

USDA (1997) Predicting soil erosion by water: a guide to conservation planning with the Revised Universal Soil Loss Equation (RUSLE). Agriculture Handbook No. 703

Walker WE, Harremoes P, Rotman J et al (2003) Defining uncertainty: a conceptual basis for uncertainty management in model-based decision support. Integr Assess 4(1):5–17

Wischmeier WH, Smith D (1978) Predicting rainfall erosion losses: a guide to conservation planning. USDA-ARS Agriculture Handbook, Washington, DC

WWF [World Wildlife Fund] (2012) Scenarios for InVEST: A primer. http://www.naturalcapitalproject.org/decisions/scenarios.html

Yang W, Li F, Wang R, Hu D (2011) Ecological benefits assessment and spatial modeling of urban ecosystem for controlling urban sprawl in Eastern Beijing, China. Ecol Complex 8(2):153–160, ISSN 1476-945X, http://dx.doi.org/10.1016/j.ecocom.2011.01.004

Chapter 11
Quantifying the Stabilizing Effect of Forests on Shallow Landslide-Prone Slopes

Luuk Dorren and Massimiliano Schwarz

Abstract Shallow landslides are natural hazards that can affect human life and infrastructure both directly and indirectly. Such landslides usually involve low-cohesion soil mantles less than a few meters deep. As shown by evidence worldwide, the presence of forests can lead to increased slope stability, due to mechanical and hydrological mechanisms, and therefore significantly reduce the landslide risk in many locations. Therefore, the nationwide project SilvaProtect-CH, which provided data and defined uniform criteria for protection forest delimitation in Switzerland, has also included shallow landslide protection forests. According to the modelling results of SilvaProtect-CH, approximately 27 % of the Swiss protection forests provide a protective function against shallow landslides. To facilitate a quick quantitative evaluation of the slope stabilizing effect of such forests, we developed the tool SlideforNET, which is described in this chapter.

Keywords Shallow landslides • Protection forests • Quantitative tool: SlideforNET • Slope stability

11.1 Introduction

Shallow landslides can pose significant risks to human livelihoods and infrastructure (Sidle and Ochiai 2006) by directly impacting buildings and traffic ways (Fig. 11.1). In addition, shallow landslides and soil loss in the upper part of stream catchments can lead to high sediment yields downstream (e.g. Benda and Dunne 1997). Consequently, shallow landslides may indirectly amplify problems such as debris flows in

L. Dorren (✉)
Bern University of Applied Sciences, Zollikofen, Switzerland
e-mail: luuk.dorren@ecorisq.org

M. Schwarz
International EcorisQ Association, Geneva, Switzerland

© Springer International Publishing Switzerland 2016
F.G. Renaud et al. (eds.), *Ecosystem-Based Disaster Risk Reduction and Adaptation in Practice*, Advances in Natural and Technological Hazards Research 42,
DOI 10.1007/978-3-319-43633-3_11

255

Fig. 11.1 Problems caused by shallow landslides (debris in the village, as well as road burial and closure downslope) in Curaglia triggered by a rainstorm in Sept. 1991. This landslide occurred one and a half years after tree removal (both uprooting and breaking) by the windstorm Vivian (Feb. 1990). Earlier, wooden barriers had been constructed out of fear for snow avalanches (Photo: A. Sialm)

residential areas, reservoir sedimentation, as well as accelerated underscouring of bridge pillars of important traffic ways.

Shallow landslides usually involve low-cohesion colluvial soils (less than a few meters deep) and are often translational, falling along a sliding plane parallel to the slope surface (Milledge et al. 2014). The actual initiation of shallow landslides depends on local topography, slope material and its physical properties, as well as hillslope hydrology. The presence of forests has an influence on the latter two factors. Evidence of the stabilizing effect of the presence of trees on slopes that are prone to shallow landslides can be found worldwide (Haigh et al. 1995; Bathurst et al. 2007; Rickli and Graf 2009; Bischetti et al. 2009; García-Ruiz et al. 2010; Peduzzi 2010; Kim et al. 2013; Okada and Kurokawa 2015).

The actual degree of stabilization depends very much on the forest characteristics – amongst others the spatial distribution of trees and more importantly their roots (cf. Schwarz et al. 2012; Hwang et al. 2015). Forest characteristics can be influenced by humans – although an improvement of the state of forests can only be attained over longer periods of time, a worsening of conditions can be caused rapidly (for example by excessive timber harvesting: Sidle and Wu 1999; Jakob 2000).

The national natural hazard risk management strategy of Switzerland, defined by PLANAT (2005), considers forests equal to technical or civil engineering measures regarding prevention of natural hazards. The presence of forests can, for example, lead to increased slope stability and therefore reduce the landslide risk in many locations to an acceptable level. Where forests are present, the implementation of technical measures for landslide risk reduction are often redundant or cheaper (lesser installation and maintenance costs) on slopes that are prone to shallow landslides.

Since the presence of forests can reduce the risk posed by shallow landslides, the nationwide project SilvaProtect-CH, which defines uniform criteria for protection forest delimitation in Switzerland, also included so-called 'shallow landslide protection forests' (Losey and Wehrli 2013). As a consequence, the Swiss Confederation also pays service-based compensations to the cantons which are responsible for carrying out silvicultural measures in forests that function as protection against shallow landslides (the 26 cantons of Switzerland are the member states of the Swiss Confederation). This is done under the condition that these measures aim at reaching a defined target profile according to the criteria of the Swiss guideline "Sustainability and Success Monitoring in Protection Forests" (Nachhaltigkeit im Schutzwald or in short NaiS guideline; Frehner et al. 2005). Due to the lack of a suitable tool for the practice, the currently used NaiS criteria for shallow landslide protection forests are rather general and do not account for site-specific topography and physical soil properties. The current version of the shallow landslide protection forest target profile in the guideline defines a desired forest structure, a minimum forest cover, a maximum gap size, as well as a required degree of regeneration and site-indigenous trees. Important factors such as the slope gradient, the potential slide plane (or shearing plane) depth and the soil material are not taken into account in the definition of the target profile. Based on the current knowledge on the effect of forests on shallow landslides, the existing guidelines could be improved by including these aforementioned factors.

This also holds for the current practice of assessing the hazard posed by shallow landslides, as being undertaken by most natural hazard engineers today. They encounter difficulties when accounting for the stabilizing effect of forests, not the least because the quantification of the slope stabilizing effect of forests remains complicated without suitable and accessible tools. In addition, many engineers consider forests to be excessively vulnerable to disturbances such as storms, pests and diseases, and therefore they often opt for a worst case scenario without forests. This is the present-day reality, despite the fact that examples of the stabilizing effect of forests, as well as the scientific knowledge and methods for its quantification exist.

Our work focuses on developing simple, but suitable and robust tools for quantifying the effect of forests on slope stability, which account for site-specific topography, soil physical properties and forest characteristics. This chapter firstly resumes the current knowledge on shallow landslides and vegetation. Secondly, it briefly describes the SilvaProtect-CH project and the results for shallow landslide protection forests. Finally, it presents one of the tools we developed called SlideForNET, as well as three real case applications.

11.2   Short Background on Shallow Landslides and Vegetation

Many landslide inventories show that rainfall-triggered shallow landslides usually have typical depths of 0.5–1.5 m, scar areas of 50–1000 m^2 and volumes ranging from a few to several thousand cubic meters (Moser and Schoger 1989; Malamud et al. 2004; Markart et al. 2007; Rickli and Graf 2009). According to the definition of BRP/BWW/BUWAL (1997), the maximum depth of the shearing plane of shallow landslides in Switzerland is 2 m. There is, however, no universal criterion for the maximum shearing plane depth and therefore, landslides with a shearing plane depth of 5 m are in some countries still considered as shallow landslides.

Soil hydrology is one of the main drivers of shallow landslides. It is the change in pore water pressures – and therefore a weakening of the soil shear strength (Fredlund 1979) – often linked to precipitation events, which cause a slope to fail or in other words, triggers a landslide (Toll et al. 2011; Lehmann and Or 2012). Both high intensity rainstorms of short duration, or low intensity rainfalls of long duration can trigger a landslide (Rickli and Graf 2009).

As already mentioned in the introduction, presence of forests has a beneficial influence on slope stability through hydrological and mechanical mechanisms (Greenway 1987; Sidle and Ochiai 2006). The main beneficial hydrological mechanisms are:

- Intercepting rainfall, controlling both the amount and timing of precipitation reaching the potentially unstable soil mantle;
- Altering hydraulic conductivity through physical transformation of the soil by roots;
- Root water uptake and evaporation – in short evapotranspiration – as remover of water from the soil layers. In general, evapotranspiration rates in temperate regions are 5–10 times higher from forests, compared to bare soil.

The main beneficial mechanical mechanism is root reinforcement. Many authors agree (e.g. Wu et al. 1979; Greenway 1987; Phillips and Watson 1994; Sidle and Ochiai 2006) that root reinforcement is more important in stabilizing slopes than the hydrological mechanisms.

According to Sidle (1992), reinforcement by tree roots may provide the difference between stability and instability during storms or snowmelt when hillslope soils are in a tenuous state of equilibrium. Many field studies on steep forested slopes worldwide have observed a significant increase (up to 10-fold) in mass erosion and landsliding 3 to 15 years after timber harvesting (e.g. Bishop and Stevens 1964; for additional references see Sidle and Ochiai 2006).

As shown in Fig. 11.2, root reinforcement can be grouped in three stabilizing mechanisms (Giadrossich et al. 2013):

1. Anchoring unstable soil mantle into more stable substrate, also called basal root reinforcement;

Fig. 11.2 Illustration of three different mechanisms of root reinforcement: (1) basal root reinforcement along a potential shearing plane – *dark line*, (2) stiffening of the soil mass, (3) lateral root reinforcement at the margins of the landslide (Modified from Giadrossich et al. 2013)

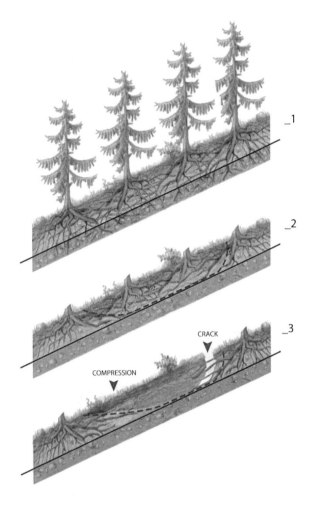

2. Stiffening the unstable soil mantle, increasing the stability through buttressing and arching by trees (Stokes et al. 2008);
3. Reinforcing the potential unstable soil mantle by roots under shearing, tension and compression acting on the lateral edge of the landslide body (Schwarz et al. 2013), also called lateral root reinforcement.

An increasing number of studies differentiate between these three types of root reinforcement (Schmidt et al. 2001; Roering et al. 2003; Schwarz et al. 2010a). Most roots in a forest stand are confined within the first meter of soil and vertical roots only occasionally reach the depth (usually 1–2 m) of potential shear planes of shallow landslides (Schmidt et al. 2001; Danjon et al. 2008). Hence, including lateral roots is critical for realistic stability analyses of shallow landslides.

For the quantitative analysis of slope stability, the three main methods available are (1) the limit equilibrium (LE) method, (2) the finite element (FE) method and

(3) the discrete element (DE) method. Currently, most slope stability analyses in practice involve LE analysis (Yu et al. 1998), amongst others due to its simplicity. LE methods consist of cutting the slope into fine slices or cells and applying equations that calculate the equilibrium of the forces and/or moments of each individual slice or cell (cf. Milledge et al. 2014). The equilibrium is mostly expressed as a factor of safety, which is the ratio between retaining (resisting) and driving (sliding) forces. A main shortcoming is that progressive failure cannot be accounted for by LE methods (Schwarz et al. 2010b).

FE methods are numerical techniques based on partial differential equations. These methods subdivide a larger problem into simpler parts called finite elements, which are connected to each other by nodes. This is done by mesh generation techniques. All nodes are defined by multiple characteristics and all connections are described by a set of equations. Using FE methods for slope stability has several advantages: slopes can be modelled with a high degree of realism (complex geometry), the presence of material for reinforcement can be included as additional strength and the action of infiltrating water can be integrated (Matthews et al. 2014). FE approaches allow for the modelling of the propagation of the failure surface and the progressive failure of the slope (Wu et al. 2015). The phenomena that remain difficult to be properly modelled by the numerical methods are local deformations and breakages in the sliding mass (Wu et al. 2015).

DE methods (re. Cundall 1971) are numerical techniques for computing the motion of and force interaction between a number of discretised particles. Nowadays, DE methods are frequently used for solving engineering problems, particularly in granular and discontinuous materials. An advantage of DE methods is the possibility to realistically implement stress-strain relationships between particles, which allow reproducing phenomena such as breakage and local deformations of the sliding mass. A disadvantage is that DE methods are computationally intensive and limit therefore the number of discretized particles or the total number of simulations.

In most existing modelling approaches, root reinforcement is considered a constant, homogeneously distributed factor (e.g. Sidle and Ochiai 2006; Bathurst et al. 2007; Simoni et al. 2007). The three aforementioned mechanisms of root reinforcement are generally simplified as an increased soil cohesion term of the Mohr–Coulomb failure criterion (e.g. Wu et al. 1979; Mao et al. 2014). Despite considerable progress in understanding of root reinforcement (e.g. Schwarz et al. 2010a and Cohen et al. 2011), realistic descriptions of the spatial distribution, as well as the different mechanisms of root reinforcement in vegetated hillslopes are rarely implemented in slope stability analyses. One of the reasons is the characterization of temporal and spatial root distribution, which is a function of tree species, tree size and location in the landscape (Schwarz et al. 2012).

11.3   Defining Protection Forests for Shallow Landslides in CH

In 2004, the Swiss Federal Office for the Environment (FOEN) launched the SilvaProtect-CH project (for details see Losey and Wehrli 2013), with the following aims:

1. to formulate a protection forest index – required for the allocation of Federal subsidies to the cantons for protection forest management;
2. to formulate a damage potential index – required for the allocation of Federal subsidies to the cantons for the management of technical protective measures;
3. to provide the cantons with uniform criteria and objective data for delimiting protection forests in Switzerland.

A large part of the project is based on process simulation and data modelling. This part was organized in five modules with the following objectives (Losey and Wehrli 2013):

1. EVENT: to model the run-out areas of the hazard processes, rockfall, torrential processes (debris flow and overbank sedimentation), avalanches and shallow landslides using pessimistic scenarios without the protective role of forests;
2. DAMAGE: to define the relevant damage potential and to collect the required geodata;
3. INTERSECT: to select runout zones of hazard processes that reach relevant damage potential (so-called damage-relevant process areas – DRPA);
4. SILVA: to provide the forest area of Switzerland;
5. SYNTHESE: (1) to extract the damage-relevant process areas in the forest by intersecting the DRPA with the Swiss forest area; (2) To calculate the protection forest index, as well as (3) the damage potential index.

In the EVENT module, the different previously mentioned natural hazard processes were modelled nationwide, using different process simulation models. These simulations were carried out by natural hazard engineering and consultancy offices. Pessimistic shallow landslide scenarios (without the protective effect of forests) were modelled using two models, one for the disposition (SLIDISP) and another for the runout of shallow landslides (SLIDESIM). The model SLIDISP is based on an LE analysis and calculates the probabilities for potential shallow landslide release in every cell of the defined raster covering the study area. The major model input data for the EVENT module were slope gradients derived from a 10×10 m digital elevation model (DEM), geotechnical parameters derived from the geotechnical map of Switzerland (1:200'000 SGTK), as well as assumptions on the depth and water-saturation of the subsoil (Liener 2000). Such a high DEM resolution is not a prerequisite, but with resolutions larger than 25×25 m, the modelling results are probably too imprecise, which could become a problem in some parts of the world. Based on the starting areas, the runout zones (consisting of single trajectories) of shallow landslides were simulated with the model

SLIDESIM. This model simulates the runout paths of slope-type debris flows using a random walk approach (Gamma 2000) for determining the flow direction and the Voellmy approach (Voellmy 1955) for calculating the runout distance. In addition, a set of pre-defined spread parameters determine the final runout zone. For the entire area of Switzerland, approximately 48 million shallow landslide trajectories were simulated. All simulated trajectories were finally converted from single falltrack lines in run-out areas by enlarging all individual trajectories with a 10 m buffer and dissolving the adjacent ones.

The damage potential included important roads (highways, cantonal roads, communal roads and even local roads ensuring a connection with residential areas), all railways, residential buildings, public buildings (hospitals, stations, community halls, etc.), industrial buildings, churches, important agricultural buildings and important installations, such as reservoir dams, harbors, power plants, etc. For details see Losey and Wehrli (2013).

After the validation and finalization of the delimitation of nation-wide protection forests by the cantonal authorities on the basis of the data and criteria of the SilvaProtect-CH project, the total protection forest area added up to 585,000 ha, which corresponds to 49 % of the forest area in Switzerland. Based on the modelled results of the project only, Losey and Wehrli (2013) calculated that approximately 27 % of the modelled protection forests provide a protective function against shallow landslides (Fig. 11.3), the remaining 2/3 of the protection forests mainly protect against rockfall, torrential processes and snow avalanches. Table 11.1

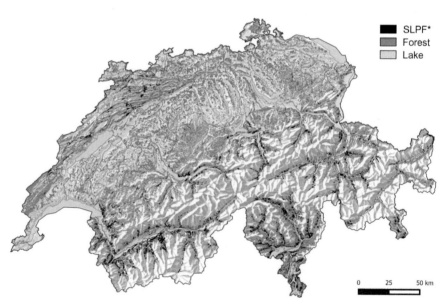

Fig. 11.3 The shallow landslide protection forests (SLPF) based on the modelling results of SilvaProtect-CH (Data source: Federal Office for the Environment FOEN) and all other forested areas in Switzerland (Data source: Vector25-Swisstopo). The background is a relief map of Switzerland (Image: Swisstopo)

Table 11.1 Slope gradients covered by shallow landslide protection forests following the SilvaProtect-CH data

Slope gradient class	Frequency (%)	Area (ha)
0–10°	0	0
>10–20°	10	15,795
>20–25°	18	28,431
>25–30°	26	41,067
>30–35°	17	26,852
>35–40°	13	20,534
>40–45°	10	15,795
>45–55°	5	7898
>55°	1	1580
Total	100	157,950

shows that shallow landslide protection forests mainly cover slopes with gradients between 20° and 40°. On flatter slopes, shallow landslides are rarely triggered and on steeper slopes, sufficient thickness of the soil mantle for shallow landslides is generally missing.

Although the initial objectives of the SilvaProtect-CH project have been achieved, the project still continues since additional analyses need to be conducted to answer secondary questions on the basis of the available data, including calculating the costs for protection forest management that should be assumed by the Federal Roads Office (FEDRO) or by the Swiss Federal Railways, for example.

Silvicultural interventions in protection forests aim at improving and/or ensuring their protective function over time. These interventions are being financed by the direct beneficiaries. If these interventions occur in a protection forest perimeter defined by the canton, a large part of the costs are covered by the federal and cantonal governments. The cantonal forest services are responsible for the prioritization and realization of these interventions. Hereby, two types of protection forests should be favoured. Firstly, those which currently provide a high level of protection but require interventions to sustain this level over the long term, and secondly, those forests where a significant improvement of the level of protection can be reached with minimal investments. To facilitate the definition of a set of key characteristics of these two types of protection forests, we propose the use of a simple, but robust tool, which implements the state-of-the-art in slope stability and root reinforcement.

11.4 SlideforNET – A Simple Tool for Quantifying the Effect of Forests on Slope Stability

SlideforNET (www.slidefor.net) is a freely available web tool developed for estimating root reinforcement effects on slope stability and for comparing the protective effect of different forest types and different slope characteristics (slope and soil material). The stability calculation in SlideforNET is based on a 3D force balance that assumes an elliptical shape (Rickli and Graf 2009) of shallow landslides. Analogue to existing landslide modelling approaches (e.g. Milledge et al. 2014), the landslide mass is considered perfectly rigid, allowing all the lateral forces to act simultaneously. In the stability calculation, the compression forces acting on the downslope margin of the landslide are not subtracted from the maximum tensile forces, since we assume that they will not occur simultaneously during sliding. For a detailed description of the equations we refer to Schwarz et al. (2010b).

Root reinforcement is implemented in the calculation by considering (1) the roots crossing the upper margin of the landslide (lateral root reinforcement along the potential tension crack) and (2) the roots crossing the basal shearing plane (basal root reinforcement). The latter is calculated using an exponentially decreasing cumulative density function, approaching a basal root reinforcement of 0 kPa at a depth of 2 m. Based on the input parameters stand density, mean stem diameter at breast height (DBH) and species composition, the model calculates the minimum lateral root reinforcement assuming a mean tree distance based on a tree distribution following a regular grid. Recent data of root distribution of the main alpine tree species (Norway spruce – *Picea abies*, silver fir – *Abies alba*, and European beech – *Fagus sylvatica*) allow the attribution of a root reinforcement value (5, 10, or 15 kN/m), based on the mean tree distance, the mean stem diameter and trees species (Schwarz et al. 2013). Stiffening of the unstable soil mantle is not considered.

In order to consider the effects of lateral root reinforcement on different dimensions of shallow landslides, a gamma probability function is used to describe the frequency-magnitude distribution of potential shallow landslide volumes, as described by Malamud et al. (2004). The resulting number of unstable landslides is not related to a specified rainfall event magnitude or return period, rather it represents the partial probability that landslides with a certain area may occur under fully saturated conditions (pore water pressure = soil depth*9.81).

The input parameters shearing plane depth as well as effective soil cohesion are calculated using a stochastic approach; the mean value has to be defined by the user for both parameters, and subsequently a random function is used to create a normal distribution array of values based on a fixed standard deviation. For each site, 10,000 slope stability calculations with different combinations of the randomly generated scar size, shearing plane depth and effective soil cohesion are carried out. For each combination, SlideforNET analyses if a landslide can be triggered or not. Figure 11.4 shows the main graphical output of the tool. Another output is the 'degree of protection', which is the reduction of the total number of landslides due to the presence of forests expressed as a percentage. Finally, the model also accounts for the surcharge of the trees. The tool provides information to the user in which the tree weight is compared to the weight of the subsoil in order to demonstrate the negligible role of this force component on slope stability (cf. Fan and Lai 2014).

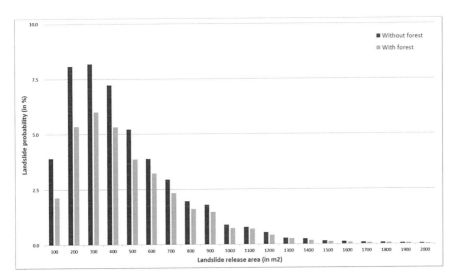

Fig. 11.4 Graphical output of SlideforNET showing the probability density function of the occurrence of shallow landslides under current forested and non-forested conditions at the Schangnau site

11.4.1 Real Case Applications

We applied SlideforNET on three forested sites in the Swiss Prealps where shallow landslides have occurred, as described in Schwarz et al. (2013). Table 11.2 summarizes the main characteristics of the sites. The soil depth indicates the depth of potential shearing plane, which can be estimated by observing previous events. The most difficult parameters to be estimated are the effective friction angle and the effective soil cohesion. However, soil classification tables (e.g. the Unified Soil Classification System USCS) provide an aid for obtaining plausible values. Species compositions of the forest stand are characterized in terms of coverage percentage. Here we compared the 'present' conditions with the 'optimal' target profile according to the NaiS guidelines (Frehner et al. 2005). For all cases, a safety factor F of 1 was used. F is the ratio of the forces that make a slope fail to those that prevent a slope from failing. $F < 1$ means unstable slope conditions, $F = 1$ means that the slope is at the point of failure and $F > 1$ means stable slope conditions.

For the three cases, SlideforNET reproduces our field observations, in the sense that the tool predicted that a landslide occurrence probability under the present forest conditions exists even though the calculated probability at the Gantrisch site is very low. According to the results, the slope stabilising effect at the Spisibach site does not influence the landslide occurrence probability, but with an optimal species composition this stabilizing effect would decrease the occurrence probability with 31 % (i.e. (68–47)*100/68). On both other sites, these values are similar (Schangnau 38 %; Gantrisch 27 %). The histogram in Fig. 11.4 shows that lateral

Table 11.2 Input and output data of SlideforNET for the three sites

Site	Slope gradient [°]	Soil depth [m]	Effective friction angle [°]/Effective soil cohesion [kPa]	Species present [%]	Species optimal* [%]	Mean DBH [cm]/Stand density [N/ha]	Landslide probability without forest [%]	Landslide probability present forest [%]	Landslide probability optimal species [%]
Gantrisch	20	1	27/1.0	Sp:80 Fi:20 Be:0	Sp:30 Fi:70 Be:0	41/300	11	3	0
Schangnau	25	1.5	25/0.5	Sp:80 Fi:20 Be:0	Sp:0 Fi:40 Be:60	39/350	45	33	16
Spisibach	35	2	29/0.5	Sp:40 Fi:50 Be:10	Sp:0 Fi:40 Be:60	40/250	68	68	47

Sp = Norway spruce, Fi = silver fir, Be = European beech
*Species composition according to the "optimal" target profile of the NaiS guidelines

root reinforcement is especially effective for landslides with release areas up to 500 m^2.

It is important to note that these forests, in addition to the slope stabilizing effect, may reduce both surface erosion and the mobilization of non-consolidated material covering the slope surface, which might otherwise be transported by debris flows.

Since December 2014, SlideforNET is available to practitioners worldwide; an important restriction, however, is the root distribution data, which are currently only based on measurements around Norway spruce, Silver fir and European beech trees in the Alps. For other broadleaved and coniferous species, we made very general assumptions.

In Switzerland, foresters are currently getting used to online tools that help in defining 'optimal' target profiles since the current rockfall protection forest guideline in NaiS is based on a similar tool (http://www.gebirgswald.ch/de/anforderungen-steinschlag.html). To facilitate the use of these tools, courses have to be offered to practitioners and examples of applications will have to be published in professional magazines. Therefore, a proper introduction of tools like SlideforNET in the daily practice of protection forest management is a process of several years. Practitioner feedback and the continuous application of the tool in real cases as part of our research work will enable us to further improve the tool over the coming years.

11.5 Conclusion

Forests play an important role in stabilizing shallow landslide-prone slopes and therefore also in natural hazard risk management in Switzerland. The modelled results of the SilvaProtect-CH project showed that approximately 27 % of the protection forests, which corresponds to 13 % of all forests, provide a protective function against shallow landslides in Switzerland. These forests mainly cover slopes with gradients between 20° and 40°.

The SlideforNET tool allows practitioners to perform a quick quantitative evaluation of the protective effect of forests in regions susceptible to shallow landslides. This tool may be used to discuss the protective function of present forest conditions, or to study the effects of future management scenarios. Furthermore, the tool serves as complementary instrument for defining an 'optimal' target profile on the basis of forest, slope and soil characteristics. SlideforNET can be easily and directly used in the field via smartphones by practitioners worldwide, which will hopefully increase its recognition in the coming years.

References

Bathurst JC, Moretti G, El-Hames A et al (2007) Modelling the impact of forest loss on shallow landslide sediment yield, Ijuez river catchment, Spanish Pyrenees. Hydrol Earth Syst Sci 11:569–583

Benda L, Dunne T (1997) Stochastic forcing of sediment supply to channel networks from land sliding and debris flow. Water Resour Res 33:2849–2863

Bischetti GB, Chiaradia EA, Epis T, Morlotti E (2009) Root cohesion of forest species in the Italian Alps. Plant Soil 324:71–89

Bishop DM, Stevens ME (1964) Landslides in logged areas in Southeast Alaska. U.S.D.A. Forest Service Research Paper NOOR-1, Northern Forest Experimental Station, Juneau, AK: 18 p

BRP/BWW/BUWAL (1997) Berücksichtigung der Massenbewegungsgefahren bei raumwirksamen Tätigkeiten – Empfehlung. BRP/BWW/BUWAL, Bern: 42 p

Cohen D, Schwarz M, Or D (2011) An analytical fiber bundle model for pullout mechanics of root bundles. J Geophys Res 116:F03010

Cundall PA (1971) A computer model for simulating progressive, large scale movement in blocky rock system. In: Symposium ISRM. Nancy, France, Proc. 2:129–136

Danjon F, Barker DH, Drexhage M, Stokes A (2008) Using three-dimensional plant root architecture in models of shallow-slope stability. Ann Bot 101(8):1281–1293

Fan CC, Lai YF (2014) Influence of the spatial layout of vegetation on the stability of slopes. Plant Soil 377:83–95

Fredlund DG (1979) Second Canadian geotechnical colloquium: appropriate concepts and technology for unsaturated soils. Can Geotech J 16(1):121–139

Frehner M, Wasser B, Schwitter R (2005) Nachhaltigkeit und Erfolgskontrolle im Schutzwald – Wegleitung für Pflegemassnahmen in Wäldern mit Schutzfunktion. BUWAL, Bundesamt für Umwelt, Wald und Landschaft, Bern

Gamma P (2000) dfwalk – Ein Murgang-Simulationsprogramm zur Gefahrenzonierung, Geographica Bernensia G 66, Univ. Bern: 144 p.

García-Ruiz JM, Beguería S, Alatorre LC, Puigdefábregas J (2010) Land cover changes and shallow landsliding in the flysch sector of the Spanish Pyrenees. Geomorphology 124 (3–4):250–259

Giadrossich F, Schwarz M, Pirastru M, Niedda M (2013) Stabilization mechanisms of hillslopes due to root reinforcement. Quaderni di Idronomia Montana 31:353–362

Greenway DR (1987) Vegetation and slope stability. In: Anderson MF, Richards KS (eds) Slope stability. Wiley, New York

Haigh MJ, Rawat JS, Rawat MS et al (1995) Interactions between forest and landslide activity along new highways in the Kumaun Himalaya. For Ecol Manag 78(1–3):173–189

Hwang T, Band LE, Hales TC, et al. (2015) Simulating vegetation controls on hurricane-induced shallow landslides with a distributed ecohydrological model. J. Geophys. Res. Biogeosci. n/a

Jakob M (2000) The impacts of logging on landslide activity at Clayoquot Sound, British Columbia. CATENA 38:279–300

Kim D, Im S, Lee C, Woo C (2013) Modeling the contribution of trees to shallow landslide development in a steep, forested watershed. Ecol Eng 61:658–668

Lehmann P, Or D (2012) Hydromechanical triggering of landslides: from progressive local failures to mass release. Water Resour Res 48:8250–8262

Liener S (2000) Zur Feststoffflieferung in Wildbächen, Geographica Bernensia G 64, Univ. Bern. 91 pp

Losey S, Wehrli A (2013) Schutzwald in der Schweiz. Vom Projekt SilvaProtect-CH zum harmonisierten Schutzwald. Federal Office for the Environment FOEN, Bern: 29 pp

Malamud BD, Turcotte DL, Guzzetti F, Reichenbach P (2004) Landslide inventories and their statistical properties. Earth Surf Process Landf 29:687–711

Mao Z, Bourrier F, Stokes A, Fourcaud T (2014) Three-dimensional modelling of slope stability in heterogeneous montane forest ecosystems. Ecol Model 273:11–22

Markart G, Perzl F, Kohl B, et al. (2007) 22nd and 23rd august 2005 – analysis of flooding events and mass movements in selected communities of Vorarlberg. BFW-Dokumentation 5/2007: 45 pp

Matthews C, Farook Z, Helm P (2014) Slope stability analysis – limit equilibrium or the finite element method? Ground Eng 2014:22–28

Milledge DG, Bellugi D, McKean JA et al (2014) A multidimensional stability model for predicting shallow landslide size and shape across landscapes. J Geophys Res Earth Surf 119:2481–2504

Moser M, Schoger H (1989) Die Analyse der Hangbewegungen im mittleren Inntal anlässlich der Unwetterkatastrophe 1985. Wildbach- Lawinenverbau 53(110):1–22

Okada Y, Kurokawa U (2015) Examining effects of tree roots on shearing resistance in shallow landslides triggered by heavy rainfall in Shobara in 2010. J For Res 20:230–235

Peduzzi P (2010) Landslides and vegetation cover in the 2005 North Pakistan earthquake: A GIS and statistical quantitative approach. Nat Hazards Earth Syst Sci 10:623–640

Phillips CJ, Watson AJ (1994) Structural tree root research in New Zealand: a review, Landcare Research Science Series N°7: 70 pp

PLANAT (2005) Strategie Naturgefahren Schweiz – Synthesebericht. Bundesamt für Wasser und Geologie BWG, Biel: 81 pp

Rickli C, Graf F (2009) Effects of forests on shallow landslides – case studies in Switzerland. For Snow Landsc Res 82(1):33–44

Roering JJ, Schmidt KM, Stock JD et al (2003) Shallow landsliding, root reinforcement, and the spatial distribution of trees in the Oregon Coast Range. Can Geotech J 40:237–253

Schmidt KM, Roering JJ, Stock JD et al (2001) The variability of root cohesion as an influence on shallow landslide susceptibility in the Oregon 690 Coast Range. Can Geotech J 38:995–1024

Schwarz M, Lehmann P, Or D (2010a) Quantifying lateral root reinforcement in steep slopes – from a bundle of roots to tree stands. Earth Surf Process Landf 35:354–367

Schwarz M, Preti F, Giadrossich F et al (2010b) Quantifying the role of vegetation in slope stability: the Vinchiana case study (Tuscany, Italy). Ecol Eng 36:285–291

Schwarz M, Cohen D, Or D (2012) Spatial characterization of root reinforcement at stand scale: theory and case study. Geomorphology 171–172:190–200

Schwarz M, Feller K, Thormann J-J (2013) Entwicklung und Validierung einer neuen Methode für die Beurteilung und Planung der minimalen Schutzwaldpflege auf rutschgefährdeten Hängen, Final report «Wald- und Holzforschungsfonds», Swiss Federal Office for the Environment FOEN

Sidle RC (1992) A theoretical model of the effects of timber harvesting on slope stability. Water Resour Res 28:1897–1910

Sidle RC, Ochiai H (2006) Landslides: processes, prediction, and land use. American Geophysical Union, Washington, DC

Sidle RC, Wu W (1999) Simulation effects of timber harvesting on the temporal and spatial distribution of shallow landslides. Z Geomorphol NF 43:185–201

Simoni S, Zanotti F, Bertoldi G, Rigon R (2007) Modelling the probability of occurrence of shallow landslides and channelized debris flows using GEOtop-FS. Hydrol Process 22:532–545

Stokes A, Norris JE, Beek LPH (2008) How vegetation reinforces soil on slopes. In: Norris JE, Stokes A, Mickovski SB et al (eds) Slope stability and erosion control: ecotechnological solutions. Springer, Dordrecht, pp 65–118

Toll DG, Lourenco SDN, Mendes J et al (2011) Soil suction monitoring for landslides and slopes. Q J Eng Geol Hydrogeol 44:23–33

Voellmy A (1955) Über die Zerstörungskraft von Lawinen. Schweizerische Bauzeitung 73:159–162, 212–217, 246–249, 280–285

Wu TH, McKinnell WP, Swanston DN (1979) Strength of tree roots and landslides on Price of Wales Island. Alaska Can Geotechnol J 16:19–33

Wu W, Switala BM, Acharya MS (2015) Effect of vegetation on stability of soil slopes: numerical aspect. In: Wu W (ed) Recent advances in modeling landslides and debris flows. Springer International Publishing, Cham, pp 163–177

Yu HS, Salgado R, Sloan SW, Kim JM (1998) Limit analysis versus limit equilibrium for slope stability. J Geotech Geoenviron Eng 124:1–11

Chapter 12
Integrating Landscape Dimensions in Disaster Risk Reduction: A Cluster Planning Approach

Ritesh Kumar, Munish Kaushik, Satish Kumar, Kalpana Ambastha, Ipsita Sircar, Pranati Patnaik, and Marie-Jose Vervest

Abstract Appreciation of the interplay between societal development, the changes in ecosystem functioning and services, and the creation and transformation of disaster risk is fundamental for the identification of ecosystem-based interventions and options for disaster risk reduction. While the need to integrate ecosystem services and ecosystem management has received increased attention as pathways for reducing disaster risk, little attention has been given to the actual methods and approaches that can enable such integration in practice. This chapter proposes a cluster planning approach for disaster risk reduction planning, building on the understanding of the relationship between landscape-scale drivers of disaster risk and community vulnerability and capacity. Including a cluster scale approach in risk reduction planning, which comprises smaller landscape units of communities facing similar risks, helps bridge administrative and ecological boundaries for reaching effective risk reduction outcomes. In the Mahanadi Delta, a landscape exposed to multiple hazards, applying this approach has helped to delineate three clusters wherein distinct ecosystems-based options for risk reduction can be applied. Embedding administrative planning units within ecological planning units has enabled a more realistic integration of ecosystem services in the context of disaster risk reduction. The cluster approach will be particularly useful for planners responsible for developing risk reduction plans across administrative and ecological boundaries.

Keywords Cluster planning approach • Disaster risk • Ecosystem services • Landscapes

R. Kumar (✉) • S. Kumar • K. Ambastha • I. Sircar • P. Patnaik
Wetlands International South Asia, A-25, Second Floor, Defence Colony, New Delhi 110024, India
e-mail: ritesh.kumar@wi-sa.org

M. Kaushik
Cordaid, c/o Caritas India, 9-10, Bhai Vir Singh Marg, New Delhi 110001, India

M.-J. Vervest
Wetlands International Headquarters, Ede-Wageningen, The Netherlands

© Springer International Publishing Switzerland 2016   271
F.G. Renaud et al. (eds.), *Ecosystem-Based Disaster Risk Reduction and Adaptation in Practice*, Advances in Natural and Technological Hazards Research 42,
DOI 10.1007/978-3-319-43633-3_12

12.1 Linking Ecosystems Within Community Risk Reduction Planning – The Challenge of Scale

Ecosystems provide a range of benefits; they help reduce, buffer, and in certain circumstances, mitigate disaster risk, as well as assist societies in adapting to increasing risk. Explicit acknowledgement of ecosystem-based approaches to disaster risk reduction in the Sendai Framework for Disaster Risk Reduction (2015–2030), which resulted from the Third World Conference on Disaster Risk Reduction held in March 2015, signals a clear advancement in the way disaster risk is understood and tackled (UNISDR 2015; see also Renaud et al., Chap. 1).

The science and practice of disaster management in the last two decades have progressively evolved towards a greater focus on disaster risk rather than on the disaster itself, and towards the importance of a prospective and preventive approach to reducing *risk* compared to a mere reactive approach (UNISDR 2009). Increased recognition of underlying social drivers of hazards, exposure and vulnerabilities also reflect an emerging understanding that while physical conditions potentiate disaster risks, disasters are inherently 'social constructions' and attributable to societal choices, constraints and actions (ICSU-LAC 2009; Lavell et al. 2012). This shift in the understanding of disasters helps to factor in the role of skewed socio-economic development in creating and increasing disaster risk, but also the possibility of influencing social decisions that favour risk reduction. This new understanding also puts greater impetus on 'people-centred' approaches, such as community-managed disaster risk reduction which call for bringing people in a community together to collectively address shared disaster risks and collectively pursue a common framework for disaster risk reduction (Maskrey 2011; IIRR and Cordaid 2013).

The way communities relate to and manage their ecosystems within the wider developmental context is an important constituent of the social construction of risk. Similar to the new trends in disaster management discussed above, the last four decades have also seen changes in the science and practice of managing ecosystems, shifting away from solely conservation-centred approaches and towards recognition of the role of ecosystems in ensuring human well-being, encapsulated in the concepts of 'ecosystem services' and 'coupled socio-ecological systems' (Mace 2014). Ecosystem services describe a framework for structuring and synthesizing biophysical understanding of ecosystem processes in terms of human well-being (Brauman et al. 2007). The continued degradation of ecosystems have raised concerns about the ability of current development pathways to ensure maintenance of the critical ecosystem processes which underpin delivery of ecosystem services, including those for disaster risk reduction, while also recognizing limits to their substitution by human or manufactured capital (Barbier 1994; Daily 1997). The concept of coupled socio-ecological systems underlines the complex and adaptive linkages that exist between environment, social actors and institutions (Folke et al. 2010; Vogel et al. 2007). Understanding the interplay between societal development, the changes in ecosystem structure, functions and services, and the creation and transformation of disaster risk is fundamental to identifying potential ecosystem-based interventions and options for disaster risk reduction. While the need to integrate sustainable ecosystem management within disaster risk reduction planning processes has lately received increased attention, there is little elaboration of methods and approaches to enable multi-scale interventions that support such integration within a given landscape.

Community-managed disaster risk reduction planning involves a nuanced, participatory process of assessing interaction of a potentially damaging event and vulnerable conditions of a society or element exposed, through a range of qualitative and quantitative measures. However, by their very nature of being community-based, these assessments are organized on the basis of political and administrative scales (for example along villages, districts and states) and do not accommodate the dimensions of spatial and functional scales relevant for ecosystem management (for example, at levels of river basins or coastal zones). Such limitations have consequences for risk reduction programming in landscapes, for instance in a coastal delta, wherein geophysical elements, such as water and sediment fluxes, have significant bearing on hazards, vulnerabilities and capacities.

This chapter summarizes the lessons from applying a 'cluster planning approach' developed to address the issues of landscape change and increasing disaster risk in the Mahanadi Delta region, which lies within the coastal state of Odisha, India. The first section provides the local context, while the second section describes the cluster planning approach. Results of the suggested cluster planning approach are discussed in the third section. The concluding section summarizes the lessons learned and provides recommendations for further replication and improvements.

12.2 Cluster Planning Approach

The cluster planning approach is an intuitive, multi-scale framework aimed at identifying ecosystem-based intervention needs and opportunities to support disaster risk reduction planning within communities. Such an approach builds on knowledge of landscape-scale drivers of disaster risks and organises risk reduction interventions in 'clusters', or smaller landscape units comprising communities facing similar patterns of risks from natural hazards. The underlying assumption is that community-level risk reduction actions need to be complemented by multi-scale interventions to harness and benefit from ecosystem services that reduce disaster risk.

Cluster planning approach enables blending of administrative scales (along which disaster risk reduction planning is conventionally organized) with landscape scales (which is relevant for managing ecosystem services). Ecosystem services are delivered at multiple spatial and temporal scales which vary from being short term, site level to long term and global level (Holling et al. 2002; Millennium Ecosystem Assessment 2003; Tacconi 2000; Levin 1992; Limburg et al. 2002). For instance, communities may harvest fish from their village pond, but can benefit from flood protection provided by floodplains within the upstream reaches of the river, and are buffered from the impacts of cyclones by a vegetated coastline extending well beyond the village boundaries. The physical scale at which ecosystem services are delivered also has an influence on stakeholders who are likely to benefit from ecosystem services. Institutions organized along political and geographical hierarchies mediate the extent and manner in which communities influence the resource use and development decisions within a given socio-ecological system (Becker and Ostrom 1995). At the lowest level, a household can influence the decisions of its members, wherein state and national governments influence developmental programming at state and national scales, respectively. The interactions between ecosystems and societies are realized in functional landscapes wherein communities integrate ecosystem services within their livelihood strategies, including those for disaster risk reduction.

Compared with a conventional disaster risk reduction planning approach which derives information from community assessments, a cluster planning approach begins with locating the community within its landscape, and subsequently identifying different spatial scales at which ecosystem services can be managed to reduce disaster risk. While recognizing that landscapes can be large, the approach allows for incorporation of intermediate planning level(s) comprising smaller landscape entities, which due to factors related to the geomorphic setting and land-use create a hazard context that is shared between adjacent communities. This enables the bridging between administrative and ecological scales within an intermediate 'cluster' scale, at which landscape management and developmental programming can converge to facilitate application of ecosystem-based approaches for disaster risk reduction.

The method for cluster planning modifies the hazard-capacity-vulnerability assessment methods conventionally used in disaster risk reduction planning (Fig. 12.1). A 'risk context analysis' is undertaken at the beginning of the process. Knowledge of the landscape setting, specifically status and trends in geomorphological features, and developmental programming contexts, are used to delineate

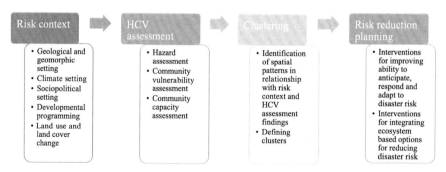

Fig. 12.1 Method for cluster planning (Source: Authors) (HCV = Hazard, Capacity and Vulnerability)

spatial patterns in relation to relevant land-use and development patterns. The next step in cluster planning is the analysis of community vulnerability and capacity, using established participatory assessment and appraisal tools. Information from the risk context analysis and vulnerability-capacity assessment are analysed to identify spatial patterns and homogeneity (using spatial planning and statistical tools) to help define the appropriate clusters. In the final stage, response options for integrating ecosystem-based approaches for disaster risk reduction at multiple scales are identified.

12.3 Applying Cluster Planning Framework: The Case of Mahanadi Delta

The cluster planning framework was applied in the Mahanadi Delta, in the state of Odisha, India, as a part of the 'Partners for Resilience' project aimed at building community resilience to increasing disaster risk using approaches integrating ecosystem management, disaster risk reduction and climate change adaptation. The risk context analysis for the Mahanadi Delta was based on a review and synthesis of available information on geology, geomorphology, developmental planning interventions, and trends in land-use and land cover. Data for community vulnerability and capacity analysis was elicited through a survey of households and community institutions. Of the 1760 villages in the delta, 100 villages (covering 5 % of total rural households) geographically distributed and located in distinctive geophysical and livelihood settings were selected for analysis. Baseline information on demography, livelihood capitals and access to early warning information were collected through a set of structured questionnaires at household and *panchayat* levels (the primary unit of formal village-level governance in India) involving 2306 households. Information on exposure, risk planning and coping and adaptation mechanisms was obtained through participatory appraisals in these villages. In each village, questionnaires were administered to 10 % of the households, which were identified using selection criteria related to occupation and ownership of social and economic assets. The Vulnerability-Capacity Indicator framework as developed by Mustafa et al. (2010) was applied considering the ability of the method to provide a quantitative index for measuring vulnerability and capacity. The indicator framework captures four major dimensions of measuring vulnerability and capacity, namely social, economic, environmental, and institutions and governance. Identification of spatial patterns was done using statistical data reduction techniques. The results are discussed according to the sequence of steps taken within the proposed method, namely risk context analysis, vulnerability-capacity assessment, clustering and risk reduction planning.

12.3.1 Risk context

The delta of the Mahanadi River spans 13,871 km^2 (including the area of Chilika lagoon and its direct catchment) between 19°40′–20° 35′ N latitude and 85°40′–86°45′ E longitude around the confluence of the Mahanadi River within the Bay of Bengal, on the east coast of India. Mahanadi is one of the major east flowing peninsular rivers of the country, flowing for 958 km and draining a basin of 139,681 km^2 across the states of Chattisgarh, Maharashtra, Jharkhand and Odisha before flowing into the sea (CWC and NRSC 2014). The arc-shaped delta has an apex at Naraj near Cuttack about 100 km from the sea wherein Mahanadi divides into two main distributaries, Mahanadi to the north and Kathjodi to the south. Each of these branches further divide into a number of distributaries giving rise to eight *doabs* (deltaic land between two distributary channels) (Maejima and Mahalik 2000). The delta is believed to have been formed during a tectonic down wrap of the Gondwana graben (Jagannathan et al. 1983) with major delta building processes placed between 6000 years BP and 800 years BP and the shoreline shifting seaward during the last 800 years (Mohanti 1993). The interaction between freshwater and coastal processes gives rise to some unique geomorphological features along the delta coastline. Along the Mahanadi mouth, strong longshore drift particularly during the rainy season has led to formation of a number of hooks. The high tidal prism keeps the mouths of the Devi, the Mahanadi, the Brahmani, the Baitarani and the Rushikulya open to form estuaries. The huge complex spit formed north of the Mahanadi Estuary has given rise to Hukitola Bay. The deltaic climate is monsoonal, with an average annual rainfall of 1572 mm; nearly 70 % is precipitated during the south-west monsoon between June to October.

Highly fertile soils and abundance of water have made the Mahanadi Delta a hub of economic activities. As per population census of 2011 (Census of India 2011), 7.96 million people inhabit the delta constituting 19 % of the total population of the state, while constituting only 9 % of the state's total geographical area. There are 1760 rural and 14 urban centres, and agriculture and associated activities form the primary occupation for 46 % of the population. Highly productive fisheries, as found in Chilika lagoon (e.g. nearly 12,000 MT of fish are harvested by 23,000 fisher households), and over 30 distinct tourism centres (including the Sun Temple of Konark and Jagannath temple of Puri) make the Mahanadi Delta a significant source of revenue for the state. Paradeep is a natural harbour and an important centre for export of minerals, coal and metallurgical products. Salt is manufactured at Humma, whereas the dunes near Gopalpur and Aryapalli are a source of limonite. Such highly productive economic assets co-exist with high conservation value areas. Of the 18 protected areas in the State of Odisha, six are located in the delta region. Chilika lagoon and Bhitarkanika mangroves have been designated as Ramsar sites or Wetlands of International Significance. Gahirmatha beach, Devi mouth and Rishikulya Estuary support extensive rookeries of globally vulnerable Olive Ridley Turtle (*Lepidochelys olivacea*). Chilika is one of only two lagoons in the world that support Irrawaddy dolphin populations and supports over a million wintering waterbirds during their migration within the Central Asian Flyway (Kumar and Pattnaik 2012).

The existence of deltas is closely linked to the amount of sand and mud that the rivers discharge to compensate the currents that wash them away (Giosan et al. 2014). At the delta head, the Mahanadi River receives water from its several tributaries building up 66,640 million m^3 discharge by the time the river flows into in the Bay of Bengal. Sediment brought in with this discharge (nearly 30 million tonnes per annum (GoO 1986)) spreads along the floodplains and shoreline, thereby helping delta build up.

Historically, human development in the region evolved in alignment with the dynamic hydrological regimes and ecological setup. Cultivators developed farming systems that adequately distributed crop failure risks from the recurrent floods and droughts (D'Souza 2006). Cropping cycles were distributed across the year so that even if one crop was negatively affected by floods or droughts, the other two would provide sufficient production to compensate for the loss. In addition to evolving rice strains especially suitable to withstand the impacts of floods or droughts, cultivators also adopted other risk distributing strategies, such as adopting varying cropping strategies for different soil types and locations. Most importantly, the strategy recognized the flood regime and its regular inundation patterns as beneficial for crops due to the resulting natural fertilization of agriculture lands.

The onset of the nineteenth century marked extensive efforts for hydrological regulation of the water regimes, with an intent to primarily raise revenues by controlling water supply to irrigators. A series of hydraulic structures were constructed to harness flows of various delta tributaries. The Naraj spur was constructed in 1856–1863 to maintain discharge distribution between various branches of the Mahanadi River. Simultaneously, the Mahanadi anicut at Jobra (Cuttack) and Birupa anicut at Jagatpur were constructed along with canal systems to irrigate the deltaic land. Hirakud Dam was constructed in 1958 as a multi-purpose project supporting irrigation, hydropower and flood control. A weir was constructed at Mundali at the delta head to support irrigation in lower parts of the delta. The Naraj spur was subsequently converted to a barrage. Embankments were constructed to supply the stored water to agriculture farmers, as a revenue recovery measure but against the wishes of communities who were apprehensive of the implications of waterlogging and decline in agricultural productivity (D'Souza 2002).

Later development in the delta emphasized the extension of hydrological regu-lating activities without reviewing their long-term implications nor taking into consideration views of the communities. In the later part of the twentieth century, hydrological regime fragmentation was coupled by loss of forest cover and exten-sive conversion of wetlands for permanent agriculture and settlements. Analysis of remote sensing imageries of 1975 and 2010 indicate that the total area of wetlands has undergone a 32 % decline (WISA 2010). Along the coastline, extensive con-version of mangroves for aquaculture and settlements took place during the 50s and 70s (Ravishankar et al. 2004).

Floods, droughts and tropical cyclones are the major natural hazards in the delta. There is a high hydrological build-up at the head of Mahanadi Delta as the river drains a huge catchment. Hirakud Dam with a gross storage capacity of 8136 million cubic metre controls 59 % of the catchment (CWC and NRSC 2014). The unregulated part of the catchment experiences heavy rainfall during the southwest monsoon, creating flows beyond the 29,500 m^3/s discharge capacity at Naraj Barrage, which severely affect the low-lying delta tracts (*ibid*). Situations can worsen significantly if the Hirakud Reservoir is operated to release water simultaneously as rains are being experienced downstream. Tropical depressions in the Bay of Bengal are frequently formed during the pre-monsoon (May-June) and post-monsoon (October-November) months, building into severe storms and cyclones (such as super cyclonic storm *Kalinga* of 1999 and extremely severe cyclonic storm *Phailin* of October 2013), damaging the deltaic tracts.

Extensive hydrological fragmentation and land use changes have gradually converted a 'flood dependent' landscape into a 'flood vulnerable' landscape. The deltaic build up processes have been impeded with as much as a 67 % reduction observed in the river sediments reaching downstream (Gupta et al. 2012), leading to a shrinkage of the delta. Extensive waterlogging is experienced in 0.27 million ha of the delta area, mainly as a result of drainage congestion created by embankments which disconnect floodplains with river channels (Khatua and Patra 2004). Degradation of wetlands has further diminished the natural flood buffering capacity of the delta. During super storm *Kalinga,* villages with a lower mangrove width along the coastline reportedly suffered higher number of deaths as compared to those villages with a higher width of mangroves (Das and Vincent 2009). On the other hand, the contribution of healthy ecosystems can make towards buffering against extreme events, in addition to their critical role in providing livelihood security for dependent communities, is well reflected in the outcomes of integrated lake-basin management implemented in Chilika lagoon over last two decades (Kumar and Pattnaik 2012). Healthy fisheries and resurgent ecotourism provide the basis of livelihood security for nearly 0.2 million people living around this lagoon.

Climate change is increasingly having a major influence on hazard dynamics in the delta. Odisha State's assessments of monsoonal patterns indicate declining dry season rainfall and intensifying monsoon rainfall (Ghosh and Mujumdar 2006). Variability in river flows, particularly increasing monsoon season flows, is also predicted along with high probability of extreme hydrological events (Gosain et al. 2006). The Bay of Bengal has recorded the maximum annual sea level rise within the Indian coast (Unnikrishnan and Shankar 2007). Such changes are likely to increase intensity as well as variability in extreme events.

Based on synthesis of the available information on the risk context, the Mahanadi Delta can be characterized into three sub-regions, namely upper delta, central delta and coastal plains (Fig. 12.2). The upper delta region formed along the delta margin has a fanning distributary system with intervening alluvial plains used for agriculture. The region constitutes 29 % of the delta's area and has elevations ranging between 20 and 50 m.a.s.l., with 43 % of the area under forests and 41 % under cultivation. There are a number of hillocks in this part of the delta, limiting

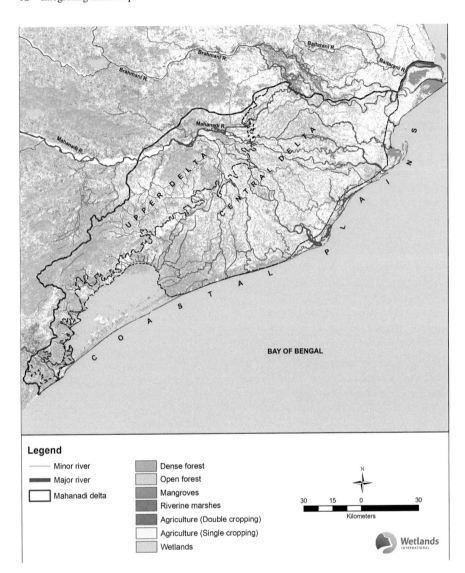

Fig. 12.2 Mahanadi Delta with its three sub-regions (Source: Authors)

extension of agriculture. Although agriculture is the key economic activity, it is constrained by lack of irrigation, leading to a high dependence on wage labour for livelihoods.

The central delta, constituting 53 % of the total delta area is flat land with elevations ranging between 5 and 20 m.a.s.l. This region is predominantly clayey with much of the area intensely cultivated. A dense network of embankments impedes drainage, creating waterlogging conditions during post-monsoon and winter months.

The narrow coastal plains run parallel to the coastline constituting 15 % of the total delta area, a major part (68 %) of which is comprised by wetlands. The narrow coastline is used for cultivation of groundnuts and other salt-tolerant crops. Elevations in this predominantly saline, sandy soil range between 0 and 5 m.a.s.l. This zone is prone to multiple hazards in the form of tropical cyclones, salinity ingress and waterlogging.

12.3.2 Vulnerability-capacity assessment

Data from the questionnaire survey and participatory appraisals indicate that, in general, high dependence on local natural resources and weak economic and physical asset base of the delta communities make them highly vulnerable to the impact of disasters. However, a distinct spatial variability in livelihood systems and disaster risk coping and preparedness mechanisms could be discerned between upper, central and coastal delta sub-regions.

Wage labour forms the primary occupation of the majority (52 %) of rural households living in the delta, followed by agriculture farming (28 %) and fishing (16 %). The average household incomes at Rs. 33,000 are nearly half the state average for Odisha (Rs. 61,000). Access to basic amenities (safe drinking water and sanitation, clean source of energy for domestic use, and access to fair-weather roads) is marginal at best. Piped water is accessed by only 2 % of the sampled households and safe drinking water by 48 % of the sampled households, compared to 92.7 % stated to have access to safe drinking water at state level (GoO 2015). Of the surveyed households, 75 % reported sourcing water for domestic use directly from rivers, creeks and village ponds. The situation in terms of access to sanitation facilities is equally dismal, with < 3 % of the households with access (as compared to ~10 % access at state level).

In terms of occupational diversity, coastal communities have relatively higher sources of incomes per household (2.15) as compared to the upper (1.94) and central delta (1.8). Communities in the central delta, which have abundant water and fertile alluvial soil, primarily depend on agriculture (41 %) and operate riverine fisheries (25 %) as a secondary livelihood option. In the upper delta, despite the fact that all households own land, limited irrigation facilities make these communities depend on wage labour (56 %) as the primary income source. Communities in the coastal plains depend mainly (35 %) on fisheries and related activities for livelihoods. Aquaculture made up of ponds provides an income source to 26 % of the households living in the central region and 24 % in coastal regions of the delta. Community grazing lands are key natural assets for upper and central delta communities. Forests and plantation are a source of fuelwood for nearly all the villages. In terms of asset ownership (occupational assets, housing type, livestock, savings and means of transportation), central delta communities have a higher diversity as compared with the rest. The majority (34 %) of households living in the central delta and coastal plains source their loans from Self Help Groups (SHGs).

In terms of seasonality, monsoon is a stress period for fisheries, as inland fishery is banned and the marine fishing is hazardous due to choppy sea conditions. In the central delta, extensive waterlogging in the post-monsoon season encourages local migration for wage labour. For households in the upper delta, the summer season with low water availability is a stressful period. Membership to different community institutions plays a critical role to overcome periods of water stress. For instance, in coastal villages, the majority of households are members of primary fishermen cooperative societies and SHGs to support fisheries and associated activities. Similarly, in the central delta, the majority of respondent households are members of SHGs that support diversified livelihood options. The prevalence of SHGs is limited in the upper delta region. Mostly farmer clubs exist in this region, which help households in gaining access to various welfare schemes of the government.

Cyclones, floods, droughts and waterlogging were reported as the main natural hazards for the Mahanadi Delta communities. A close relationship exists between the hazard, the geophysical setup of the delta and impact of developmental interventions. While the upper delta communities reported drought as the major hazard, waterlogging was reported to be the main hazard in the central delta. Most interestingly, within the central delta, floods of moderate intensity are not seen as a hazard, but as beneficial to farming, owing to natural nutrient enrichment. Communities living in the coastal plains reported multiple hazards, including cyclones, salinity ingress in agricultural fields and groundwater, and floods.

In terms of individual and community level preparedness measures, grain and fuel banks were used as coping mechanisms by 70 % of the surveyed households. Within the central delta, especially in areas with extended waterlogging, it is a common practice to construct houses on raised plinth levels. Investment into corpus funds for use during disasters/stress periods was reported by 45 % of the households. At the same time, use of insurance as a risk transfer mechanism was not observed to be popular. Life insurance was subscribed by only 12 % of the respondents. In the central delta region, 9 % of households reported use of crop insurance. None of the households reported the use of livestock, accident or asset insurance.

Coastal communities, being repeatedly exposed to hazards, have strengthened their early warning systems over a period of time. Village institutions have also developed mechanisms to interpret complex weather-related information from the Block administration (blocks being the next higher administrative authority to village *Panchayats*). The majority (89 %) of central delta communities reported accessing flood-related early warning information from radio and television. However, the reach of more sophisticated forecast/early warnings information as available through the Indian Meteorological Department or the State Disaster Management Authority was observed to be limited. Amongst the three sub-regions, the upper delta communities reported the least usage of any form of early warning system. This may be attributed to the fact that drought as the primary hazard in this region is characterized as slow onset, and drought early warning systems are not yet readily available.

There is emerging emphasis on disaster risk reduction planning through the state administration. Odisha was one of the first states to constitute a State Disaster Management Authority for this purpose. Overall, 16 % of the villages reported having a Disaster Management Plan, with the proportion highest in coastal villages. However, in no case was fund allocation made to implement the disaster management plans. In 23 % of the villages, developmental work was reported to be carried out which had an impact on the reduction of disaster risk in the target villages.

12.3.3   Clustering

The social, economic, environmental and institutional vulnerability indicator scores derived from surveys and participatory appraisals are presented in Table 12.1. These scores derived for each of the 100 villages were subjected to Principal-Components Analysis. This is a statistical procedure for converting a set of observations of possibly correlated variables into a set of linearly uncorrelated variables, thereby helping to identify underlying factors. Kaiser-Meyer-Olkin measure of sampling adequacy (0.623 and 0.644 for vulnerability and capacity indices, respectively) and Barlett's test of sphericity (Vulnerability indices: Chi square $= 269.5$, df $= 78$, Sig. $= 0.000$; Capacity indices: Chi square $= 129.8$, df $= 36$, Sig. $= 0.000$) indicate factorability. Varimax rotation provided the maximum variation of factors.

The factor plot of vulnerability indices (Fig. 12.3a) isolated four components explaining 54 % of the variability. The first two components explaining 30 % of the variance isolated indices related to institutional (Vew, Vir) and environmental settings (Vph, Vex and Vmh). The third and fourth components, explaining 24 % of the variance were related to economic (Vis) and social (Vba and Vem) indices. Vulnerability indices related to incomes, asset diversity, and coverage of organizational membership were located almost near the centre of the distribution indicating their relatively weaker influence on the variance.

The factor plot of capacity indices (Fig. 12.3b) isolated three components explaining 56 % of the variance. Component 1 (accounting for 22 % of variance) included institutional indices (Ccr, Cid and Cis), whereas Component 2 (accounting for 18 % of variance) is a mix of economic (Cod) and environment (Cnr) indices. Component 3 (accounting for 16 % of variance) comprises only one social index (Cte).

The clustering of villages resulting from the factors extracted in the analysis presented in Fig. 12.4 nearly aligns the villages as per the three sub-regions. Villages located in coastal plains are totally included in two clusters: 1(a) and 1 (b). Central delta villages are clustered primarily in 2(a) and 2(b), whereas Cluster 3 is a mix of upper and central delta villages. In terms of vulnerability indices, exclusion of coastal villages can be explained due to lower scores on account of existence of relatively better early warning systems and individual risk management measures. However, within this cluster, villages around Bhitarkanika which have relatively lower vulnerability index score on use of early warning systems are

Table 12.1 Vulnerability-capacity indicators for Mahanadi Delta communities

		Indicator score+		
		Coastal plains (N = 37)	Central delta (N = 41)	Upper delta (N = 22)
Indicator	Indicator description			
Illiteracy [V*ill*]** (F = 7.300)	Measures the absence of formal education.	0.60 (0.23)	0.44 (0.11)	0.51 (0.20)
Lack of access to basic amenities [V*ba*]** (F = 15.932)	Measures the degree of lack of access to five basic amenity types (safe drinking water, sanitation, energy for cooking, roads and electricity). A higher value indicates pre-existing vulnerability due to reduced access to basic infrastructure.	0.89 (0.11)	0.77 (0.10)	0.88 (0.07)
Physical disability [V*pd*] ** (F = 5.785)	Measures the pre-existing indisposition due to reduced mobility.	0.04 (0.05)	0.03 (0.03)	0.01 (0.02)
Membership of ethnic minority [V*em*] (F = 2.033)	Measures the proportion of households within a village belonging to an ethnic monitory with a higher risk of historical marginalization.	0.17 (0.29)	0.17 (0.35)	0.34 (0.46)
Working population dependency ratio [V*wp*] (F = 0.551)	Measures the proportion of households which depend on the working population for income. A high ratio value is an indicator of vulnerability.	0.44 (0.16)	0.46 (0.12)	0.44 (0.10)
Proportion of household employment dependent on local sources [V*ls*] (F = 0.684)	Measures the extent to which the employment is sourced from local sources (agricultural land, employment in wage labour within village, fisheries in local ponds and streams etc.). A high value indicates higher probability of being affected in the event of a disaster.	0.88 (0.13)	0.90 (0.10)	0.86 (0.15)
Income insufficiency [V*is*] ** (F = 10.207)	Measures the sufficiency of income to meet household, occupational and medical needs. A high degree of income insufficiency indicates pre-existing economic vulnerability within the household.	0.34 (0.18)	0.22 (0.13)	0.17 (0.16)
Multiplicity of hazards [V*mh*] (F = 2.434)	Indicates the types of hazards which are known to affect communities. A higher value indicates increased vulnerability due to multiplicity of hazards.	0.54 (0.16)	0.49 (0.17)	0.45 (0.16)

(continued)

Table 12.1 (continued)

Indicator	Indicator description	Indicator score+		
		Coastal plains (N = 37)	Central delta (N = 41)	Upper delta (N = 22)
Frequency of prime hazards [V*ph*] (F = 9.879)	Measures the frequency of prime hazards. A higher value indicates higher risk of being affected due to a disaster induced by a hazard.	0.80 (0.28)	0.71 (0.28)	1.00 (0.00)
Exposure to prime hazard [V*ex*] (F = 2.761)	Measures the degree of exposure of the community to a prime hazard. A higher value indicates higher probability of life and assets being affected due to a disaster induced by a hazard.	0.53 (0.12)	0.48 (0.07)	0.50 (0.09)
Degree of coverage of organizational membership [V*mm*] (F = 0.017)	Organizations help manage risks by enabling collective action and increasing level of information. A lower coverage however indicates that the community on an overall is oriented towards individual measures for risk reduction and does not operate collectively.	0.49 (0.43)	0.48 (0.29)	0.49 (0.36)
Lack of use of individual risk management measures [V*ir*] (F = 5.380)	Lower use of individual risk management measures such as fuel banks/grain banks and other local solutions increases vulnerability.	0.01 (0.04)	0.17 (0.37)	0.00 (0.00)
Lack of use of early warning systems [V*ew*] (F = 7.525)	Measures the degree to which communities do not use an early warning system as a risk reduction strategy. A high value indicates lack of, limited utility or ineffective early warning system.	0.57 (0.50)	0.61 (0.49)	1.00 (0.00)
Occupational diversity [C*od*] (F = 0.752)	Measures the different work skills being practised by a household for livelihood. A higher value indicates capacity to diversity income sources and thereby reduced risk of lost income from a single source.	0.24 (0.05)	0.23 (0.05)	0.24 (0.06)
Asset diversity [C*ad*] (F = 2.283)	Measures the range of assets available within the family for productive use. Higher value indicates the possibility for livelihood diversification and thereby reduced vulnerability.	0.37 (0.07)	0.41 (0.10)	0.42 (0.11)

(continued)

Table 12.1 (continued)

Indicator	Indicator description	Indicator score+		
		Coastal plains (N = 37)	Central delta (N = 41)	Upper delta (N = 22)
Functional diversity of local institutions [Cid]** (F = 6.109)	A higher functional diversity of local institutions provides an opportunity to benefit from collective action on multiple livelihood aspects and is therefore an indicator of higher capacity.	0.20 (0.18)	0.31 (0.22)	0.15 (0.10)
Use of collective risk transfer mechanisms [Ccr]* (F = 3.414)	Increased use of risk transfer mechanisms as insurance is an indicator of sophistication in risk reduction planning. A higher value is indicative of capacity.	0.01 (0.05)	0.09 (0.21)	0.03 (0.08)
Technical education [Cte] (F = 2.350)	Measures the proportion of household members with technical education/training on risk management, environment education, and climate change adaptation. A high value indicates the probability of the household using the acquired information to cope/adapt better.	0.02 (0.07)	0.03 (0.08)	0.00 (0.00)
State of natural resources [Cnr]** (F = 9.441)	Measures the state of natural resources (forests, soils, wetlands) within and in immediate surroundings of the village relative to dependence. A higher value indicates an increased ability of natural resources to reduce risks as well as support livelihoods during the event of a disaster.	0.45 (0.14)	0.53 (0.13)	0.38 (0.11)
Complexity of information sources used for early warning [Cis]** (F = 31.016)	Provides an insight into the type of information that is used for early warning systems. Higher rank indicate use of multiple and complex information sources.	0.29 (0.19)	0.12 (0.10)	0.02 (0.06)
Sufficiency of risk reduction planning and implementation [Crr]* (F = 4.774)	Better risk reduction planning and implementation reduces disaster vulnerability.	0.25 (0.14)	0.20 (0.00)	0.20 (0.00)

+mean scores, standard deviations indicated in parentheses
*difference of means significant at 95 % confidence level
**difference of means significant at 99 % confidence level

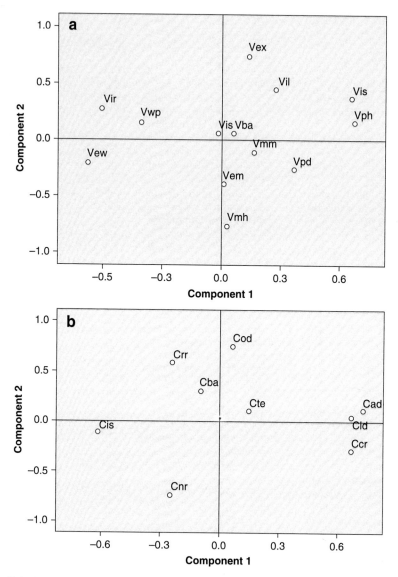

Fig. 12.3 Component plot of (**a**) vulnerability indicators and (**b**) capacity indicators

isolated as Cluster 1(a). Amongst the villages in the central delta, a set of communities with prevalence of individual risk management measures is grouped together
with upper delta villages (Cluster 3). Central delta villages are separated into cluster
2(a) and 2(b), the first cluster having a high vulnerability index score due to lower
scores for use of individual risk management measures and higher scores for
frequency of the main hazard, as compared with the second cluster.

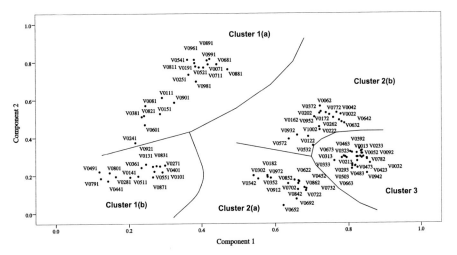

Fig. 12.4 Clustering of villages as per vulnerability and capacity indices (First three numerals of village codes are identifiers, and the last digit refers to village location within the three sub-regions of the delta: 1 – coastal plains; 2 – central delta and 3 – upper delta)

Villages of the central delta have relatively higher capacity index scores as compared to the rest, particularly standing out in terms of the use of insurance. Villages located in coastal plains and upper delta have invariably lower capacity scores for all significant indices, but the coast stands out in terms of high scores for the indicator related to complexity of information sources used for early warning. Central delta villages have the highest scores on functional diversity of local institutions and use of collective risk transfer mechanisms. Furthermore, villages located around Chilika, which report higher functional diversity of institutions as compared to the rest of the coastal villages, have been segregated as Cluster 1(b). The majority of the social and economic vulnerability and capacity indices did not significantly influence variability.

12.3.4 Risk reduction planning

Information from the previous three steps indicates that interventions for disaster risk reduction can be spatially organized at four levels. The bottom two levels are administrative entities wherein conventional disaster risk reduction interventions have been organized. These entities are located within a large landscape unit, the Mahanadi Delta. The cluster analysis indicates the possibility of considering an intermediate planning unit, which includes the three sub-regions of the delta, namely the upper delta, central delta and coastal plains, respectively. Existing and planned disaster risk reduction measures, the possibility of influencing

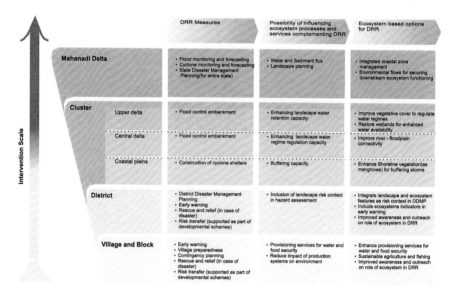

Fig. 12.5 Response options for integrating ecosystem-based options for disaster risk reduction in the Mahanadi Delta (Source: Authors)

ecosystem functions and services, and ecosystem-based disaster risk reduction options are presented in Fig. 12.5.

As can be seen, the existing and planned disaster risk reduction measures are mainly designed for action at district and village levels. At the cluster level, the interventions are more on structural measures that are intended to reduce the level of impacts on communities (through flood defence measures or cyclone shelters). Use of flood embankments has been historically observed to promote waterlogging and impact communities in the long term. At the delta level, efforts are mostly concentrated on monitoring and forecasting to enable action at lower administrative levels, i.e. districts and ultimately blocks and villages. Planning along these scales also presents opportunities for influencing ecosystem processes that underpin ecosystem services which could be harnessed for disaster risk reduction. Thus at a community level, ecosystems can be managed to enhance local food, water and livelihood security. Within the upper delta region, wetlands can be rejuvenated to improve water holding capacity, whereas within the central delta region, the flood buffering capacity of floodplains can be targeted. Within the coastal plains, creation of vegetative buffers can augment protection provided by cyclone shelters. At the delta level, sediment and water dynamics can be influenced by integrated coastal zone planning and accommodating environmental flows for maintenance of downstream ecosystem functions.

The multi-scale planning also highlights the need for institutional partnerships at different levels to deliver various interventions. While the village and block level interventions can be delivered mostly through engagement with village-level community institutions, at the cluster and delta levels, it is pertinent to connect with water infrastructure operators and agencies capable of managing coastal zones, for example, the Odisha Coastal Zone Management Authority. The Coastal Zone Management Authority is putting in place an integrated coastal zone management plan in order to secure the various ecosystem services while promoting balanced economic development in the region.

12.4 Conclusions

The cluster planning approach provides a useful multi-scale planning framework for identifying ecosystem-based interventions for disaster risk reduction within a landscape. Through the use of this approach in the Mahanadi Delta, it was possible to identify a range of ecosystem-based interventions for disaster risk reduction, building on the knowledge of delta-wide water and sediment dynamics related processes which underpin functioning of healthy ecosystems. The analysis also highlighted the limitations of structural interventions such as flood embankments. Such structural interventions, when viewed in the context of landscape processes, have the potential of aggravating vulnerability in the medium to long term, by impeding water, sediment and nutrient exchange between river channels and floodplains. The approach also enables capturing of landscape scale information, related to geomorphological as well as developmental programming, and the analysis of risk contexts, thereby supporting more robust disaster risk reduction planning. By embedding administrative planning units within ecological planning units, a more realistic integration of ecosystem services within disaster risk reduction is made possible. Finally, the clusters form an important intermediate planning stage between district level (which does not fully encompass ecosystem services scale) and delta level (at which heterogeneity created by geomorphological and developmental programming processes tend to merge). The approach will be particularly useful for authorities responsible for disaster risk reduction plans across several administrative and political boundaries.

Knowledge of functioning of the landscape and its relationship with ecosystem services is an important prerequisite for applying the cluster planning framework. Such knowledge helps to focus on a few critical landscape processes, such as the role of water and sediment flux in the case of the Mahanadi Delta. Equally important is the need to place developmental programming in the context of landscape processes. Future application of this approach can make use of more complex spatial modelling techniques to demonstrate the interlinkages between landscape processes and ecosystem services.

Acknowledgements Comments made by two anonymous reviewers have greatly helped to improve the clarity of various arguments made in this paper. The conceptual framework for risk assessment was developed under the International Development Research Center (IDRC) supported 'Chilika Climate Change' project (Grant ID: 106703-001). Data collection and collation were supported by the Partners for Resilience Programme implemented under MFS II funding of the Dutch Ministry of Foreign Affairs. Kamal Dalakoti prepared the maps and designed the illustrations.

References

Barbier EB (1994) Valuing environmental functions: tropical wetlands. Land Econ 70:155–173

Becker CD, Ostrom E (1995) Human ecology and resource sustainability: the importance of institutional diversity. Annu Rev Ecol Syst 26:113–133

Brauman KA, Daily GC, Duarte TK, Mooney HA (2007) The nature and value of ecosystem services: an overview highlighting hydrologic services. Annu Rev Environ Res 32:67–98

Census of India (2011) Provisional population totals – Orissa New Delhi: Office of the Registrar General and Census Commissioner, Ministry of Home Affairs, Government of India

CWC – Central Water Commission, NRSC – National Remote Sensing Centre (2014) Mahanadi basin. New Delhi: Central Water Commission (CWC), Ministry of Water Resources and Hyderabad: National Remote Sensing Centre (NRSC), ISRO, Department of Space, Government of India

D'Souza R (2002) Colonialism, capitalism and nature: debating the origins of Mahanadi delta's hydraulic crisis (1803–1928). Econ Polit Wkly 37(13):1261–1272

D'Souza R (2006) Drowned and dammed: colonial capitalism and flood control in Eastern India. Oxford University Press, New Delhi

Daily G (ed) (1997) Nature's services: societal dependence on natural ecosystems. Island Press, Washington, DC

Das S, Vincent RV (2009) Mangroves protected villages and reduced death toll during Indian super cyclone. Proc Natl Acad Sci 106(18):7357–7360

Folke C, Carpenter SR, Walker B et al (2010) Resilience thinking: integrating resilience, adaptability and transformability. Ecol Soc 15(4):20

Ghosh S, Mujumdar PP (2006) Future rainfall scenario over Odisha with GCM projections by statistical downscaling. Curr Sci 90(3):396–404

Giosan L, Syvitski J, Constantinescu S, Day J (2014) Protect the world's deltas. Nature 516:31–33

GoO – Government of Odisha (2015) Odisha economic survey 2014–2015. Planning and Coordination Department, Directorate of Economics and Statistics, Government of Odisha, Bhubaneswar

GoO – Government of Orissa (1986) Delta development plan Mahanadi Delta command area: geology geomorphology and coast building, Volume IV. [Unpublished Report with Engineer in-Chief]. Irrigation Department, Government of Orissa

Gosain AK, Rao S, Basuray D (2006) Climate change impact assessment on hydrology of Indian river basins. Curr Sci 90(3):346–353

Gupta H, Kao S, Dai M (2012) The role of mega dams in reducing sediment fluxes: a case study of Large Asian Rivers. J Hydrol 464–465:447–458

Holling CS, Gunderson LH, Peterson GD (2002) Sustainability and Panarchies. In: Gunderson LH, Holling CS (eds) Panarchy, understanding transformations in human and natural systems. Island Press, Washington, DC. Ch.3

ICSU-LAC (2009) Understanding and managing risk associated with natural hazards: an integrated scientific approach in Latin America and the Caribbean. International Council for Science Regional Office for Latin America and the Caribbean, Rio de Janeiro/Mexico City

IIRR, Cordaid (2013) Building resilient communities. A training manual on community managed disaster risk reduction. Philippines: International Institute of Rural Reconstruction (IIRR) and Cordaid

Jagannathan CR, Ratnam C, Baishya NC, Dasgupta U (1983) Geology of the offshore Mahanadi Basin. Petrol Asia J IV(4):101–104

Khatua KK, Patra PC (2004) Management of high flood in Mahanadi and its Tributaries below Naraj. Published in 49th Annual Session of IEI (India) 2nd Feb.2004. Bhubaneswar: Orissa State Center

Kumar R, Pattnaik AK (2012) Chilika – an integrated management planning framework for conservation and wise use. Wetlands International South Asia/Chilika Development Authority, New Delhi/Bhubaneswar

Lavell A, Oppenheimer M, Diop C et al (2012) Climate change: new dimensions in disaster risk, exposure, vulnerability, and resilience. In: Field CB, Barros V, Stocker TF (eds) Managing the risks of extreme events and disasters to advance climate change adaptation. A special report of working groups I and II of the Intergovernmental Panel on Climate Change (IPCC). Cambridge University Press, Cambridge, UK/New York

Levin SA (1992) The problem of pattern and scale in ecology. Ecology 73(6):1943–1967

Limburg KE, O'Neill RV, Costanza R, Farber S (2002) Complex systems and valuation. Ecol Econ 41:409–420

Mace GM (2014) Whose conservation? Science 345(6204):1558–1560

Maejima W, Mahalik NK (2000) Geomorphology and land use in Mahanadi Delta. In: Mahalik NK (ed) Mahanadi delta: geology, resources and biodiversity. Asian Institute of Technology Alumni Association, New Delhi (India Chapter) Ch 5

Maskrey A (2011) Revisiting community-based risk management. Environ Hazards 10(1):1–11

Millennium Ecosystem Assessment (MEA) (2003) Ecosystems and human well-being: a framework for assessment. A report of the conceptual framework working group of the millennium ecosystem assessment. Island Press, Washington, DC

Mohanti M (1993) Coastal processes and management of the Mahanadi River Deltaic complex, East Coast of India. In: Kay R (ed) Deltas of the World. American Society of Civil Engineers, New York

Mustafa D, Ahmed S, Saroch E, Bell H (2010) Pinning down vulnerability: from narratives to numbers. Disasters 35(1):62–86

Ravishankar T, Navamuniyammal M, Gnanappazham L et al (2004) Atlas of mangrove wetlands of India, Part – 3 Orissa. M. S. Swaminathan Research Foundation, Chennai

Tacconi L (2000) Biodiversity and ecological economics: participation, values and resource management. Earthscan Publications, London

UNISDR (2009) Global assessment report on disaster risk reduction: risk and poverty in a changing climate – invest today for a safer tomorrow. United Nations Office for Disaster Risk Reduction (UNISDR), Geneva

UNISDR (2015) Sendai framework for disaster risk reduction 2015–2030. United Nations Office for Disaster Risk Reduction (UNISDR), Geneva

Unnikrishnan AS, Shankar D (2007) Are sea-level-rise trends along the North Indian Ocean Coasts consistent with global estimates? Glob Planet Chang 57(3-4):301–307

Vogel C, Moser SC, Kasperson RE, Dabelko GD (2007) Linking vulnerability, adaptation, and resilience science to practice: pathways, players, and partnerships. Glob Environ Chang 17:349–364

WISA – Wetlands International – South Asia (2010) Wetlands and livelihood project technical report. Wetlands International-South Asia, New Delhi

Part III
Innovative Institutional Arrangements and Policies for Eco-DRR/CAA

Chapter 13
Progress and Gaps in Eco-DRR Policy and Implementation After the Great East Japan Earthquake

Naoya Furuta and Satoquo Seino

Abstract In 2011, Japan experienced a huge earthquake followed by a tsunami and a nuclear power accident known as the Great East Japan Earthquake (GEJE). This chapter focuses on impacts of the tsunami and the reconstruction of coastal zones affected by GEJE, with a greater emphasis on sea wall reconstruction. The main question addressed in this chapter is how ecosystems played and are playing a role in GEJE and the reconstruction process from both policy and implementation perspectives. In this respect, it reviews how sea walls, coastal forests, traditional knowledge and protected areas played out during the GEJE. The chapter also provides a review of policy responses after GEJE to promote ecosystem-based disaster risk reduction (Eco-DRR) both at the global as well as national level by the government of Japan. It then reviews reconstruction activities on the ground with a particular focus on coastal areas such as reconstruction of sea walls and coastal forest. Finally, policy-implementation gaps and lessons are discussed from an Eco-DRR point of view based on these practical experiences.

Keywords Earthquake • Tsunami • Coastal zone • Sea walls • Japan

N. Furuta (✉)
Japan Liaison Office, IUCN (International Union for Conservation of Nature), Gland, Switzerland

Institute of Regional Development, Taisho University, 3-20-1, Nishisugamo, Toshima-ku, Tokyo 170-8470, Japan
e-mail: naoya.furuta@iucn.org

S. Seino
Ecological Engineering Laboratory, Department of Urban and Environmental Engineering, Graduate School of Engineering, Kyushu University, Fukuoka, Japan

© Springer International Publishing Switzerland 2016
F.G. Renaud et al. (eds.), *Ecosystem-Based Disaster Risk Reduction and Adaptation in Practice*, Advances in Natural and Technological Hazards Research 42,
DOI 10.1007/978-3-319-43633-3_13

13.1   Introduction

In 2011, Japan experienced a huge earthquake followed by a tsunami and a nuclear power accident. Known as the Great East Japan Earthquake (GEJE), the earthquake that occurred on 11 March, 2011 had a magnitude of 9.0 on the Richter scale, making it the largest in recorded history in Japan, and the fourth largest in the world. It triggered a huge tsunami which struck the Pacific Coast of Japan with waves as high as 40.1 m. The area affected by tsunami waves of more than 10 m stretched 530 km from north to south along the coastline facing the Pacific Ocean (The 2011 Tohoku Earthquake Tsunami Joint Survey Group 2013). The number of people dead or missing from GEJE was more than 18,000 and the number of damaged or collapsed houses and other buildings was more than 400,000 (National Police Agency of Japan 2015). It is important to note that most of this damage was caused by the tsunami and more than 230,000 people were still living in temporary housing as of December 2014 (Reconstruction Agency of Japan 2014). The Government of Japan estimated its economic loss at 16 trillion Japanese Yen (Cabinet Office of Japan 2011).

With this background, this article will focus on impacts of the tsunami and the reconstruction of coastal zones affected by GEJE, with a greater emphasis on sea wall reconstruction, based on an examination of various policy documents prepared by the Government as well as other sources including newspaper articles addressing case studies related to ecosystem-based disaster risk reduction (Eco-DRR). Case studies cited in this article were selected by the authors based on their expert judgment supported by information provided by local experts and Government officials. This approach allows for objectivity through the selection of well-documented cases using various sources including newspapers, as there are still very few peer-reviewed articles published on this topic at the time of writing.

The main question addressed in this chapter was how ecosystems played and are playing a role in GEJE and the reconstruction process from both policy and implementation perspectives. To this end, this chapter first briefly reviews how sea walls, coastal forests, traditional knowledge and protected areas played a role during the GEJE. It then reviews recent national and international policy development from the Eco-DRR point of view in relation to the GEJE. It also reviews reconstruction activities after the GEJE with a particular focus on sea wall reconstruction projects, which are controversial in many places. It also reviews coastal forest restoration, followed by a discussion on the policy-implementation gap in Eco-DRR after the GEJE.

13.1.1   Sea Walls, Coastal Forests, Traditional Knowledge and Protected Areas During the Great East Japan Earthquake

The coastline affected by GEJE was one of Japan's most well-prepared regions for tsunamis as it is also one of the most disaster-prone regions in the country. It has experienced many earthquakes and tsunamis in the past (Suppasri et al. 2013). For example, Taro town in Iwate prefecture, a coastal town, was very famous for its 2.4 km long dual sea walls built to a height of 10.45 m to prevent tsunami impacts. However, this great sea wall could not protect the town from the GEJE tsunami with its 16 m high waves at that location. About 200 people out of a population of 4434 were lost in Taro. It was said that people had a false sense of security and many did not run away because they felt safe behind this sea wall (Asahi Shinbun 2011, Iwate Nippou 2011). Although it was argued that existing sea walls along the coastline delayed the intrusion of the tsunami and reduced its height (Ministry of Land, Infrastructure, Transport and Tourism of Japan 2011a), most of these sea walls could not prevent damage during GEJE as was the case in Taro. One study also demonstrated that the death ratio caused by the tsunami was lower in areas where residents could see the ocean, in comparison to areas where residents could not see the ocean (Tanishita and Asada 2014).

Another common way to prepare for coastal natural hazards has been planting coastal forests. For more than four centuries Japan has been developing coastal forests along its coastline (Shaw et al. 2011). These forests reduce the impact of coastal hazards such as airborne sand, salt winds, high tides, and tsunamis (see also Takeuchi et al., Chap. 14). Japan's Forest Law stipulates that forests for disaster risk management should be planted in coastal areas to prevent damage from wind, airborne sand, and tsunamis. Another benefit of coastal forests is their scenic beauty – a green forest along a white sandy beach is considered particularly beautiful in Japan. Rikuzen-takata city, Iwate prefecture, was well-known for its beautiful coastline of pine forests, called Takata-matsubara. According to a local historical record, pine tree planting along the coast was initiated in the seventeenth century in order to prevent damage from strong winds and sand and protect against high tides. The area was also designated as one of Japan's premier places of scenic beauty by the Government of Japan. These coastal pine trees also protected its residential quarters from tsunamis which manifested themselves in this area many times, such as the Meiji Sanriku Tsunami in 1896, Showa Tsunami in 1933 and Chilean Tsunami in 1960 (Ministry of Land, Infrastructure, Transport and Tourism et al. 2014). The tsunami caused by GEJE, however, was too big for these forests to play a significant mitigation role. Tsunami waves higher than 10 m uprooted and washed away all but one pine tree in this region.

In other parts of the area affected by GEJE, it was reported that coastline forests reduced the impacts of the tsunami, delayed its arrival time, and protected houses by capturing debris (see also Takeuchi et al., Chap. 14). In Hachinohe city, Aomori prefecture, a forest caught 20 ships washed inland by 6 m tsunami waves, thereby protecting the houses located behind the trees. In Fudai village in Iwate Prefecture, it was reported that the combination of a 15.5 m high sea wall and a forest belt behind it prevented the residential area from being affected by the tsunami. Although the tsunami overtopped the sea wall, the forest belt trapped concrete debris. It was also reported that a forest belt in Iwaki city, Fukushima prefecture trapped several cars washed away by a 7 m high tsunami, thereby preventing them from being deposited on adjacent farm lands. Another example was Oharai town in

Ibaragi prefecture, where sand dunes reduced the energy of the tsunami waves, thereby reducing direct negative impacts on roads and houses (Experts' working group on restoration of coastal protection forest at Great East Japan Earthquake 2012).

On the other hand, in other areas, it was reported that uprooted pine trees from the coastal forests increased wave impacts and caused more damage as they were the first debris to hit houses. Many of these trees were planted in areas (such as on shallow mounds) where they could not establish comprehensive root systems. Therefore, proper planting and management of trees as well as species selection are critical factors in this context (Renaud and Murti 2013).

GEJE was also an eye-opening event for recognizing the importance of traditional knowledge and historical records handed down from the past. After GEJE, it was found that many historical records of past tsunamis existed in the region, such as stone monuments, place names, and old historical literature that were not taken seriously or simply ignored before GEJE (Fig. 13.1). After GEJE, the Ministry of Land, Infrastructure, Transport and Tourism of Japan (MLITJ) surveyed stone monuments related to past tsunamis and found some 300 stone monuments in three prefectures of the affected Tohoku region. These stone monuments relay lessons from past tsunami disasters, such as tsunami predictive information, how to escape from a tsunami, warnings about the location of houses, records of damage by tsunamis, and so on. (Ministry of Land, Infrastructure, Transport and Tourism, Tohoku Regional Bureau 2014).

Fig. 13.1 Stone monuments in GEJE affected area tell stories about past tsunamis (Photo: Kiyotatsu Yamamoto reproduced with permission)

The coastline affected by the tsunami was also famous for its beautiful landscape, particularly in the north where its saw tooth shaped coast characterized the landscape. The major industry in this area was fisheries, with small towns and cities scattered along the coastline. Several protected areas including a national park (Rikuchu-kaigan National Park), a quasi-national park, and several prefectural natural parks existed along the coastline. It was reported that islands designated as a prefectural natural park reduced the energy of tsunami waves, thereby reducing its impact on the residential/commercial area in Matsushima city (Renaud and Murti 2013). More generally, it can be argued that greater limitation of development activities within these protected areas could have reduced the potential damage to lives and assets otherwise present along the coastline from GEJE.

13.2 Policy Response After GEJE to Promote Eco-DRR at the Global Level

Policy responses and reconstruction activities in the wake of GEJE show mixed results from an Eco-DRR point of view. One of the positive and interesting policy responses at the national level was the creation of the Sanriku Fukko (Reconstruction) National Park along the coastline (for details see Takeuchi et al., Chap. 14). The Ministry of the Environment Japan (MOEJ) is also planning to integrate DRR elements into other national parks in Japan (Ministry of the Environment Japan 2012).

MOEJ has also started to advocate the role of protected areas for disaster risk reduction nationally and internationally, particularly since 2012 when IUCN (International Union for Conservation of Nature), UNEP (United Nations Environment Programme) and UNU (United Nations University) organized an experts' workshop on "Ecosystem-based Disaster Risk Reduction for Resilient and Sustainable Development" in Sendai, Japan, where the concept and global discussion on Eco-DRR were introduced in Japan for the first time. Some 35 Japanese experts from the national and local governments, academia and NGOs participated (IUCN et al. 2012). In November 2013, the first Asia Parks Congress was organized by MOEJ and IUCN in Sendai, Japan, where about 800 participants from 40 countries got together and discussed various topics relating to protected areas (Ministry of the Environment of Japan 2013). One of the six themes highlighted at the Asia Parks Congress was disasters and protected areas, and various examples of how protected areas contribute to DRR, including the example of Sanriku Fukko National Park, were presented and discussed. In addition, MOEJ, IUCN and UNU organized an international symposium on Ecosystem-based Disaster Risk Reduction in Tokyo leading up to the Asia Parks Congress, and MOEJ and IUCN organized an international workshop on DRR and Protected Areas immediately after the Asia Parks Congress in Sendai (Ministry of the Environment of Japan and IUCN 2013).

Based on the interest shown and success of the Asia Parks Congress, MOEJ and IUCN also organized a dozen sessions on disaster risk reduction and protected areas at the IUCN World Parks Congress in Sydney, Australia in November 2014 (IUCN et al. 2014). In these sessions, various aspects of DRR and protected areas were discussed from the view points of science, policy and practice and a new case study publication was launched (Murti and Buyck 2014). The World Parks Congress is a global conference of protected area experts and has been held by IUCN almost every 10 years since 1962. More than 6000 people from over 170 countries participated in the Congress in Sydney. Ms Margareta Wahlström, United Nations Special Representative of the Secretary-General for Disaster Risk Reduction also participated in one of the high-level events, the World Leaders' Dialogue. A two-day training course on disaster risk reduction and protected areas was held by IUCN, the Secretariat of Convention on Biodiversity (CBD) and United Nations Office for Disaster Risk Reduction (UNISDR) leading up to the Congress (IUCN and CBD 2014). Based on these various inputs, the "Promise of Sydney" the outcome document from the Congress, recognized the role of protected areas for disaster risk reduction for the first time in the history of the Congress (World Parks Congress 2014).

An important milestone was also reached in the 12th meeting of the Conference of Parties (COP12) of the CBD in October 2014 in Pyongchang, Korea. Parties endorsed a decision, initially proposed by MOEJ, known as "Biodiversity, Climate Change and Disaster Risk Reduction" which explicitly mentioned disaster risk reduction in its title for the first time in CBD history and encourages Parties to promote and implement ecosystem-based approaches to climate change related activities and disaster risk reduction in the context of the Hyogo Framework for Action 2005–2015 (CBD Decision XII/20). In March 2015, the 3rd UN World Conference on Disaster Risk Reduction (WCDRR) was held in Sendai, Japan and the Sendai Framework for Disaster Risk Reduction 2015–2030 was adopted in which environment and ecosystems received stronger emphasis than in the previous framework document, the Hyogo Framework for Action. MOEJ actively participated in this conference and organized a side event on Eco-DRR where a new handbook on DRR and protected areas was launched (Dudley et al. 2015). Also, a new capacity development project on Eco-DRR for developing countries was officially announced by the Minister of the Environment of Japan.

13.2.1   Domestic Policy Development in Japan on Eco-DRR

In addition to the establishment of the Sanriku Fukko National Park by MOEJ, several important domestic developments have occurred regarding Eco-DRR in the wake of GEJE. In September 2014, the Science Council of Japan published a recommendation on Encouragement of the Use of Ecological Infrastructure in Reconstruction and National Resilience (Science Council of Japan 2014). This recommendation refers to the green infrastructure movement that started in the

USA in the 1990s and the Green Infrastructure Strategy adopted by the European Commission in May 2013 and proposes to promote the use of ecological infrastructure and Eco-DRR for post-disaster reconstruction and for enhancing resilience in Japan.

GEJE also triggered a policy reaction resulting in the development of a new national law on resilience in Japan (Cabinet Secretariat of Japan 2014). The "Basic Act for National Resilience Contributing to Preventing and Mitigating Disasters for Developing Resilience in the Lives of the Citizenry" (Act No. 95 of 2013) was adopted in 2013 by the National Diet and its Fundamental Plan and Action Plan were approved by the Cabinet in June 2014. The basic goal of this Act is to enhance the resilience of the nation against future large-scale disasters. In order to achieve this goal, the Act was designed to supersede any other sectional national plans managed by other related ministries. Each sectional national plan will now have to be revised in accordance with this Act and its basic plan. In addition, each local government can develop its own local resilience action plan.

Some elements related to Eco-DRR were incorporated in the text of this Act such as reference to symbiosis with nature, harmony with the environment in accordance with the characteristics of each region, and maintenance of landscape beauty. Promoting land-use planning using ecosystem functions that act in favor of disaster risk reduction and the need for quantitative evaluation of ecosystem functions for DRR were also mentioned in the 'Fundamental Plan'. MOEJ is also developing technical guidelines for local governments on Eco-DRR to support implementation of this Act. There is, however, also a major concern that this Act may actually accelerate more reliance on engineering solutions due to the various challenges to implementing Eco-DRR on the ground (see Discussion section below).

13.2.2 Reconstruction Activities on the Ground After GEJE

In response to the massive impacts caused by the GEJE, the Government of Japan promptly initiated recovery and reconstruction. The basic policy orientation was set out in a policy document "Towards Reconstruction: Hope beyond the Disaster" issued in June 2011 (Reconstruction Design Council 2012). One of the most important lessons learned from the GEJE was that no amount of preparation can allow society to completely avoid the destruction caused by disasters of this magnitude. As such, a paradigm shift in disaster management policy that moves away from an almost exclusive reliance on structural countermeasures and moves towards strengthening disaster risk reduction measures is being encouraged. These include approaches such as escape strategies, land use planning/management and establishing multiple defenses.

Based on the experience of the GEJE, a basic principle for future earthquake and tsunami preparedness was also revisited. In Sep 2011, Japan's Central Disaster Prevention Council published a report, "Technical Investigation on Countermeasures for Earthquakes and Tsunamis based on the lessons learned from the 2011

Tohoku Pacific Coast Offshore Earthquake" (Central Disaster Management Council of Japan 2011). This report established a basic principle that distinguishes two levels of tsunami disaster management measures:

- For largest-possible tsunamis with extremely low probabilities of occurrence but with devastating consequences once they do occur – give first priority to protection of human life and establish comprehensive tsunami countermeasures embracing every possible instrument, placing evacuation of residents at the core (Level 2).
- For tsunamis that occur frequently but cause major damage despite relatively low tsunami height – develop coastal protection facilities from the point of view of protecting human life and residents' assets, stabilizing the regional economy and securing essential industrial bases (Level 1).

Based on these principles, the Tsunami Resilient Cities Act was also introduced in December 2011. This Act allows each prefectural governor to estimate and publish the location of areas of inundation which could be caused by the largest possible tsunamis in the future. The prefectural governor can then designate "Tsunami Disaster Security Zones" and "Tsunami Disaster Special Security Zones". For "Tsunami Disaster Special Security Zones," also referred to as orange and red zones, land use regulation can be imposed by restricting the construction of homes, hospitals and schools, and other critical infrastructure (Ministry of Land, Infrastructure, Transport and Tourism 2015).

Although these guidelines and acts were developed to deal with future earthquakes and tsunamis and can be applied anywhere in Japan, their basic ideas were also applied in planning for reconstruction after the GEJE in the affected areas. In July 2011, MLITJ and Ministry of Agriculture, Forestry and Fisheries of Japan (MAFFJ) jointly provided a set of technical standards to calculate the appropriate heights of sea walls to be reconstructed based on Level 1 criteria mentioned above (Ministry of Land, Infrastructure, Transport and Tourism of Japan 2011b). Prefectural governments calculated heights of sea walls to be reconstructed based on these technical standards and then decided the height of each sea wall. For some local communities, reconstruction of sea walls according to the technical guidance provided by the Central Government was perceived as the inviolable, non-negotiable condition for the reconstruction process, and this has created tensions among local community members and local authorities.

Sea walls extending some 300 km in length, most of them under the jurisdiction of prefectural governments, had been constructed along the coastline affected by the GEJE before the earthquake; of these 190 km were destroyed. The basic plan for reconstruction was to rebuild these sea walls within 5 years after the GEJE with a budget of 1 trillion JPY (Nikkei Shinbun 2014). Although the technical standards provided by MLITJ and MAFFJ called for consideration of nature conservation, harmonization with the surrounding landscape, and cost effectiveness when deciding sea wall heights, there are many cases in which the locations and heights of sea walls chosen have been questioned by local residents, particularly in the area of Kesennuma city, Miyagi prefecture (Yokoyama 2011). In these cases, local residents often questioned the necessity of constructing huge sea walls while destroying assets such as important spaces for fisheries, beautiful landscapes, and sandy beaches. In some cases, sea wall reconstruction is still planned, even though all local residents have already relocated to higher elevations, and there is nothing to protect other than a few rice paddies. Koizumi district in Kesennuma city is one such example.

In Koizumi, 43 people out of a population of 1810 were lost, and 266 out of 518 homes were washed away or completely destroyed by GEJE (Mori 2013). After GEJE, the residents of Koizumi decided to relocate their homes to higher elevations in order to reduce their exposure to future tsunamis. Although nobody will live in the inundated area, the prefectural government has planned to build a sea wall with a height of 14.7 m, and requiring a budget of 22 million JPY (Fig. 13.2). Some local residents and experts from outside the area have been questioning this plan from the perspectives of cost effectiveness and negative impacts on the natural environment (Nikkei Shinbun 2014). Residents have even proposed an alternative plan. An economic evaluation estimated the net benefit of building the sea wall at minus 20.7 million JPY; thus, there is no rationale for the project from an economic point of view (Kaku et al. 2014). The Nature Conservation Society of Japan also sent an opinion letter to Prime Minister Shinozo Abe on this issue (NACS-J 2014).

Another controversial example is a sea wall reconstruction plan at Gamo tidal flat in Sendai city, Miyagi prefecture. Gamo tidal flat is located in the coastal zone of Sendai city and is designated as a special national wildlife sanctuary zone. Gamo tidal flat had been one of the hot spots for nature conservation activities in the region and was also designated as a nature restoration project site under the Nature Restoration Promotion Act before the GEJE. This tidal flat was heavily affected by the tsunami during the GEJE and experienced large scale geological modification followed by dynamic change in its land forms including reformation of a coastal lagoon (Fig. 13.3). After the GEJE, Miyagi prefecture proposed to reconstruct a

Fig. 13.2 Coastal sea wall construction plan at Koizumi, Kesennuma (Source: Katsuhide Yokoyama reproduced with permission)

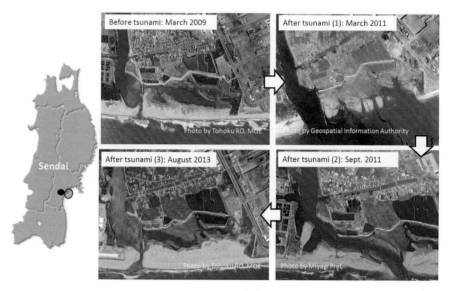

Fig. 13.3 Changes in the natural environment of Gamo tidal flat, Miyagi by GEJE (Source: MOEJ reproduced with permission)

7.2 m high sea wall by replacing the old 4 m high one. Some local residents and conservation organizations, however, have been opposing this plan as the new sea wall would destroy some parts of the tidal flat. Miyagi prefectural government revised its plan with a 20–30 m setback of the sea wall but consensus with the local residents has not been reached at the time of writing (Kahoku Shinpo 2014). Local high school students drew up an alternative plan to set back the sea walls more drastically and designate the area as a natural park, submitting their proposal to the prefectural government (Kahoku Shinpo 2014).

Several factors that contributed to these disputes during the reconstruction process. First of all, sea wall construction projects are executed as part of reconstruction activities mandated by the Basic Act on Reconstruction in Response to the GEJE. Sea walls can be relatively easily constructed under this Act, which does not require an environmental impact assessment or cost-benefit analysis. Other common reasons why some local communities have accepted such seemingly too-large sea walls have been "to avoid delaying the reconstruction process" and "land will be bought by the Government at a good price" (Yokoyama 2011). Yokoyama (2011) also pointed out much better reconstruction plans would have been possible if the budget for reconstructing sea walls had also been used for buying subsided land in lowland areas to conserve as brackish water wetlands for nature conservation. It was also pointed out that one of the reasons for these disputes about sea wall construction was the insufficiency of participatory processes during the reconstruction planning phase (Nikkei Shinbun 2014).

Fig. 13.4 Historical changes of tidal flats in Otomoura, Rikuzen-Takada City, Iwate Prefecture (Source: MOEJ reproduced with permission)

There are several places where opinions from local community in favor of setting back or removing sea walls were reflected in reconstruction plans such as Kodanohama beach and Tanakahama beach in Kesennuma Ohshima island, Miyagi prefecture (see Takeuchi et al., Chap. 14). Another such example can be found in Otomoura in Rikuzen-takada city, Iwate prefecture. Otomoura used to be a tidal flat used for fishing activities and recreation for local residents. In the 1950s, the prefectural government initiated a 30 ha landfill project to create farmland on this tidal flat, but the resulting farmland has been abandoned over the last 15 years. Tsunami and land subsidence caused by the GEJE turned this farmland back into the original tidal flat (Fig. 13.4). According to the reconstruction plan for Otomoura developed by Rikuzen-takada city government, the coastal levee will be reconstructed but located about 215 m further inland and the tidal flat will be restored. The goals of this restoration project are to restore the tidal flat for recreation and nature education, to provide an ideal site for clam digging, and to create a tidal flat to nurture biodiversity. The local community is highly enthusiastic about this reconstruction plan (Tokai Shinpo 2013).

In some other areas, sea wall reconstruction was completely rejected following consultations with local communities such as in Moune district in Kesennuma city. Instead of building a 9.9 m high sea wall as indicated by the prefectural government, local communities and the city government decided not to build a sea wall at all despite the fact that during the GEJE, 44 households were damaged in the district (Yomiuri Shinbun 2012). One of the reasons for this decision was that all the residents in this district will relocate to higher ground. Another reason was consideration of the landscape and natural environment which are critical as this district's main industry is oyster farming, and fishermen here have been planting trees in the upstream mountains to enrich the nutrient cycle from the mountain to the sea in order to improve their oyster farms (Mori wa umi no koibito 2015).

13.3 Reconstruction of Coastal Forests

As mentioned above, Japan has been developing coastal protection forests since the 15th and sixteenth centuries. Prior to the GEJE, the Pacific coastlines of Aomori, Iwate, Miyagi, Fukushima, Ibaraki and Chiba prefectures had a total of about

230 km of disaster-prevention forests but the 2011 tsunami destroyed about 140 km of these (Renaud and Murti 2013). The Forestry Agency of Japan established an expert group to review the damage to and effectiveness of coastal forests during the GEJE. The expert group produced a report with recommendations on how to restore coastal forests (Experts' working group on restoration of coastal protection forest at Great East Japan Earthquake 2012). This report pointed out that coastal forests reduced the energy of tsunami waves and trapped debris in some places. A numerical simulation also showed that a 600 m long coastal forest could significantly reduce the washout area of coastal residential zones from a 10 m high tsunami (Tanaka et al. 2014). Nonetheless, it was also reported that in low areas with high ground water levels, coastal tree roots could not develop well and so trees were uprooted by the tsunami (Experts' working group on restoration of coastal protection forest at Great East Japan Earthquake 2012).

Based on these lessons learned, the Government of Japan is planning to complete infrastructure preparations, for example by establishing mounds for reforestation, within 5 years and complete the plantation process within 10 years after the GEJE (Forestry Agency of Japan 2013). For example, along the coastline in Sendai city, a plantation project is being implemented by creating mounds of more than 2 m height (Fig. 13.5).

A recent study suggested that this kind of coastal forest development can be problematic from a biodiversity point of view (Onza et al. 2015). The study revealed that plant diversity was much enhanced by the disturbances created by the GEJE. This result poses a question about the way coastal reforestation is being implemented, which could result in destroying these rich natural habitats created by geological changes such as tsunamis and land subsidence. MOEJ has also been conducting detailed monitoring studies along the coastline affected by the GEJE and identified many places that are important in terms of natural habitat (Biodiversity Center of Japan 2014). Although the usefulness and importance of coastal forests or forests in general for disaster risk reduction are generally acknowledged and recognized in many policy documents, including the Action Plan for implementing the Basic Act for National Resilience (National Resilience Promotion Headquarters 2014), there are also controversies and discussions surrounding how to design and implement protection forests on the ground.

A unique public/private partnership project was recently implemented to establish a coastal defense system in Hamamatsu city, Shizuoka prefecture. This is not part of the reconstruction post-GEJE but a project in preparation for a future earthquake and tsunami predicted in that region. A private company founded in Hamamatsu donated 30 billion JPY and the prefectural government is designing and constructing a coastal defense facility of 17.5 km in length (Shizuoka Prefecture 2015). What is unique about this project is its design, which integrates a coastal forest system with artificial mounds supported by a new construction method. Usually, coastal levees and coastal forests are designed and constructed by different departments of a prefectural government with separate technical standards and

Fig. 13.5 Coastal Forest Plantation and coastal sea wall in Sendai (Photo: Naoya Furuta)

subsidy systems provided by different ministries. As this project is funded by a private company, the prefectural government has much more freedom in designing and implementing the project: sea walls and forests can be designed and implemented in an integrated manner. Local residents have also been participating in the planning process from the beginning through committees on landscape design, nature conservation and forest design.

13.4   Discussion

GEJE was the greatest natural disaster ever recorded in the history of Japan. The event taught the country many lessons. Massive reconstruction activities are still ongoing and many policy responses have been triggered, not only addressing the GEJE, but also preparing for future natural hazards. As discussed above, progress has been achieved in terms of promoting Eco-DRR at global and domestic policy levels. However, it is also clear that there remain some gaps and challenges with respect to the implementation of these policies on the ground. There are several very interesting examples of Eco-DRR initiatives such as the establishment of Sanriku Fukko National Park and coastal protection facilities such as Tanakahama and Kodanohama in Oshima Island and Otomoura in Rikuzen Takata (Takeuchi et al., Chap. 14). In the meantime, there are still concerns being discussed in other locations such as Koizumi and in coastal forestation projects from an biodiversity point of view.

Challenges and recommendations for addressing policy and implementation gaps in Eco-DRR from various experiences both in Japan as mentioned above and in other countries have been discussed and shared through several occasions over the past few years in Japan. For example, at the International Workshop on Disaster Risk Reduction and Protected Areas in Sendai, Japan November 2013 (Ministry of the Environment of Japan and IUCN 2013), various reasons were identified why this policy/implementation gap exists. These include lack of communication among scientists from different disciplines such as engineering or ecology and with holders of traditional ecological knowledge, as well as a shortage of scientific evidence on implementing ecosystem-based solutions on the ground. Also, no ecological scientists were invited into the field during the rescue and recovery phases just after the GEJE as humanitarian activities were prioritized, and no ecological scientists were involved in developing reconstruction policies at the national level or in developing reconstruction plans at the local level.

After the GEJE, responsibility for the reconstruction planning process was delegated to each local government and had to be bottom-up. However, difficulties with consensus building through this participatory reconstruction planning process have been observed in many places. Due to the disruption, shock, loss of many lives and much property, together with the huge pressure for early recovery, it was not easy for all the local authorities to carry out the reconstruction process in a participatory way. In addition, many local government officials were also killed or went missing. Lack of experience in participatory planning and lack of human resources in many local governments hampered the process. In this situation, most of the local governments tended to apply guidelines prepared by the central government without sufficient consultation with local community members. It was also pointed out that ecological experts were not usually invited to take part in such consultation processes.

Another stock-taking occasion took place during the international conference on Disasters and Biodiversity organized by International Union for Biological Science

(IUBS) and Biodiversity Network Japan (BDNJ) in Sendai, Japan in September 2014. This conference identified key lessons from various experiences in the recovery and reconstruction process from disasters around the world (including the GEJE) which could serve to achieve a biodiversity-harmonious disaster recovery process (IUBS and BDNJ 2014). Identified lessons included:

- Importance of research and studies:

It is important to conduct and take into account research and studies covering various dimensions, not only limited to biodiversity but also regarding traditional culture, which often includes biodiversity-harmonious DRR practices, and other social dimensions and relationships between disasters and these elements.

- Importance of sharing knowledge and information:

There are already enough positive and negative experiences from around the world on disaster recovery processes and it is very important to share these lessons, knowledge and information more effectively and widely in order to avoid repeating past mistakes.

- Importance of pre-disaster preparation:

In a post-disaster situation society has to address many urgent needs, and policy makers are under tremendous pressure to effect speedy recovery and reconstruction. Under these circumstances, biodiversity tends to be forgotten or marginalized and often decisions are taken that are inappropriate from a long-term perspective. In order to avoid these failures, it is important to prepare for post-disaster recovery and reconstruction procedures that include appropriate biodiversity perspectives to be followed when disaster strikes.

- Importance of social capital:

The importance of social capital for realizing a biodiversity-harmonious disaster recovery and reconstruction was also commonly observed in past disaster experiences. It is very important to respect communities' human relationships in the recovery and reconstruction process. Community activities such as festivals can often strengthen social capital and thus help make recovery and reconstruction more effective should a disaster happen. The importance of sharing knowledge among stakeholders in the community during the recovery and reconstruction process was also emphasized as a way to promote better choices, as there is often a large gap in knowledge and information between government officials and community members.

- Environmental impact assessment before reconstruction:

It has been commonly observed around the world that most of the destruction to biodiversity does not happen directly as a result of the actual disaster but as a result of inappropriate reconstruction projects following the disaster. The lack of environmental impact assessments (EIA) in the reconstruction process can be one of the causes of this kind of failure. It might be unrealistic to conduct a full EIA for all recovery and reconstruction projects, but it was strongly suggested that at least a rapid EIA or Strategic Environment Assessment (SEA) should be conducted. In addition, in planning reconstruction projects, it is also important to take into account and respect the "biological legacy" created by disasters such as newly emerged micro habitats and ecological processes.

• Importance of the philosophy and technology of green (natural) infrastructure:

The importance of understanding and applying green (natural) infrastructure philosophy and technology during the reconstruction process was recognized. Also, the importance of careful examination of quality of the green/nature such as species being planted as green (natural) infrastructure technology was noted; sometimes inappropriately planned green infrastructure can also destroy local ecosystems and biodiversity.

• Need for capacity building and empowerment:

In order to realize biodiversity-harmonious disaster recovery and reconstruction, the importance of capacity building at all levels was emphasized, as was the importance of empowerment of local community members during the recovery and reconstruction process so that they can make better-informed choices.

It is hoped that these lessons will be shared widely and will be further examined and enriched in the course of the post-GEJE reconstruction phase.

13.5 Conclusions

This chapter reviewed tsunami impacts, analyzed from both policy and implementation points of view the reconstruction activities in the coastal zone affected by the GEJE which have emphasized the (re)building of sea walls, and tried to draw some lessons. It documents interesting progress in Eco-DRR on the policy side while mixed results are observed on the ground when it comes to implementation. Various aspects which created this policy/implementation gap were discussed in previous sections but there are some other elements which may also contribute to this gap. Such elements include:

• GEJE destroyed facilities, infrastructure and industries such as fisheries and the fishery product processing industry as well as tourism-related businesses. In the reconstruction phase after initial recovery, it seems that construction projects have become a major local industry in order to substitute affected industries and create jobs.
• Even though many areas experienced subsidence after the GEJE, the government tried to maintain the location of the original coastline by constructing engineered infrastructure to protect the original outline of the national territory.
• Immediately after the GEJE, construction of giant infrastructure was expected by society because of the fear of another big disaster. During the implementation

stage, many people questioned this decision because of feasibility and sustainability concerns. However, people felt unable to change their decisions for fear this would delay the reconstruction process.

The Japanese archipelago is located along the ring of fire, facing the Pacific Ocean and is a hotspot for disasters arising from various natural hazards including earthquakes, tsunamis, volcanic eruptions, flooding, landslides, and heavy snow fall. In March 2014, the Government of Japan revealed an estimate of damage which would probably result from the next Nankai megathrust earthquake, which occur every 90–200 years offshore off the southwestern coast of the main island of Japan. The number of deaths from this earthquake is estimated at 330,000 and the economic loss could reach 220 trillion JPY (Cabinet Office of Japan 2013). Therefore, Japan has to continuously prepare for future disasters.

The GEJE reminded Japanese people that no amount of preparation can completely prevent the destruction caused by hazards of this magnitude. In this respect, it highlighted both the usefulness and limitations of engineering- and ecosystem-based solutions and reminded people of the importance of traditional knowledge. It also highlighted the importance of preparedness before disasters occur because there might not be enough time to revise existing systems and institutions once a disaster strikes. We believe that these lessons of the reconstruction process after the GEJE, including both successes and failures, need to be widely shared with the rest of the world in order to make our world more resilient.

References

Asahi Shinbun (2011) 20 March 2011. http://www.asahi.com/shimbun/nie/kiji/kiji/20110328.html (in Japanese) Accessed 9 Jan 2015

Biodiversity Center of Japan (2014) Impact of the Great East Japan Earthquake on the natural environment in Tohoku coastal regions

Cabinet Office of Japan (2011) http://www.bousai.go.jp/2011daishinsai/pdf/110624-1kisya.pdf (in Japanese) Accessed 9 Jan 2015

Cabinet Office of Japan (2013) http://www.bousai.go.jp/jishin/nankai/nankaitrough_info.html (in Japanese) Accessed 9 Jan 2015

Cabinet Secretariat of Japan (2014) http://www.cas.go.jp/jp/seisaku/kokudo_kyoujinka/index_en.html Accessed 3 April 2015

Central Disaster Management Council of Japan (2011) Report of the committee for technical investigation on countermeasures for earthquakes and tsunamis based on the lessons learned from the 2011 Tohoku Pacific Coast Offshore Earthquake

Dudley N, Buyck C, Furuta N et al (2015) Protected areas as tools for disaster risk reduction, A handbook for practitioners. MOEJ/IUCN, Tokyo/Gland. 44 pp

Experts' working group on restoration of coastal protection forest post- Great East Japan Earthquake (2012) Restoration of coastal forest in the future (in Japanese)

Forestry Agency of Japan (2013) Annual report on forest and forestry in Japan. Fiscal Year 2013

IUBS, BDNJ (2014) Message from WS3: biodiversity-harmonious disaster recovery process (DRAFT)

IUCN, CBD – Convention of Biological Diversity (2014) Report of the workshop on understanding and strengthening the role of protected areas for disaster risk reduction and climate change adaptation pre-congress workshop for the IUCN World Parks Congress 10–11 November 2014, Sydney

IUCN, UNU, UNEP (2012) Report of the workshop on "Ecosystem-based Disaster Risk Reduction for Resilient and Sustainable Development" ecosystem based disaster risk reduction in Japan

IUCN, Parks Australia, NSW National Parks and Wildlife Service (2014) http://www.worldparkscongress.org/Accessed 9 Jan 2015

Iwate Nippou (2011) 5 May 2011, http://www.iwate-np.co.jp/311shinsai/saiko/saiko110505.html (in Japanese) Accessed 9 Jan 2015

Kaku J, Kato D, Sakurai Y et al. (2014) Cost benefit analysis of coastal levee construction project in Koizumi, Kesennuma. March 2011. University of Tokyo

Ministry of Land, Infrastructure, Transport and Tourism of Japan (2011a) White Paper on Land, Infrastructure, Transport and Tourism in Japan 2011 (in Japanese)

Ministry of Land, Infrastructure, Transport and Tourism of Japan (2011b) http://www.mlit.go.jp/common/000149772.pdf (in Japanese) Accessed 9 Jan 2015

Ministry of Land, Infrastructure, Transport and Tourism Tohoku Regional Bureau (2014) http://www.thr.mlit.go.jp/bumon/b00045/road/sekihijouhou/index.html (in Japanese) Accessed 9 Jan 2015

Ministry of Land, Infrastructure, Transport and Tourism, Iwate Prefecture and Rikuzen-takata city (2014) Basic plan for reconstruction of the Takata-matsubara Tsunami Memorial Park (in Japanese)

Ministry of Land, Infrastructure, Transport and Tourism of Japan (2015) http://www.mlit.go.jp/sogoseisaku/point/tsunamibousai.html (in Japanese) Accessed 9 Jan 2015

Ministry of the Environment of Japan (2012) Green reconstruction: creating a new National Park

Ministry of the Environment of Japan (2013) 1st Asia Parks Congress website http://asia-parks.org/Accessed 9 Jan 2015

Ministry of the Environment of Japan, IUCN (2013) Report of international workshop on "Disaster Risk Reduction and Protected Areas" 18 November 2013, Sendai

Mori S (2013) Relocation to higher ground and continuation of the community of Koizumi district in Kesennuma (summary) Urban Study 56:1–18. 2013–06 (in Japanese)

Mori wa umi no koibito (2015) http://mori-umi.org/english/Accessed 9 Jan 2015

Murti R, Buyck C (eds) (2014) Safe havens: protected areas for disaster risk reduction and climate change adaptation. IUCN, Gland

NACS-J (2014) An opinion letter requesting to secure natural environmental conservation in Koizumi. Nature Conservation Society of Japan, Kesennuma

National Police Agency of Japan (2015) http://www.npa.go.jp/archive/keibi/biki/higaijokyo.pdf (in Japanese) Accessed 9 Jan 2015

National Resilience Promotion Headquarters (2014) Action plan for National Resilience 2014, June 3, 2014

Nikkei Shinbun (2014) 12 July 2014, http://www.nikkei.com/article/DGXNZO74004420Z00C14A7X93000/ (in Japanese) Accessed 9 Jan 2015

Onza N, Ishida I, Tomita M et al (2015) Plant diversity and environmental heterogeneity in coastal forests affected by the tsunami. Jpn J Conser Ecol 19(2):177–188

Reconstruction Agency of Japan (2014) http://www.reconstruction.go.jp/topics/main-cat2/sub-cat2-1/20141226_hinansha.pdf (in Japanese). Accessed 9 Jan 2015

Reconstruction Design Council (2012) Towards reconstruction – "Hope beyond the Disaster". Report to the Prime Minister of the Reconstruction Design Council in response to the Great East Japan Earthquake, 25 Jun 2011

Renaud F, Murti R (2013) Ecosystems and disaster risk reduction in the context of the Great East Japan Earthquake and Tsunami – a scoping study. UNU-EHS Publication Series No. 10

Science Council of Japan (2014) Encouragement of the use of ecological infrastructure for reconstruction and national resilience http://www.scj.go.jp/ja/info/kohyo/pdf/kohyo-22-t199-2.pdf (in Japanese) Accessed 9 Jan 2015

Shaw R, Noguchi Y, Ishiwatari M (2011) Green belts and coastal risk management. Knowledge Note 2–8, Cluster 2: Nonstructural Measures, World Bank

Shizuoka Prefecture (2015) http://www.pref.shizuoka.jp/kensetsu/ke-890/bouchoutei/(in Japanese) Accessed 9 Jan 2015

Suppasri A, Shito N, Imamura F et al (2013) Lessons learned from the 2011 Great East Japan Tsunami: performance of Tsunami countermeasures, coastal buildings, and tsunami evacuation in Japan. Pure Appl Geophys 170:993–1018

Tanaka N, Yasuda S, Limura K, Yagisawa J (2014) Combined effects of coastal forest and sea embankment on reducing the washout region of houses in the Great East Japan tsunami. J Hydro Environ Res 8:270–280

Tanishita M, Asada T (2014) Factors associated with district victim rate by Tohoku earthquake Tsunmi in Minamisanriku Town, J Jpn Soc Civil Eng, Ser. A1 (Structural Engineering, Earthquake Engineering (SE/EE)) 70(4):I66-I70 (in Japanese)

The 2011 Tohoku Earthquake Tsunami Joint Survey (TTJS) Group (2013). http://www.coastal.jp/ttjt/index.php (in Japanese). Accessed 9 Jan 2015

Tokai Shinpo (2013) 8 Mar 2013 (in Japanese)

World Parks Congress (2014) The promise of Sydney vision http://worldparkscongress.org/about/promise_of_sydney_vision.html Accessed 9 Jan 2015

Yokoyama K (2011) http://www.47news.jp/47gj/furusato/2014/03/post-1058.html (in Japanese) Accessed 9 Jan 2015

Yomiuri Shinbun (2012) 8 Jun 2012 (in Japanese)

Chapter 14
Ecosystem-Based Approaches Toward a Resilient Society in Harmony with Nature

Kazuhiko Takeuchi, Naoki Nakayama, Hiroaki Teshima,
Kazuhiko Takemoto, and Nicholas Turner

Abstract Ecosystem-based approaches have proven effective and efficient in reducing disaster risks while ensuring continued benefits to people from ecosystem services. In this article, a new concept of Ecosystem-based Disaster Risk Reduction (Eco-DRR) for enhancing social-ecological resilience is proposed, based on analysis of several case studies. Field studies in developing countries such as Ghana and Myanmar have shown the benefits of Eco-DRR as implemented by local communities. These projects improve local livelihoods and social-ecological resilience. In Japan, after the massive damage from the 11 March 2011, Great East Japan earthquake and tsunami, ecosystem-based approaches were an important element of the national government's DRR efforts. Analysis of these cases shows that Eco-DRR is a socially, economically and environmentally sustainable tool for DRR that creates new value for a region. It also shows the importance of multi-stakeholder participation in the process of promoting Eco-DRR. It is likely to become even more important in the future, as a means for addressing the increase in disasters resulting from climate and ecosystem change as well as demographic change. The contribution of Eco-DRR to maintaining and restoring ecosystems is particularly valuable for countries where there is reduced capacity for land management, as currently occurring in Japan due to rapid population decline and aging.

Keywords Geophysical and meteorological hazards • 11 March 2011 • Great East Japan earthquake and tsunami • Multi-stakeholder engagement • Social-ecological resilience

K. Takeuchi
United Nations University (UNU), Tokyo, Japan

IR3S, The University of Tokyo, Tokyo, Japan

N. Nakayama (✉) • H. Teshima
Ministry of the Environment, Government of Japan, Tokyo, Japan
e-mail: naoki_nakayama@env.go.jp

K. Takemoto • N. Turner
Institute for the Advanced Study of Sustainability, United Nations University, Tokyo, Japan

© Springer International Publishing Switzerland 2016   315
F.G. Renaud et al. (eds.), *Ecosystem-Based Disaster Risk Reduction and Adaptation in Practice*, Advances in Natural and Technological Hazards Research 42,
DOI 10.1007/978-3-319-43633-3_14

14.1   Introduction

The region around the rim of the Pacific Ocean is situated on the "Ring of Fire", an area where a large number of earthquakes and volcanic activities occur. Disasters are particularly concentrated in this region, which is home to over 75 % of the world's volcanoes, and where 90 % of the world's earthquakes occur (USGS 2015).

According to the Emergency Events Database (Guha-Sapir et al. 2015), the number of disasters and the resulting economic damage around the world have generally been rising over recent decades, although there was a decrease in the early years of the twenty-first century (Fig. 14.1). While the number of geological hazards has remained constant, meteorological, hydrological and climatological hazards are on the rise, and are expected to increase further in the future due to the effects of climate change (IPCC 2014). Meanwhile, the number of deaths caused by natural hazards is on a downward trend. This is believed to be the result of disaster risk reduction (DRR) efforts and enhanced knowledge of natural hazards. In contrast, associated economic losses are increasing drastically, now estimated at USD 314 billion per year in the built environment alone (UNISDR 2015). Moreover, if we look at natural hazard occurrences by region, Asia accounts for an overwhelmingly large percentage in terms of the number of hazards and resulting deaths, as well as the magnitude of related economic losses (Fig. 14.1).

While the "Ring of Fire" carries the threat of calamities, it also brings valuable natural capital or "blessings" of nature such as the scenic beauty of volcanic landforms and plentiful hot springs (Fig. 14.2). However, natural hazards such as earthquakes, tsunamis and volcanic eruptions strike suddenly, causing tremendous damage to human lives and livelihoods. But as well as causing damage, these disasters, depending on their scale and type, can also help provide water resources through rain and preserve biodiversity by disturbing the natural environment and forming vegetation landscapes such as riparian forest floodplain.

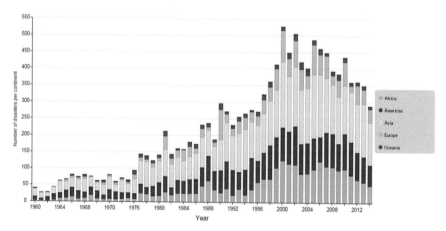

Fig. 14.1 Total number of reported natural hazards from 1960 to 2014 by region (Reproduced from: Guha-Sapir D, Below R, Hoyois P (2015) EM-DAT: International Disaster Database – www.emdat.be – Université Catholique de Louvain – Brussels – Belgium)

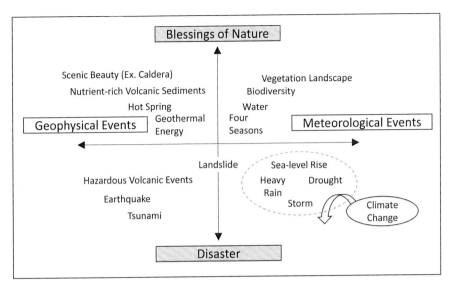

Fig. 14.2 Blessings and threats of nature from natural hazards (Source: authors)

This ambiguity therefore needs to be recognised. The National Biodiversity Strategy of Japan 2012–2020 states that: "Great East Japan Earthquake (GEJE) was an opportunity to recognize once again that nature which provides us with plentiful benefits also becomes a threat on occasion causing disasters and that we have to live with these two opposing characteristics of nature" (MOEJ 2012, p. 2). The strategy, which was revised after the 11 March 2011, GEJE, is based on the vision of approaching nature with a mentality of gratitude and reverence, and achieving a society in harmony with nature by rebuilding relationships between people and nature (MOEJ 2012).

In this chapter, through the analysis of case studies which includes the GEJE, we consider how ecosystems can provide services not only for our daily lives but also for mitigating impacts from natural hazards. We explore how ecosystems can generate new value for communities and enhance social-ecological resilience as a result, and also emphasize how these multiple benefits could be enhanced through multi-stakeholder participation, and through virtuous cycles of supply of and demand for ecosystem services and the return of funds and labor in exchange for these services, which we term the "socio-ecological sphere".

14.2 Eco-DRR for Enhancing Social-Ecological Resilience

Ecosystem-based disaster risk reduction (Eco-DRR) is "the sustainable management, conservation and restoration of ecosystems to reduce disaster risk, with the aim of achieving sustainable and resilient development" (Estrella and Saalismaa 2013, p.30). The ecosystem-based approach is an important tool for building a society in harmony with nature, as it enables us to effectively utilize DRR functions and other blessings of nature.

In Japan after the GEJE, it was recognized that disaster countermeasures including conventional hard engineering were insufficient, and that it was necessary to build social-ecological resilience against future risks and shocks (Takeuchi et al. 2014; Furuta and Seino, Chap. 13). Resilience in this context means "the capacity of a system to absorb disturbance and reorganize while undergoing change so as to still retain essentially the same function, structure, identity, and feedbacks" (Walker et al. 2004). Based on the concept of Eco-DRR, it is vital to reduce disaster risk while enhancing social-ecological resilience, with the aim of building a resilient society in harmony with nature that can respond flexibly to various natural hazards

Ecosystems can reduce physical exposure to natural hazards such as landslides, floods, avalanches, storm surges and droughts (Sudmeier-Rieux et al. 2013; World Bank 2010). Therefore, in addition to reducing risks related to earthquakes, volcanic activities and tsunamis in the short-term, they are effective in reducing risks of longer-term climate change impacts and increases in hydrological and climatological hazards. For this reason, an ecosystem can contribute to building a social-ecological system for local communities to respond to both short-term and long-term risks in an integrated manner. Lowlands in coastal areas are vulnerable to damage from disasters such as tsunamis as well as sea level rise caused by climate change. DRR measures that take advantage of ecosystems can be particularly useful in this context.

Eco-DRR also contributes both to responding to threats from nature, and to enjoying the blessings of nature. Moreover, it offers further benefits by providing various ecosystem services not only when disasters occur, but also in the reconstruction phase and in ordinary times, contributing to enhancing the resilience of regions and maintaining the livelihoods of people. Nevertheless, since there is a limit to the functions of ecosystem-based approaches, it is necessary to combine and integrate them with various other DRR measures to effectively strengthen the social and economic resilience of regions, as the need for long-term measures is expected to increase. In this way, ecosystem-based approaches can contribute greatly to building a resilient society that is in harmony with nature.

Eco-DRR is not only beneficial in reducing disaster risks; it can also be more cost-effective than conventional approaches (IFRC 2002; Sudmeier-Rieux et al. 2013). It has the additional advantage of providing multiple benefits through ecosystem services. This was evident in southern Louisiana when the region was hit by Hurricane Katrina and the ensuing storm surges in 2005. A study by Wamsley et al. (2009) showed that salt marshes in the area were effective in reducing damage resulting from storm surges caused by the hurricane in inland areas. In addition, it was observed through model simulations that the salt marshes functioned to reduce the scale of the disaster under certain conditions (Wamsley et al. 2009). These DRR effects of the coastal marshes in Louisiana were estimated before Hurricane Katrina to have an annual economic value of USD 940 per hectare (Costanza et al. 1989). With the addition of ecosystem services, this amount increased to USD 12,700 per hectare (Costanza et al. 1997). On this basis, restoring 480,000 ha of marshes led to the revival of ecosystem services providing a value of USD 600 million annually, and a total of USD 20 billion as of 2006 (Costanza et al. 2006). This is a large figure even when compared to the estimated USD 2.5 billion that is required to restore the marshes and repair some embankments, and it demonstrates the positive economic effects of marshland regeneration. We should also note that utilization of the ecosystem is generating new value for the community.

14.3   Multi-Stakeholder Participation and the "Socio-Ecological Sphere"

It is important that local communities participate in the process of promoting Eco-DRR because the maintenance and recovery of ecosystems requires management, and these communities will benefit from various ecosystem services and enhanced social and economic resilience (see also Lange et al.). Also, local communities should take decisions about the future of their regions and choose policies that they will not regret in future. Strengthening the link between local residents and landscapes will contribute to enhancing social-ecological resilience through conservation and restoration of ecosystems (Takeuchi et al. 2014). For example, the participation of local communities is essential for the maintenance and management of the greenbelt of coastal forests. It has also been emphasised that it is crucial to utilise ecosystems so that they contribute to the livelihoods of people in local communities (Shaw et al. 2014).

This idea is similar to the concept of the "socio-ecological sphere" outlined in the National Biodiversity Strategy of Japan 2012–2020, which aims to set up a natural zone of symbiosis, consisting of the supply of and demand for ecosystem services and the return of funds and labor in exchange for these services (Fig. 14.3). "Satoyama" in the Fig. 14.3 is defined as dynamic mosaics of managed socio-ecological systems, which produce a bundle of ecosystem services fo human well-being (Saito and Shibata 2012, p.26). According to this policy, in order to achieve a society that exists in harmony with nature, it is fundamentally important to build decentralized, self-sustaining communities that use and circulate ecosystem services in their regions. But if this faces challenges, we must expand the circulation of ecosystem services to incorporate other areas, including urban areas and other countries, to build relationships between the supply and demand for ecosystem services and the return of funds and labour, which are mutually supportive within these spheres. Although local residents should be at the center of such efforts, a wide range of stakeholders including governments, research and educational institutions, the private sector and non-governmental organizations (NGOs) can help strengthen connections in the socio-ecological sphere by providing resources and support. This concept could be applied to local Eco-DRR activities. To make Eco-DRR activities socially, economically and environmentally sustainable, it is

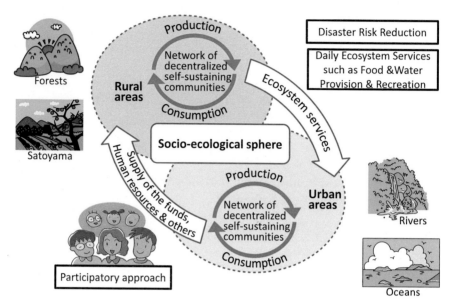

Fig. 14.3 The concept of "socio-ecological sphere" and Eco-DRR (Reproduced and modified from MOEJ 2012, with permission)

important that regions benefiting from ecosystem services contribute human and financial resources to the regions need Eco-DRR.

The following three case studies, from both developing and developed countries, further illustrate how ecosystems have been utilized to contribute to the livelihoods of people in the communities.

14.4   Case Study Analysis

In this chapter, three case studies were selected that provide strong examples of Eco-DRR for both geophysical and meteorological hazards, from sites where the authors have conducted research. The case studies in Ghana and Myanmar focus on meteorological hazards, while the case study of the GEJE deals with the tsunamis caused by geophysical events. At all of these sites, Eco-DRR is closely linked with the daily livelihoods of local communities, creating new value to the regions. Another common feature of these cases is multi-stakeholder participation, which ensures the sustainability of Eco-DRR activities.

14.4.1 Mangrove Forest Rehabilitation and Community Resilience Building in the Coastal Regions of Myanmar

Cyclone Nargis, which devastated coastal regions of Myanmar in 2008, is deemed the country's worst environmental catastrophe. The official death toll in Myanmar from Nargis is 84,537 with 53,836 people still missing, which is 1000 times higher than that from the 2004 Indian Ocean Tsunami in the country (Post-Nargis Joint Assessment – Myanmar 2008). The hardest-hit area was the south-west coast, the Ayeyarwady region in particular. It also devastated the mangrove forests on which people living in the area largely relied upon for their livelihoods. Recognizing this, international and national resources were used to rehabilitate these mangroves and the services they provide to local communities. These services include DRR against storm surges (Post-Nargis Joint Assessment – Myanmar 2008) and contribute to community resilience building (see below). This interrelationship can be considered as an example of the "socio-ecological sphere" illustrated in Fig. 14.3. To build a network of decentralized, self-sustaining communities, the capacity of local communities for sustainable forest management, agriculture and fisheries has been developed through the participation of multiple stakeholders.

The cyclone's impact on the mangrove forests and communities in the region is described in the Post-Nargis Joint Assessment – Myanmar (2008). According to the report, land use conversion had reduced the mangrove area to half of its original size and the clearing of mangrove and other coastal vegetation to create rice fields had increasingly exacerbated the damage inflicted by natural hazards. When Nargis struck, the mangrove forest provided protection to coastal communities (Fig. 14.4), but 38,000 ha of it were destroyed or badly damaged, dealing a massive economic blow to the region, which relies on the forests for livelihoods such as fishing. The report estimated the economic loss associated with the forest's destruction to be USD 4.3 million. The impact of deforestation on local villages is substantial because many of the residents are dependent on the forests for all or part of their livelihoods, and because they use mangrove resources—which involve no monetary transactions—for food and other daily necessities.

To address this issue, a community resilience-building project targeting the region was implemented by the Japan International Cooperation Agency (JICA) with the goal of restoring the mangrove forests surrounding the villages. Efforts were also made as part of the project to identify wind and salt-water tolerant crops and promote use of the region's traditional home gardens. Prior to the project commencing, JICA began working with local villagers in 2002 to develop a master plan for mangrove rehabilitation and launched a technical cooperation project in 2007 based on this plan. The project aimed to enhance the forest management capability of the Myanmar Forestry Ministry and local residents and thereby help restore the forest; it also assisted the villages with recovery from the cyclone's damage. Under this project, JICA, in collaboration with the Myanmar Forestry Ministry and local NGOs, helped plant mangrove trees over an area of 8000 ha after

Fig. 14.4 Mangrove forest and home garden in Ayeyarwady region, Myanmar (Photo: Akira Nagata, reproduced with permission)

the Nargis disaster. In addition, community forest management and agriculture and fisheries revitalization strategies were implemented to improve community livelihoods. These approaches were intended to enhance both DRR and residents' livelihoods through the use of ecosystem services.

14.4.2   Ecological Adaptation to Climate Change and Capacity Building of Villagers in Northern Ghana

The case study of Northern Ghana provides another good example of the "socio-ecological sphere", whereby community resilience to meteorological hazards is enhanced through agricultural ecosystems. This project focused on capacity building for national researchers and engineers as well as local farmers, which is expected to strengthen the human resources contributing to ecological adaptation in rural areas in the future.

Located in Sub-Saharan Africa, Northern Ghana faces challenging climate conditions. There are concerns that the acceleration of climate change may further reinforce the region's vulnerability. It is predicted that a rise in temperature of 2.3 °C to 4.2 °C caused by climate change will push the region's temperatures even higher during the dry season and exacerbate the impact of droughts (Tachie-Obeng et al. 2014). Climate change is also expected to trigger more floods in the region's Volta River basin (Sawai et al. 2014). In short, climate change is projected to cause rising temperatures and pronounced extreme weather events.

A number of measures can be effective in reducing the impacts of climate change, such as building irrigation dams or installing flood defense infrastructure, but in the poverty-stricken region of northern Ghana, projects requiring large amounts of capital are not feasible options. It is therefore important to promote disaster preparedness and risk mitigation strategies that utilize agricultural ecosystems, and to support capacity building at the community level to enable local people (farmers) to implement these strategies. In fact, coping strategies used by households in communities during droughts and floods rely heavily on ecosystem services, including food provision (Lolig et al. 2014).

The University of Tokyo and the United Nations University, together with several research institutions in both Japan and Ghana, are currently implementing the research project Enhancing Resilience to Climate and Ecosystem Changes in Semi-Arid Africa: An Integrated Approach (CECAR-Africa) in northern Ghana, as part of the Science and Technology Research Partnership for Sustainable Development (SATREPS) program administered by JICA and the Japan Science and Technology Agency (JST).

Figure 14.5 compares the diversity of crops and rice yields from the two districts of Wa West and Tolon (Takeuchi 2015). Wa West enjoys advantages over Tolon in terms of crop diversity. A broader variety of crops implies greater resilience because farmers are better able to adapt to frequent and intense extreme weather conditions by selecting crops with a short vegetation period or greater drought resistance (see Kloos and Renaud, Chap. 9). On the other hand, Tolon's rice yields far exceed those of Wa West. Production volume is essential for farmers to secure sustainable livelihoods. This study concluded that resilience of agricultural

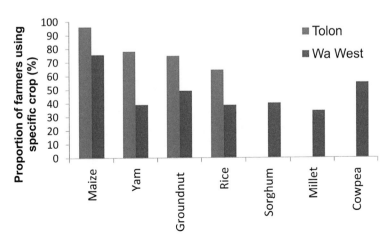

Fig. 14.5 Diversity of crops and rice yields from the two districts of Wa West and Tolon (Source: Takeuchi 2015)

production can be achieved by maintaining crop diversity while at the same time boosting production (Takeuchi 2015)

Shea trees (*Butyrospernum Parkii*) are drought- and flood-tolerant trees common to this region. Some are found in cultivated settings, but the trees grow naturally in the wild in the savannah grasslands. Shea butter, a product extracted from shea tree nuts, is used to produce soaps and cosmetics that are exported worldwide. Research on shea butter production in northern Ghana found that processors in rural areas consume large amounts of water and energy, particularly for the production of high quality butter. With the introduction of a more resource-efficient production process, butter-producing communities could maximize the value of shea trees as a key resource for building community resilience (Jasaw et al. 2015). Given that it is mostly women engaged in shea nut collecting and processing, improved production practices could also lead to skill enhancement and the diversification of women's livelihoods, which in turn would help strengthen community resilience (Otsuki et al. 2014).

As part of the CECAR-Africa project, researchers and engineers from Ghana took part in a skill enhancement training program in Japan and in Ghana. The project also sent a team of Japanese and Ghanaian researchers to target villages, where they held seven community workshop sessions to present their research findings and hold discussions with the villagers and stakeholders (Fig. 14.6). These initiatives provided capacity development opportunities for engineers and local communities alike to promote DRR strategies that utilize agricultural ecosystems and contributed to building a more resilient social system.

Fig. 14.6 Community workshop in Ghana in August 2014 (Photo: Osamu Saito, reproduced with permission)

14.4.3 Building a Resilient Society in Harmony with Nature in Japan after the GEJE

Japan is part of the "Ring of Fire" and its geographical, geomorphological and meteorological conditions make it especially vulnerable to natural hazards (MLIT 2014). As such, Japan's rate of exposure to natural hazards is ranked amongst the highest in the world, along with the Pacific Island States (Alliance Development Works 2014). In particular, the GEJE in 2011 caused the greatest damage Japan has suffered in recent years (see Furuta and Seino, Chap. 13).

This devastating experience has, however, led to the accelerated promotion of DRR measures as well as other initiatives to enhance the resilience of the country. Reports of coastal forests reducing damages during the GEJE also underlined the importance of Eco-DRR. For example, the Basic Act for National Resilience was enacted in 2013, and related measures are being promoted to advance DRR and to expedite reconstruction in a comprehensive and systematic manner based on lessons learned from the GEJE. Harmony with nature and the environment was put forth as one of the basic policies of this undertaking. The Fundamental Plan for National Resilience, which is based on the Basic Act for National Resilience, outlines steps to evaluate the ecosystem functions of natural ecosystems such as coastal forests and marshlands under both extraordinary and ordinary conditions and to actively utilize such functions in DRR measures.

Moreover, authors of this chapter have worked to promote the mainstreaming of the importance of Eco-DRR worldwide through international conferences such as the 1st Asia Parks Congress in 2013 in Sendai, Japan, the Twelfth Meeting of the Conference of the Parties to the Convention on Biological Diversity (CBD COP 12) in 2014 in Pyeongchang, Republic of Korea, the 6th World Parks Congress in Sydney, Australia and the 3rd UN World Conference on Disaster Risk Reduction in 2015 in Sendai (Furuta and Seino, Chap. 13 for details). A notable achievement was the adoption of a decision on DRR and biodiversity at CBD COP 12 based on a proposal by Japan (COP 12 Decision XII/20). The decision encourages Parties to the Convention to promote and implement ecosystem-based approaches to disaster risk reduction.

In this manner, the GEJE became a turning point in Japan and prompted significant progress in domestic and international measures related to Eco-DRR. In addition, reconstruction measures that take advantage of the natural environment of each region have been launched in areas affected by the GEJE with the aim of realizing a sustainable society. Such efforts are being led by Fukushima, Miyagi and Iwate prefectures, the three which suffered the heaviest damages (MOEJ 2013). In the following sections we introduce two initiatives for strengthening the resilience of local communities and contributing to DRR. These initiatives provide best practices of the "socio-ecological sphere" (Fig. 14.3) for adapting to meteorological events. As with the other case studies, these projects utilize multiple ecosystem services with multi-stakeholder participation. Another noteworthy characteristic of both projects is their consideration of the ecosystem and biodiversity in their regions, which is important for ensuring sustainability.

14.5 Restoration of Coastal Forests Along the Area Devastated by GEJE

In Japan, coastal disaster-prevention forests have been developed for centuries for the purpose of preventing or mitigating disasters (see Furuta and Seino, Chap. 13). The tsunami after the GEJE caused flood damage in coastal forests, the total area of which was approximately 3660 ha. An assessment was conducted using aerial photographs and other materials to measure the extent of woodland that was washed away, inundated or destroyed. The result revealed severe damage, with about 30 % of the area rated at a damage level of 75 % or more, and over 20 % rated at a damage level of 25–75 % (Forestry Agency of Japan 2012). Although many coastal disaster-prevention forests were devastated by the tsunami, there were reports of coastal forests helping to dampen the energy of tsunamis and delay their arrival time (Forestry Agency of Japan 2012). It is conceivable that the coastal disaster-prevention forests that were devastated were also effective in this way. Moreover, in coastal disaster-prevention forests where woodland remained, cases were reported in which floating wreckage was trapped and damage to houses and other properties behind the woodland was reduced (Forestry Agency of Japan 2012; Furuta and Seino, Chap. 13). In addition, they can defend against blowing sand and wind and have other disaster prevention functions that play an important role in protecting the region (Forestry Agency of Japan 2012). Based on this, coastal disaster-prevention forests are now being restored in various regions. However, at the same time, there is a fact that many seawalls are still being erected along the coast as a conventional DRR measure (see Furuta and Seino, Chap. 13).

The Greenbelt Project launched in Watari Town in Miyagi Prefecture brings together various stakeholders in the region to work together with the aim of strengthening resilience and reducing the effects of disasters in the region, and restoring coastal forests while giving consideration to the local ecosystem. In Watari Town there are endangered plants and a distinctive forest of black pines along the coastline, covering an area that is 4 km long and 400 m wide. For over 100 years, the scenic forest was a symbol of the town and protected the residents from sea winds and blowing sand (MOEJ 2013). As a result of the GEJE, however, 77 ha of the 120-ha coastal forest were washed away, and the damage also extended to houses (Fig. 14.7). In the eastern district of Yoshida in Watari Town, there were about 230 households before the earthquake, but only 23 families are returning to the area. The restoration and management of the coastal forest and the farmland behind it has become a major issue in this district. Here, we see two characteristics that can contribute to building resilience in the region.

Fig. 14.7 The coastal forest in Watari Town before and after the GEJE (Photo: Tohoku Regional Development Association, reproduced with permission)

One is the adoption of an approach that gives consideration to the ecosystem and biodiversity in the region. Using guidelines compiled by the Forestry Agency of Japan entitled "Future Restoration of Coastal Disaster-Prevention Forests" as a reference, local tree species were selected for cultivation and planting in the region. Also, a broad-leaved forest is being restored behind the pine forest in a 200 m-wide greenbelt. Steps were taken to preserve endangered plant species as well as native broad-leaved trees, while invasive alien species such as desert false indigo (*Amorpha fruitcosa*) and black locust (*Robinia pseudoacacia*) were exterminated (Fig. 14.8). In this manner, efforts are being made to conserve biodiversity while promoting the use of ecosystem services, such as those for DRR.

The other characteristics for resilience-building is the reconstruction plan, which was compiled with the participation of a wide variety of stakeholders, most of which were community-based (Fig. 14.9). The community had a strong desire to realize sustainable development of the region for the next generation. Therefore, in addition to the local government, local residents are also playing an active role in planning reconstruction efforts in the area. Five workshops were held with the participation of a wide variety of stakeholders, including some from outside the town, to compile a master plan for the restoration of coastal disaster-prevention forests (MOEJ 2013). The plan also includes steps to grow seedlings and conduct eco-tours, taking advantage of capacity inside and outside the region and

Fig. 14.8 Eradication of invasive alien tree species in Watari Town (Photo: Takao Ogawara, reproduced with permission)

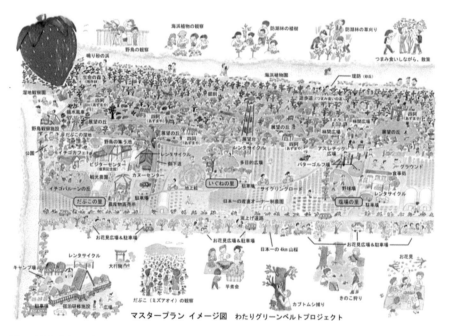

Fig. 14.9 Map of reconstruction of coastal areas in Watari Town, produced through multi-stakeholder participation (Reproduced from: http://watarigbpj.sakura.ne.jp/dl/masterplan_image with permission)

developing businesses that take advantage of local resources (MOEJ 2013). The approach utilises ecosystem services and gives consideration to the livelihoods of the people in the reconstruction process.

14.5.1 Sanriku Fukko (Reconstruction) National Park

Another project that was launched in a region affected by the GEJE takes advantage of protected areas to build a resilient society in harmony with nature. Based on an analysis of the disaster-stricken areas, Takeuchi et al. (2014) proposed that a combination of ecosystems and social and economic resilience, such as a transformation to sustainable agriculture, provides a variety of options for flexible responses to future disaster risks and improved quality of life. They presented as an example the efforts to create the Sanriku Fukko (Reconstruction) National Park (Fig. 14.10). This is the central focus of the Green Reconstruction Project, whose three basic policies are: (1) make the most of the blessings that nature provides, (2) study the threats from nature, and (3) strengthen interconnections between the forests, farmlands, rivers and coasts. This new national park was established by combining several existing protected areas and upgrading it into one integrated national park. The purpose of this initiative was to support the reconstruction of the

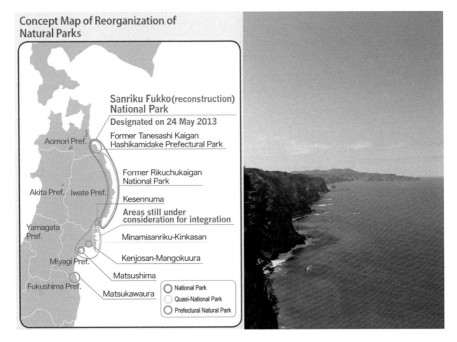

Fig. 14.10 Concept Map of Sanriku Fukko National Park and its landscape (Source: MOEJ, reproduced with permission)

affected areas by creating a 700 km long hiking trail along the coastline, promoting eco-tourism in collaboration with fishermen and disaster education for visitors (MOEJ 2012). It was very symbolic that the name of the national park includes the term "reconstruction" and that it is committed to serve reconstruction activities.

At the Sanriku Fukko (Reconstruction) National Park, efforts are being made to maximize the blessings of nature during ordinary times and protect the people from natural threats when disasters strike. An example of an initiative that attempts to combine disaster prevention measures that strengthen the resilience of local communities with measures to conserve the natural environment can be found on Kesennuma Oshima Island in Kesennuma City, Miyagi Prefecture.

Before the GEJE in 2011, Kesennuma Oshima in the Sanriku Fukko (Reconstruction) National Park attracted many tourists. It was a major center of tourism, providing visitors with a natural environment, nature walks, ocean swimming and fishing. At the time of the GEJE, the island suffered heavy damage from a 12-m tsunami, which took the lives of about 30 people on an island of approximately 3000 residents.

After the disaster, a reconstruction plan was compiled based on discussions between Miyagi Prefectural Government and the local residents of Tanaka-hama beach on Kesennuma Oshima Island. The plan places top priority on the security of the residents and aims to achieve coexistence with the natural environment.

Originally, there was a proposal to construct a sea wall with a height of 11.6 m. However, there was opposition to this from the local community, because it may have damaged coastal landscapes and ecosystem functions. As a result the agreed plan for Tanaka-hama beach includes a sea wall with a height of 3.9 m above Tokyo Pail (Fig. 14.11). The local administrative organ decided to purchase the land inside the embankment, including farmland that was damaged by the disaster, and fill the land so that the maximum height will be 11.8 m above Tokyo Pail. It will create a vegetation base and develop coastal disaster-prevention forests to prepare for tsunamis that occur relatively frequently in the area. It was possible to take such an approach in Tanaka-hama because of geographical features such as the level of the ground, which is higher in inland areas, as well as the fact that several evacuation routes existed, fulfilling requirements for the security of the residents.

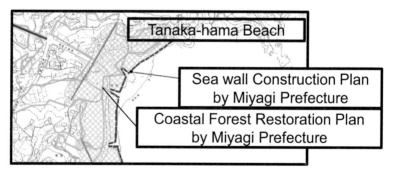

Fig. 14.11 Reconstruction plan in Tanaka-hama Beach (Source: Kesennuma City Reconstruction Plan, http://www.city.kesennuma.lg.jp/www/contents/1387874115071/files/ooshima.pdf)

Fig. 14.12 Evacuation exercise for students on an evacuation route in Tanaka-hama beach (Photo: MOEJ, reproduced with permission)

In addition to these disaster prevention facilities, the disaster prevention functions of coastal forests were used to prepare for future tsunami disaster risks.

The Ministry of the Environment, which is responsible for the management of national parks in Japan, restored facilities near the beach to promote nature experience programs in Tanaka-hama, and prepared for tsunami disaster risks by building an evacuation route that will allow people to quickly evacuate to the top of a hill (Fig. 14.12). During ordinary times the escape route will also be used as a promenade leading to accommodation on top of the hill.

As a result, the scenic coastal landscape was preserved in the process of restoration and reconstruction, and it was possible to continue promoting eco-tourism featuring a nature experience program that takes advantage of various ecosystem services that the national park provides. The plan satisfies the needs of local residents for reconstructing the area as a sustainable community in harmony with the natural environment (Dudley et al. 2015).

14.6 Conclusion

Eco-DRR is a socially, economically and environmentally sustainable tool for DRR that creates new value for a region. It is effective against both geophysical and meteorological hazards, while also providing important benefits in both ordinary and extraordinary situations through ecosystem services. The case studies introduced in this chapter are all efforts to strengthen resilience and promote ecosystem-based DRR with the participation and involvement of various stakeholders, including local communities. In addition to multi-stakeholder participation, it is important to create a virtuous cycle of ecosystem services and the human and financial resources needed to maintain and enhance these services. From these case studies, we conclude that the concept of the "socio-ecological sphere" can be a key element for ensuring Eco-DRR, sustainability and social-ecological resilience.

Such efforts to strengthen the resilience of communities and contribute to DRR are likely to become even more important in the future. In Japan, rapid population decline and aging is expected to cause a shortage of people involved in land management and reduced standards of land management around the country. Meanwhile, as the utilisation of land decreases and more land is abandoned, the costs of maintaining and updating existing social capital are expected to rise (MLIT 2014). If the map of Japan is drawn on grid paper with 1-km squares, it is predicted that in more than 60 % of the currently populated areas, the population will decrease to less than half by 2050 (MLIT 2014). In this situation, it will become increasingly important to actively maintain and restore the functions of ecosystems, especially where measures utilising ecosystems are effective and economical and provide various services to a region. An example of this would be to take an artificial forest of conifers that was abandoned by its managers and regenerate it as a native broad-leaved forest that is more disaster-resistant. Such efforts are also vital from the standpoint of responding and adapting to an increase in natural hazards resulting from climate change. This point of view is reflected in the National Land Grand Design Plan 2050, which proposes that land that has become abandoned as a result of demographic changes and shifts in the distribution of the population should be returned to its natural state (MLIT 2014). Such approaches may serve as a useful reference for other countries where the population is expected to decline in the future.

References

Alliance Development Works (2014) World risk report. UNU-EHS, Bonn

Costanza R, Farber SC, Maxwell J (1989) The valuation and management of wetland ecosystems. Ecol Econ 1:335–361

Costanza R, d'Arge R, de Groot R et al (1997) The value of the world's ecosystem services and natural capital. Nature 387:253–260

Costanza R, Mitsch WJ, Day JW (2006) A new vision for New Orleans and the Mississippi delta: applying ecological economics and ecological engineering. Front Ecol Environ 4(9):465–472

Dudley N, Buyck C, Furuta N et al (2015) Protected areas as tools for disaster risk reduction. A handbook for practitioners

Estrella M, Saalismaa N (2013) Ecosystem-based disaster risk reduction (Eco-DRR): an overview. In: Renaud FG, Sudmeier-Rieux K, Estrella M (eds) The role of ecosystems in disaster risk reduction. United Nations University Press, Tokyo

Forestry Agency of Japan (2012) Future restoration of coastal disaster-prevention forest. http://www.rinya.maff.go.jp/j/press/tisan/pdf/120201-01.pdf Accessed 25 Apr 2015

Guha-Sapir D, Below R, Hoyois P (2015) EM-DAT: International Disaster Database – www.emdat.be – Université Catholique de Louvain – Brussels

IFRC (2002) Mangrove planting saves life and money in Viet Nam. International Federation of Red Cross. IUCN, Geneva

IPCC (2014) Climate change 2014: impacts, adaptation, and vulnerability: summary for policy makers. International Panel on Climate Change WGII (report) Available online at http://ipcc-wg2.gov

Jasaw GS, Saito O, Takeuchi K (2015) Shea (Vitellaria paradoxa) butter production and resource use by urban and rural processors in Northern Ghana. Sustainability 7:3592–3614

Lolig V, Donkoh SA, Obeng FK et al (2014) Households' coping strategies in drought- and flood-prone communities in Northern Ghana. J Disast Res 9(4):542–553

MLIT – Ministry of Land, Infrastructure and Transportation, Japan (2014) National land grand design plan 2050

MOEJ – Ministry of the Environment, Japan (ed) (2012) National biodiversity strategy of Japan 2012–2020

MOEJ – Ministry of the Environment, Japan (ed) (2013) Annual report on the environment, the sound material-cycle society and the biodiversity

Post-Nargis Joint Assessment – Myanmar (2008) Tripartite Core Group: Government the Union of Myanmar, Association of Southeast Asian Nations, United Nations country team in Myanmar http://reliefweb.int/report/myanmar/myanmar-post-nargis-joint-assessment Accessed 7th May 2015.

Otsuki K, Jasaw GS, Lolig V (2014) Framing community resilience through mobility and gender. J Disast Res 9:554–562

Saito O, Shibata H (2012) Satoyama and satoumi, and ecosystems services: a conceptual framework. In: Duraiappah AK et al (eds) Stoyama-satoumi ecosystems and human well-being. Springer, Tokyo

Sawai N, Kobayahsi K, Takara K et al (2014) Impact of climate change on River Flows in the Black Volta River. J Disast Res 9:432–442

Shaw D, Scully J, Hart T (2014) The paradox of social resilience: how cognitive strategies and coping mechanisms attenuate and accentuate resilience. Glob Environ Chang 25:194–203

Sudmeier-Rieux K, Jaquet S, Derron M-H et al (2013) A neglected disaster: landslides and livelihoods in Central-Eastern Nepal. Glob Environ Chang 4:169–176

Tachie-Obeng E, Hewitson B, Gyasi EA et al (2014) Downscaled climate change projections for Wa District in the Savanna Zone of Ghana. J Disast Res 9:422–431

Takeuchi K (2015) Annual implementation report 2014 of SATREPS project "Enhancing Resilience to Climate and Ecosystem Changes in Semi-Arid Africa: an Integrated Approach", Science and Technology Research Partnership for Sustainable Development, Environment and Energy (Global-scale environmental issues). http://www.jst.go.jp/global/kadai/pdf/h2302_h26.pdf. Accessed 24 Nov 2015

Takeuchi K, Elmqvist T, Hatakeyama M et al (2014) Using sustainability science to analyze social-ecological restoration in NE Japan after the Great Earthquake and Tsunami of 2011. Sustain Sci 9:513–526

UNISDR (2015) Global assessment report on disaster risk reduction 2015. United Nations International Strategy for Disaster Reduction http://www.preventionweb.net/english/hyogo/gar/2015/en/home/index.html Accessed 7 May 2015

USGS (2015) Earthquake glossary – ring of fire. U.S. Geological Survey http://earthquake.usgs.gov/learn/glossary/?term=Ring%20of%20Fire Accessed 25 Apr 2015

Walker B, Holling CS, Carpenter S, Kinzig A (2004) Resilience, adaptability and transformability in social-ecological systems. Ecol Soc 19:2

Wamsley TV, Cialone MA, Smith JM et al (2009) Nat Hazards 51:207–224

World Bank (2010) Convention solutions to an inconvenient truth: ecosystem-based approaches to climate change. World Bank, Washington, DC

Chapter 15
Potential for Ecosystem-Based Disaster Risk Reduction and Climate Change Adaptation in the Urban Landscape of Kathmandu Valley, Nepal

Simone Sandholz

Abstract This chapter elaborates on the potential for applying ecosystem-based solutions for urban disaster risk reduction in developing countries, based on the case study of the Kathmandu Valley in Nepal. The high level of mainly informal urbanization in the Kathmandu Valley has led to severe environmental problems and loss of ecosystem services. As a consequence, the livelihoods of the 2.5 million inhabitants in the almost entirely built-up Kathmandu Valley are increasingly at risk, as seen in the aftermath of the 2015 earthquakes that caused widespread damage in the urban area. Combined risks from natural hazards and unsuitable urban planning are likely to be exacerbated by climate change. Due to political instability during the past decades, the poor execution of existing plans and policies as well as the enactment of new ones remain challenging, without real signs for improvement. In addition, the complex governance system involving local, national and international actors is another challenge being faced in this urban agglomeration. Understanding of human-nature interactions, including values attached to natural assets by local communities, is crucial for the development of successful long-term strategies for risk reduction that integrate ecosystem-based solutions.

Keywords Kathmandu • Urban risk • Ecosystem services • Livelihoods • Cultural values • Disaster risk reduction

S. Sandholz (✉)
Institute of Geography, University of Innsbruck, Innsbruck, Austria
e-mail: simone.sandholz@th-koeln.de

© Springer International Publishing Switzerland 2016
F.G. Renaud et al. (eds.), *Ecosystem-Based Disaster Risk Reduction and Adaptation in Practice*, Advances in Natural and Technological Hazards Research 42,
DOI 10.1007/978-3-319-43633-3_15

15.1   Introduction

The potential for applying ecosystem-based approaches to disaster risk reduction in urban areas, especially in developing countries, is barely tapped. Compared to any other ecosystem, urban environments are even more complex, as both natural and built environment[1] and the cultural and socio-economic settings interact in a very limited space (Dizdaroglu et al. 2012: 5). Although aspects of ecological planning are gaining importance, they are mainly restricted to urban administrative boundaries. Only seldom are they extended to the urban surroundings, e.g. by including benefits of watershed management or protected areas management into urban planning strategies (Trzyna 2014). Research on the linkages between urban areas, ecosystem services and their potential to reduce urban risks from disasters is still scarce.

Rapid urbanization and population growth are increasing overall vulnerability of urban areas to disasters (Briceño 2015). Disaster impacts in urban areas of developing countries have the potential to be especially high due to the large number of inhabitants. This is in particular true for the urban poor that are highly concentrated in vulnerable areas such as floodplains, often with inadequate and unenforced building codes, and comprising a large share of marginal settlements. Many cities are severely threatened by the impacts of climate change, which will very likely impact negatively on risks and vulnerabilities, particularly in informal settlements (Bigio 2003; Quarantelli 2003; IPCC 2012; Joint UNEP/OCHA Environment Unit 2012).

Needs for an urban transition to adapt to climate change impacts as well as man-made and natural hazards are especially high in developing countries. Cities in developing countries are confronted with the need to solve several problems simultaneously, while experiencing high urbanization rates. However, urban policies and governance are not always well-equipped to tackle such complex challenges. For example, policies related to climate change in cities and urban environmental protection are often seen as a separate policy sector, with little or no interaction with other urban policies or processes (Bulkeley 2010). This lack of urban policy integration makes it difficult to implement long-term strategies, such as ecosystem-based disaster risk reduction, that require longer time horizons to demonstrate effective protection against hazard impacts, e.g. a protection forest needs time to grow before it can stabilize a slope to protect people settling further down the Valley from landslides.

In addition, hard engineering or 'grey' infrastructure solutions are still often preferred over 'green' or ecosystem-based solutions in disaster management, even though in many cases, conventional engineered measures have been demonstrated to increase the severity of flooding and harm the ecological balance of water flows (Quarantelli 2003). Local knowledge that has emerged over generations contributes

---

[1]defined as 'the totality of humanly created, modified or constructed spaces and places' in World Disasters Report (2014: 121).

to the potential and capacities of local communities to deal with disasters. Yet, local knowledge has often been overlooked in favor of pure engineered solutions to cope with disasters (Mercer et al. 2012a). Attempts to control nature through dams, levees, and reclamation of swamps and wetlands were and still are popular.

Environmental challenges in the developing world require different policies, as urban agglomerations are comparably more diverse, dynamic and complex than cities in Europe or North America. They face a high level of informal economy and housing, a lack of or weakened governance (Simone 2010; Dahiya 2012) and a relatively more limited financial resource or tax base. Particularly cities in Asia have to face serious 'risk overlaps' as the urban transition "occurs rapidly over one long wave", preventing cities from managing environmental problems over time and sequentially as is the case in the Western world (Marcotullio 2006: 42). These conditions call for more integrated urban policies to tackle cross-scale issues, such as environmental management, especially in growing urban agglomerations such as Kathmandu in Nepal.

This chapter is based on a combination of an extensive literature research and interviews and field research carried out in July–September 2013 as primary data sources (see also Takeuchi et al., Chap. 14). Data gathered from twelve in-depth semi-structured expert interviews has been used to support the research arguments derived from literature. Interviews were conducted with Nepalese experts in disaster management, and urban planning and built environment-related disciplines to validate the case study findings. Experts were representatives from ministries, urban planning authorities, non-governmental organizations and researchers (cf. list in Appendix).

15.1.1 Potential of Ecosystem-Based Disaster Risk Reduction and Climate Change Adaptation for Urban Areas

Understanding the interactions of urban activities and ecosystems is essential for a sustainable urban development (Dizdaroglu et al. 2012). Linking urban planning with sustainable ecosystem management that takes into account potential climate change impacts requires a holistic approach. Such an approach would have to include analyses of urban and environmental policies and related policy actors, as well as the local, socio-cultural aspects that shape behavior and perception. At the same time, it would require the consideration of ecosystem services in urban contexts along with the risk and livelihoods profiles of urban dwellers, who are often not a homogenous group but experiencing different levels of risk.

Addressing climate change at the urban scale is not taking place within a social, political, economic, or material vacuum. Instead, it is influenced and shaped by the willingness and capacity of officials and those at risk to take action and reduce exposure and susceptibility to climate-change-associated hazards in a specific place (Bulkeley 2010; Pelling 2011). Accelerated urbanization, the unsustainable use of natural resources and subsequent ecosystem degradation are altering the natural environment and leading to changes in the micro- and macroclimate. This vicious cycle calls for improved urban planning and urban governance (Abbate 2010; Jones et al. 2014). The same processes are triggering vulnerability to disasters, resulting in recurrent and increasingly costly disasters (Fra Paleo 2013). Pelling (2003) talks about a 'coevolution of urbanization and risk', emphasizing that there is no simple one-way line of causality in the production of urban risk. Urban risk is the result of numerous "feedback loops and thresholds and competing ideas, mechanisms and forms" (Pelling 2003: 7). Consequently, a minor hazard event could trigger major disasters with huge impacts throughout the city. Climate change and other urban development dynamics interact in the city, bringing to light different visions of urbanization and associated risk management preferences of the actors involved. In the best case scenario, this dynamic situation opens up the possibility of re-negotiating priorities, both within single policy areas and on a broader scale among different policy areas concerned with the topic (Pelling 2011).

In recent years, an increasing number of studies and papers have discussed the benefits of ecosystem services for sustaining livelihoods and reducing disaster risk. In fact, research has mostly focused on rural communities, despite the fact that on a global scale the majority of dwellers are now urban (United Nations 2014). However, there is growing awareness of the role of ecosystems and their importance to the urban context. In urban areas, the linkages between livelihoods and ecosystem services are comparably less immediate and less visible compared to rural areas, where the direct dependency of people on agriculture and natural resources is more obvious. Urban population density is much higher, and consequently the urban built environment has been modified tremendously, easily resulting in the misperception that neither nature nor natural hazards are of major concern (Grove 2009; Krasny et al. 2014).

Ecosystems provide important services, namely supporting services (e.g. soil dynamics and nutrient flux regulation), provisioning services (e.g. production of freshwater and food) and regulating services (e.g. regulation of the urban climate or hydrology). Cultural ecosystem services comprise recreational benefits and spiritual values attached to the urban natural environment, which in turn support urban inhabitants' social identity (Grove 2009). In particular, the regulating services of ecosystems can support disaster risk reduction, such as slope stabilization through suitable vegetation or green spaces such as wetlands that decelerate rainwater runoff and minimize flood peaks (Joint UNEP/OCHA Environment Unit 2012; Secretariat of the Convention on Biological Diversity 2012). However, disasters can also impact negatively on ecosystem services, increasing people's vulnerability to current and future disasters, e.g. hills deforested by landslide are more likely to erode and slide down again (Lange et al. 2013). Another example is the deterioration of cultural ecosystem services, e.g. if a disaster interrupts or harms traditional customs, such as festivals, markets, or craft production (Taboroff 2003).

Berkes et al. (2009: 129) claim that the conservation of ecosystem services requires "maintaining cultural connections to the land and at times restoring and cultivating new connections". One of the recommendations identified to sustain the linkage between people and land includes maintenance of local and traditional knowledge, of cultural legacies, social institutions and networks, which all play a critical role in sustaining the use of ecosystem services. Such recommendations have been geared mainly for rural contexts. It is worth asking why this type of analysis should not apply to urban areas similarly, e.g. for the maintenance of urban rivers, such as the case of the Australian Aboriginal people attaching cultural and religious values to the Darling river, and for linking physical safety with cultural health and well-being, as described by Gibson (2012). Urban green spaces and urban protected areas can contribute to mitigate disaster impacts, support climate change adaptation and, at the same time, improve human well-being (Beck 2012; Trzyna 2014).

Unfortunately, the significance and linkages of ecosystem services to urban dynamics and urban risk reduction are often poorly understood. The same could be said with respect to the interactions between different ecosystem services and urban livelihoods, as preferences may vary among social groups in urban settings. Social identity, knowledge, spirituality, recreation, and aesthetics attached to ecosystems are also very likely to differ between different cities (Grove 2009). 'Place' is a social concept and the ways in which such a physical place or space is perceived, experienced, imagined and ultimately maintained is tied to cultural values and beliefs (Gibson 2012).

This chapter argues that traditional cultural ties to different ecosystem services could have the potential to support the maintenance (and conservation) of ecosystem services in urban areas. This premise has high potential for application in agglomerations like the Kathmandu Valley in Nepal, where established community-based societal groups and/or where communal stewardship of natural resources can be found. Rapid urbanization, however, has the likelihood of degrading cultural and social ties to particular ecosystem services. As a consequence, awareness and subsequent concerted action of actors at different levels is needed, ranging from local communities to national authorities and international organizations operating on the ground.

15.2 The Kathmandu Valley – Risks, Policies and the Potential Role of Ecosystems

Kathmandu is one out of five municipalities located in the wider 665 km^2 Kathmandu Valley (see Fig. 15.1) and is in the center of the Nepal Central Development Region. It is situated in the Nepalese 'central hills' zone at an average elevation of 1350 m above sea level. The surrounding mountains of up to around 3000 m elevation form a natural barrier which limits both access and further urban expansion. The area has a humid subtropical/subtropical-highland climate, with a monsoon period from June to August, enabling its inhabitants to plant water-intensive crops like rice.

Fig. 15.1 Kathmandu Valley (by author)

The Valley itself has nourished its inhabitants for centuries. Until the middle of the 20th century, more than half of the agglomeration's inhabitants used to practice farming on the fields surrounding the densely built residential areas (Gutschow and Kreutzmann 2012), leaving a maximum of fertile land for agricultural production. Forests served as a source for firewood and burning material for cooking, heating and production of construction materials. The soil itself was and is used for brick production, and the rivers crossing the Valley provided a source for drinking and irrigation water. Water resources from stone spouts (cf. Fig. 15.2) were managed by local communities (Historical Stone Spouts and Source Conservation 2007). The local population was rooted deeply in their natural environment, as can be seen for instance in the variety of cultural ecosystem services derived, such as temples and shrines for religious functions constructed on hilltops or riverbanks, or recreational sites near springs. However, urbanization has fundamentally changed the natural environment towards a densely populated and almost totally built-up urban land-scape (Ellingsen 2010; Thapa and Murayama 2012).

Nepal currently is the fastest urbanizing country in South Asia (Muzzini and Aparicio 2013), with 29 million inhabitants (UN-HABITAT 2010b). What makes Nepal unique is that most of the urban growth is concentrating solely in the capital of Kathmandu, which is growing annually by 3,94 % (calculated for 2010–2015, United Nations, Department of Economic and Social Affairs, Population Division 2014). Around 2.5 million dwellers are settling in the Valley (Muzzini and Aparicio 2013), which is prone to multiple natural hazards, in particular earthquakes, floods, and landslides.

Extreme natural hazards pose an obstacle to urban, social and economic development in the Valley, which is also highly susceptible to climate change. In 2011,

Fig. 15.2 Examples for stone spouts and their use by local communities (by author)

Nepal was ranked the fourth country in the world most vulnerable to climate change (GFDRR 2012). Climate change is expected to exacerbate urban disaster risks, namely from annual droughts and flooding during the monsoon season and landslides triggered by heavy rainfall (Shrestha and Aryal 2011; GFDRR 2012; Jha and Shrestha 2013; Ginnetti and Lavell 2015). Rainfall patterns are predicted to change, leading to more frequent and intense summer floods and winter droughts; rising temperatures are predicted to increase health risk, especially related to waterborne diseases. (Regmi et al. 2009; Ministry of Environment 2010; Shrestha and Aryal 2011). Important opportunities for development, such as in the agriculture sector, are highly susceptible to climate change and to extreme events such as droughts and floods (GFDRR 2012).

High vulnerability is amplified by population growth and migration to the urban area, social exclusion of different societal groups, and the unstable political situation after ten years of civil war between Maoists and the Government which only ended in 2006 (Titz 2012; Jones et al. 2014). Nepal was among the first countries to create a policy and legal framework for disaster risk management, but due to the ongoing political transition, full implementation is still lacking. Various interviewees mentioned the negative consequences of the prolonged, unstable political situation, with one saying: "If politics were more stable, maybe the city would look different and the conservation might be different, too", referring to both, conservation of culture and nature.

The ongoing degradation of ecosystems, in combination with the growing population and poor construction, is putting urban livelihoods at risk (UN-HABITAT 2015). Forests on the surrounding hills are cut, cultivated land is vanishing, and urban river ecosystems are suffering from pollution, built-up infrastructure along river banks and sinking water levels. Such degradation is exacerbating the urban dwellers' vulnerability to disaster impacts, at different scales but throughout all phases of the disaster cycle. For instance, relief after any disaster would be hindered, for instance by lack of clean water. The same is true in recovery and reconstruction especially after devastating disasters such as earthquakes: there is lack of land for growing food or available building materials. Fig. 15.3 depicts the key features of urban development in Kathmandu Valley which result in increasing

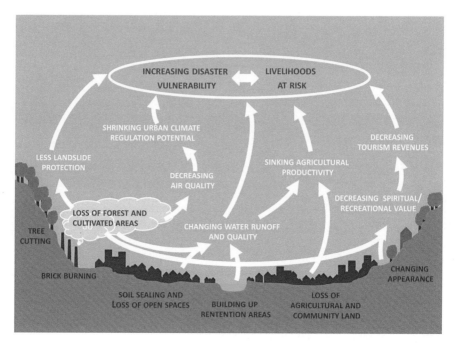

Fig. 15.3 Degradation of ecosystem services and consequences on urban livelihoods and urban risk in Kathmandu Valley (author's own figure)

environmental degradation and vulnerability to disasters, putting urban dwellers and their livelihoods at risk.

15.2.1 Environment

The forest area of the watersheds surrounding Kathmandu Valley decreased by 40 % from 1955 to 1996, with the remaining forests in a mainly regenerating stage. The 2007 Kathmandu Valley Environment Outlook still found 32.7 % of the total Valley area covered by forests, although increasingly threatened by deforestation (ICIMOD et al. 2007). Trees are cut down for settlements, firewood and cultivation. Urban flood retention areas along rivers, open lands and fertile agricultural lands are being built-up, leading to more severe annual flooding. In turn, flooding is exacerbated by deforestation in upstream areas due to urban sprawl and the use of wood for cooking or construction (Nehren et al. 2013). Deforestation and land degradation in hilly areas in turn result in soil erosion, landslides, and siltation in and around the catchment area. Marginal segments of the population construct on hazard prone areas like river banks and steep slopes, further reducing flood retention capacities and limiting agricultural lands in the watershed (Ellingsen 2010; UN-HABITAT 2010a).

Fig. 15.4 Environmental problems in Kathmandu Valley – waste disposal in urban river (*left*), urban sprawl and loss of agricultural areas, brick burning kilns in background (*right*)

The largely uncontrolled process of land conversion has become an environmental problem. Agricultural productivity for the growing population is decreasing; consequently, the Valley is annually becoming less self-sustaining. Moreover, water quality is deteriorating, resulting in severe consequences to the aquatic ecosystem and on the health of urban dwellers. Groundwater recharge has decreased, harming the quality and quantity of groundwater sources. The main response to the urban water shortfall is simply extracting more groundwater in an unsustainable manner (GoN/NTNC 2009; British Red Cross et al. 2014). Access to piped water fell from 68 to 58 % from 2003 to 2010 (Muzzini and Aparicio 2013), forcing people, especially informal settlers, to use surface water. However, rivers and streams are largely used as dumping sites for all types of wastes (cf. Fig. 15.4). It is estimated that half of Kathmandu's daily 150 tonnes of waste are dumped into the urban rivers (CFE-DMHA 2015).

In addition, the rich cultural heritage along the river and its tributaries, such as traditional monuments, shrines and temples, and values attached to the waterways are gradually eroding (GoN/NTNC 2009). "In this sense, the notion of the holiness of the river Bagmati is rendered paradoxically by the massive pollution, even for many devout Hindus" (Ellingsen 2010: 6). Ecosystem services that forests had provided, such as clean water for Valley residents, spiritual as well as recreational benefits for locals and tourists, are decreasing (ICIMOD et al. 2007).

Kathmandu's urban rivers have their sources in the Himalayan Mountains. Aside from being one of the main drinking water sources, these rivers are also strategically important for the country's energy supply. Like the whole country, Kathmandu depends to a large extent on hydropower. However, poor infrastructure, a growing energy demand and climate change, which is predicted to intensify summer floods and winter drought, are contributing to a severe electricity crisis (Sovacool et al. 2011; Surendra et al. 2011; GFDRR 2012). Already today, there are regular power cuts of up to 18 h per day in Kathmandu, hampering development. As a consequence, hydropower plants may become inefficient in providing a regular power supply throughout the year.

The ongoing dependence on firewood as an energy source (Sovacool et al. 2011) is one reason for the rapidly worsening air quality, which is further exacerbated by the Valley's topography. Economic activities such as the brick-burning industry are lowering air quality even further, turning the atmosphere hazy throughout the dry season (ICIMOD et al. 2007; Ellingsen 2010). Urban air pollution in Kathmandu is one of the worst globally, impacting negatively on humans, animals and vegetation (Bhattarai and Conway 2010; Muzzini and Aparicio 2013; British Red Cross et al. 2014).

The most important policies are the Environment Protection Act (1996) and the Forest Act (1993/amended in 1999). The Environment Protection Act, and the related Environment Protection Rules (1997) and Environment Impact Assessment (EIA) Order are the most important policies concerning both environmental protection and sustainable development. Although these policies do not consider DRR as such, there is ample scope for the consideration of disaster risk reduction when assessing the potential environmental impacts of major projects, which could increase disaster risk (International Federation of Red Cross and Red Crescent Societies 2011).

The Forest Act regulates the designation of forests areas and the type of permissible forest uses. It authorizes the designation of "Community Forests", which are formally handed over to "user groups" for the further development, protection and sustainable utilization of forests in the common interest of the community. This includes exploitation of timber, fruits, and animals (International Federation of Red Cross and Red Crescent Societies 2011).

Water resources are protected by different policies, in particular the Soil and Watershed Conservation Act (1982), followed by the Water Resources Act (1993) and subsequently the National Water Resources Strategy (2002), the National Water Plan (2005), and the Water Induced Disaster Management Policy 2006.

However, poor execution of policies is hindering the effective protection of the Valley's natural resources. One respondent interviewed reported: "When Ring Road was built, in the initial plan, trees and green areas around the road were foreseen. The plan was perfect, but in the end the belt is a dump yard; there are no trees and the land is gone." Another respondent talked about poor law enforcement of existing legislation that sought to keep riverbanks free of buildings: "The initial intention of the law to protect the riverbank was skipped for the political pressure." Ecosystem services, namely the provision of wood, water, and soil are exploited in an unsustainable manner, at the expense of ecosystem regulating and supporting services, for instance climate regulation, flood regulation or soil formation, which in turn impact negatively on urban livelihoods in multiple ways (c.f. Fig. 15.2).

15.2.2 Urban Development

Urban vulnerability in the Kathmandu Valley is high due to both the occurrence of natural hazards and societal conditions such as poor housing standards (Bhattarai and Conway 2010). Rapid and mostly unplanned urbanization, fostered by weak institutional arrangements and an administration system recovering from years of conflict, is shaping land use and settlement patterns. Urban sprawl and middle class preferences for 'bungalow' style housing have led to increasing pressure on agricultural land and skyrocketing land prices. Garbage disposal and water supply and sanitation are working partly at best (Muzzini and Aparicio 2013). Considerable concerns concerning the limits of urbanization in relation to the availability of space, water or building materials are warranted. The need for construction materials, especially sand and bricks from local clay, are already exceeding production supplies within the Valley's borders. Presently, there are almost one hundred brick kilns in the Valley, converting fertile sand and soil into building materials for houses, "sealing" even more land with built structures (Gutschow and Kreutzmann 2012).

Open spaces are being lost rapidly, lowering urban retention capacities for absorbing excess flood waters as well as limiting public evacuation areas in case of emergency situations. Agricultural land and open spaces are built up and divided into small plots of up to 15-45 square meters, hindering sustainable land-use planning (Bhattarai and Conway 2010; Muzzini and Aparicio 2013; British Red Cross et al. 2014).

Urban development in the Valley currently does not consider environmental aspects or potential risk reduction services provided by ecosystems. Informal development often takes over formal urban development planning processes, changing the urban landscape even before urban plans can be implemented. In a study on the urban housing sector in Nepal, UN-Habitat (2010a) notes the need for balancing housing needs and ecological carrying capacity through adequate policies.

A large number of regulations and policies on urban planning and construction exist, including the 1976 Physical Development Plan of the Kathmandu Valley and the 1988 Town Development Act that informs the Kathmandu Valley Master Plan, which is renewed every 5 years. The Nepal National Building Code of 1994 was meant to regulate building construction; however, it is rarely enforced (ICIMOD et al. 2007; British Red Cross et al. 2014), although recently some improvement was made in simplifying the process of acquiring building construction permits (World Bank 2014). Like the other urban policies before it, the Building Code stipulates the minimum amount of open spaces or minimum width of roads, but today's Nepali townscape is dominated by contiguous buildings with few open spaces and narrow and encroached roads (UN-HABITAT 2010a).

Asked about the state of urban planning in the Valley, interviewees responded with statements that describe the current reality: "Which urban planning?", "There is no urban planning since 1976", or "In theory all the rules, regulations and laws are there, but in practice no one wants to take the responsibility for it." Informal (illegal) settlements are constructed along ecologically-sensitive riverbeds and lowlands, with numbers growing from 17 settlements in 1985 to 40 in 2010 (UN-HABITAT 2010a). According to those interviewed, urban planning in Nepal has failed, evidenced by the expansion of informal settlements.

Currently, urban land-use planning is not clearly regulated, and institutional responsibilities between different authorities at different levels are unclear. Unfortunately, there is no consistent legal mechanism for the relocation of people away from high-risk land. Relocation mainly occurs in response to disasters, instead of through a more organized, planned approach.

The five municipalities in the Valley cooperated once in 2014 under the umbrella of the Kathmandu Valley Development Authority, to jointly tackle problems of disaster risk, proper land use planning and inadequate open spaces. However, most interviewees claimed a lack of interaction between the different actors at the various levels. As one interviewee mused, "The key limitation is the lack of interaction between the municipalities and the Kathmandu Valley Development Authority, and a lack of enforcement, as the municipalities always neglect the plans coming from the Development Authority".

Nevertheless, it is in the urban land-use planning context that ecosystem components could be embedded on a large scale. One promising activity being implemented by the Ministry of Physical Planning and Works is a system of voluntary 'land pooling' in the Valley, aiming towards a more planned and sustainable urban planning, including the creation of public open spaces (International Federation of Red Cross and Red Crescent Societies 2011).

15.2.3   Disaster Risk Reduction and Climate Change Adaptation

Kathmandu Valley is prone to a multitude of natural hazards, namely earthquakes, landslides, annual flooding, and, increasingly, strong winds due to storms (c.f. Fig. 15.5[2]) (Ministry of Local Development/GoN and Disaster Risk Reduction at the National Level in Nepal/UNDP 2011). Unplanned urban development, encroachment of buildings in open spaces, and the depletion of the water table are increasing people's exposure and vulnerability to different types of hazards (UN-HABITAT 2015). Critical infrastructure such as for water supply and the road network as well as essential services like schools or hospitals are also extremely vulnerable to hazards. The Valley has only limited access by road or plane, contributing to its high risk level.

Kathmandu Valley is located in the most at-risk seismic urban area worldwide (British Red Cross et al. 2014; GFDRR 2014). After the last devastating earthquake in 1934, with a magnitude of 8.3 on the Richter scale, the country was hit by two major earthquakes with magnitudes of 7.8 and 7.3 on April 25, and May 12, 2015, respectively. These earthquakes and their aftershocks have led to almost 8900 casualties and significant damages to buildings and infrastructure (UN OCHA 2015a). Kathmandu itself was among the most affected areas. Overall, however,

---

[2]Although fire is also another major hazard which could potentially result in significant losses to homes and infrastructure. It is not further considered here as fire in an urban agglomeration is mostly human-induced, e.g. due to electrical short circuits or poor wiring (Ministry of Local Development/GoN and Disaster Risk Reduction at the National Level in Nepal/UNDP 2011)

Houses destroyed/damaged
Nepal Central Region, 1971-2011

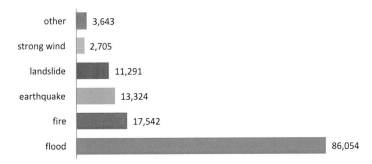

other — 3,643

strong wind — 2,705

landslide — 11,291

earthquake — 13,324

fire — 17,542

flood — 86,054

Fig. 15.5 Houses destroyed or damaged by natural hazards from 1971-2011, Nepal Central Region (Data source: DesInventar 2014)

the earthquakes' impacts have been far below the worst case scenarios, in terms of human losses and impacts on infrastructure (Knight 2013; British Red Cross et al. 2014; National Planning Commission 2015).

It has been predicted that another 8.3 magnitude earthquake could have killed more than 100,000 people, injured 300,000 and displaced 1.8 million. In such an event, limited road access and lack of heavy equipment to remove rubble will be serious barriers to an effective large-scale response (British Red Cross et al. 2014). The World Food Programme, which functions as the global coordination lead on post-disaster logistics in Nepal, estimates that if the airport was closed and all three main access roads were impassable, their organization would only be able to provide enough food to feed 100,000 people for a week – and even with a functioning airport initial supplies would only be available for 10 % of the assumedly displaced people (UN OCHA 2013).

Landslide risk, which is already high in Nepal's mountain and hill areas, is predicted to be higher in the post-2015 earthquake phase, especially at the peak of the monsoon season (Faris and Wang 2014; Yua et al. 2014; Nepal Earthquake Assessment Unit 2015). As of 21 May 2015, approximately 3600 landslides have already been identified all over Nepal, triggered or reactivated by the April 2015 earthquake only (Earthquakes without Frontiers 2015). Such impact of the earthquakes on the number and intensity of landslides is likely to continue for years (Witze 2015). This situation poses a direct risk to local residents with potential damages to homes, agricultural land, critical infrastructure, but could also indirectly result in increasing flood risk as landslides block rivers (Nepal Earthquake Assessment Unit 2015).

Ecosystem services obviously do not have the potential to directly reduce risks from earthquakes; however, healthy, well-maintained ecosystems which contribute to agricultural production would provide much-needed food supplies and serve to lower post-disaster relief needs in the Valley (Muzzini and Aparicio 2013; British Red Cross et al. 2014). In addition, proper urban planning that provides open spaces would help reduce disaster impacts by maintaining water supply and evacuation spaces. Maintaining vegetation cover on hillsides can stabilize steep slopes, and has the potential to reduce earthquake- as well as rainfall-induced landslides (Sudmeier-Rieux and Ash 2009; Peduzzi 2010; Sudmeier-Rieux et al. 2011; Faris and Wang 2014)

With the National Calamity Act of 1982, disaster preparedness and relief are regarded as a national issue. Since then Nepal has made great strides in terms of improving disaster preparedness (CFE-DMHA 2015). The National Disaster Management Plan endorsed in 1996 emphasizes the need to link natural resource management, climate change and development with disaster management. In 1999 the Local Self Governance Act enabled local authorities to take action on a regional (sub-national) and local scale, including for disaster risk reduction (GFDRR 2012; British Red Cross et al. 2014).

In 2008, the Government of Nepal began to shift its focus from a mainly disaster response approach to more *ex-ante* disaster prevention and mitigation. The National Strategy for Disaster Risk Management (NSDRM), approved in 2009 and based on the Hyogo Framework for Action (2005–2015), appears to be widely accepted and supported at the national level. District governments have already established disaster management plans under this strategy, and the next stage will be at local government level (International Federation of Red Cross and Red Crescent Societies 2011; GFDRR 2012; British Red Cross et al. 2014).

In 2011, the Government of Nepal announced a revised draft Bill for a new Disaster Management Act, aiming at a holistic approach and going beyond the reactive approach to disasters. The Act seeks to integrate disaster risk reduction in national, regional (sub-national) and local development processes. However, the Act is not yet legislated, but the January 2015 national progress report on the implementation of the Hyogo Framework for Action mentions the Act. In the same year, the Nepal Risk Reduction Consortium (NRRC) was launched, bringing together members of the national government, international financial institutions, international development partners and donors, including the Red Cross and Red Crescent Movement and the United Nations. The NRRC established the 'Flagship 5' Programme, which outlined different priority areas for pursuing long-term disaster risk reduction (The Nepal Risk Reduction Consortium 2011; Taylor et al. 2013).

To deal with the impacts of climate change, the country developed a National Adaptation Programme of Action (NAPA) in 2010 (Ministry of Environment). The NAPA mentions the challenging urban planning process and calls for cross-cutting solutions, but without being more specific. Nevertheless, this could be an effective entry point for ecosystem-based disaster risk reduction and climate change adaptation measures, integrating and going beyond the disperse strategies and policies that are already in place. At a local scale, ecosystem-based DRR and CCA measures could support the implementation of the 2011 National Climate Change Policy which ensures that 80 % of climate finance should be allocated to support local level implementation.

So far, integration of DRR and CCA in Nepal is still needed. Ecosystem-based management approaches potentially support stronger linkages between DRR and CCA, especially with respect to addressing water-related and landslide risks. For example, maintaining forest cover in the catchment areas and on hill slopes would reduce water runoff and thus provide flood regulation as well as lower landslide risks.

15.2.4 Social Ties and Values

The natural environment and urban structures within Kathmandu Valley are closely associated with legends, rituals and festivals (Government of Nepal; Department of Archaeology 2007). In some areas, forest patches remain untouched because they are considered to be sacred. These sacred places reflect important cultural and religious values, are protected and conserved, and often serve as recreational sites for picnics or hiking (ICIMOD et al. 2007). While the traditional urban layout of Kathmandu Valley was "a shining example of energy and space-efficient building techniques with a distinct community harmonization component (UN-HABITAT 2010a: 99)", it has been replaced by western-style constructions, paying much less attention to disaster and climate risks.

Before 1982, maintenance of natural and cultural sites was done by the '*Guthi*', local organizations or associations based on caste and locality. Disaster preparedness and relief was regarded as social works of local communities. The '*Guthis*' are also, at the same time, a system of community land ownership, responsible for endowing land for religious purposes and charity. Such community-based maintenance and ownership of land decayed over the years for different reasons. Among the most apparent consequences is the deteriorating traditional water supply of the Valley, often fed from springs coming from the surrounding hills (UN-HABITAT 2008; GoN/NTNC 2009). With access to urban water supply becoming limited, people are increasingly dependent on traditional, community-based water supply systems, which are also under pressure. However, more than one fourth of the 400 community-based traditional water spouts have already disappeared, with more expected to be lost in the near future due to deterioration, construction and infrastructure development (UN-HABITAT 2008). The deterioration of the traditional water supply system reduces disaster response capacities, given that the traditional water supply system functioned as a reliable water source during emergency situations when electricity – needed for water pumping – could not be relied on (Jigyasu 2014). Moreover, much of the community land which was formerly managed as a commons has been lost. Local community-based maintenance of land resources and waterways (e.g. cleaning of springs and wells), often linked to religious values, is no longer in place or has been replaced by government policies that have turned out to be less efficient (Ellingsen 2010).

Different respondents expressed concern that "people are forgetting the traditional knowledge", "there is no more community devotion as in my grandparents' generation" and "society is highly globalized and does not accept the tradition; they want 'modern' things". The loss of community values and community-managed activities is taking place together with a deteriorating environment, impacting on ecosystem services which could support people in case of a disaster – hence resulting in a vicious cycle of ecosystem degradation and increasing disaster risk. Deteriorating cultural and social values attached to ecosystems such as rivers and streams also reduce opportunities for recreation and spiritual expression for locals and tourists, which are also a major source of income for Nepal.

One way forward for Nepal could be the (re)appreciation of community-based values and traditional cultural ties attached to different ecosystem services. Community ties still play an important role in risk reduction strategies of urban dwellers, particularly in countries where public services are not reliable or fully developed. As a part of traditional knowledge systems, such ties evolved over time and are embedded in the particular natural and cultural environment. Related skills, crafts and cultural practices can provide mutual support and can be part of coping mechanisms for community members after a disaster has struck (Jigyasu 2014).

In reality, people are putting more importance to everyday survival issues than to potential extreme events which may only occur once in a few generations (International Federation of Red Cross and Red Crescent Societies 2014). As a consequence, daily life and also knowledge systems may be adapted to 'normal' hazards, such as annual rainfalls and floods in Nepal, but not to major events such as the recent earthquakes, although such extreme events are largely anticipated. This situation is particularly true for the most vulnerable societal groups that focus on daily survival. In a study on the perception of climate change impacts carried out in nine riverine communities within Kathmandu Valley, Nehren et al. (2013) found that people gave higher importance to everyday risks, such as health-related problems, than flooding or predictions of future climate change impacts (c.f. Fig. 15.6).

The 2014 World Disasters Report identified that people generally give a very low priority to serious hazards that are considered within the mandate of public policy and authorities. "They apply much higher significance to problems of everyday life and issues that they have to confront for normal survival, most of which are linked to their livelihoods" (International Federation of Red Cross and Red Crescent Societies 2014). Therefore, one main challenge is to find long-lasting, socially compatible, and environmentally sustainable solutions to everyday development challenges that are generating local risks. Such strategies can also help to lessen disaster impacts, for instance through proper land use planning that pays attention to environmental and social concerns (Pelling 2012). The challenge is to break down the cultural barriers between the 'natural' and the 'urban' (Trzyna 2014), which have only emerged in Kathmandu's recent history.

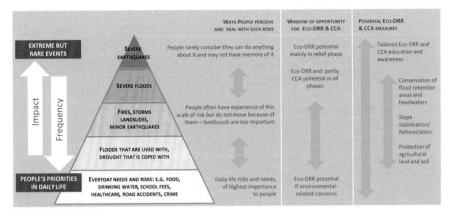

Fig. 15.6 Risk hierarchies in urban Kathmandu (Source: Cannon T. in: World Disasters Report 2014, pp. 68; modified)

15.3 Ways Forward

Correlations between Kathmandu Valley's evolving land use, land tenure, degrading environmental conditions, local culture and living conditions are complex. Global climate change is likely to increase vulnerabilities (Ellingsen 2010; British Red Cross et al. 2014), urgently requiring an integration of climate change, disaster risk reduction, and sustainability, as stressed by Kelman et al. (2015). Successful risk reduction strategies therefore should focus on concerted action, such as the recent Post Disaster Need Assessment (National Planning Commission 2015) when stating: "Specific strategies will be needed to address the complexities involved in the recovery of urban environments", asking for recovery activities that are environmentally sustainable and appropriate to the region.

Fortunately, the integration of DRR has been recognized by the National Government of Nepal as part of its National Development Planning, National Policy on Adaptation to Climate Change, and National Strategy for Disaster Risk Management (International Federation of Red Cross and Red Crescent Societies 2011). However, the effective implementation of disaster management legislation and of other related policies mentioned, and maximizing their inter-linkages, are lacking. The GFDRR report on Disaster Risk Management in South Asia (2012) cites the lack of legislation enforcement in Nepal. Existing capacities and institutions are not working to their full potential, and policies and legislation are implemented on an *ad-hoc* basis rather than through a systematic approach. Short-term thinking and planning potentially hinders the adoption of long-term approaches, including ecosystem-based management approaches. One major constraint is resources: "Plagued by an inadequate budget, the municipality has to set priorities for its engagements (Ellingsen 2010: 119)." As a consequence, state activities or responsibilities are given to non-government actors which have the necessary budget – a shift from government to a multi-stakeholder governance in DRR (Jones et al. 2014) and CCA. Jones et al. (2014) talked about an ongoing proliferation of NGOs engaging in DRR around Nepal already before the recent earthquakes. It remains to be seen how far the number and composition of actors will change.

In Nepal, any disaster risk reduction strategy or urban planning project is heavily influenced by development assistance coming from international donors. The country is one of the major global recipients of development assistance (ODA). In 2012, foreign aid represented 26 % of the national budget, with more than 40 donors (Government of Nepal et al. 2013). Regarded as both a blessing and a curse, international donor agencies support national and urban development; however, the down-side of international development assistance is now widely acknowledged. For instance, long-term, strategic planning needed especially in such dynamic urban areas as Kathmandu is constrained by the limited timeframes and the lack of local and cultural sensitivity of projects being supported by donor countries. Dependency on foreign aid and changing priorities of donor organizations over time are major challenges (cf. Sovacool et al. 2011, in their study on social and technical barriers to hydroelectric power plants in Nepal). Such dependency is likely to further grow in the aftermath of the 2015 earthquakes.

Various respondents expressed concern with the assistance being provided by international organizations in Nepal: "You foreigners don't know about Nepalese culture". Another person interviewed asked for a "Nepali way" considering local customs: "The urban planning in Kathmandu should be made by Nepali planners, not by borrowing Western things". Another respondent worried that "If foreign people come to Nepal it is difficult for them to internalize the local things, therefore their advices sometimes fail as they are not aware of the local customs."

Nonetheless, important opportunities exist to enhance and improve ODA. For example, the NRRC Flagship 5 program discussed earlier has prioritized integrated, community-based disaster risk reduction as well as policy and institutional support for disaster risk reduction (The Nepal Risk Reduction Consortium 2011). These priority interventions are crucial for the sustainable urban development of Kathmandu, and there is opportunity to include aspects of ecosystem-based disaster risk reduction and climate change adaptation (cf. Fig. 15.4). In particular, Flagship 4 on "Integrated Community Based DRR" to reduce community vulnerability and Flagship 5 on "Policy/Institutional Support for DRM" focusing on the integration of DRM in plans, policies and programs at national, district and local levels could be potential entry points. Under Flagship 4, a coordination mechanism for urban DRR is underway, and among the expected outcomes are mechanisms and tools for community-based disaster risk reduction/management, based on nine minimum characteristics of disaster resilient communities. Eco-DRR/CCA could and should be considered under characteristic no. 8 of disaster resilient communities which call for local level risk/vulnerability reduction measures. Flagship 5 comprises the risk-sensitive land use plan for Kathmandu and its surroundings. Eco-DRR/CCA could also be integrated to support risk-sensitive land use planning and explicitly mainstream DRR and CCA in the development planning process, a key priority under Flagship 5 (The Nepal Risk Reduction Consortium 2011; Taylor et al. 2013; Jones et al. 2014).

In Kathmandu itself, the most recent undertaking is the Kathmandu Metropolitan City Risk Sensitive Land Use Plan, which is based on an integrated sectoral approach and on cooperation between urban, national and international actors (Earthquakes and Megacities Initiative 2010). In this Plan, environmental deterioration in its various facets is mentioned as the main cause of urban challenges. The Plan identifies the available environmental resources as crucial for achieving safer environments. It focuses in particular on flood-prone areas, cropland crucial for food supply, as well as urban parks and open spaces due to their importance for recreation and shelter. Proposed interventions for such areas would allow for ecosystem-based measures, but specific proposals listed still focus on infrastructure projects such as drainage for flood-prone riverbanks or legal aspects, such as review of by-laws, increased penalties or land-use policies.

The challenge is to introduce up-to-date approaches that take into account local culture and knowledge, which also should not be 'over-romanticised' (Mercer et al. 2012b) but taken seriously. In reference to introducing urban agriculture, one respondent criticized the concept, claiming that such solutions were "low-tech" and do not match the needs of a modern city: "Do British want to have paddy fields around the Buckingham Palace?". Traditional knowledge systems – particularly (land) management systems – have evolved over time and can play a significant role in disaster risk reduction and community resilience (Jigyasu 2014). Healthy ecosystems and their services have the potential to play a vital role in reducing climate and disaster risk and providing opportunities for sustainable urban development (Mercer et al. 2012b). Mainstreaming the ecosystem perspective and approach in cities still needs time (Pickett et al. 2008), recognizing that man-made components are essential parts of the urban ecosystem (Tanner et al. 2014).

Recognition of ecosystem services is slowly appearing on the policy agenda. For example, a Payment for Ecosystems (PES) feasibility study was carried out in the Shivapuri-Nagarjun National Park, a major water source for Kathmandu Valley. Mature hardwood forests are now confined to protected parks and sacred areas such as Nagarjun (Raniban), Gokarna and Shivapuri watersheds and the Wildlife Conservation forest, and Bajrabarahi forest (ICIMOD et al. 2007). In a study on protected areas in India and Nepal, Karanth and Nepal (2011) found that local residents perceived ecosystem-related benefits such as access to fuel wood, fodder and economic benefits from tourism. At the same time, they found a local appreciation for conservation that ecosystem-based measures could be based upon – not only in protected areas, but potentially in urban areas as well as in surrounding urban fringes. Promising initiatives are, for instance the 2015 introduction of a 'one house, two trees' policy in Lahan Municipality in southern Nepal, requiring two trees to be planted in order to formally complete a new building construction (Direction Kathmandu 2015) and the 'Green Homes: Promoting Sustainable Housing in Nepal' project. This undertaking promotes eco-friendly homes with greenery to solve problems of water scarcity, waste management and environmental pollution in Lalitpur Municipality in Kathmandu Valley (UN Habitat Nepal 2014). Likewise, in the aftermath of the recent earthquakes, greener approaches in rebuilding are emphasized, to reduce pressure on natural resources and carbon emissions

(UN Habitat Nepal 2015). At the same time, environmental education is gaining importance, starting from primary schools to university-level education. As one respondent stated, "It is the right time to think about 'eco-conservation'".

It remains to be seen how disaster reconstruction after the recent 2015 earthquakes will take place. The latest assessments speak volumes about the extent of reconstruction required in the Valley: More than 65,000 buildings are partially damaged, and more than 68,000 fully damaged (ICIMOD 2015). Water supply has been affected seriously; access to sanitation and hygiene is interrupted for a large share of the population. Monsoon-induced floods are expected to further damage water systems (UN OCHA 2015b, 2015c). In particular, informal camps located in low-lying areas of Kathmandu Valley will be prone to flooding; hydropower plants, highways and agricultural lands are also likely to be affected by landslides (Nepal Earthquake Assessment Unit 2015).

A sustainable urban reconstruction process will require both a focus on the natural as well as the built environment. Natural resources constitute an integral part of human livelihoods (Joint UNEP/OCHA Environment Unit 2012; International Federation of Red Cross and Red Crescent Societies 2014). Thus, efforts to reconstruct a built environment that reduces disaster risk but at the expense of natural resources or cultural considerations, such as social norms and sensitivities, will not be considered sustainable over the long-term (International Federation of Red Cross and Red Crescent Societies 2014; Rutherford 2015). There is a window of opportunity to mainstream ecosystem-based approaches into the reconstruction debate, as both environmental and urban planners could be brought together.

15.4 Outlook and Conclusions

In Kathmandu Valley, one witnesses, on the one hand, growing awareness of the potential of ecosystem services to sustain or improve urban livelihoods and resilience and, on the other hand, ongoing ecosystem deterioration. This reality becomes an allegory for ongoing development processes in Nepal: promising small-scale or community-based approaches that seek to preserve or restore ecosystem services for a growing urban population increasingly at risk, while at the same time formal urban planning remains a challenge, particularly when it comes to environmental protection.

Kathmandu is at a crossroads between becoming a modern, international city and the challenging reality of being unable to provide basic infrastructure and services. On the way to 'modernity', many of the traditional practices, habits and cultural roots have been eroded, including community-based maintenance of public areas, water supply and agricultural systems that supported disaster preparedness and mitigation of past generations. Ecosystem services and their sustainable use were a major source of people's livelihoods, which are now eroding and changing fundamentally for the worse; blaming the inhabitants comes up short. As one respondent put it: "I do not blame the poor and illiterate people, but the rich people and the authorities for what is going wrong."

The potential for integrating Eco-DRR solutions in different policy sectors do exist, but have to be mainstreamed into ongoing planning and governance schemes. The Kathmandu Valley case highlights very well the comparably higher challenges in urban contexts than in rural areas. At the same time, it shows the potential for mainstreaming ecosystem-based solutions, for instance through improved land-use planning, reforestation and slope stabilization and conservation of flood retention areas, into an improved urban governance scheme.

Urban areas in developing countries are facing comparably more significant problems than cities in developed countries, such as budget constraints, highly dynamic formal and informal development and weak governance, which hinder adequate planning. In a country such as Nepal that relies heavily on donor funding, international organizations and international financial institutions are actors involved in setting the Nepalese development agenda. Only a few countries are comparable to Nepal with respect to ODA dependency. Unfortunately, the short- and medium-term timeframe associated with development assistance are not conducive to fostering ecosystem-based approaches for disaster risk reduction, which often require longer time horizons to yield tangible protective functions (although potentially providing already short-term benefits at the community level with respect to livelihoods). Nonetheless, there are opportunities. The NRRC, in particular, through the Flagship 5 program, has the mandate to establish long-term strategic planning and could be an entry point for mainstreaming ecosystem-based DRR and CCA into urban risk management.

On a larger geographic scale, closer cooperation with the neighboring countries of China and particularly India may offer opportunities for transboundary watershed initiatives, potentially under the umbrella of the 2014 hydropower investment deal (CFE-DMHA 2015). Ecosystem-based measures, for instance to reduce flood risk, implemented in Nepal could benefit India in the upper courses of Ganga River catchment. In addition, exchanging experiences with other Himalayan countries prone to comparable disasters such as Pakistan (as e.g. analysed by Sudmeier-Rieux et al. 2011, in the case of the 2005 earthquake) or Bhutan (Sovacool and Meenawat 2011; Sovacool et al. 2012) should be encouraged.

Like Kathmandu, many cities in developing countries are facing comparable challenges to overcome risks, exacerbated by environmental degradation. Most urban environmental challenges, such as urban heat waves, flooding, landslides or water shortages, are expected to worsen as a result of global climate change. Ecosystem-based measures, such as maintenance or recovery of flood retention areas, the reforestation of hills for stabilization and reduction of river sedimentation, and the selective establishment or maintenance of urban green areas for recreation, local climate regulation and emergency evacuation areas, potentially reduce disaster and climate change-related risks. While none of these measures are very innovative or new, the major challenge is that they require a long-term approach and concerted action between government, private sector, civil society and urban dwellers. There is need for improved urban governance mechanisms including a broad range of stakeholders, from communities to international donors, and linking natural and social processes. A long-term sustainable vision is centrally-based on the consideration of community-based action and their ties to natural assets.

Appendix: Institutions Interviewed for the Study

- Kathmandu Valley Development Authority (KVDA)
- Kathmandu Valley Preservation Trust (KVPT)
- Ministry of Culture, Tourism and Civil Aviation, Government of Nepal
- Ministry of Urban Development, Government of Nepal
- National Society for Earthquake Technology, Nepal (NSET)
- Center for Disaster Studies, Tribhuvan University, Kathmandu
- Department of Architecture and Urban Planning, Tribhuvan University, Kathmandu
- Central Department of Geography, Tribhuvan University, Kathmandu

References

Abbate C (2010) A Vicious Circle. In: Plunz R, Sutto MP (eds) Urban climate change crossroads. Ashgate, Farnham, pp 129–134

Beck H (2012) Understanding the impact of urban green space on health and wellbeing. In: Atkinson S, Fuller S, Painter J (eds) Wellbeing and place. Ashgate, Farnham, pp 35–52

Berkes F, Kofinas GP, Chapin SF III (2009) Conservation, community, and livelihoods: sustaining, renewing, and adapting cultural connections to the land. In: Chapin SF III, Kofinas GP, Folke C (eds) Principles of ecosystem stewardship: resilience-based natural resource management in a changing world. Springer, New York, pp 129–148

Bhattarai K, Conway D (2010) Urban vulnerabilities in the Kathmandu valley, Nepal: visualizations of human/hazard interactions. J Geogr Inf Syst 2(2):63–84. doi:10.4236/jgis.2010.22012

Bigio AG (2003) Cities and climate change. In: Kreimer A, Arnold M, Carlin A (eds) Building safer cities: the future of disaster risk. The World Bank, Washington, DC, pp 91–99

Briceño S (2015) Looking back and beyond Sendai: 25 years of international policy experience on disaster risk reduction. Int J Disas Risk Sci 6:1–7. doi:10.1007/s13753-015-0040-y

British Red Cross, Nepal Red Cross Society, Groupe Urgence – Réhabilitation – Développement (2014) Urban preparedness – lessons from the Kathmandu Valley. British Red Cross, London

Bulkeley H (2010) Cities and governance. In: Plunz R, Sutto MP (eds) Urban climate change crossroads. Ashgate, Farnham, pp 29–38

CFE-DMHA (2015) Nepal disaster management reference handbook, Disaster Management Reference Handbook Series. Center for Excellence in Disaster Management & Humanitarian Assistance, Hickam

Dahiya B (2012) Cities in Asia, 2012: demographics, economics, poverty, environment and governance. Cities 29(2):S44–S61

DesInventar (2014) Nepal central region – composition of disasters. UNISDR. http://www.desinventar.net/DesInventar/profiletab.jsp. Accessed 30 Nov 2014

Direction Kathmandu (2015) Lahan municipality to enforce 'one house-two tree' policy. http://directionkathmandu.com/information/lahan-municipality-to-enforce-one-house-two-tree-policy/. Accessed 13 Sep 2015

Dizdaroglu D, Yigitcanlar T, Dawes L (2012) A micro-level indexing model for assessing urban ecosystem sustainability. Smart Sust Built Environ 1(3):291–315

Earthquakes and Megacities Initiative (2010) Risk-sensitive land use plan: Kathmandu Metropolitan City, Nepal. German Federal Foreign Affairs Office and EMI, Quezon City

Earthquakes without Frontiers (2015) Nepal: UPDATED (28 May) landslide inventory following 25 April Nepal earthquake. Earthquakes without Frontiers. http://ewf.nerc.ac.uk/2015/05/28/nepal-updated-28-may-landslide-inventory-following-25-april-nepal-earthquake/. Accessed 05 June 2015

Ellingsen W (2010) Ethnic appropriation of the city. The territoriality of culture in Kathmandu. LAP Lambert, Saarbrücken

Faris F, Wang F (2014) Stochastic analysis of rainfall effect on earthquake induced shallow landslide of Tandikat, West Sumatra, Indonesia. Geoenviron Disasters 1(1):1–13. doi:10.1186/s40677-014-0012-3

Fra Paleo U (2013) A functional risk society? Progressing from management to governance while learning from disasters. In: SC and UNESCO (ed) Changing global environments, World social science report 2013. OECD Publishing/UNESCO Publishing, Paris, pp 434–438

German Federal Foreign Affairs Office, EMI (2010) Risk-sensitive land use plan: Kathmandu Metropolitan City, Nepal. German Federal Foreign Affairs Office and EMI, Quezon City

GFDRR (2012) Disaster risk management in South Asia: a regional overview. The World Bank, Washington, DC

GFDRR (2014) Nepal – country program update May 2014. GFDRR, Washington, DC

Gibson L (2012) 'We are the river': place, wellbeing and aboriginal identity. In: Atkinson S, Fuller S, Painter J (eds) Wellbeing and place. Ashgate, Farnham, pp 201–215

Ginnetti J, Lavell C (2015) The risk of disaster-induced displacement in South-Asia – technical paper. Internal Displacement Monitoring Centre, Geneva

GoN/NTNC (2009) Bagmati action plan (2009–2014). Government of Nepal & National Trust for Nature Conservation, Kathmandu

Government of Nepal; Department of Archaeology (2007) Kathmandu World heritage site: integrated management framework. Government of Nepal, Ministry of Culture, Tourism and Civil Aviation, Department of Archaeology in close collaboration with the World Heritage Centre and UNESCO-Kathmandu Office, Kathmandu

Government of Nepal, Ministry of Finance, International Economic Cooperation Coordination Division (2013) Foreign aid in Nepal (FY 2011–12). Government of Nepal, Ministry of Finance, International Economic Cooperation Coordination Division, Singhadurbar, Kathmandu

Grove JM (2009) Cities: managing densely settled social-ecological systems. In: Chapin SF III, Kofinas GP, Folke C (eds) Principles of ecosystem Stewardship: resilience-based natural resource management in a changing world. Springer, New York, pp 281–294

Gutschow N, Kreutzmann H (2012) Handlung schlägt Plan: Stadtentwicklung im Kathmandu-Tal (Nepal). Geographische Rundschau 64/4 (April 2012):42–49

Historical Stone Spouts and Source Conservation Association (HSSCA) (2007) National convention on stone spouts. Kathmandu

ICIMOD (2015) Nepal earthquake 2015. http://www.icimod.org/v2/cms2/_files/images/e92e3b0202d11e51262a6e2cb1ed6f2d.jpg. Accessed 09 June 2015

ICIMOD, MoEST, UNEP (2007) Kathmandu valley environment outlook. International Centre for Integrated Mountain Development (ICIMOD), Kathmandu

International Federation of Red Cross, Red Crescent Societies (2011) Analysis of legislation related to disaster risk reduction in Nepal. IFRC, Geneva

International Federation of Red Cross and Red Crescent Societies (2014) World disasters report 2014 – focus on culture and risk. International Federation of Red Cross and Red Crescent Societies, Geneva

IPCC (2012) Managing the risks of extreme events and disasters to advance climate change adaptation, A special report of working groups I and II of the intergovernmental panel on climate change. Cambridge University Press, Cambridge/New York

Jha PK, Shrestha KK (2013) Climate change and urban water supply: adaptive capacity of local government in Kathmandu City. Nepal J Forest Livelihood 11(1):62–81

Jigyasu R (2014) Fostering resilience: towards reducing disaster risks to World Heritage. World Heritage 74:4–13

Joint UNEP/OCHA Environment Unit (2012) Keeping up with megatrends – the implications of climate change and urbanization for environmental emergency preparedness and response. Joint UNEP/OCHA Environment Unit, Geneva

Jones S, Oven KJ, Manyena B, Aryal K (2014) Governance struggles and policy processes in disaster risk reduction: a case study from Nepal. Geoforum 57:78–90

Karanth KK, Nepal SK (2011) Local residents perception of benefits and losses from protected areas in India and Nepal. Environ Manag 49:372–386

Kelman I, Gaillard JC, Mercer J (2015) Climate change's role in disaster risk reduction's future: beyond vulnerability and resilience. Int J Disas Risk Sci 6:1–7. doi:10.1007/s13753-015-0038-5

Knight K (2013) Imagining a major quake in Kathmandu. United Nations/OCHA. http://www.irinnews.org/report/97925/imagining-a-major-quake-in-kathmandu. Accessed 25 Apr 2015

Krasny ME, Russ A, Tidball KG, Elmqvist T (2014) Civic ecology practices: participatory approaches to generating and measuring ecosystem services in cities. Ecosyst Serv 7:177–186

Lange W, Cavalcante L, Dünow L, Medeiros R, Pirzer C, Schelchen A, Valverde Y (2013) HumaNatureza^2 = Proteção Mútua. Percepção de riscos e adaptação à mudança climática baseada nos ecossistemas na Mata Atlântica, Brasil. Schriftenreihe des Seminars für Ländliche Entwicklung. SLE, Berlin

Marcotullio PJ (2006) Comparison of urban environmental transitions in North America and Asia Pacific. In: Tamagawa H (ed) Sustainable cities – Japanese perspectives on physical and social structures. United Nations University Press, Tokio, pp 7–49

Mercer J, Gaillard JC, Crowley K, Shannon R, Alexander B, Day S, Becker J (2012a) Culture and disaster risk reduction: lessons and opportunities. Environ Haz 11(2):74–95. doi:10.1080/17477891.2011.609876

Mercer J, Kelman I, Alfthan B, Kurvits T (2012b) Ecosystem-based adaptation to climate change in Caribbean small island developing states: integrating local and external knowledge. Sustainability 4:1908–1932

Ministry of Environment (2010) National Adaptation Programme of Action (NAPA) to climate change. Government of Nepal, Kathmandu

Ministry of Local Development/GoN, Disaster Risk Reduction at the National Level in Nepal/UNDP (2011) A needs and a capacity assessment of fire preparedness in the municipalities of Nepal. Ministry of Local Development/GoN/Disaster Risk Reduction at the National Level in Nepal/UNDP, Kathmandu

Muzzini E, Aparicio G (2013) Urban growth and spatial transition in Nepal: an initial assessment. Directions in development. World Bank, Washington, DC. doi:10.1596/978-0-8213-9659-9

National Planning Commission (ed) (2015) Nepal earthquake 2015 post disaster needs assessment, vol. A: key findings. Government of Nepal, Kathmandu

Nehren U, Subedi J, Yanakieva I, Sandholz S, Pokharel J, Chandra Lal A, Pradhan-Salike I, Marfai MA, Sri Hadmoko D, Straub G (2013) Community perception on climate change and climate-related disaster preparedness in Kathmandu Valley, Nepal. J Nat Res Dev 04:35–57

Nepal Earthquake Assessment Unit (2015) Pre-monsoon overview Nepal earthquake. https://www.humanitarianresponse.info/sites/www.humanitarianresponse.info/files/assessments/150526_pre-monsoon_brief_final.pdf. Accessed 05 June 2015

Peduzzi P (2010) Landslides and vegetation cover in the 2005 North Pakistan earthquake: a GIS and statistical quantitative approach. Nat Hazards Earth Syst Sci 10(4):623–640. doi:10.5194/nhess-10-623-2010

Pelling M (2003) The vulnerability of cities. Natural disasters and social resilience. Earthscan, London

Pelling M (2011) Adaptation to climate change: from resilience to transformation. Routledge, London/New York

Pelling M (2012) Hazards, risk and urbanisation. In: Wisner B, Gaillard JC, Kelman I (eds) The Routledge handbook of hazards and disaster risk reduction. Routledge, London/New York, pp 145–155

Pickett STA, Cadenasso ML, Grove JM et al (2008) Beyond urban legends: an emerging framework of urban ecology, as Illustrated by the Baltimore Ecosystem Study. Bioscience 58(2):139–150. doi:10.1641/B580208

Quarantelli EL (2003) Urban Vulnerability to Disasters in Developing Countries: Managing Risks. In: Kreimer A, Arnold M, Carlin A (eds) Building safer cities: the future of disaster risk. The World Bank, Washington, DC, pp 211–231

Regmi BR, Pandit A, Pradhan B et al (2009) Climate change and health country report – Nepal. LI-BIRD, Pokhara

Rutherford N (2015) How to turn awareness into action on disaster preparedness. http://www.euractiv.com/sections/sustainable-dev/how-turn-awareness-action-disaster-preparedness-314413. Accessed 11 May 2015

Secretariat of the Convention on Biological Diversity (2012) Cities and biodiversity outlook. Secretariat of the Convention on Biological Diversity, Montreal

Shrestha AB, Aryal R (2011) Climate change in Nepal and its impact on Himalayan glaciers. Reg Environ Chang 11(1):65–77

Simone A (2010) City life from Jakarta to Dakar. Movements at the crossroads. Routledge, New York

Sovacool BK, Meenawat H (2011) Improving adaptive capacity and resilience in Bhutan. Mitig Adapt Strateg Glob Chang 16(5):515–533. doi:10.1007/s11027-010-9277-3

Sovacool BK, Dhakal S, Gippner O, Bambawale MJ (2011) Halting hydro: a review of the socio-technical barriers to hydroelectric power plants in Nepal. Energy 36:3468–3476. doi:10.1016/j.energy.2011.03.051

Sovacool BK, D'Agostino AL, Meenawat H, Rawlani A (2012) Expert views of climate change adaptation in least developed Asia. J Environ Manag 97:78–88. doi:10.1016/j.jenvman.2011.11.005

Sudmeier-Rieux K, Ash N (2009) Environmental guidance note for disaster risk reduction: healthy ecosystems for human security, Revised Edition. IUCN, Gland

Sudmeier-Rieux K, Jaboyedoff M, Breguet A, Dubois J (2011) The 2005 Pakistan earthquake revisited: methods for integrated landslide assessment. Mt Res Dev 31(2):112–121

Surendra KC, Khanal SK, Shrestha P, Lamsal B (2011) Current status of renewable energy in Nepal: opportunities and challenges. Renew Sust Energ Rev 15:4107–4117. doi:10.1016/j.rser.2011.07.022

Taboroff J (2003) Natural Disasters and Urban Cultural Heritage: A Reassessment. In: Kreimer A, Arnold M, Carlin A (eds) Building safer cities: the future of disaster risk. The World Bank, Washington, DC, pp 233–240

Tanner CJ, Adler FR, Grimm NB et al (2014) Urban ecology: advancing science and society. Front Ecol Environ 12(10):574–581. doi:10.1890/140019

Taylor G, Vatsa K, Gurung M, Couture E (2013) Review of the Nepal Risk Reduction Consortium (NRRC). NRCC, Kathmandu

Thapa RB, Murayama Y (2012) Scenario based urban growth allocation in Kathmandu Valley, Nepal. Landsc Urban Plan 105:140–148

The Nepal Risk Reduction Consortium (2011) Disaster risk reduction in Nepal. Flagship programmes. NRCC, Kathmandu

Titz A (2012) Naturgefahren als Entwicklungshemmnis. Das Beispiel Nepal-Himalaya. Praxis Geographie 42(9):34–38

Trzyna T (2014) Urban protected areas: profiles and best practice guidelines. Best practice protected area guidelines series No. 22. IUCN, Gland

UN-HABITAT (2008) Water movements in Patan with reference to traditional stone spouts. UN-HABITAT Water for Asian Cities Programme Nepal, Kathmandu

UN-HABITAT (2010a) Nepal. Urban housing sector profile. UN-HABITAT, Nairobi

UN-HABITAT (2010b) The State of Asian Cities 2010/11. UN-HABITAT, UN-HABITAT Foundation, UN-ESCAP, Fukuoka

UN-HABITAT (2015) Kathmandu valley, Nepal: climate change vulnerability assessment. United Nations Human Settlements Programme (UN-HABITAT), Nairobi

UN Habitat Nepal (2014) Green homes Nepal – Dharan sub-metropolis to promote green homes. http://unhabitat.org.np/media-center/news/dharan-sub-metropolis-to-promote-green-homes/. Accessed 13 Sep 2015

UN Habitat Nepal (2015) Green homes Nepal – 'Greening the Re-building'. http://unhabitat.org.np/media-center/news/greening-the-re-building/. Accessed 13 Sep 2015

United Nations (2014) World urbanization prospects – highlights. United Nations, New York

United Nations, Department of Economic and Social Affairs, Population Division (2014) World urbanization prospects: The 2014 revision, custom data acquired via website

UN OCHA (2013) Nepal: preparing for an earthquake in the Kathmandu Valley. http://www.unocha.org/top-stories/all-stories/nepal-preparing-earthquake-kathmandu-valley. Accessed 30 Nov 2014

UN OCHA (2015a) Humanitarian bulletin. Nepal earthquake. UN OCHA. http://reliefweb.int/sites/reliefweb.int/files/resources/OCHANepalEarthquakeHumanitarianBulletinNo3%28Aug2015%29_Final.pdf. Accessed 12 Sep 2015

UN OCHA (2015b) Nepal flash appeal revision: Nepal earthquake April – September 2015. UN Office for the Coordination of Humanitarian Affairs, UN Resident and Humanitarian Coordinator for Nepal, Kathmandu

UN OCHA (2015c) Nepal: earthquake 2015 – situation report No. 20 (as of 3 June 2015). Office for the Coordination of Humanitarian Affairs in collaboration with the Office of the Resident and Humanitarian Coordinator and humanitarian partners. http://reliefweb.int/sites/reliefweb.int/files/resources/OCHANepalEarthquakeSituationReportNo.20%283June2015%29_Final.pdf

Witze A (2015) Mappers rush to pinpoint landslide risk in Nepal. Nature 521:133–134. doi:10.1038/521133a

World Bank (2014) Doing business 2015: going beyond efficiency. World Bank, Washington, DC. doi:10.1596/978-1-4648-0351-2

World Disasters Report 2014 (2014) World disasters report 2014 – focus on culture and risk. International Federation of Red Cross and Red Crescent Societies, Geneva. doi:ISBN 978-92-9139-214-8

Yua B, Wub Y, Chuc S (2014) Preliminary study of the effect of earthquakes on the rainfall threshold of debris flows. Eng Geol 182(Part B):130–135. doi:10.1016/j.enggeo.2014.04.007

Chapter 16
Towards Anticipatory Management of Peat Fires to Enhance Local Resilience and Reduce Natural Capital Depletion

Johan Kieft, Talia Smith, Shiv Someshwar, and Rizaldi Boer

Abstract Greenhouse gas emissions from peat lands are key sources of overall emissions in Indonesia. These emissions are mainly caused by fires and to a lesser extent decomposition of degraded peat lands which have been cleared for either food crop or palm productions. Land clearing has taken place since the colonial times; however, it had accelerated dramatically since the mid-90s, fuelled by palm oil expansion and poorly planned efforts to open peat land for food production.

Fire activity is driven by clearing peat forest lands. Risk of fires increases significantly during drier than normal years, often linked to El Nino phenomena. Once fires are ignited in peat areas, they tend to be submerged, making them difficult to extinguish. This prompts the need for an anticipatory approach to fire management. Such an approach would enact anticipatory risk reduction actions 1–3 months ahead of an anticipated fire outbreak. These actions would be integrated into existing standard operating procedures for fire prevention, while at the same time mainstreaming fire risk reduction into spatial and development planning to address long-term fire vulnerability.

The collaboration between the Earth Institute at Columbia University and the *Institut Pertanian Bogor's* (IPB) Centre for Climate Risk and Opportunity Management in Southeast Asia Pacific (CCROM SEAP), with support from the National REDD+ Agency and facilitated by the United Nations Office for REDD+ Coordination in Indonesia (UNORCID), has resulted in the development of a seasonal fire early warning system, known as the Fire Risk System (FRS), for managing fires at the provincial and district level, with particular focus on Central Kalimantan and

J. Kieft (✉) • T. Smith
UN Office for REDD+ Coordination in Indonesia (UNORCID), Jakarta, Indonesia
e-mail: johan.kieft@unorcid.org; talia.sara.smith@gmail.com

S. Someshwar
Center for Globalization and Sustainable Development, The Earth Institute,
Columbia University, New York, NY, USA

R. Boer
Director of Climate Risk and Opportunity Management in Southeast Asia Pacific,
Institut Pertanian Bogor, Bogor, Indonesia

© Springer International Publishing Switzerland 2016
F.G. Renaud et al. (eds.), *Ecosystem-Based Disaster Risk Reduction and Adaptation in Practice*, Advances in Natural and Technological Hazards Research 42,
DOI 10.1007/978-3-319-43633-3_16

Riau provinces. The system is designed to enhance capacity of national, provincial and local stakeholders to prevent fires and addressing underlying fire vulnerably by integrating anticipatory actions into planning processes.

Keywords Peat • Indonesia • Fire • Anticipatory early warning

16.1 Introduction

Forest and peat land fires are, in terms of financial damage and health impact, one of the most damaging humanitarian disasters. Recent assessments of these fires in Indonesia have identified that peat land fires in particular are the key source of haze, which results in detrimental health and economic impacts (Betha et al. 2013; Forsyth 2014). Alarmingly high levels of air pollution have been measured both in terms of small particle density as well as heavy metals (Goldammer 1999; Betha et al. 2013). The frequency of these fires is on the rise; and in contrast to the past, fires are occurring in years with normal rainfall (Gaveau et al. 2014). However, while fires affect millions in Indonesia and neighbouring countries, recent research has shown that the majority of these fires are located in a limited number of districts in Sumatra and Kalimantan, and are linked to peat land disturbance (i.e. drainage) and increased access to peat land areas.

Such fires in Indonesia severely undermine the sustainable use of peat lands and their ability to deliver key environmental services. Peat lands act as a sponge, absorbing water during the wet season and gradually releasing water during the dry season. These functions provide critical ecosystem services, such as: stabilizing critical coastal and lowland ecosystems, reducing land subsidence, supporting fresh water access to coastal cities located in lowlands, and allowing for navigation of key river systems. In addition, peat lands serve as an important carbon sink, storing over 550 Gt of carbon worldwide (Wetlands International 2014; Jaenicke et al. 2008). When peat is burned, the result in terms of carbon emissions is staggering: currently, peat fires produce an estimated 40–45 % of total greenhouse gas emissions (GHG) in Indonesia (Hooijer et al. 2014). These figures underline that Indonesian peat lands contain significant natural capital that is essential for the country's sustainable development.

Tropical peat lands are characterized by a layer of organic matter stored under water logged circumstances which can reach up to 20 meters. In Indonesia, peat is mainly fibric in nature and consists of forest biomass which as consequence easily ignites. Peat lands store large amounts of organic matter, estimated at around 55 ± 10 Gt of carbon by Jaenicke et al. (2008). When peat lands are drained, the layer above the water table becomes fuel for fires during periods of low rainfall. Thus, peat land characteristics make traditional forest fire management approaches, which are generally based on reducing the fuel load and avoiding fire use, inappropriate for peat lands fire management. In contrast, anticipatory fire management of Indonesia's peat lands requires that such lands remain under a sustainable forest management system where drainage efforts are avoided in order to keep fuel wet. As a result of reduced drainage, maintaining and managing a sufficiently high water table level based on peat land depths will have a significant impact on reducing fire incidence, accumulating natural capital and increasing local resilience of communities dependent on peat land ecosystems (Someshwar et al. 2010; Hooijer et al. 2012; Moore et al. 2013a; Hooijer et al. 2014).

Recent fire episodes, such as in Riau, Sumatra, in February–March 2014, have shown that despite significant gains in hotspot monitoring and fire suppression, the current fire management system in Indonesia cannot reduce the risk of fire due to its inherent reactionary nature, meaning that actions are only taken after fires have been ignited. The recent fire events have pushed the Government to take action through a combination of new policy initiatives. Of these, the most significant one is the development of a Standard Operating Procedure (SOP), which seeks to institutionalize prevention actions for fire risk reduction. Adopting an anticipatory-based approach to fire management is therefore being advocated, as utilized in the Central Kalimantan Province by local governments through support provided by CARE (an international, humanitarian non-governmental organization (NGO)), Bogor Agricultural University and Earth Institute Columbia University (Sommershar et al. 2010).

Through these initiatives, it is expected that fire can be more effectively suppressed and controlled, particularly if such an anticipatory fire management approach is better mainstreamed in district development planning and budgeting. Other measures should include the adoption of a climate-based, seasonal fire early warning system (in addition to a fire danger rating system), ensuring that fire prevention efforts are better guided through a centralized fire prevention system and better target vulnerable areas, such as degraded peat lands, land under small-scale and medium-sized palm oil plantations. Additionally, prevention efforts must be aligned with a more centralized hotspot management and fire suppression system.

In this chapter, we review the dominant trends of peat land usage and the impact on fire and discuss the unique nature of peat fires in Indonesia, in order to make an argument for ecosystem-based, anticipatory management of peat fires. Ultimately, in the long term, such anticipatory fire management will require an improved water management approach. This is a critical action needed to reduce risk of peat fires and shift towards sustainable land-use in the Indonesian lowlands, which would then improve the overall management of natural capital stored in these peat lands.

16.2 The Challenging Nature of Peat Fires

16.2.1 The Magnitude of Peat Land Fires in Indonesia

Forest and peat fires in Indonesia have significant impacts on local and regional ecosystems, human health, and the economy of Indonesia and its neighbouring countries. Health impacts alone are estimated to affect 20–40 million people living in western Indonesia, the Malayan Peninsula and the island of Borneo. Increasingly, Indonesian fires are linked with what is called the Asian Brown cloud, a haze phenomenon linked to climate change impacts and public health deterioration in major Asian cities (Goldammer 1999). The magnitude of peat fires is underlined in the Indonesian Second National Communication under the United Nations Framework Convention on Climate Change. Figure 16.1 identifies peat land fires as the second single largest source of GHG emissions in Indonesia (Ministry of Environment 2011).

Moreover, the public health, environmental and economic risks resulting from these fires are of increasing concern as fire risks have risen significantly. A major peat fire episode occurring in October 2009 in Central Kalimantan resulted in the highest concentration of particulate matter ($PM_{2.5}$). The concentrations measured far exceeded what is permissible within a 24-h standard, on several occasions during that specific episode. Furthermore, it took months before particle matter levels returned to acceptable background levels (Betha et al. 2013). The risk analysis, which was part of the study undertaken by Betha et al. (2013), indicated that four or five individuals out of 1000 exposed to smoke haze may be affected by

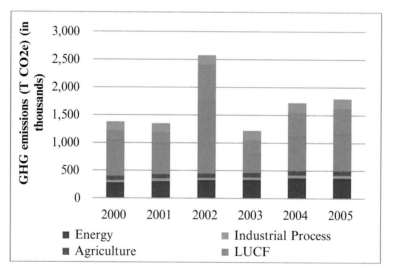

Fig. 16.1 Indonesian GHG emissions by sector (Source: Ministry of Environment 2011. Note: LUCF stands for GHG emissions from Land Use Change and Forestry)

cancer after prolonged exposure to high concentrations of carcinogenic metals in $PM_{2.5}$ emitted from peat fires.

Additionally, other costs incurred from fires include damages to the economy. Specifically, increased flooding, land subsidence and peat degradation are incredibly costly, leading to a loss of land and depreciation of the value of land-based assets. Such investment risks are important to investors, although studies have not yet calculated the extent of these costs. Particularly, the issue of land subsidence is severe (Hooijer et al. 2012). For example, one study of Southeast Asian peat lands shows that cumulative subsidence from peat land drainage over a depth of three metres results in an average of 1–1.5 m subsidence within the first year after drainage (Hooijer et al. 2012). The same study shows that draining peat lands might cause prolonged flooding across 30–69 % of coastal peat land surface within fifty years; as these lands are currently used for agriculture, flooding comes at a significant cost to the agricultural sector. Most of these lands are planted with palm oil (*Elaeis guineensis Jacq.*) and Acaccia (*Acacia mangium Willd*) (Hooijer et al. 2012). In addition, there are immeasurable impacts on local communities who are highly dependent on peat ecosystems; for example, in Kalimantan, communities impacted by floods are primarily located close to peat lands and are therefore vulnerable to the flooding due to subsidence of surrounding peat lands.

16.2.2 Past and Current Uses of Peat Lands in Indonesia and Their Impact on Fire Vulnerability

Current practices of unsustainable peat management are relatively new. Historically, peat lands have been used for agriculture in Indonesia, starting with the Banjarese people, who introduced an approach of gradual land clearing. Shallow hand-dug channels (*handil*), in which land was cleared and planted with rice initially, while raised beds were created for upland and perennial crops, which resulted in a gradual conversion to perennial crops (Watson 1987). Their success triggered the interest of the Netherlands Indies Government which subsequently started to clear peat lands for transmigration purposes in the late 1930s, using agriculture systems based on Banjarese practices. Additionally, during the colonial period, the Dutch constructed channels connecting big rivers. For example, the first channel built in Kalimantan, named Anjirserapat (where *anjir* means channel), was completed in 1890 and connects Kapuas river and Barito river, spanning a distance of approximately 28 km (Watson 1987; Alpian et al. 2013).

During the early New Order government which lasted from 1967 to 1998, various relatively small-scale peat land development sites were established where rice and food crops were cultivated by transmigrants. These were located mainly in the south-eastern part of Central Kalimantan. New approaches were developed, namely fish bone-based drainage and irrigation systems.[1] At the same time, loss of fertile agricultural lands in Java took place due to urbanization, with losses averaging around 100,000 ha annually. Consequently, the Government decided to increase the scope and magnitude of agricultural use of peat lands to compensate the loss of agricultural land, which was known as the one million hectare Mega Rice Project in 1995. Over 1.4 million hectares of peat were cleared during this period, as these lands were sparsely populated, still covered by primary forest and seen as under-utilized (Boehm and Siegert 2001). Previously, a small-scale project in Sumatra had proven that to some extent, peat could be used for rice cultivation. However, these schemes were small-scale and on shallow peat (IRRI 1984). The Mega Rice Project was abandoned in 1999 (Gamma et al. 2010), but the clearing and draining have resulted in a degraded landscape of scrubs and ferns, which now burn during El Niño years. The Government initiated efforts to rehabilitate the area in 2007 (GOI 2007). These efforts were challenged by technical difficulties, as initial approaches were based on simply blocking waterways that have not led to the anticipated results. In addition, social acceptance of rewetting remains challenging despite the fact that communities are exposed to annual fire outbreaks (Jaenicke et al. 2010). Hence, progress has so far remained limited, as there are only small-scale efforts implemented by NGOs with very limited success.

16.2.3 Special Characteristics of Peat Fires

Peat has unique features that exacerbate fires when drained, compacted and then exposed to prolonged period of drought. Fallow peat land is primarily covered by shrubs and trees. Drained peat land consists of 95-99 % organic matter and can exceed a depth of 15 m. Tropical peat lands have developed under humid conditions and are completely rain-fed.

In contrast to peat fires, bush and forest fires occur when there are sufficient fuel loads, an ignition source and dry weather (Usup et al. 2004). On the other hand, peat fire is influenced by soil bulk density (which affects the availability of air in soils), water table depth, rainfall and ignition sources (e.g. fire use). As Fig. 16.2 depicts, these characteristics make peat fires much harder to extinguish, mainly due to the ignition of fire that persists underneath the peat surface. Fire depth depends on soil moisture and can exceed over 0.5 m, which depends on various factors, of which the dynamics are still not well understood.

Bushfires can be controlled by reducing exposure to ignition and decreasing the fuel load (e.g. through fire breaks, controlled burning of undergrowth, etc.). However, for peat fires, the fuel load is not variable, but rather constant; for example, it

---

[1]Fish bone drainage and irrigation systems are tidal driven drainage systems of peat and acid-sulphate soils which are dug in a fish bone shape to allow maximum peneration of fresh water sources during high tide while allowing optimal irrigation during low tide.

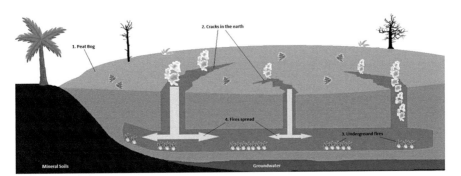

Fig. 16.2 The proccess of submerging peat fires in a peat landscape (Reproduced by UNORCID from Hooijer et al. 2014)

is not possible to create a fire break in peat due to the nature of peat itself as organic matter. Furthermore, the groundwater table in peat is completely rainfall dependent (i.e. peat build-up depends on rainfall), unless degradation is advanced enough that the peat reaches the lower groundwater table (Jaenicke et al. 2010).

As explained above, in the southeast Asian context, peat lands are an integrated element of coastal and lowland ecosystems. When peat lands are burnt, the burnt-over peat lands are susceptible to erosion, when they are exposed to rain. This process leads to loss of organic matter to adjacent seas, as the ash washes into the waterways and rivers. When such loss of organic matter and carbon from these environments (particularly along rivers) is included in the assessments of carbon loss – something often ignored in peat land budgets – the amount of total estimated carbon loss in peat lands increases by 22 % (Moore et al. 2013b).

Another complex feature of peat is the social dimension of peat land communities. To avoid conflicts over land with local communities, plantations are typically cleared on peat land away from settlements, with communities allowed to continue operating around the edges of concession boundaries. As a result of this situation and due to land tenure issues, there are frequent disagreements and conflicts over boundaries, where communities have lost access to lands as a result of oil palm and pulpwood plantations (Chokkalingam et al. 2007).

Peat land ecosystems in Indonesia are rich in globally important biodiversity, which further prompts the need for an improved conservation effort, particularly amongst the more notable flagship species, such as the Sumatran tiger and orang-utan. Protection of lowlands, therefore, depends on the ability to maintain healthy ecosystems that can deliver critical environmental services. Ultimately, the clearing and drainage of peat land forests alter the peat hydrological functions, reducing the ability of peat to provide regulating services and the delivery of other key ecosystem, cultural and social services (Hooijer et al. 2012). Put more simply, peat land clearing could lead to the collapse of peat ecosystems across Indonesia. The next section will explore peat dynamics with a closer look at two of the most vulnerable provinces in Indonesia, Riau and Central Kalimantan.

16.3   Peat Fire Dynamics in Riau and Central Kalimantan

16.3.1   Peat Characteristics in Riau and Central Kalimantan

Since the first large scale fire in 1982–1983, fire outbreaks in Indonesia have become increasingly frequent and severe, in terms of area affected and impacts. After large opening of peat lands was initiated during the mid-1990s, most fire hotspots were observed on degraded peat lands. The uncontrolled spread of fires occurs mainly in the islands of Sumatra and Kalimantan, with Riau and Central Kalimantan provinces showing some of the highest fire hotspot densities in the world (Yulianti et al. 2013). This section discusses the specific peat dynamics of Riau and Central Kalimantan, including their differences and changes in peat land use, as well as how these different land use development paths have led to widespread use of fire.

The circumstances in Riau vary significantly from Central Kalimantan. In Riau, there are significant challenges in developing a climate-based, early warning system due to the great variation of geographical and climatic conditions in Sumatra Island. While research programmes have shown that in Central Kalimantan fire events usually occur during drought years induced by climate anomalies from the Pacific (the El Niño Southern Oscillation – ENSO) and Indian Ocean (Indian Ocean Dipole – IOD), in Riau, the association of fire events with ENSO and IOD are not prevalent (Gaveau et al. 2014; Velde 2014).

Research by Gaveau et al. (2014) show that due to higher human pressures, peat fires are happening even after relatively short periods of lower than average rainfall during relatively wet years, as fires are linked to conflict over resources. Research results have shown that over the past decades, peat land utilization in Riau has intensified, thus resulting in enhanced vulnerability to fire. In addition, conflict and poor land demarcation leads to ineffective fire management (Loffler et al. 2014). By contrast, in Central Kalimantan, pressures on peat lands have remained lower, as most of the cleared peat lands are not yet used intensively, and relatively limited areas, around 160,000 ha of the 3 million ha (5 %), are used for palm oil production (BPS 2014). However, palm oil plantations have started to open estates on peat lands in Central Kalimantan.

Land use change trends in Sumatra can be explained by the combination of policy-driven, large-scale palm oil development, local land management systems to produce food crops (like *sonor* or swamp rice cultivation), and smallholder palm oil development (Chokkalingam et al. 2007). In Sumatra, smallholder engagement in palm oil production has increased rapidly, based on similar social-economic drivers as witnessed during the early 20th century for rubber. The agronomy of smallholder palm oil production makes it difficult to adopt mixed cropping arrangements, which has profound environmental impacts due to the need to clear forests for production (Budidarsono et al. 2012). Hence, the development of palm oil has led to a widespread loss of forest vegetation in Sumatra.

From 2004 to 2009, the oil palm area across Indonesia increased by 21 %, as a result of the growth of smallholder plantations (Directorate General Estate Crop – Ministry of Agriculture 2010). Smallholder plantations in Riau involve around 380,000 families, producing around 5.9 million tonnes of fresh fruit bunches (FFB) annually from around 1 million hectares of land. These FFB are processed in 144 palm oil mills in the province (Dinas Perkebunan Provinsi Riau 2010). It is expected that Central Kalimantan, where oil palm production is not yet as developed as in Riau, will be following the trends witnessed in Sumatra.

Deforestation has been aggravated over the past few decades due to a policy allowing the establishment of palm oil processing units without access to feed stock, which then trigger local demand for palm oil and is in turn often met by smallholders clearing forests on peat land. The impact on palm oil development is significant; palm oil production in Riau currently contributes up to 24 % of total national production. Today, in Riau, the registered oil palm plantations (including state-owned, smallholder and private enterprises) occupy an estimated 1.9 million hectares of land, or around 21 % of the total area of Riau province (Dinas Perkebunan Provinsi Riau 2010).

Aside from oil palm, pulpwood production in Riau, in particular, has led the province to be the largest producer of pulpwood in Indonesia, with Riau hosting the two largest pulpwood mills in the country. Pulpwood covers 5.02 million hectares across Indonesia and is still expanding both on mineral soils and peat lands (Ministry of Forestry 2011). The impact on peat lands is severe, not only due to the use of fire during land clearing and replanting by contractors, but also the effects on land subsidence (Hooijer et al. 2014).

As already indicated, research suggests that human activities are responsible for a large proportion of the fire events (Dennis et al. 2005). Figure 16.3 provides an overview of the impacts and causes of peat land fires. As the figure explains, clearing peat forests for palm oil is leading to the increased use of fire and generates drained peat lands. Palm oil production requires drainage; the drainage then further enhances fire vulnerability, in particular as smallholders and even larger-sized companies cannot control drainage on their individual parcels. Figures 16.4 and 16.5 present the varying degrees of vulnerability to fire, which incorporates: (1) current trends in clearance of peat forest, (2) drainage pattern following forest clearance (3) tenure insecurity, and (4) subsequent conflicts over land. Conflicts arise between concessionaires and communities as well as between spontaneous settlers and communities. Recent social dynamics trends have increased conflicts in which fire is used as a tool to secure tenure over land or signal that tenure is contested, which in turn have led to the uncontrolled clearing of land (Loffler et al. 2014).

Such dynamics in both Central Kalimantan and Riau are the challenges that the institutions, such as the Provincial Planning Agency, Disaster Risk Management agency and Environmental/Forestry services have had to face to implement spatial and forestry plans. These dynamics are further complicated by rapid demographic change, fuelled by the economic opportunities gained from the expansion of palm oil and, to a lesser extent, pulpwood production.

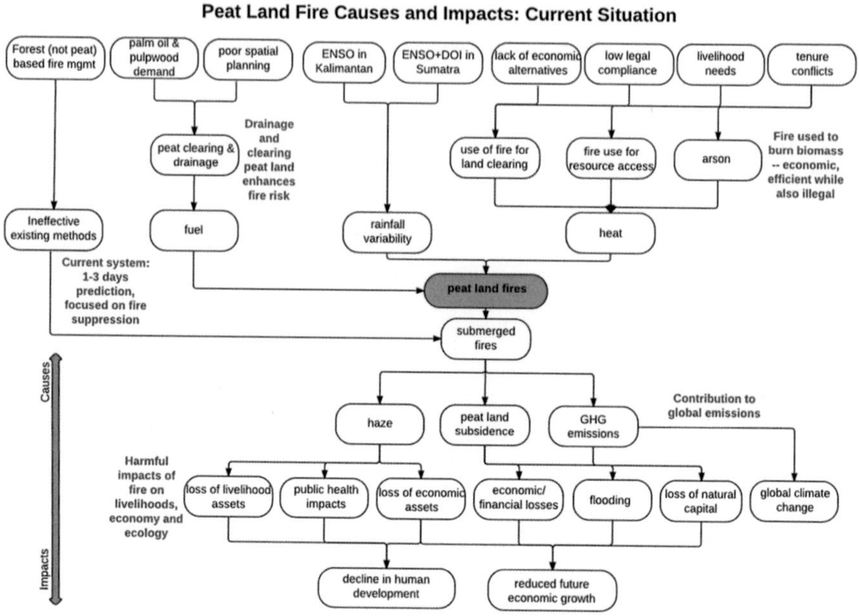

Peat Land Fire Causes and Impacts: Current Situation

Fig. 16.3 Peat land fires in Indonesia: causes and impacts of peat land fires (Own figure)

Rapid changes were initiated by past policies, which often date back to the colonial period but aggravated by the more recent transmigration and especially plantation-driven palm oil extension programmes. These socio-economic changes have undermined the ability of indigenous communities to adhere to *adat*,[2] or customary land use practices. Environmental change due to deforestation and drainage of peat lands makes that local knowlegde on fire use for land clearing and old fallow management practices have lost their value. The shift towards palm oil, which requires some form of more permanent land tenure, and lead to significant changes in land value, have created an environment where *adat* land use is often manipulated to secure access to land.

Despite the impacts of palm oil on land tenure, smallholder palm oil production has had a positive impact on poverty alleviation due to the profitability of palm oil. At the same time, this has simultaneously resulted in the intensification of forest clearance in peat lands and the use of fire, resulting in accelerated peat degradation and increased vulnerability to fire and other hazards (including land subsidence, erosion and flooding).

---

[2]*Adat* refers, in a broader sense, to the customary norms, rules, interdictions, and injunctions that guide individual's conduct as a member of the community and the sanctions and forms of address by which these norms and rules, are upheld (Keat Gin 2004). Adat also includes the set of local and traditional laws and dispute resolution systems by which society was regulated.

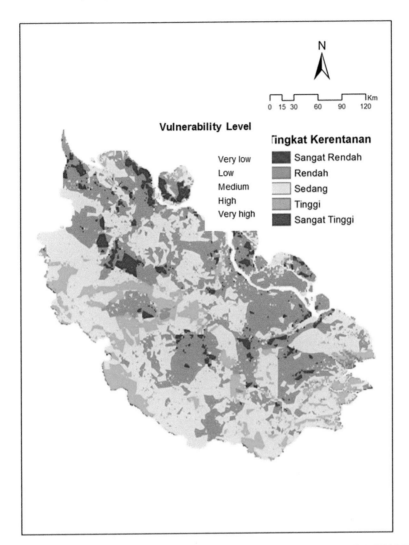

Vulnerability Level

	Tingkat Kerentanan
Very low	Sangat Rendah
Low	Rendah
Medium	Sedang
High	Tinggi
Very high	Sangat Tinggi

Fig. 16.4 Fire vulnerability map of Riau Province (Reproduced from IPB CCROM-SEAP Fire Risk System)

16.3.2   The Development of an Early Warning System in Riau and Central Kalimantan

The complex ecological dynamics of peat land in Central Kalimantan and Riau, alongside the political-economic dimensions of palm oil development, call for the need of new approaches towards addressing peat land fires. These approaches must consider geographical specificities, particularly the unique nature of peat and fire in Kalimantan and in Sumatra. In both provinces, the Government's institutional focus

Fig. 16.5 Fire vulnerability map of Central Kalimantan Province (Source: IPB CCROM-SEAP Fire Risk System)

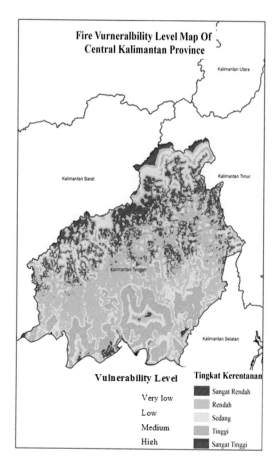

on fire suppression has limited efficacy in reducing vulnerability to fire risk. This has prompted the need for an early warning system to support a critical shift within fire management institutions to work towards an anticipatory system. An anticipatory fire management system would make it possible to identify and perform critical actions, months before a fire event starts, to mitigate the risk of an outbreak.

Collaboration between the Earth Institute at Columbia University and the Institut Pertanian Bogor's (IPB) Centre for Climate Risk and Opportunity Management in Southeast Asia Pacific (CCROM SEAP), with support from the National REDD+ Agency and facilitated by the United Nations Office for REDD+ Coordination in Indonesia (UNORCID), has resulted in the development of a seasonal fire early warning system, known as the Fire Risk System (FRS), for managing fires at the provincial and district level, with particular focus on Central Kalimantan and Riau province. This system informs local government and communities of the level of fire risk within a 1–3 month response time, providing them the opportunity to develop proactive measures for high fire risk when weather patterns indicate drier than normal conditions (Someshwar et al. 2010). Trainings for local government staff have also been conducted on how the dynamic land use module, which can be adjusted to evaluate different land use options can be used. As these outputs can be translated into determining levels of fire risk in Central Kalimantan and possible measures to manage the risk. The final section will describe how an ecosystem-based, anticipatory management approach, which would include such an early warning tool as developed by CCROM SEAP and Columbia University, can successfully contribute to the mitigation of peat fires across Indonesia.

16.4 Towards Ecosystem-Based Anticipatory Peat Fire Management

The main factors limiting the widespread and formal operational use of seasonal early warning information are the challenges associated with the current institutional focus on reactive or response measures (i.e. after the fires occur). Political unwillingness combined with administrative hurdles limit the availability of incentives and sufficient enforcement of penalties to change behaviour. A complicated regulatory structure and a system-based approach to fire suppression means that Indonesian institutions do not yet have the requisite authority to undertake measures that anticipate and reduce fire risk based on available risk information. Neither are they yet in a position to work with stakeholders to change behaviour on fire use. The newly-enacted Standard Operating Procedure (SOP) (National REDD+ Agency 2014) on forest and land fires requires an anticipatory approach to addressing peat land and forest fire disasters. However, the work undertaken in Central Kalimantan and in Riau indicates that a user-friendly, fire early warning system integrated into a revised, more comprehensive land and forest fire SOP can significantly improve the capacity of local governments to effectively manage fire risks.

16.4.1 Towards an Anticipatory Fire Management Approach

As already described above, peat fires are difficult to control, and they behave differently than forest fires or fires on mineral soils. Recognizing this, the Government of Indonesia initiated a policy, specifically to address peat land degradation. Government Regulation No. 71/2014 set clear standards for land that has to be conserved (three metres of depth or greater) and established the maximum drainage depth (40 cm). Unfortunately, the approved Regulation has not yet been implemented at the time of writing; moreover, the Regulation has been challenged by numerous palm oil and pulp wood industry leaders. Nevertheless, regulatory action is just one component of activities required to mitigate peat fires.

The way forward requires a multi-pronged, ecosystem-based, anticipatory approach for a sustainable reduction in fire. Peat fire management requires improved and comprehensive peat land and water management, and ultimately, peat land rehabilitation will be critical to allow for its continued sustainable use. Characteristics of an anticipatory fire management approach include proper assessment of risks to ecosystem services, determination of the level of fire vulnerability at the ecosystem level and administrative unit, a climate-based early warning system to inform policymakers of high-risk areas, an appropriate incentive and/or penalty scheme to change behaviour towards responsible fire use, and sustainable peat land management practices in high risk areas. In addition, a requisite component of such an incentive and/or penalty scheme will be a REDD+ (Reducing Emissions from Forest Degradation and Deforestation) province-based reference emission level (REL) to serve as a baseline for a reward or penalty. REDD+ can provide payment for GHG emissions related to controlling forest degradation and deforestation, including in Indonesian peat lands. Thus, REDD+ provides resources for local governments to provide incentive for land managers to reduce emissions from peat land fires and thus reduce fire vulnerability.

A key function of an anticipatory approach is also an orientation towards ecosystems, which appreciates traditional, indigenous knowledge of communities that have depended on such ecosystems for centuries. For example, the gradual clearing of peat dome edges by communities, indicative of the Banjar approach described above, along with the use of *handils* for deeper access, is one way to more sustainably manage peat land. Additionally, alternatives to oil palm or pulpwood plantation should be explored. For example, through the use of *sorjan* cropping systems – *i.e.* inter-cropping on alternating raised beds and deep sinks – rice and other higher value crops could be utilized in peat land ecosystems. Other options include a gradual transition to other crops. First, farmers can start growing rice with trees through inter-cropping and then gradually transitioning towards a mixed fruit crop/tree crop system. Lastly, studies have shown that the *Melaluca cajaputi* species can be successfully used in the rehabilitation of peat swamp forests (Alpian et al. 2013).

16.4.2 A Paradigm Shift to Reverse Peat Degradation

In line with the ongoing efforts of the Government of Indonesia, it is critical to effectively address peat fires and their detrimental impacts, and reverse peat land degradation. Through a new anticipatory-focused paradigm, peat land rehabilitation will be the ultimate goal required to mitigate fire risks in the long run. With this understanding, the next question is how to shift towards this paradigm. Specifically, what concrete steps are required to address the challenges of mitigating peat land fires described above?

Institutional structures could potentially support the approach outlined above, based on a review of the current SOP for Forest and Peat fires (National REDD+

Agency 2014) and underscored by the need for a comprehensive SOP to support local governments in adopting anticipatory-based interventions. The SOP might require the inclusion of guidelines on adopting a climate-based early warning system, such as that developed by IPB CCROM-SEAP (2015) and Columbia University. This scientific system, which provides a probabilistic forecast of fire risk at the village level, is a powerful tool for policymakers at all levels of the Indonesian Government. Fire risk information is particularly important to local government officers, who are seeking to improve the preparedness of their communities to deal more effectively with fire hazards and also invest more directly in sustainable peat land management. Using this probabilistic forecast, policymakers can access funding for prevention activities and peat rehabilitation efforts, which may range from short-term activities such as blocking canals and banning the use of fire in high-risk activities to long-term, full-scale peat rehabilitation. Alternatively, such forecasting of risk can be used to monitor changes in risk over time, information that can inform the medium and long-term development planning of local governments. Streamlining fire prevention efforts at multiple time scales – short, medium and long-term – in the context of development and spatial planning is a critical component of an anticipatory fire management approach, which is required for sustainably managing the natural capital stored in Indonesia's peat land ecosystems. Ultimately, such a comprehensive approach to fire management will not only mitigate the ecological and socio-economic impacts of peat fires, but it will support socially inclusive, economic growth-oriented and environmentally sustainable policies across Indonesia.

References

Alpian, Prayitno TA, Gentur JP Sutapa, Budiadi (2013) Biomass Distribution of Cajuput Stand in Central Kalimantan Swamp Forest. Jurnal Manajem Hutan Tropika 19(1):1

Betha R, Pradani M, Lestari P et al (2013) Chemical speciation of trace metals emitted from Indonesian peat fi res for health risk assessment. Atmos Res 122 (May 1998): 571–578. Available via http://dx.doi.org/10.1016/j.atmosres.2012.05.024

Boehm HDV, Siegert F (2001) Ecological impact of the one million hectare rice project in Central Kalimantan, Indonesia, using remote sensing and GIS. Paper presented at the 22nd Asian conference on remote sensing 5:9

BPS (2014) Kalimantan Tengah Dalam Angka. BPS-Statistics of Kalimantan Tengah Province, Palangka Raya

Budidarsono S, Dewi S, Sofiyuddin M, Rahmanulloh A (2012) Socioeconomic impact assessment of palm oil production. Technical brief No. 27: palm oil series. Bogor. World Agroforestry Centre (ICRAF) Southeast Asia Program

Chokkalingam U, Suyanto, Permana RP et al (2007) Community fire use, resource change and livelihood impacts: the downward spiral in the wetlands of Southern Sumatra. Mitig Adapt Strateg Glob Chang 12:75–100

Dennis RA, Mayer J, Applegate G et al (2005) Fire, people and pixels: linking social science and remote sensing to understand underlying causes and impacts of fires in Indonesia. Hum Ecol 33 (4):465–504. doi:10.1007/S10745-005-5156-Z

Dinas Perkebunan Provinsi Riau (2010) Data Statistik Perkebunan., Pekanbaru

Directorate General Estate Crop – Ministry of Agriculture (2010) Agricultural statistics database. Ministry of Agriculture, Jakarta

Forsyth T (2014) Public concerns about transboundary haze : a comparison of Indonesia. Glob Environ Chang 25:76–86. Available via, http://dx.doi.org/10.1016/j.gloenvcha.2014.01.013

Gamma G, Van Noordwijk M, Suyanto et al (2010) Hot spot of emission and confusion: land tenure insecurity, contested policies and competing claims in the Central Kalimantan Ex-Mega Rice Project Area. Working Paper No. 98. World Agroforestry Centre. Bogor. 34 p

Gaveau DLA, Salim MA, Hergoualc'h K et al (2014) Major atmospheric emissions from peat fires in Southeast Asia during non-drought years: evidence from the 2013 Sumatran fires. Sci Rep 4:6112, http://dx.doi.org/10.1038/srep06112

GOI (2007) Instruksi Presiden Republik Indonesia-Nomor 2/2007. Tentang Percepatan Rehabilitasi Dan Revitalis Kawasan Pengembangan Lahan Gambut Di Kalimantan Tengah. Available via http://www.setneg.go.id/components/com_perundangan/docviewer.php? id=1530&filename=Instruksi_Presiden_no_2_th_2007.pdf

Goldammer JG (1999) Health guidelines for vegetation and fire events. WHO/UNEP/WMO. International Forest Fire News (IFFN) No. 31 (July – December 2004, 132–139)

Hooijer A, Triadi B, Karyanto O et al (2012) Subsidence in drained coastal peatlands in Southeast Asia: implications for sustainability. Paper submission presented to the International Peat Society, June 2012

Hooijer A, Page S, Navratil P et al (2014) Carbon emissions from drained and degraded peatland in Indonesia and emission factors for Measurement, Reporting and Verification (MRV) of peatland greenhouse gas emissions

IPB CCROM-SEAP (2015) Fire risk system. www.kebakaranhutan.or.id. Accessed April 2015

IRRI (1984) Workshop on research priorities in Tidal Swamp rice. Available via: http://books.irri. org/9711041022_content.pdf

Jaenicke J, Rieley JO, Mott C et al (2008) Determination of the amount of carbon stored in Indonesian peatlands. Geoderma 147:151–158

Jaenicke J, Wotsen H, Budiman A, Siegert F (2010) Planning hydrological restoration of peatlands in Indonesia to mitigate carbon dioxide emissions. Mitig Adapt Strateg Glob Chang 15:223–239

Keat Gin O (ed) (2004) Southeast Asia: a historical encyclopedia, from Angkor Wat to East Timor, ABC-CLIO

Loffler H, Afiff S, Burgers P et al (2014) Agriculture beyond food: experiences from Indonesia. NOW/WOTRO, The Hague

Ministry of Environment (MoE) (2011) Indonesia second national communication under the UNFCCC. MoE-Govermment of Indonesia, Jakarta

Ministry of Forestry (2011) Statistik Kehutanan 2010. Ministry of Forestry, Jakarta

Moore S, Gauci V, Evans CD, Page SE (2013a) Increased losses of organic carbon and destabilising of tropical peatlands following deforestation, drainage and burning. In: AGU Fall Meeting, 9–13 Dec 2013, San Francisco

Moore S, Evans CD, Page SE et al (2013b) Deep instability of deforested tropical peatlands revealed by fluvial organic carbon fluxes. Nature 493:660–663

National REDD+ Agency (2014) Empat Mentri sahkan POSNAS Karhutla. Available via http:// www.reddplus.go.id/berita/berita-redd/2054-empat-kementerian-sahkan-posnas-karhutla

Someshwar S, Boer R, Conrad E (2010) World resources report case study. Managing Peatland fire risk in Central Kalimantan, Indonesia. World Resources Report, Washington, DC. Available via http://www.worldresourcesreport.org

Usup A, Hashimoto Y, Takahashi H, Hayasaka H (2004) Combustion and thermal characteristics of peat fire in tropical peatland in Central Kalimantan, Indonesia. Tropics (14)1

Velde BV (2014) Forests news update: Q&A on fires and haze in Southeast Asia' Forests News. A blog by the Center for International Forestry Research. Available via http://blog.cifor.org/ 23000/forests-news-update-qa-on-fires-and-haze-in-southeast-asia#.U__hj8WSxch. Accessed on 18 Jun 2014

Watson G (1987) Settlement in the coastal wetlands of Indonesia: an argument for the use of local models in agricultural development. Crosscurrents 1:18–32

Wetlands International (2014) Carbon emissions from peatlands. Wetlands International. Available via www.wetlands.org/whatarewetlands/peatlands/carbonemissionsfrompeatlands/tabid/2738/default.aspx. Accessed 2 Mar 2015

Yulianti N, Hayasaka H, Sepriando A (2013) Recent trends of fire occurrence in Sumatra (Analysis using MODIS hotspot data): a comparison with fire occurrence in Kalimantan. Open J Forest 3:129–137. doi:10.4236/ojf.2013.3402

Chapter 17
Protected Areas, Biodiversity, and the Risks of Climate Change

Jeffrey A. McNeely

Abstract Protected areas are becoming a major land use, approaching 15 % of the Earth's terrestrial surface and a growing percentage of coastal waters. These sites are popular for visitors, but face many management challenges, including how to adapt to climate change. Often established for biodiversity conservation, scenic beauty, or tourism objectives, protected areas should become a major part of national strategies to address climate change and the disasters that may come in the form of extreme climatic events. Protected areas often contain the ecosystems that are the most effective in storing carbon and make major contributions to adapting to climate change. But these sites need to be managed more effectively, and linking them to the growing public concern about climate change could be one means of doing so. Management approaches that should be supported include establishing protected area complexes that expand their influence to a landscape scale, incorporating climate change issues into protected management at both site and system scales, identify the multiple ecosystem services that protected areas provide as a means of building broader support for them, and many others. Protected areas can also contribute to recovery from extreme hazard events, for example by working with local communities to restore natural vegetation. To date, protected areas have been largely ignored by the Clean Development Mechanism established by the Climate Change Convention. This should change, and protected areas should be recognized for the many contributions they make to climate change mitigation and adaptation, thereby contributing to reducing disaster risks.

Note that a useful decision-support tool is now available to help protected area managers identify climate risks and integrate them into site management. See www.iisd.org/cristaltool/

J.A. McNeely (✉)
Department of National Parks, Wildlife and Plant Conservation Hua Hin Thailand, 1445/29 Petchkasem Road, Saitai, Cha Am, Petchburi 76021, Thailand
e-mail: jeffmcneely2@gmail.com

© Springer International Publishing Switzerland 2016
F.G. Renaud et al. (eds.), *Ecosystem-Based Disaster Risk Reduction and Adaptation in Practice*, Advances in Natural and Technological Hazards Research 42,
DOI 10.1007/978-3-319-43633-3_17

379

A relatively simple step would be to incorporate protected area agencies more actively in the preparation of the national reports called for by the Framework Convention on Climate Change. Protected areas should also become eligible for support under the REDD+ programme.

Keywords Climate change • Adaptation • Protected areas • Biodiversity • REDD+

17.1 Introduction

Weather-related disasters have long been a fact of life. The evidence strongly supports the view that their increasing frequency is linked to global climate change (IPCC 2014), and that the extreme climatic events are likely to become more frequent as the climate warms (Cai et al. 2015). It is therefore not surprising that many governments are making larger investments in climate-related disaster risk reduction, at least at the planning level, and the issue is being actively discussed at meetings of the UN Framework Convention on Climate Change. However, an important element in climate-related disaster risk reduction has been receiving inadequate attention: the role of protected areas.

This may be changing. The previous focus of concern on the threats of climate change to the integrity of protected areas has been joined by a broader consideration of how protected areas can contribute to addressing climate-change related problems. Protected areas are increasingly being recognized for their important roles in mitigating greenhouse gasses, reducing the effects of extreme climatic events, building resilience to changing conditions, and adapting to climate change (Barber et al. 2004; Murti and Buyck 2014). Protected areas are globally approaching the Convention on Biological Diversity's Aichi Target of 17 % of the land and freshwaters and 10 % of territorial seas (Juffe-Bignoli et al. 2014), so they are becoming major components of national responses to the risks posed by climate change. The real challenge now is the rest of Aichi Target 11, which calls for these protected areas to be "conserved through effectively and equitably managed, ecologically representative and well-connected systems of protected areas and other effective area-based on conservation measures, and integrated into the wider landscape and seascape." This chapter is a contribution to this operational part of the Aichi Target 11.

Climate change will affect all ecosystems, including those found within protected areas. Impacts will be felt on the coastline (e.g. rising sea levels, acidification, changing distribution of harvest fisheries and storm surges) (Woodruff and Woodruff 2008); on lands growing food and fiber (e.g. changing rainfall patterns and temperatures will affect seasonality and crop production); systems of surface transport (e.g. due to threats of floods); delivery of water supply (e.g. due to changing rainfall patterns and changing snowfall in temperate regions); destinations for tourism (e.g. affecting the times when tourists will find protected areas appropriate for visiting); many aspects of human health (e.g. increasing likelihood of emerging infectious diseases and invasion of harmful non-native species); and many others (Groves et al. 2012).

That said, the impacts of climate change should be seen as the most dramatic, though not necessarily the most urgent, among the numerous anthropogenic threats to the environment. Problems such as land-use change, pollution, desertification, human population growth, increasing demand for resources, over-exploitation of harvested species, spread of invasive alien species, and many others may be higher on the agenda for most protected area agencies (MEA 2005). So while climate change will play the leading villain role in this chapter, the other environmental problems remain important gang members looming in the background and serving as force multipliers (House et al. 1996) to the disastrous negative impacts of climate change. Another way of look at this is that climate change will accelerate these other environmental problems, though they are already sufficiently worrisome by themselves. The potential benefits of climate change for some species, ecosystems, and economies are well worth considering, but will not be further explored here.

This chapter will briefly review the negative impacts of climate change on biodiversity, summarize the steps being taken to address these impacts, and then indicate how protected areas are making a major contribution to both mitigation of climate change and adaptation to the changes that are likely to come irrespective of any mitigation measures that might be adopted. It will conclude with policy recommendations on how protected areas can become a more significant supporter of reducing disaster risks posed by climate change.

Experience from various parts of the world illustrate that environmental degradation contributes to the negative impacts of extreme climatic events, which become disasters when they affect human interests. For example, the impacts of the 2005 Hurricane Katrina, which caused nearly US$ 150 billion in damage to the US state of Louisiana and devastated New Orleans, were increased by the deterioration of the ecosystems of the Mississippi Deltaic Plain (e.g. Burby 2006; Tibbetts 2006). Restoration efforts therefore include significant investments in restoring the coastal wetlands as a means of protecting New Orleans (Day et al. 2007). As a low-lying coastal and delta city, the government of New Orleans learned lessons which could be applied to geographically similar cities such as Shanghai, Guangzhou, Bangkok, Yangon, Jakarta, Manila, Ho Chi Minh City, Miami, Tokyo, Dacca, Cairo, London, Washington D.C., and thousands of other coastal communities in all parts of the world. This chapter will suggest ways of doing so.

17.2 The Negative Impacts of Climate Change on Biodiversity, and Ways to Address Them

Climate change is having major impacts on species and ecosystems though its effects on temperature, seasonality, ocean chemistry (Orr et al. 2005), and rainfall – major determinants of the population and distribution of most species and the ecosystems of which they are part (Groom et al. 2006; Rosenzweig 1995). When superimposed on anthropogenic habitat change and increasing demand for natural resources, the future of species and ecosystems is fraught with risks of extinction (Dickson et al. 2014). If current trends continue, over a third of the world's species may be committed to premature extinction (Thomas et al. 2004). This problem is being addressed by the Parties to the Convention on Biological Diversity through its Strategic Plan for Biodiversity 2011–2020 and its 20 targets. But very few of these targets show signs of being reached. One such exception is Aichi Target 11 on protected areas which appears to be on track (Tittensor et al. 2014), with regards to percentage coverage, although not necessarily in terms of management effectiveness (Le Saout et al. 2013).

Information from many sources are required to identify which species and ecosystems are most vulnerable to the changes that climate change will bring (Dawson et al. 2011). Multiple approaches are available to assess the vulnerability of species (Pacifici et al. 2015), and active monitoring is needed to keep track of how species and ecosystem are changing as their vulnerabilities are exposed (Spellerberg 2005). New kinds of ecosystems may evolve to replace or supplement the existing ones, or variants of them with novel assemblages of species (Young 2014). These changes may be so profound that some scientists are concluding that Earth is entering a new Epoch, referred as the "Anthropocene" (Steffen et al. 2007; Monastersky 2015; Lewis and Maslin 2015).

These new ecosystems are likely to have fewer species to provide the multiple functions that they carry out, losing connectedness and even their capacity to provide some ecosystem services (Gitay et al. 2002; Root et al. 2003). The latest updating of the IUCN Red List of Threatened Species assessed 76,199 species and found that 22,176 were Threatened (IUCN 2014). Of the latter, climate change was specifically identified as one of the threatening factors for 2334 species (10.5 %), though many others may also have been affected (Akçakaya et al. 2014). More generally, the IUCN Red List and other such assessment tools need to be improved so that species particularly vulnerable to climate change can be identified at a sufficiently early stage, enabling management measures to be designed, tested, and implemented before the status of the species reaches a critical stage (Akçakaya et al. 2014).

In a report prepared for the Intergovernmental Panel on Climate Change (IPPC), Gitay et al. (2002) concluded that changes in biodiversity at the landscape scale, such as those driven by forest fires or deforestation, are highly likely to further affect regional climates by changing forest-related ecosystem services such as evapotranspiration and carbon sequestration. Greater efforts to conserve biodiversity are therefore justified on climate change grounds, adding to the long list of significant reasons to conserve the biological systems that provide so many benefits to humanity in addition to their intrinsic values (see, for example Duffy 2009; Ninan 2009; Wilson 1988; and Ghilarov 2003).

Approaches being considered or implemented for conserving species in the face of climate change include:

- Assisted colonization, which involves intentionally moving species of plants or animals to new habitats that biologists predict may be more suitable for them, and that the species may not be able to reach without active assistance (for example, if potential dispersal habitats have been converted to agriculture) (Hoegh-Guidberg et al. 2008; Hobbs et al. 2011). Guidelines for such assisted movements have been provided (IUCN/SSC 2013);
- Greater attention is being given to identifying and conserving climate refugia, areas that are likely to maintain their traditional species or offer landscapes to which species can retreat even as other areas are modified (Reside et al. 2013). Such refugia can be given high priority as new protected areas when necessary, and may be the destination of assisted colonization for some species;
- Linking protected areas through conservation corridors that enable species to move in response to the effects of climate change is being more widely applied, in what could be considered a sort of unassisted colonization through habitat management (Lombard et al. 2010; Hess and Fischer 2011; Damschen et al. 2006);
- Establishing protected areas or improving their management, including enrichment planting, creating waterholes, and controlling invasive alien species (the latter may be an issue in assisted colonization);
- Linking protected areas through conservation corridors that enable species to move in response to the effects of climate change (Lombard et al. 2010; Hess and Fischer 2011; Damschen et al. 2006); and
- Modifying protected area management to anticipate climate changes and respond to them, especially through developing new forms of governance that involve people more directly in managing the risks of climate change, and ensure that local communities are involved in preparing contingency plans for providing emergency sources of timber, shelter, freshwater, and other goods and services in case of extreme natural events affecting people living in and around protected areas (see also Harmáčková et al., Chap. 5; Erisman et al. 2015).

The last four points highlight the important role of protected areas for conserving biodiversity in a time of rapid climate change, and modifying governance to enable them to do so. The following section will address how protected areas are also helping to address climate change issues, and thereby helping to address at least some of the risks posed by the changing climate.

17.3 Protected Areas and Climate Change

Protected areas are contributing to climate change mitigation and adaptation and thereby help contribute to reducing climate-related, disaster risks. While most of the investment from the UNFCCC's Clean Development Mechanism has gone to mitigation (especially seeking to reduce emissions from fossil-fuel energy facilities), more attention is now being given to adaptation. Here, mitigation and adaptation are considered as part of the same package of responses to climate change, though they are separated for ease of exploration. But it is worth keeping in mind that the best methods of using natural ecosystems to sequester carbon are also the most likely to contribute to climate change adaptation as well as conserving biodiversity. This leads to environmental, economic, social, and cultural benefits (SCBD 2009).

17.3.1 Protected Areas Mitigate Climate Change and its Effects

The role of protected areas in conserving biodiversity, as reflected in Articles 8 (1) and 8(2) of the Convention on Biological Diversity (UNEP 1992), also helps to address climate change mitigation. For example, deforestation globally is the second largest anthropogenic source of carbon dioxide to the atmosphere, after fossil fuel combustion, accounting for as much as 20 % of the CO_2 emissions (though some estimates are as low as 12 %, especially at times when the fossil fuel emissions are responsible for a very high percentage of carbon emissions) (UNEP 2014). And since protected areas are intended to prevent deforestation, this makes them critical elements of national climate change mitigation programmes even if protected areas are not explicitly recognized for their contributions.

The ecosystem service of carbon sequestration is of considerable value (Luyssaert et al. 2008), and its value is likely to increase as governments become more aware of the urgency of mitigating climate change (IPCC 2014) and as protected areas expand. Forests are estimated to store about 2.4 billion metric tons of carbon per year (Pan et al. 2014), and forest protected areas store more carbon than other land uses, with the global system of protected areas storing over 15 % of the terrestrial carbon stock (Campbell et al. 2008). Southeast Asia has the highest carbon density, storing 267 tonnes per hectare of old-growth forest; the total value of the carbon stored within Southeast Asia's protected areas was estimated at between US$250 billion and over $500 billion (depending on the price of carbon, a volatile figure).

Old-growth forests, even up to 800 years old, continue to accumulate carbon at an increasing rate, in both trees and the soil, that is far higher than the new forests that are being advocated as climate change mitigation measures (Stephenson et al. 2014; Luyssaert et al. 2008; Pan et al. 2011). More important, much of the stored carbon is likely to return to the atmosphere if the old-growth forests are harvested, and the more intensive the logging, the more carbon is returned to the atmosphere. This is a powerful argument for conserving the old-growth forests contained within protected areas.

Lewis et al. (2009) found that old-growth tropical forests in Africa are absorbing nearly 20 % of the CO_2 produced by the burning of fossil fuels, with mature, intact tropical forests sequestering nearly 5 billion tonnes of CO_2 annually, yielding an economic benefit of US$18.7 billion (again, subject to the variable price of carbon). They concluded that the carbon storage of these old-growth forests is increasing, as has been found in other parts of the tropics. If these forests are logged, about 40 % of their carbon will be released (with the figure even higher if they are intensively logged). Perhaps more interesting, Schimel et al. (2014) found that topical forests are increasing their uptake of CO_2 as more of this pollutant enters the atmosphere, judging from in situ, atmospheric, and simulation estimates. While this finding awaits further confirmation, it remains apparent that conserving old-growth forests such as those found in tropical protected areas is an important strategic aim of protected areas and a major contributor to climate change mitigation.

Conserving forests provides climate effects that go beyond the long-term global public good of carbon sequestration. Through evapotranspiration, forests also help to generate cloud cover, a phenomenon demonstrated by deforestation in tropical lowlands. Clearing the lowland forests reduces evapotranspiration, so fewer clouds are formed and less rain falls on mountain cloud forests. These cloud forests historically have provided important water supplies downstream during dry seasons, but this ecosystem service is declining. And as the cloud forests decline, they also lose species that are endemic to them (Lawton et al. 2001; Pounds et al. 1999). More locally, the shade and air movement generated by forests help to reduce local temperatures by several degrees, providing some relief to forest communities from global warming by the cooling effect they provide locally (Bonan 2008).

Mature vegetation supported by protected areas makes additional contributions to climate change mitigation, helping to moderate the effects of heavy rainfall, storm surges along the coast, and landslides accompanying earthquakes or rainstorms wherever they may occur. For example, Mt. Elgon National Park in Uganda has far fewer landslides following heavy storms than do the more vulnerable hillsides outside the park (Bintoora 2014). Along many tropical coastlines, healthy mangrove forests help to control storm surges, but if they are to be effective in protecting against the most extreme events, such as tsunamis, they need to be at least 150 m wide (Dahdouh-Guebas et al. 2005). So protection from tsunamis provided by coastal vegetation and fringing reefs is only partial; while it is very helpful with extreme climatic events, no coastal vegetation or fringing coral reefs could have saved Banda Aceh from the disastrous effects of the 2004 tsunami in the Andaman Sea (Borrero 2005).

Despite such limitations (tsunamis are not climate-related), protected areas mitigate climate change impacts in both the long term, through carbon sequestration, and in the short term, through providing a variety of disaster risk reducing ecosystem services. For example, one study has found that the loss of each hectare of coastal wetlands in the US led to an increase in storm damage, especially to cities, of an average of US$33,000 (Costanza et al. 2008), a good indication of how valuable protecting coastal wetlands can be. A study in Barbados supported by Swiss Re, a major reinsurance company, found that each dollar invested in conserving the mangroves of Folkstone Marine Park provided benefits of $20 in hurricane protection. It also found that healthy coral reefs reduce the damage of storm surges by about 50 % (Mueller and Bresch 2014). In Sri Lanka, the 3000 ha

Muthurajawella Marsh provides flood protection values estimated at over US$5 million per year (Emerton and Kekulandala 2003). In New Zealand, the Whangamarino Ramsar Wetland contributes flood control benefits estimated at over US$600,000 per year, increasing to as much as US$4 million when serious flooding affects the region (Schuyt and Brander 2004). As a final example, the Pantanal Wetland of Brazil provides annual economic values of US$181 million in greenhouse gas regulation, $120 million in climate regulation, and $4.7 billion in disturbance regulation, in addition to numerous other ecosystem services that give it an economic value of well over $15 billion per year (Seidl and Moraes 2000). Many other such figures can be cited (see, for example Kumar 2010 and Ninan 2009).

It therefore should come as no surprise that the remaining natural wetlands that are included in protected areas that conserve coastal vegetation, floodplains, and coral reefs are beginning to receive higher priority in view of their role in disaster risk reduction (Keddy et al. 2009). Many of these ecosystems are given additional protection by being included on the List of Wetlands of International Importance, under the Ramsar Convention (www.ramsar.com), though much more remains to be done.

As one well-known example, the US National Park Service is allocating over US $50 million to help restore the Gulf Islands National Seashore and the Jean Lafitte National Historic Park and Preserve, to help mitigate future hurricane damage to New Orleans (Ford 2014). But of course this does little to replace the many coastal ecosystems that have been destroyed by creating sea transport routes, introducing invasive alien species that destroy native vegetation, and digging petroleum exploration canals through the Mississippi River Delta.

At least theoretically, climate change mitigation can reduce the need for adaptation, but substantial risks from climate change are still expected (IPCC 2014). Even if all carbon sequestration targets are met, extreme climatic events have become the new normal, so the role of protected areas in helping societies and ecosystems adapt to climate change has become increasingly important.

17.3.2 Managing Protected Areas to Help Adapt to Climate Change and Reduce Climate-Related Risks of Disaster

Though most protected area managers give higher priority to urgent management issues such as encroachment, poaching, invasive alien species, human-animal conflict, and tourism, many protected area agencies are encouraging their site managers to give greater attention to the implications of climate change. In order to address disaster risk reduction and climate change adaptation more effectively, protected area agencies have several management options available and no doubt others will be explored in the future. Current tactical responses that are available include:

- Use protected area complexes, including both legally protected sites and surrounding lands managed by other government agencies and private landholders, as the basis for larger ecosystem restoration and climate change adaptation, giving particular attention to watersheds, wetlands, and coastal/marine ecosystems. This may require new approaches to governance at both site and system levels to include local stakeholders as active participants in planning and management for climate change. These approaches should seek good governance models that can help people recover better from disasters, and use the nearby protected areas to support resilience in the face of extreme climatic events and adapt to them when they do occur;

- Design national systems of protected areas with climate change mitigation and adaptation in mind. To mitigate climate change, conserve natural vegetation (especially old-growth forests, coastal vegetation such as mangroves, wetland ecosystems, and marine ecosystems such as coral reefs). To adapt to climate change, identify and protect climate refugia (Reside et al. 2013), especially through expanding the size of protected areas and the elevation variability of them, enhance ecosystem connectivity to expand the effective size of protected areas, and link them to the surrounding lands (Bennett 2003). This increase in the effective size of protected areas also enhances their primary duty to conserve biodiversity, but may increase the hazard of encouraging the establishment and spread of invasive alien species because larger areas have more potential entry points;
- For marine protected areas, the consensus is that refugia need to be established in the more resilient habitats. Protected areas should be located in areas that are less likely to be affected by land-based stresses. Multiple examples of each major habitat should be protected, and critical areas (e.g. fish spawning aggregations, nesting and feeding areas, and breeding areas) should be given particular attention (Green et al. 2014). Realistically, much of this will need to take place outside strictly protected areas and involve management at the seascape scale;
- Quantify and qualify the ecosystem services of protected areas, namely carbon sequestration, watershed protection, genetic diversity, and disaster risk reduction and adaptation to climate change, as a means of providing stronger support to old-growth forests, coastal vegetation, coral reefs, and wetlands; and communicate these multiple benefits to Ministries of Finance as public goods worthy of greater budgetary support;
- Incorporate a regular programme of measuring, reporting, and verifying the carbon sequestration and other ecosystem benefits as a standard part of PA management;
- Use protected area complexes, including both legally protected sites and surrounding lands managed by other government agencies and private landholders, as the basis for larger ecosystem restoration and climate change adaptation, giving particular attention to watersheds, wetlands, and coastal/marine ecosystems;
- Link protected areas in north-south complexes that will enable greater movement of species in response to climate change, not least by expanding the genetic variability in the species that occur (Peters and Lovejoy 1992). For species that are threatened by climate change, this tactic is likely to be especially important;

- Use the protected area complexes as models of adaptation for larger landscapes and seascapes, ultimately seeking to demonstrate how a country's entire territory can be managed in a sustainable manner that is low in carbon emissions and is adaptive to changes in climate and other dynamic factors of modern economies (Brown et al 2005);
- Strengthen research and monitoring to assess impacts of climate change, guide management responses to these changes, and provide a baseline against which mitigation and adaptation in anthropogenic systems can be assessed;
- Capitalize on opportunities emerging to enhance the role of protected areas as part of local, national, regional, and global responses to climate change and the increasing risks of climate-related disasters (see Sect. 17.3.4 for an example);
- Include climate change adaptation and mitigation in site management plans, with specific activities identified to involve local communities, restore degraded ecosystems, establish connectivity between protected areas, and integrate the protected area with the wider landscape and seascape (including transboundary connections where appropriate);
- Base any protected area management intervention designed to address climate change on the principle of adaptive management, with the results of management being carefully monitored so that lessons learned can be applied quickly and work toward constantly improving management;
- Use the "captive audience" of visitors to protected areas to communicate the major issues of climate change, including practical steps that visitors can take to help address the issues (see also Harmáčková et al., Chap. 5).

17.3.3   Protected Areas and Post-Disaster Recovery

Adapting to extreme climatic events also includes converting immediate disasters into longer-term benefits. For example, protected areas that include wetlands are adapted to flooding, a natural process that helps to provide nutrients to the system, enables fish to migrate, and recharges aquifers. Changing climates may make it possible to enlarge the wetlands within some protected areas, such as those along coastlines that are experiencing sea level rise or inland areas that are receiving significantly increased rainfall. This could enhance the effectiveness of protected areas, and help them carry out their process of recharging the ground water and aquifers that provide a significant proportion of the water consumed by humans. Protected areas more generally provide healthy ecosystems that can help restore floodplains after a heavy rainfall and provide emergency resources if required (Stolton et al. 2008).

Protected area managers need to be prepared to respond quickly to climate-related disasters (and indeed to any disasters). Appropriate steps can include:

- Seek to restore the natural vegetation as soon as possible after a destructive natural hazard such as a flood, a storm surge, a hurricane/typhoon, a landslide, or a catastrophic fire; avoid wherever possible the common response, especially along coasts, of hard engineering measures such as sea walls since the natural vegetation is both less costly and more sustainable, as well as maintaining the character of the protected areas;

- Rehabilitate green spaces, using native species, which may have been lost (for example, to pave a parking lot in the protected area before the disaster struck). This provides an opportunity to "build back better", using ecological rehabilitation and reconstruction (Fan 2013).
- Engage full and informed community participation in post-disaster recovery programmes, and consider ways to enable these communities to use some of the ecosystem services of the protected area in the recovery process;
- Use the disaster as a means to help communities prepare for future disasters (which seem likely to come, judging from recent history of climate-related extreme natural events). Such preparation can enable communities to become less vulnerable and more resilient to future disasters. This aspect of recovery may be led by other organizations, but active participation by the protected area site managers will demonstrate the added value of protected areas in facilitating recovery of affected communities.
- Develop disaster preparedness and contingency planning for protected areas, including training of staff and making any structures in the protected area as disaster resilient as possible.

More broadly, protected areas should be recognized as part of wider disaster risk reduction and recovery strategies, so the agencies with the larger mandate of coordinating risk reduction and response need to be shown the many contributions protected areas provide to reducing future risks and the services they provide to support recovery.

A more innovative step is to respond to a disaster by creating a new protected area, a step which could be taken most easily along coastlines that are likely to come under increasing threat. An outstanding example is the Sanriku Fukko Reconstruction National Park in Japan (see also Furuta and Seino, Chap. 13 and Takeuchi et al. Chap. 14; Takeuchi et al. 2014). While this was a response to a non-climate disaster, the new protected area will also provide numerous climate benefits, in terms of both mitigation and adaptation.

17.3.4 Building Protected Areas into the UNFCCC Clean Development Mechanism

Emerging opportunities in response to climate change can potentially provide financial support for the ecosystem services that protected areas provide in sequestering carbon, protecting watersheds from extreme climatic events, and maintaining the capacity of ecosystems to adapt to changing conditions (UNEP 2014). For example, Parties to the UNFCCC adopted in 2005 an initiative known as Reducing

Emissions from Deforestation and Forest Degradation in Developing Countries (REDD). It was originally focused solely on carbon, but due to pressure from some developing country governments, in 2010 a "+" was added to include the role of conservation, sustainable management of forests, inclusive governance, equitable sharing of costs and benefits, and enhancement of forest carbon stocks in the initiative, known hereafter as REDD+. Disaster risk reduction has not yet been an explicit part of the discussion, but could be implied though the acceptance of ecosystem-based adaptation as relevant to REDD+. Whether this will actually receive REDD+ funding remains to be seen.

While REDD+ is relatively new, the idea that the Kyoto Protocol should support protected areas in tropical forests because of their values for carbon sequestration remains essentially sound. This idea was raised already at the turn of the century (Kremen et al. 2000), though many governments object to it. This objection is apparently based on the assumption that governments of tropical countries have already established protected areas and pay for their effective management. This assumption has proven to be false repeatedly ever since. For example, only 24 % of protected areas are currently being effectively managed, according UNEP's World Conservation Monitoring Centre (Juffe-Bignoli et al. 2014). However, donor governments, apparently unaware of the serious problems facing protected area management (even in their own countries), have yet to change their position. This chapter has shown that protected areas are highly valuable for climate change adaptation and mitigation, and should be eligible for financial support to pay for the climate-related ecosystem services they provide to the larger community (Bietta et al. 2013).

The Annex 1 countries (the industrialized countries that are members of OECD) under the UNFCCC may also be giving insufficient attention to more general problems linked to the Clean Development Mechanism (CDM) developed as part of the Kyoto Protocol. Most of the investment in forests as forms of carbon storage has been given to "new" forests, "afforestation" (planting trees where no forests had occurred in recent times) instead of "reforestation" (planting trees to re-establish forests that had been cleared for other uses, such as agriculture). Afforestation is misdirected investment, because the young fast-growing trees do not approach the carbon storage capacity of old-growth forests for at least 200 years (Harmon et al. 1990), and CDM payments cover only a few years, far short of the residency time of carbon in the atmosphere. As a result, concern about "carbon farming" by planting new forests is coming under increasing criticism (Becker and Lawrence 2015), recognizing protection of the old-growth forests in protected areas as the most viable option (Smith et al. 2000).

The CDM's efforts to increase carbon sequestration by supporting afforestation also has a significant problem in relation to water flows. Jackson et al. (2005) found that such forests, often using eucalyptus or other non-native species, reduce stream flow by about 40 %, increasing to 50 % in older plantation forests. Their study of 114 planted forests in 16 countries found that about 10 % resulted in completely drying out streams and increased the soil acidity in 85 % of the sites. On the other hand, they found that reforestation of cropland that had replaced forest improved both the quantity and quality of water flow. Such reforestation could become new protected areas or could be added to existing ones, thereby helping to assure long-term benefits by contributing to their permanence.

To date, 38 "REDD+ Countries" are receiving funds from 16 "Funder Countries", with over US$4 billion in funding already pledged (UNEP 2014). Optimists expect this level of funding to increase, and it could provide a useful source of funding to forest protected areas. The World Bank's Carbon Fund (which held US $2.5 billion in 2014) agreed in December 2013 to 38 criteria that will allow forested countries to sell REDD+ emission reductions to the Carbon Fund from 2015 to 2020, though protected areas have still not received specific mention (World Bank 2014).

Even if a country is not part of REDD+, the carbon stored in its protected areas is still valuable because in its reports to the UNFCCC the country can indicate the public good value of the carbon stored in this form. While this seems unlikely to yield funds directly for protected areas, it can be used to justify to governments an increase in budgets for protected areas to maintain the increasingly valuable old-growth forests, wetland, coral reefs, and other biotopes contained within protected areas.

17.4 Building a Stronger Constituency for Including Protected Areas in Climate Change and Disaster Risk Reduction Programmes

The designers of national and international responses to climate change have yet to pay adequate attention to the role of protected areas. This may be due to a lack of information about the multiple contributions protected areas make to climate change issues. To the extent that this is the case, those promoting greater recognition of the role of protected areas need to provide more effective messages. These messages need to be sent to the relevant government agencies (including Ministries of Foreign Affairs that will be leading international climate negotiations, Ministries of Finance who may be interested to learn the dimensions of the public goods protected areas provide, Ministries of Agriculture that will be addressing climate issues on the ground, Ministries of Environment that might be preparing national reports for submission to the Conference of Parties of the UNFCCC, Ministries of Energy that will be seeking low-carbon approaches to development, and many others). Political leaders and relevant private businesses may also be interested, provided civil society is sufficiently mobilized to push towards this direction. The recent dramatic increase in divestments from fossil fuel companies indicates the kind of move that is required. The kinds of messages that might generate such an effort could include:

- Climate changes are already affecting vulnerable populations in many parts of the world (IPCC 2014), with the various weather-related disasters such as heat waves in Australia, droughts in California, disastrous rainfall in India and Pakistan, Typhoon Haiyan in the Philippines, and so forth hitting hardest those who are already disadvantaged. Such extreme natural events are indicators of how the future might well look, indicating that the ecosystem services being provided by protected areas will be under even higher demand. This clearly calls for greater investments in protected areas to ensure that their benefits can continue to be provided.

- Old growth forests continue to increase their capacity to store carbon as they age, and since protected areas hold most of a country's old-growth forest, they provide an important means of storing up to 20 % of the country's carbon emissions. Most coral reefs, mangroves, and marine grass beds are also in protected areas, and they too are effective carbon sinks. Therefore, protected areas need to be included when discussing carbon sequestration at national and international level.

- The increasing threat of extreme climatic events, such as storm surges, can be addressed in part by protected areas that are located along the coastline (especially mangroves, coastal wetlands, and coral reefs). They therefore need to be included as part of the discussion on extreme natural events and disaster risk reduction. This coastal vegetation is also effective in protecting coastlines for high-frequency, low-magnitude events that are likely to occur with sea level rise.

- Climate change may be felt especially through changes in rainfall and seasonality that affect the flow of water to farms, cities, and industries, and sometimes lead to floods. Protected areas, which contain most of the old-growth forest of the country, are important providers of water through watershed protection, and help to limit the effects of floods. Since every human depends on water, the water-related ecosystem services provided by protected areas deserve strong support from all.

- Protected areas can play a strong social role in disaster risk reduction in relation to climate change, giving local communities confidence that the natural ecosystems in nearby protected areas are helping to protect them against floods, landslides, and any disastrous effects of storms.

- Because protected areas are so effective in sequestering carbon and adapting to climate change, they should form a foundation for national plans for climate-related disaster risk reduction and adaptation to climate change. And because they support ecosystems such as forests and coral reefs that are the most effective at sequestering carbon, they should be given highest priority for investments under REDD+.

17.5 Conclusions and Recommendations

This chapter has indicated some of the multiple values of protected areas to disaster risk reduction in relation to climate change. These sites have many other advantages that make them unique contributors (Dudley et al. 2013; Buyck et al. 2015).

Protected areas are legally established, have trained staff on site, and are well familiar with the surrounding lands and the kinds of climate-related disasters that pose risks to them (and to the protected areas). They may have a legal mandate to provide support in the case of disasters from extreme hazard events and are accustomed to providing such support. While climate change may be just one more management issue to be addressed, the increasing priority this issue is receiving will lead to stronger responses in many protected area sites. The steps outlined in Sect. 17.3.2 above may be helpful to them.

Even if the Aichi Biodiversity Target 11 is fully implemented, 83 % of the land and 90 % of the sea will still need attention. Addressing climate change-related risk reduction needs to involve far more than the formal legally protected areas, and in addition involve the many "other effective area-based conservation measures" that can include community conserved areas, lands where indigenous peoples are rights-holders, and other land and ocean management approaches that may be able to apply some of the management measures suggested in this chapter.

This chapter has presented evidence demonstrating that protected areas are an important part of disaster risk reduction in relation to climate change and the many other events and processes that affect human interests but can be ameliorated by protected areas. But protected areas can play their role in disaster prevention and mitigation only if other agencies, local communities, businesses, and others also manage their ecosystems with hazard events in mind.

Managing rivers in as natural a form as possible has been shown to be far more effective than attempting to guide them through engineering approaches, though of course this may not be possible in areas with high-density human populations. Keeping floodplains available for fulfilling their natural functions is usually far better than using what some might perceive as "wastelands" to build new factories or other infrastructure. Zoning should avoid construction in areas vulnerable to disasters, such as along coastlines or in and around wetlands. Roads constructed in mountain areas, especially in protected areas, need to be designed to minimize their vulnerability to landslides. In other words, disaster risk reduction is a team effort, and it is time to give protected areas their well-earned opportunity to contribute, and to receive the necessary support to enable them to do so.

While the specific impacts of a changing climate remain essentially unpredictable, national systems of protected areas need to include measures to recognize the role of protected areas in mitigating climate change and adapting to it, sustaining ecosystem processes and functions, and capitalizing on new sources of support for effective management of protected areas. To contribute to the latter point, governments should continue negotiating REDD+ as an important part of the UNFCCC implementation, and direct its international financial support to protected areas that conserve old-growth forests, coral reefs, wetlands, mangroves, and other ecosystems that make important contributions to climate change mitigation and adaptation. This would be a major improvement in the effectiveness of the Clean Development Mechanism (CDM).

However, for the contribution of protected areas to be fully realized, they need to be brought into local, regional, and national climate change mitigation and adaptation programs. This chapter has provided both the justification for doing so and some practical steps that can be taken, often at minimal cost, to enable protected areas to take their proper prominent role in climate-related disaster risk reduction.

References

Akçakaya HR, Butchart SHM, Watson JEM, Pearson RG (2014) Preventing species extinctions resulting from climate change. Nat Clim Chang 4:1048–1049

Barber CV, Miller KR, Boness M (eds) (2004) Securing protected areas in the face of global change: issues and strategies. IUCN, Gland

Becker K, Lawrence P (2015) Carbon farming: the best and safest way forward? Carbon Manage 5 (1):31–33

Bennett A (2003) Linkages in the landscape: the role of corridors and connectivity in wildlife conservation. IUCN, Gland

Bietta F, Chung P, Massai L (2013) Supporting international climate negotiators: lessons learned by the coalition for rainforest nations. Coalition for Rainforest Nations, New York

Bintoora AK (2014) Initiatives to combat landslides, floods and effects of climate change in Mt. Elgon Region. In: Murti R, Buyck C (eds) Safe havens: protected areas for disaster risk reduction and climate change adaptation. IUCN, Gland

Bonan GB (2008) Forests and climate change: forcings, feedbacks, and the climate benefits of forests. Science 320:1440–1449

Borrero JC (2005) Field data and satellite imagery of tsunami effects in Banda Aceh. Science 308:1596

Brown J, Mitchell N, Beresford M (2005) The protected landscape approach: linking nature, culture and community. IUCN, Gland

Burby RJ (2006) Hurricane Katrina and the paradoxes of government disaster policy: bringing about wise governmental decisions for hazardous areas. Ann Am Acad Pol Soc Sci 604 (1):171–191

Buyck C, Dudley N, Furuta N et al (2015) Protected areas as tools for disaster risk reduction: a handbook for practioners. Japan Ministry of Environment/IUCN, Gland

Cai W et al (2015) Increased frequency of extreme La Nina events under greenhouse warming. Nat Clim Chang. doi:10.1038/nclimate2492

Campbell A, Miles L, Lysenko I et al (2008) Carbon storage in protected areas: technical report. UNEP World Conservation Monitoring Centre, Cambridge, UK

Costanza R, Perez-Maqueo O, Martinez ML et al (2008) The value of coastal wetlands for hurricane protection. Ambio 37(4):241–248

Dahdouh-Guebas F, Jayatissa LP, Nitto D et al (2005) How effective were mangroves as a defence against the recent tsunami? Curr Biol 15(12):R443–R447

Damschen E, Haddad NM, Orrock JL et al (2006) Corridors increase plant species richness at large scales. Science 313:1284–1286

Dawson TP, Jackson ST, House JI et al (2011) Beyond predictions: biodiversity conservation in a changing climate. Science 332:53–58

Day JW, Boesch DF, Clairain EJ et al (2007) Restoration of the Mississippi delta: lessons from hurricanes Katrina and Rita. Science 315:1679–1684

Dickson MG, Orme CDL, Suttle KB, Mace GM (2014) Separating sensitivity from exposure in assessing extinction risk from climate change. Sci Rep. doi:10.1038/srep06898

Dudley N, MacKinnon K, Stolton S (2013) Reducing vulnerability: the role of protected areas in mitigating natural disasters. In: Renaud FG et al (eds) The role of ecosystems in disaster risk reduction. United Nations University Press, Tokyo

Duffy JE (2009) Why biodiversity is important to the functioning of real-world ecosystems. Front Ecol Environ 7:437–444

Emerton L., Kekulandala L. (2003). Assessment of the economic value of Muthurajawela Wetland. Occasional Papers of IUCN Sri Lanka 4: iv + 1–28

Erisman JW, Brasseur G, Ciais P et al (2015) Put people at the centre of global risk management. Nature 519:151–153

Fan L (2013) Disaster as opportunity? Building back better in Aceh, Myanmar, and Haiti. Overseas Development Institute, London

Ford M (2014) Hurricane Katrina: the role of US National Parks on the Northern Gulf of Mexico and post-storm wetland restoration. In: Murti R, Buyck C (eds) Safe havens: protected areas for disaster risk reduction and climate change adaptation. IUCN, Gland

Ghilarov A (2003) Ecosystem functioning and intrinsic value of biodiversity. Oikos 90 (2):408–412

Gitay H, Suarez A, Watson RT, Dokken DJ (2002) Climate change and biodiversity, IPCC technical paper V. Intergovernmental Panel on Climate Change, Geneva

Green AL, Fernandes L, Almany G et al (2014) Designing marine reserves for fisheries management, biodiversity conservation, and climate change adaptation. Coast Manag 42(2):143–159

Groom MJ, Meffe GK, Carroll CR (2006) Principles of conservation biology, vol 3. Sinauer Associates, Sunderland

Groves C, Game ET, Anderson MG et al (2012) Incorporating climate change into systematic conservation planning. Biodivers Conserv 21:1651–1671. doi:10.1007/s10531-012-0269-3

Harmon ME, Ferrell WK, Franklin JF (1990) Effects on carbon storage of conversion of old-growth forests to young forests. Science 247:699–702

Hess GR, Fischer RA (2011) Communicating clearly about conservation corridors. Landsc Urban Plan 55:195–208

Hobbs RJ, Hallett LM, Ehrlich PR, Mooney HA (2011) Intervention ecology: applying ecological science in the 21st century. Bioscience 61(6):442–450

Hoegh-Guidberg O, Hughes L, McIntyre S et al (2008) Assisted colonization and rapid climate change. Science 321:345–346

House TJ, Near JB, Shields WB, et al. (1996) Weather as a force multiplier: owning the weather in 2025. DTIC Online, Accession Number: ADA333462

IPCC (2014) Climate change 2014: impacts, adaptation, and vulnerability. Intergovernmental Panel on Climate Change, Geneva

IUCN (2014) The IUCN red list of threatened species. Version 2014.3. Available via http://www.iucnredlist.org. Accessed on 15 Jan 2015

IUCN/SSC (2013) Guidelines for reintroductions and other conservation translocations, Version 1.0. IUCN Species Survival Commission, Gland

Jackson RB, Jobbágy EG, Avissar R et al (2005) Trading water for carbon with biological sequestration. Science 310:1944–1947

Juffe-Bignoli D, Burgess ND, Bingham H et al (2014) Protected planet report 2014. UNEP-WCMC, Cambridge, UK

Keddy PA, Fraser LH, Solomeshch AI et al (2009) Wet and wonderful: the world's largest wetlands are conservation priorities. Bioscience 59(1):39–51

Kremen C, Niles JO, Dalton MG et al (2000) Economic incentives for rain forest conservation across scales. Science 288:1828–1832

Kumar P (ed) (2010) The economics of ecosystems and biodiversity: ecological and economic foundations. Earthscan, London

Lawton RO, Nair US, Pielke RA et al (2001) Climatic impact of tropical lowland deforestation on nearby montane cloud forests. Science 294:584–587

Le Saout S, Hoffmann M, Shi Y et al (2013) Protected areas and effective biodiversity conservation. Science 342:803–805

Lewis SL, Maslin MA (2015) Defining the Anthropocene. Nature 519:171–180

Lewis SL, Lopez-Gonzalez G, Sonké B et al (2009) Increasing carbon storage in intact African tropical forests. Nature 457:1003–1006

Lombard AT, Cowling RM, Vlok JHJ, Fabricius C (2010) Designing conservation corridors in production landscapes: Assessment methods, implementation issues, and lessons learned. Ecol Soc 15(3):7. Available via http://www.ecologyandsociety.org/vol15/iss3/art7/

Luyssaert S, Detlef Schulze E, Börner A et al (2008) Old-growth forests as global carbon sinks. Nature 455:213–215

MEA (Millennium Ecosystem Assessment) (2005) Ecosystems and human well-being: biodiversity synthesis. World Resources Institute, Washington, DC

Monastersky R (2015) The human age. Nature 519:143–147

Mueller L, Bresch D (2014) Economics of climate adaptation in Barbados: facts for decision making. In: Murti R, Buyck C (eds) Safe havens: protected areas for disaster risk reduction and climate change adaptation. IUCN, Gland

Murti R, Buyck C (eds) (2014) Safe havens: protected areas for disaster risk reduction and climate change adaptation. IUCN, Gland

Ninan KN (ed) (2009) Conserving and valuing ecosystem services and biodiversity. Earthscan, London

Orr JC, Fabry VJ, Aumont O et al (2005) Anthropogenic ocean acidification over the twenty-first century and its impact on calcifying organisms. Nature 437:681–686

Pacifici M, Foden WB, Visconti P et al (2015) Assessing species vulnerability to climate change. Nat Clim Chang 5:215–224

Pan Y, Birdsey RA, Fang J et al (2011) A large and persistent carbon sink in the world's forests. Science 333:988–993

Pan Y, Birdsey RA, Phillips LO, Jackson RB (2013) The structure, distribution, and biomass of the world's forests. Annu Rev Ecol Evol Syst 44:593–622

Peters RL, Lovejoy TE (eds) (1992) Global warming and biological diversity. Yale University Press, New Haven

Pounds JA, Fogden MPL, Campbell JH (1999) Biological response to climate change on a tropical mountain. Nature 398:611–615

Reside AE, VanDerwal J, Phillips B et al (2013) Climate change Refugia for terrestrial biodiversity: defining areas that promote species persistence and ecosystem resilience in the face of global climate change. James Cook University and National Climate Change Adaptation Research Facility, Gold Coast

Root TL, Price JT, Hall KR et al (2003) Fingerprints of global warming on wild animals and plants. Nature 421:57–60

Rosenzweig ML (1995) Species diversity in space and time. Cambridge University Press, Cambridge, UK

SCBD (2009) Connecting biodiversity and climate change mitigation and adaptation. Secretariat of the Convention on Biological Diversity Technical Series 41: 1–126

Schimel D, Stephens BB, Fisher JB et al (2014) Effect of increasing CO_2 on the terrestrial carbon cycle. Proc Natl Acad Sci 112(2):436–441

Schuyt K, Brander L (2004) The economic values of the World's wetlands. WWF, Gland

Seidl AF, Moraes AS (2000) Global valuation of ecosystem services: application to the Pantanal da Nhecolandia, Brazil 33: 1–6

Smith J, Mulongoy K, Persson R, Sayer JA (2000) Harnessing carbon markets for tropical forest conservation: towards a more realistic assessment. Environ Conserv 27(3):300–311

Spellerberg I (2005) Monitoring ecological change. Cambridge University Press, Cambridge, UK

Steffen W, Crutzen P, McNeill J (2007) The Anthropocene: are humans now overwhelming the great forces of nature. Ambio 36(8):614–621

Stephenson NL, Das AJ, Condit R et al (2014) Rate of tree carbon accumulation increases continuously with tree size. Nature 507:90–93

Stolton S, Dudley N, Randall J (2008) Natural security: protected areas and hazard mitigation. WWF, Gland

Takeuchi K, Elmqvist T, Hatakeyama M et al (2014) Using sustainability science to analyse social-ecological restoration in NE Japan after the great earthquake and tsunami of 2011. Sustain Sci 9:513–526

Thomas CD, Cameron A, Green RE et al (2004) Extinction risk from climate change. Nature 427:145–148

Tibbetts J (2006) Louisiana's wetlands: a lesson in nature appreciation. Environ Health Perspect 114(1):A40–A43

Tittensor DP, Walpole M, Hill SLL et al (2014) A mid-term analysis of progress toward international biodiversity targets. Science 346:241–244

UNEP (1992) The United Nations convention on biological diversity. United Nations Environment Programme, Nairobi

UNEP (2014) Building natural capital: How REDD+ can support a Green Economy: a report from the International Resource Panel. United Nations Environment Programme, Nairobi

Wilson EO (ed) (1988) Biodiversity. National Academy Press, Washington, DC

Woodruff DS, Woodruff KA (2008) Paleogeography, global sea level changes, and the future coastline of Thailand. Nat Hist Bull Siam Soc 56(1):1–24

World Bank (2014) The World Bank carbon funds and facilities. World Bank, Washington, DC

Young KR (2014) Biogeography of the Anthropocene: novel species assemblages. Prog Phys Geogr 38(5):664–673

Part IV
Research and Innovation

Chapter 18
Ecosystem Services of Coastal Dune Systems for Hazard Mitigation: Case Studies from Vietnam, Indonesia, and Chile

Udo Nehren, Hoang Ho Dac Thai, Muh Aris Marfai, Claudia Raedig, Sandra Alfonso, Junun Sartohadi, and Consuelo Castro

Abstract In many developing and emerging tropical and subtropical countries, coastal dune systems (CDS) are under high pressure, which leads to progressive degradation and loss of dune areas. This in turn weakens the protection function against coastal hazards. In this chapter we discuss CDS in three case studies: Thua Thien-Hue province (Central Vietnam), Parangtritis (Java Island, Indonesia), and Ritoque (Central Chile). For these CDS, we assess relevant ecosystem services (ES) with particular regard to protection services as well as the current degradation status through a rapid assessment approach. Moreover, we analyse the legal frameworks for coastal dune management and protection in the case study countries. Main results include indicator sets for assessing ES and the degradation status of CDS, which are transferable to other coastal dune areas. Based on these sets we evaluate and compare the three dune systems and provide policy recommendations for a more efficient regulation and management of CDS.

U. Nehren (✉)
Institute for Technology and Resources Management in the Tropics and Subtropics (ITT),
TH Köln, University of Applied Sciences, Köln, Germany
e-mail: udo.nehren@fh-koeln.de

H.H.D. Thai
Institute of Resources and Environment, Hue University, Hue City, Vietnam

M.A. Marfai • J. Sartohadi
Faculty of Geography, Universitas Gadjah Mada, Yogyakarta, Indonesia

C. Raedig
Technology Arts Sciences TH Köln, Institute for Technology and Resources Management in the Tropics and Subtropics (ITT), Betzdorfer Str. 2, 50679 Cologne, Germany

S. Alfonso
Technology Arts Sciences TH Köln, Institute for Technology and Resources Management in the Tropics and Subtropics (ITT), Betzdorfer Str. 2, 50679 Cologne, Germany

Institute of Geography, University of Innsbruck, Innsbruck, Austria

C. Castro
Institute of Geography, Pontificia Universidad Católica de Chile, Santiago de Chile, Chile

© Springer International Publishing Switzerland 2016   401
F.G. Renaud et al. (eds.), *Ecosystem-Based Disaster Risk Reduction and Adaptation in Practice*, Advances in Natural and Technological Hazards Research 42,
DOI 10.1007/978-3-319-43633-3_18

Keywords Coastal dune systems (CDS) • Ecosystem services (ES) • Degradation • Hazard mitigation

18.1 Introduction: The Role of Coastal Dunes for Disaster Risk Reduction and Climate Change Adaptation

Coastal dune systems (CDS) act as natural buffers between land and sea and thereby protect people and infrastructure from natural hazards (Sudmeier-Rieux et al. 2006; Prasetya 2007; Gonsalves and Mohan 2012; Hettiarachchi et al. 2013). Depending on their height and width, shape, continuity, and ecological status, CDS can reduce the physical exposure of inland areas to hazards such as tropical cyclones, storm surges, wave action, coastal floods (Gómez-Pina 2002; Dahm et al. 2005; Takle et al. 2007; Thao et al. 2014), and, to some extent, tsunamis (Liu et al. 2005; Bambaradeniya et al. 2006; Bhalla 2007; Mascarenhas and Jayakumar 2008). Moreover, they control coastal erosion (Prasetya 2007; Barbier et al. 2011) and, at least to a certain degree, mitigate climate change-related impacts such as sea level rise and saltwater intrusion (Carter 1991; Heslenfeld et al. 2004; Saye and Pye 2007). However, at the same time coastal dunes are prone to accelerated erosion processes caused by sea-level rise (Brown et al. 2013).

Despite their importance for coastal protection, in many parts of the world, coastal dunes are under severe pressure. This is particularly the case in many tropical and subtropical countries due to urbanization processes (Martínez et al. 2004), sand mining (Sridhar and Bhagya 2007; Miththapala 2008; Takagi et al. 2014), agricultural expansion (French 2001) as well as development of aquaculture (Phan and Nguyen 2006). These activities lead to different forms of degradation, such as fragmentation, soil sealing, pollution, and introduction of invasive plant and animal species (see also Senhoury et al., Chap. 19). In the worst case, dunes are completely removed.

Even though well-developed CDS are found worldwide including in tropical and subtropical regions (Martínez and Psuty 2007), most scientific research is concentrated on mid-latitude dune systems, where the most wide-ranging conservation measures have been established (Doody 2013). In contrast, little attention has been paid to tropical and subtropical CDS (Moreno-Casasola 2004). Therefore conservation and management strategies of CDS in developing and emerging countries of the tropics and subtropics need to be improved, which requires a systematic identification and valuation of the various ecosystem services (ES) that coastal dunes provide, as well as an evaluation of the current degradation status.

It is well known that dune degradation and loss have negative impacts on their buffering function against natural hazards and on other ES, as essential functions of the CDS are increasingly eroded (Grootjans et al. 2013). For instance, the impacts of the 2004 Indian Ocean Earthquake and Tsunami were less severe along the south-eastern coastline of Sri Lanka in areas where natural dunes had been preserved (IUCN 2005). Moreover, Ishikawa (1988, cited in Harada and Imamura 2005) reported that coastal sand dunes of more than 10 m height prevented tsunami damages inland in the Aomori and Akita prefectures in Japan from the Nihonkai-Chubu Earthquake and Tsunami in 1983. On the other hand, the 2011 tsunami in

Tohoku, Japan, passed through sand dunes covered with vegetation and completely destroyed houses behind the dunes at distances of up to 1.6 km from the coastline, and led to damages at distances of up to 5.2 km inland (Yasuda et al. 2012). Therefore, we must be aware that the protection function of CDS against tsunamis depends on the magnitude of the hazard event and the physical and ecological condition of the CDS. To estimate the buffering capacity of the CDS, risk models are required that take into account the probability of tsunamis of certain heights, the physical and ecological characteristics of the dune system, and the exposure of people and infrastructure. Models, mainly based on historical records, also exist to estimate the risk of tropical cyclones and storm surges. However, these models are complex and require a lot of ground-based data, for instance, time sequence data in high spatial resolution based on monitoring over decades. Since such data are rarely available for dune areas, their application is not yet widespread, which is particularly true for tropical and subtropical countries. Even though it is difficult to clearly define the hazard protection function of CDS due to the lack of sufficient experimental evidence, we can say that coastal dunes offer protection to a certain extent and strengthen the socio-ecological resilience (as defined by Adger et al. 2005) of coastal communities to disasters in the medium to long term.

Presently, 44 % of the world's population lives within 150 km of the coast (UN Atlas of the Oceans 2014), 80 % of major cities with over 10 million inhabitants are coastal (Martínez 2008), and many coastal ecosystems are continuously degraded (Lotze et al. 2006; Waycott et al. 2009). Considering that both the world's population and urbanization processes will continue to increase, we can assume that land use pressures on CDS in urbanized areas of many tropical and subtropical countries will further intensify. Moreover, the demand for sand for construction and the extraction of minerals will also grow, which will affect both urban and rural areas. Sand is a finite resource, and according to Peduzzi (2014), 40 billion tonnes of sand are used each year globally – with a rapid upward tendency. Already today, the demand cannot be met by exploitation of sand from inland deposits, such as river beds and alluvial plains. Sand from desert regions is not ideal for concrete production as sand grains rounded by wind erosion do not bind well (Zhang et al. 2006). Therefore, beach and dune sand mining will likely increase with severe consequences for the environment (Peduzzi 2014).

CDS provide a variety of other important ES, such as freshwater provision (van der Meulen et al. 2004; Barbier et al. 2011), raw material (Kallesøe et al. 2008), food and medicine (Sridhar and Bhagya 2007; Kallesøe et al. 2008; Spalding et al. 2014), carbon sequestration (Alonso et al. 2012; Spalding et al. 2014), and space for tourism, recreation, education, and research (Moreno-Casasola 2004; Doody 2013). Furthermore, CDS are important habitats for marine and non-marine animals, including nesting sites and stop-over sites for migratory birds (de Silva and Premachandra 1998; Schlacher et al. 2014) and therefore significantly contribute to biodiversity conservation of CDS and the surrounding coastal environment.

Due to the importance of CDS for the provision of multiple ES, and in particular for disaster prevention and risk reduction, conservation and restoration efforts must be intensified (Lithgow et al. 2013, 2014). In parallel, sustainable use of CDS has to be further explored, applied and monitored. To date, only few assessments of ES of coastal dunes exist (Mendoza-Gonzales et al. 2012), and there is a lack of reliable ground data on the current status of CDS in many tropical and subtropical countries. We have started to tackle this gap by providing assessment schemes for ES and the degradation status of CDS that can be applied in data-poor environments. Thereby, we provide policymakers and other stakeholders with the necessary scientific basis to define key priorities for political intervention and action.

We selected three model regions which we analysed and compared: (i) the dunes of Thua Thien-Hue, Central Vietnam, (ii) the dunes of Parangtritis, Yogyakarta, Indonesia, and (iii) the dunes of Ritoque, metropolitan region of Valparaíso, Central Chile. The three CDS provide important protection functions against coastal hazards and are at the same time facing severe threats due to unsustainable management. To make the assessment of CDS in different regions comparable, we developed evaluation matrices with relevant ES provided by the dunes and defined criteria to evaluate their degradation status. Moreover, we analysed the legal frameworks related to coastal dune utilization and conservation. The study builds on the authors' long-term research experience in the study regions and aims at providing a methodological basis to assess the status of CDS in data-poor environments which could be transferred to other tropical and subtropical coastal regions.

18.2   Case Study Areas

18.2.1   Dunes of Thua Thien-Hue, Central Vietnam

Thua Thien-Hue province is located in North Central Vietnam and shares borders with Quang Tri province in the north, Lao People's Democratic Republic in the west, and Da Nang and Quang Nam provinces in the south. The province lies within the tropical monsoon climate (Am) according to the Koeppen classification, with two main seasons: the rainy season from September to February and the dry season from March to August. The annual mean temperature is about 25.6 °C and the total annual precipitation varies between 2800 and 3300 mm (General Statistical Office 2013).

Occupying a total area of 5054 km^2, Thua Thien-Hue province has a coastline of about 127 km in length and is located in the narrowest part of Vietnam, with an average width of about 60 km. The topography of the coastal zone is flat to undulate with coastal plains, sand dunes, and lagoons, while the hinterland is hilly to mountainous with altitudes of up to 1600 m above sea level (m.a.s.l.) The province's population is 1.12 million with a density of 223 inhabitants/km^2. The population is unevenly distributed with more than 80 % living in the coastal plains and sandy areas, where income generation of the rural population is based on paddy rice cultivation, aquaculture, and fishing (General Statistical Office 2013, reference year 2011).

In Thua Thien-Hue province sandy coastal areas cover about 27,130 ha with a narrow band of sand dunes stretching along the coastline. The average height of the dunes lies between 10 and 15 m, and maximum height is about 30 m, with steep slopes towards the ocean and towards the inland (Hoang, pers. obs.; see Fig. 18.1a). There is little knowledge on the native vegetation of the CDS. In the southern parts of Central Vietnam where CDS extend up to 30 km inland, the unique coastal forest formation of Cam Ranh Bay can still be found today (Sterling et al. 2006). We assume that historically the dunes of Thua Thien-Hue were also covered by coastal forest with indigenous tree species, but due to the narrow coastal strip in the province, those forests probably had a comparatively small extent. Today, most of the dunes are either bare, or covered with shrubby vegetation or exotic tree plantations of species which can grow on nutrient-poor soil and have an elevated level of salt tolerance, such as *Casuarina equisetifolia, Acacia auriculiformis, A. crassicarpa, Eucalyptus camaldulensis,* and *E. tereticornis.*

18.2.2 Dunes of Parangtritis, Yogyakarta Special Region, Central Java, Indonesia

This small dune system with about 400 ha of inactive (i.e. stabilized) dunes and about 70 ha of active dunes (i.e. continuously moving) is located on the southern coast of Java Island in Parangtritis Coastal area, Kretek District, Bantul Regency, about 30 km south of the center of Yogyakarta urban area. It faces the Indian Ocean and consists of barchan (crescent-shaped) dunes (Fig. 18.2a) and longitudinal dunes fed by deposits from the Merapi volcano. The climate is tropical wet and dry or savanna (Aw) according to the Koeppen classification with dominance of easterly winds and annual precipitation of about 750–1100 mm in Bantul Regency (Bantul Regency 2014a). Bantul Regency covers a total area of 506.86 km^2 with a population of about 911,500, which corresponds to a population density of 1800 inhabitants/km^2 (Bantul Regency 2014b).

Geomorphologically the dunes of Parangtritis are exceptional as barchan dunes are usually formed in arid environments and not in the humid tropics (Sunarto et al. 2010). Besides coastal sand dunes, parts of the coastal area are formed by land deposits from erosion processes along the Opak River and local marine deposits. The active dunes facing the Indian Ocean are often covered with *Spinifex sp.* (Fig. 18.2b) and *Ipomoea pes-caprae* as typical pioneer vegetation on foredunes. In some areas *Casuarina equisetifolia* has been planted to stabilize the dunes. Native *Pandanus* species (Fig. 18.2a) and *Calotropis gigantea* are also frequently found.

Fig. 18.1 CDS adjacent to Hue City, Central Vietnam (**a**) View of the coastal dune system from the inland, (**b**) freshwater originating from an aquifer under the dune; the natural channel was widened by the local inhabitants for easier access to freshwater, (**c**) view of a seaward cemetery from the top of the dune; shrimp ponds are visible in the back

18.2.3 Dunes of Ritoque, Metropolitan Region of Valparaíso, Central Chile

The dunes of Ritoque are a well-developed system in the metropolitan region of Valparaíso in Chile's 5th region (Fig. 18.3) covering an area of about 2000 ha. Due to their location between the towns of Concón and Quintero, about 30 km north of Valparaíso city, the dunes of Ritoque are designated as urban to peri-urban.

The dune system is mainly fed by fine grained sands from the Aconcagua river basin, and to a lesser extent from weathering material of coastal rocks and probably also from sediments south of the Aconcagua Estuary (Toral Ibáñez et al. 1980). The climate is Mediterranean with warm and dry summers and mild winters (Csb

Fig. 18.2 Study area of Parangtritis, Central Java, Indonesia (**a**) barchan dune, (**b**) foredune with herbaceous vegetation dominated by *Spinifex sericeus* (the long trailing runners stabilize the dune sand) and *Pandanus sp.* trees

climate according to the Koeppen classification). Total annual precipitation in the commune of Quintero is about 500 mm (Wolf Eigenherr 2008).

The geomorphology of the dunes results from the interaction of winds blowing predominantly from southwestern directions, the type and coverage of the colonizing vegetation, and the wave dynamics with remarkable seasonal and inter-annual variations (Castro 1987). The active dune system consists of incipient dunes and foredunes with pioneer vegetation (Fig. 18.3a) and mainly bare rear dunes with a height of up to 30 m.a.s.l. that are dissected by an interdunal depression. Behind the rear dunes runs a valley that is connected to the Mantuaga Estuary (Fig. 18.3b). From the valley, there is a steep rise to the stabilized Early Holocene and Pleistocene longitudinal dunes that reach altitudes of 80–100 m.a.s.l. These dunes were

Fig. 18.3 Study area of Ritoque, Central Chile (**a**) Foredunes covered by clumps of herbaceous vegetation, (**b**) Wetland of the Mantagua Estuary, (**c**) Sclerophyllous woodland in a narrow valley of a stabilized dune

naturally covered by woody vegetation and scrubs, but today only small patches of the native vegetation remain (Fig. 18.3c).

18.3 Methods

The three CDS represent an urban-rural gradient from the urbanized metropolitan region of Valparaíso/Viña del Mar, to the peri-urban dunes of Thua Thien-Huen and to the rural area of Parangtritis. We analysed the three CDS with respect to (a) the provision of ES, (b) the degradation status, and (c) the legal frameworks related to the use and management of coastal dunes, to provide an easily applicable and consistent methodology for the assessment of ES and degradation status of CDS that can be used by policymakers, planners, and decision makers in data-poor environments. With the legal framework analysis, we want to draw attention to the complexity of laws that address dune management in the different countries and underpin the need for integrated dune management policies.

The following describes the three main steps undertaken for the study:

(a) ES were categorized according to the Millennium Ecosystem Assessment (MA 2005), namely supporting, regulating, provisioning, and cultural services. Supporting services are supplemented by habitats for species, as suggested by the "The Economics of Ecosystems and Biodiversity" initiative (TEEB 2014), as well as by biodiversity and geodiversity (Fig. 18.4). Biodiversity is a crucial structural component of ecosystems that directly contributes to ecosystem functioning and the variety of available ecosystem services (Naeem et al. 1999; MA 2005). Geodiversity is a relatively new concept that considers the variety of earth materials, forms, and processes (Gray 2004), which in turn links to biodiversity and ES. For instance, the parent material has an impact on species distribution and abundance (Gray 2004). This becomes particularly obvious in large and complex CDS like the dunes of Ritoque, where abiotic

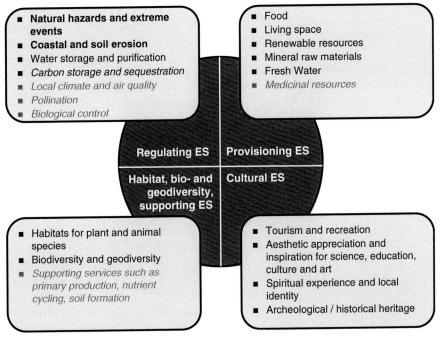

Fig. 18.4 Important ecosystem services (ES), habitats, and biodiversity and geodiversity provided by CDS (*in bold*: regulating services directly related to natural hazards and geoprocesses; *in italics*: services which could not be assessed in the case study areas on basis of available data)

factors such as parent material, topography, aspect, water balance, and micro-
climate vary at a small scale and create a mosaic of ecosystems and habitats.

For each of the four categories, we created a list of relevant services based on
a review of the scientific literature and our own field observations. Where
possible, we assessed these services in the three case study areas and rated
them using a five-step scale from 'very important' to 'not important'. The rating
is based on a set of qualitative indicators that was developed by the authors
(Table 18.1). We are aware that in further studies some of these indicators need
to be specified and others replaced by quantitative indicators to improve the
measurability and comparability. We hope that the methodologies provided by
this study will trigger both new studies in other CDS as well as the further
development of applicable quantitative indicators. The three investigated dune
systems were evaluated by the authors, all of whom have several years of
research experience in the respective dune areas.

Figure 18.4 summarizes the selected ES. Some services are specific for CDS
and require a special explanation. This is the case when distinguishing between
regrowing (biotic) and mineral (abiotic) raw materials, a distinction which we
consider useful as there is a considerable difference when using for instance
timber from plantations and sand for construction purposes. Moreover, we
included living space for humans as a provisioning service and archaeologi-
cal/historical heritage as a cultural service of CDS. All ES are valuated based on
services provided, regardless of possible destructive utilization. This means for
instance that sand extraction is considered a provisioning service, even though
its overexploitation may lead to severe degradation or even the complete loss of
the dune system.

(b) The **degradation status** of the three CDS is evaluated based on field observations and literature review in the form of a rapid assessment. For the assessment, we developed a scheme of qualitative indicators. Each indicator is ranked
from 1 = no degradation to 5 = very high degradation (Table 18.2). Apart from
the degradation status, we also assessed the total loss of dune area as a result of
sand removal. To this end, we used historical data to reconstruct the original
extent of the dune system and compared it with the current size. For our case
study areas, we found reliable historical information which allowed us to
calculate the total dune loss with good approximation.

(c) The **legal framework,** as the third step in our analysis, seeks to explore (i) the
extent to which coastal dunes are addressed in legal frameworks, (ii) how
specific the corresponding legal regulations are, (iii) if there are clearly defined
conservation categories and management strategies for CDS, and (iv) if there
are competing legislative goals and requirements. We considered the relevant
legal frameworks in the three case study countries which directly or indirectly
address coastal dunes and devoted special attention to the consideration of
different ES, in particular related to mitigation of natural hazards, extreme
events, and erosion processes.

Table 18.1 Indicators for the assessment of ecosystem services (Source: authors)

ES category	Indicators				
	Very high importance	High importance	Medium importance	Low importance	No importance
Regulating services					
Protection from natural hazards	High frequency and intensity of natural hazards + exposed people and infrastructure + high vulnerability[1]	High frequency and intensity of natural hazards + exposed people and infrastructure + medium to low vulnerability	Medium to low frequency and intensity of natural hazards + exposed people and infrastructure + medium to low vulnerability	Medium to low frequency and intensity of natural hazards + exposed people and infrastructure + low vulnerability	No people or infrastructure exposed to natural hazards
Protection from coastal erosion	High wave action and/or currents and/or winds + exposed people and infrastructure + high vulnerability	High wave action and/or currents and/or winds + exposed people and infrastructure + high to medium vulnerability	Medium to low wave action and/or currents and/or winds + exposed people and infrastructure + medium to low vulnerability	Medium to low wave action and/or currents and/or winds + exposed people and infrastructure + low vulnerability	No people or infrastructure exposed to coastal erosion
Water filtration and protection from salt-water intrusion	Dune system fulfils important filter functions and protects a coastal aquifer[2]	–	–	–	–
Provisioning services					
Living space	Larger settlement(s) within the CDS	Smaller settlement(s) within the CDS	Several individual buildings within the CDS	Few individual buildings within the CDS	No settlements or buildings within the CDS
Food production (agriculture and/or aquaculture)	Intensive production for trans-regional or international markets	Production for local markets	Small scale subsistence production with importance for local communities	Low production without existential significance for local communities	No production

(continued)

Table 18.1 (continued)

ES category	Indicators				
	Very high importance	High importance	Medium importance	Low importance	No importance
Production of renewable resources (wood, fibre, etc.)	Intensive production for trans-regional or international markets	Production for local markets	Small scale production with importance for local communities	Low production without existential significance for local communities	No production
Mineral raw materials	Intensive, market-oriented mining activities in extended parts of CDS	Intensive, market-oriented mining activities in parts of CDS	Extended sand mining for construction at local to regional scale	Occasional sand mining for construction at local to regional scale	No mining
Fresh water	Use for household consumption in communities and/or tourism, and/or high importance for agriculture	Use for household consumption in single households and/or tourism, and/or importance for agriculture	No household consumption and tourism, but some importance for agriculture	No household consumption and tourism, little importance for agriculture	No household consumption, no importance for agriculture
Cultural services					
Tourism and recreation	Site of international or national importance with tourism infrastructure	Site of regional importance with tourism infrastructure	Site of local importance without tourism infrastructure	Occasional visitors, no tourism infrastructure	No recreation site
Aesthetic appreciation and inspiration for science, education, culture and art	Site of international or national importance with regular cultural and/or social events	Site of regional importance with regular cultural and/or social events	Site of importance for local communities with occasional cultural and/or social events	Site of limited importance without cultural and/or social events	No importance
Spiritual experience and local identity	Spiritual, religious, or traditional site of international or national importance	Spiritual, religious, or traditional site of regional importance	Spiritual, religious, or traditional site with importance for local communities	Site of limited importance for local communities	No importance
Archaeological/historical heritage	Site of international or national importance with protection category	Site of supra-regional importance with or without protection category	Site of regional importance without protection category	Site of local importance without protection category	No importance

Habitat function, biodiversity and geodiversity

Habitats for plant and animal species	Habitat of international or national importance for marine and non-marine animals and/or indigenous plant species	Habitat of supra-regional importance for marine and non-marine animals and/or plant species	Degraded environment with restricted importance for marine and non-marine animals and plants	Heavily degraded environment with little importance for marine and non-marine animals and plants	Hardly suitable habitat provided
Biodiversity and geodiversity	Well-developed dune complex with different ecosystems and habitats of very high diversity	Medium dune system with various ecosystems and habitats of high diversity	Small dune system with few habitats of medium diversity	Small dune system with low degree of diversity	No visible difference in habitat or ecosystem

[a]According to the World Risk Report (2014), vulnerability "comprises the components of susceptibility, lack of coping capacities and lack of adaptive capacities [...] and relates to social, physical, economic and environmental factors which make people or systems susceptible to the impacts of natural hazards, the adverse effects of climate change or other transformation processes".

[b]We assume that dune systems generally have an important filter function and provide protection against salinization of the coastal aquifer. Therefore they are considered of very high importance. For quantitative statements, a hydrological model is required.

Table 18.2 Indicators for the degradation status of the CDS (Source: authors)

Degradation type	Degradation status				
	Very high (rank = 5)	High	Medium	Low	None (rank = 1)
Local sand extraction	Affected area of geomorphological transformation >50 %	Affected area of geomorphological transformation 20–50 %	Affected area of geomorphological transformation 5–20 %	Affected area of geomorphological transformation <5 %	No sand extraction
Soil sealing	Area of sealed soils by houses and infrastructure >50 %	Area of sealed soils by houses and infrastructure 20–50 %	Area of sealed soils by houses and infrastructure 5–20 %	Area of sealed soils by houses and infrastructure <5 %	No soil sealing
Dune fragmentation	Dunes are highly fragmented by a dense network of paved roads and/or railways and advanced urbanization processes taking place; agricultural and/or aqua-cultural activities possible	Dunes are fragmented by some paved roads and/or railways; settlements are located in the dune system; agricultural and/or aquacultural activities possible	Few paved or unpaved roads and/or trails cross the dune system; few settlements or houses located in the dunes; agricultural and/or aquacultural activities possible	No paved roads or railways, only unpaved trails within the dune system; few or no houses located in the dunes; no agricultural and/or aquacultural activities	Dunes not affected by infrastructure and agricultural and/or aquacultural activities
Degradation of vegetation cover	Original vegetation cover destroyed in >50 % of total dune area; native vegetation widely destroyed and/or replaced by exotic species	Original vegetation cover destroyed in 20–50 % of total dune area; native vegetation widely destroyed and/or replaced by exotic species	Original vegetation cover destroyed in 5–20 % of total dune area; several exotic species introduced, but native vegetation dominant	Original vegetation cover destroyed in <5 % of total dune area; few exotic species introduced, native vegetation dominant	No direct human impact visible; no exotic species
Contamination of soil and water, waste disposal	National limit values for soil and/or fresh water exceeded	National limit values for soil and/or fresh water not exceeded, but severe waste disposal visible in wide parts of the CDS	National limit values for soil and/or fresh water not exceeded; waste disposal visible in some parts of the CDS	National limit values for soil and/or fresh water not exceeded; limited waste disposal visible in some parts of the CDS	No contamination
Overuse of aquifer	Discharge exceeds recharge	–	–	–	–

18.4 Results

18.4.1 Ecosystem Services of the Studied Coastal Dune Systems

18.4.1.1 Dunes of Thua Thien-Hue

The dunes of Thua Thien-Hue provide important protection functions against typhoons, storm surges and wave action, and prevent accelerated coastal erosion by stabilizing the shoreline (Table 18.3). Due to the windbreak function, the impact of monsoonal winds and typhoons hitting the East Sea coast is reduced, and villages and agricultural land behind the dunes are protected. This has been the case when the destructive typhoons Xangsane (2006), Mekkhala (2008), Ketsana (2009), and Nari (2013) hit the coast. With its long shoreline, the coast of Thua Thien-Hue province is facing an elevated risk from natural hazards with about ten tropical typhoons hitting the coast every year (GFDRR and The World Bank 2010). Figure 18.5 shows the track lines of typhoons that have impacted on Vietnam in the period of 1945–2013. It is estimated that more than 60 % of the disaster events that occurred in Vietnam between 1953 and 1991, and 78 % of killed and missing persons, were the result of typhoons (Fritz and Blount 2007).

The dunes also play an important role for freshwater provision and protection from saltwater intrusion. Freshwater collects in small channels that run parallel to the dunes (Fig. 18.1b). This surface water is an important resource for agricultural cultivation including paddy rice, aquaculture practices, animal husbandry, as well as household consumption. Besides, sand dunes are preferred areas for planting chili, taro, and sweet potato (Chin 2008). Close to towns, dunes are used for sports and recreation, providing shade in the hot time of the day. Also, cemeteries are found along the dunes (Fig. 18.1c). Monocultural *Acacia* and *Casuarina equisetifolia* plantations cover parts of the dunes and are used for construction or as fire wood, to the extent that legislation allows. Habitat quality and biodiversity and geodiversity are considered important (Table 18.3).

18.4.1.2 Dunes of Parangtritis

Java Island is a zone of high seismic activity as it is located north of the Sunda-Java Trench, where the Indo-Australian plate is being subducted under the Eurasian plate. Bantul District has suffered from several earthquakes, of which the May 2006 Java Earthquake with magnitude 6.3 on the Richter scale was the most severe in recent years, with more than 220,000 households affected and more than 4000 casualties (Nurwihastuti et al. 2014). The coast of Bantul District is also prone to tsunamis. Historical records prove that three tsunamis hit the coast – in 1889, 1994, and 2006 – but there are no data on wave heights. The dunes of Parangtritis with a total length of about 4 km, a medium width of 1 km and heights of 6–20 m are

Table 18.3 Assessed ecosystem services provided by the studied CDS (source: authors)

ES		Thua Thien Hue, Vietnam		Parangtritis, Indonesia		Ritoque, Chile
Regulating services						
Protection from natural hazards	↑	Typhoons, storm surge, waves, small tsunamis	↑	Waves, tsunamis	↑	Waves, tsunamis, storms
Protection from coastal and soil erosion	↑	Coastal erosion, wind erosion	↑	Coastal erosion, wind erosion	↑	Coastal erosion, wind erosion
Water storage and purification	↑	Freshwater aquifer, protection from salt water intrusion, filtration	↑	Freshwater aquifer, protection from salt water intrusion, filtration	↑	Freshwater aquifer, protection from salt water intrusion, filtration
Provisioning services						
Living space	↑	Traditional villages	→	Illegal rural settlements	↑	Suburban settlements
Food	↑	Agriculture & aquaculture production	→	Agriculture & aquaculture production	→	Grazing land for cattle and horses
Regrowing raw materials	↑	*Acacia spp.* and *Casuarina equisetifolia* plantations	↓	none	↘	Coniferous plantations
Mineral raw materials	↑	Sand mining for construction and cultivation	↘	Locally sand mining for construction	→	Locally sand mining for construction
Fresh water	↑	Use for household consumption and agriculture	↑	Use for household consumption and tourism	↑	Use for household consumption and tourism
Cultural Services						
Tourism and recreation	↗	Ecotourism, beach tourism, sports, recreation	→	Ecotourism, weekend tourism, sports, recreation	↑	Beach tourism, ecotourism, sports, recreation
Aesthetic appreciation and inspiration for science, education, culture and art	→	Area of scientific and educational interest	↗	Ecological laboratory, scenery for landscape photography	↑	Scientific, educational and cultural events, location for architecture and arts
Spiritual experience and local identity	↑	Preservation of indigenous knowledge, traditional land use, religious places, social events	↗	Religious places, social events	↑	Location for meditation, poetry and reflection about local identity, social events

(continued)

Table 18.3 (continued)

ES		Thua Thien Hue, Vietnam		Parangtritis, Indonesia		Ritoque, Chile
Archaeological / historical heritage	↑	Archaeological site	↓	none	↑	Archaeological site
Habitat, biodiversity and geodiversity						
Habitats for plant and animal species	↑	Marine and non-marine animals, indigenous plant species	↑	Marine and non-marine animals, indigenous plant species	↑	Marine and non-marine animals, indigenous plant species
Biodiversity and geodiversity	↗	Various ecosystems, landforms and habitats of high diversity	↗	Various ecosystems, landforms and habitats of high diversity	↑	Various ecosystems, landforms and habitats of very high diversity

↑ Very important, ↗ Important, → medium important, ↘ less important, ↓ not important

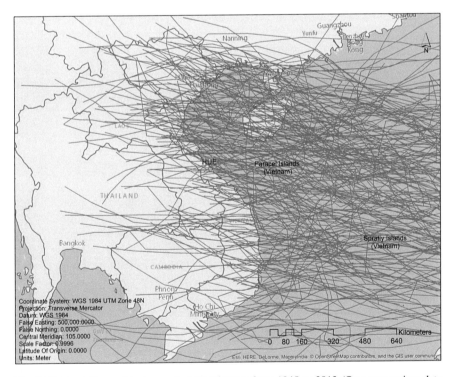

Fig. 18.5 Track lines of typhoons that hit Vietnam from 1945 to 2013 (Own processing; data source MONRE 2014)

considered as an effective buffer even against high tsunamis for the settlements located north of the dunes (Sunarto et al. 2010). For the infrastructure located within the dunes, including the Parangtritis tourist area (Fig. 18.2) as well as settlements along the mouth of the Opak River to the west of the study area, the CDS offer little protection (Table 18.3).

The dunes also mitigate coastal erosion and wind erosion (Sunarto et al. 2010). The latter applies particularly to forested dunes, but also to those areas covered by herbaceous plants such as the salt tolerant *Ipomoea pes-caprae* and *Spinifex* species, which are known as primary sand stabilizers due to their sprawling runners (Fig. 18.2b). Hydrological services include freshwater conservation, protection from saltwater intrusion, and filtration. The local aquifer provides large amounts of freshwater that are used for agriculture and household consumption.

People settle primarily in the village of Parangtritis and along the coastal road and its side streets (Fig. 18.2). According to the spatial development plan of Bantul, the barchan dunes are designated as a conservation area with limited human interference. However, although illegal, some people dwell in this area and use sand for construction purposes. While the agricultural use in the active dunes is limited to some cattle and poultry farming, there is intensive agricultural production on the inactive dunes and aquaculture in the fluvial plain.

Parangtritis is recognized as a tourism and recreation site. According to the responsible Bantul governmental agency, there are plans to further develop the tourism sector, but at the same time to protect the barchan dunes from mass tourism and infrastructural development. Other important cultural services include a religious site in Parangkusumo village, an ecological laboratory, and as a location for social events and landscape photography.

18.4.1.3 Dunes of Ritoque

The dunes of Ritoque consist of active incipient dunes, foredunes and rear dunes along the coastline and fixed Holocene and Pleistocene dunes in higher elevated areas (Fig. 18.6). According to Dura et al. (2015), the dunes of Ritoque are located within a historically active segment of the central Chilean subduction zone, where destructive earthquakes occurred within the last 500 years in a consistent recurrence interval of ~80 years. The earthquakes were often accompanied by tsunamis which varied in height. The last tsunami of 10 m or more occurred in 1730 and was triggered by an earthquake with a 9.0 magnitude. The well-documented tsunamis in 1822, 1906, and 1985 were considerably lower (<4 m) and resulted in localized damage (Moreno and Gibbons 2007).

According to a tsunami risk analysis conducted by the Chilean Navy Hydrographic and Oceanographic Service (SHOA), the rear dunes buffer high tsunamis >6 m and protect the infrastructure in the valley behind, while the foredunes would be more or less inundated depending on the height of the tsunami wave. The Mantagua Estuary and mouth of the Aconcagua south of the dune system are the

Profile: Dunes of Ritoque

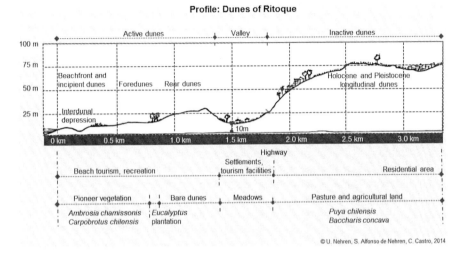

Fig. 18.6 Profile of the dunes of Ritoque. It shows that the rear dunes protect settlements and infrastructure in the valley behind the dunes (Source: authors)

main weakness zones, where a high tsunami wave would penetrate and destroy infrastructure.

Apart from its protection functions including the defense against coastal erosion and wind erosion (Table 18.3), the dunes also protect a freshwater aquifer that provides domestic water for the town of Quintero and prevents saltwater intrusion. Moreover, the dune system comprises various coastal and off-shore ecosystems and habitats, such as coastal dune vegetation dominated by *Ambrosia chamissonis* and *Carpobrotus chilensis*, coastal wetlands and marshlands, shrubland, thicket and bushes ("matorrales") with *Baccharis concava and Puya chilensis* on arid (sunny) hillsides on stabilized dunes, as well as sclerophyllous woodlands on humid (shady) slopes and in valleys of stabilized dunes (Castro 1987). This landscape mosaic makes the dunes an important habitat for rare plants and animal species and a stopover site for migratory birds.

Humans use the services of the dunes in various ways. Residential areas are mainly located on the stabilized dunes in higher elevations, but there are also some properties in the active dunes. Some parts of the marshlands and meadows are used as grazing land for cattle and horses, and few coniferous plantations have been introduced. Locally dune sand is used for construction purposes.

The beachfront and foredunes are favorable areas for beach tourism, while ecotourism with bird watching has been established in the wetlands of Mantagua. Furthermore, educational and cultural events take place in the so-called "Ciudad Abierta de Ritoque", a visionary project that was established by the school of Architecture of the Pontifical Catholic University of Valparaíso in 1970. Sport activities include hiking, motocross races, and horseback riding. Due to the dunes' vast scientific importance and aesthetic value, Castro (1987) recommended the conservation as a "site of scientific interest" that combines natural and cultural elements, archaeological heritage, and scenic beauty.

18.4.2   Degradation Status of the Studied CDS

18.4.2.1   Thua Thien-Hue Dunes

The coastal area of Thua Thien-Hue province was already inhabited in the Iron Age by the Sa Huynh Culture (*c.* 400 BC–200 AD) (Yamagata 2006). Since then, the coastal area of Thua Thien-Hue was continuously inhabited, and step-by-step dunes were converted to settlement or agricultural land.

Deforestation of coastal areas was highest during the Vietnam War. To collect information on the vegetation of the dunes and adjacent areas before and after the Vietnam War, Hoang (unpubl. data) carried out 160 interviews in coastal communities of Thua Thien-Hue province in 2013 and 2014. Some of the older interviewees could recall stories they were told from their parents and grandparents who described the coastal vegetation at the time between 1954 and 1960. Accordingly, before the Vietnam War, coastal forests in the inner slopes of the dune were already degraded, but indigenous species such as *Litsea glutinosa, Barringtonia acutangula, Ficus lacor,* and *Vatica mangachapoi* were still common.

To counteract further degradation, reforestation measures were developed to stabilize the bare dunes in the 1990s. However, since the demand for fuel wood was high, fast growing exotic tree species such as *Casuarina equisetifolia* as well as *Acacia* and *Eucalyptus* species were introduced, mainly as monocultures. Today, such ecologically poor plantations dominate, and there is hardly any near-natural coastal forest left.

With respect to protection from coastal hazards, Wolanski (2007) found that *Casuarina* stands prevented damage from small tsunami waves during the Indian Ocean Tsunami of 2004, but because of frequent uprooting, they did not serve as reliable bioshields against winds during typhoons. Even worse, Gonsalves and Mohan (2012) demonstrated that the monocultural *Casuarina* plantations that were uprooted by the 1999 super cyclone in Orissa province in India added to the cyclone's impact by damaging settlements in the hinterland.

The only remnants of natural vegetation in Thua Thien-Hue are small patches around temples. Such temple gardens were protected for centuries, and although today these miniscule forest stands are mixed with exotic species, their trees and seed banks are the only traces left to gather knowledge on diversity of historical coastal forests. The existence of these remnants is the reason for the categorization of a high level of biodiversity in Table 18.3.

In the last decades, sand extraction has become an important factor aggravating the degradation of the dunes (Table 18.4). The local coast inhabitants use sand as construction material for roads and modern houses. Increasingly, sand is also exported as building material to other areas. This leads to the fragmentation of the originally continuous dune stretches. Furthermore, sand is sold in growing quantities to extract valuable minerals such as titanium.

Another impact on the dunes is pollution: solid waste is regularly disposed in dune areas close to settlements (Fig. 18.1c). Moreover, sandy areas close to

Table 18.4 Type and degree of degradation (Source: authors)

Category	Indicators	Thua Thien-Hue, Vietnam	Parangtritis, Indonesia	Ritoque, Chile
Local sand extraction	Affected area of geomorphological transformation	3	4	3
Soil sealing	Area of sealed soils by houses and infrastructure	3	2	2
Dune fragmentation	Severity and density of fragmentation by roads, railways, settlements, agriculture, trails	4	4	4
Degradation of vegetation cover	Coverage and quality of actual vegetation compared to natural conditions, considering introduction of exotic species	4	4	3
Contamination	Contamination of soil and water, waste	3	2	2
Overuse of aquifer	Discharge exceeds recharge	No data available	No data available	No data available

1 = none, 2 = low, 3 = medium, 4 = high, 5 = very high

agricultural fields are polluted by fertilizers and pesticides. Since the year 2000, the shrimp farming sector has been expanding rapidly since the government issued Resolution 09/NQ-CP, allowing farmers to transform coastal saline rice fields into shrimp farms (Fig. 18.1c). Increasingly, aquaculture is practiced within dune areas, and the sand is polluted not only by pesticides, but also by antibiotics which are commonly used in intensive shrimp farming (Nhuong et al. 2002). We assume that the establishment of aquaculture systems also negatively affects coastal morphology and dune formation, but research in this field is scarce; thus, scientific proof is still pending.

Dunes of Parangtritis

In the 1980s, land use in the Parangtritis area was dominated by agricultural activities such as dry farming and paddy rice production in the coastal alluvial plain and some aquaculture activities in the fluvial plain, whereas huge parts of the sand dunes, in particular of the barchan type, were active and without vegetation cover. At that time, there were only few houses built on the dunes and beach ridges and along the beachfront, and tourism infrastructure did not yet exist. In the 1990s, the government promoted the development of Parangtritis as a tourist destination. Since then, tourism facilities were established and an increasing number of people moved there, increasing the settlement area.

In recent years, the area has attracted an increasing number of tourists, who, along with the growing local population, have increased the pressure on natural resources. Visible effects are the expansion of roads, settlements, and tourism facilities, which has led to increased soil sealing and dune fragmentation (Table 18.4). Moreover, dune sand is extracted for construction, vegetation is partly degraded e.g. by motorbikes, and waste and chemicals are disposed on the dunes. A critical point is also the exploitation of the local aquifer due to the increased freshwater demand. Even though there are no groundwater data available, there is a risk of over-using groundwater resources which could trigger increased saltwater intrusion. Finally, in parallel to the CDS of Thua Thien-Hue, the establishment of aquaculture systems in the coastal zone around Parangtritis is increasing, although this is not in line with the land-use plan of Parangtritis. Currently, the local government is trying to solve this problem by controlling and monitoring the aquaculture activities to avoid negative impacts on the active barchan dunes.

18.4.2.2 Dunes of Ritoque

Archaeological finds, such as fossil shells and ceramics, testify that the dunes of Ritoque were already inhabited by the pre-colonial *Bato* culture (Massone 1980). In 1822, Maria Graham, a British writer of travel books, visited the area and described the coastal dunes at the beachfront as high and free of vegetation. She explicitly referred to the rich bird life in the dunes and wetlands (Graham 1824). In 1834, Charles Darwin made a stop in the Bay of Quintero on his voyage on the HMS Beagle and expressed his admiration for the shell beds mainly formed by one bivalve species of *Erycina* (Darwin 1839).

In 1925, a railway line was constructed that runs through the foredunes over a distance of a few hundred meters parallel to the beachfront and crosses the estuary with a massive bridge (Table 18.4). Moreover, the coastal highway F-30-E crosses the longitudinal dunes and the valley behind the active dunes. Together with industrial development in the 1960s, urbanization processes have become increasingly evident. New settlements are currently in planning, which will not only exacerbate fragmentation and soil sealing processes of the dunes, but also lead to higher pressure on the freshwater aquifer.

Along the beachfront and in the foredunes, there are no larger settlements, but there are several private properties and tourism facilities, such as campgrounds. During the summer holidays, the beachfront dunes are heavily populated by tourists. Tourism creates a severe waste problem and frequently damages the vegetation due to open fires. Furthermore, the dune vegetation is negatively affected by motor rallies and numerous small footpaths.

Already in 1972 and 1976, the Chilean National Forest Corporation (CONAF) reported the introduction of the exotic species *Ammophila arenaria* (CONAF 1979). Today many other exotic species, such as *Lupinus arboreus* and *Tamarix gallica*, have established themselves (Wolf Eigenherr 2008). Moreover, *Pinus radiata, Pinus halepensis, Casuarina cunninghamiana,* as well as *Eucaplytus* and *Acacia* species have been introduced as wind breaks and a source for timber (Toral Ibáñez et al. 1980). Additionally, livestock grazing has led to the introduction of exotic grass species and a change of abiotic conditions due to soil compaction and nutrient inputs.

Sand extraction for construction purposes can be observed in several places. Recently, an area of about 1.5 km^2 of dune sand has completely been removed for infrastructural development close to the town of Concón in the southernmost tip of the dune system.

18.4.3   Synthesis

Table 18.4 summarizes the type and degree of degradation of the three studied CDS using the indicators listed in Table 18.2. Figure 18.7 shows the degradation status and total loss of the respective dune areas along an urban-rural gradient. The overall value of the degradation status (y1-axis) is the mean value of the 5 indicator values in Table 18.4. The y2-axis shows the estimated total loss of dune area in percent. For all study cases the baseline dates back to the 1950s, for which historical data (aerial views and photographs) are available. The x-axis represents the urban-rural gradient. With more study areas included, we could have more refined results with respect to establishing urban-rural degradation patterns, and we can expect to identify distinct patterns of dune loss and degradation dependent on closeness to agglomerations.

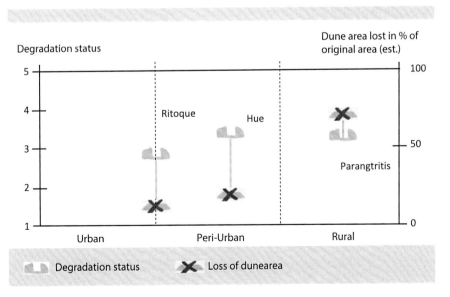

Fig. 18.7 Degradation status and total loss of the studied CDS along an urban-rural gradient

18.4.4   Legal Frameworks Addressing Coastal Dunes

18.4.4.1   Thua Thien-Hue, Vietnam

At the national level, the 'Law on Land' of 2003 and the revised 'Vietnamese Land Law' of 2013 address coastal lands with adjacent water surfaces. Both regulations determine that such areas are allocated by the state to economic organizations, households, individuals, overseas Vietnamese or foreign-invested enterprises for the purposes of aquaculture, agricultural, forestry, salt production, or non-agricultural purposes. The areas can only be used when the following requirements are met:

– The use is in agreement with the land use master plans approved by the respective state agencies (at national or provincial level)
– The use contributes to the protection of the coastal area and increases the coastal sedimentation process
– The use contributes to the protection of the ecosystem, environment, and landscape
– The use does not interfere with national security or maritime navigation

The coastal forests behind the dunes are designated as 'Protection Forest' according to Decision No 61/2005/QĐ-BNN and are defined with respect to their protection function (Table 18.5). The criteria used to classify coastal forests into zones where extraction and use are allowed or forbidden are related to the elements that are protected (e.g. settlements, infrastructure), but not in relation to the diversity, quality, or structure of the protection forest. Coastal protection forests are established and managed by the Departments of Agriculture and Rural Development (DARD) of the respective province.

Even though criteria for the use or non-use of protection forest have been defined, large stretches of the coastal dunes have nonetheless been converted into agricultural land or shrimp ponds, and remaining parts of the dunes have been used for sand extraction. Besides the lack of an effective monitoring system, regulations from other laws are likely to interfere, allowing the use of high priority protection forest. Moreover, there are no clearly defined conservation categories and management strategies for CDS at the national level.

18.4.4.2   Parangtritis, Indonesia

In Indonesia, the use and management of coastal systems are regulated at the national level by two main legal frameworks: Act 26-2007 and Act 27-2007. Act 26-2007 is a general regulation about spatial planning. It addresses coastal zones as buffers and local protection areas that should be conserved. Act 27-2007 regulates the management of coastal areas and small islands and aims at sustainable

Table 18.5 Specific protection of the coastal dunes of Thua Thien-Hue based on Decision No 61/2005/QĐ-BNN (2005), and of Parangtritis according to the current spatial plan of Bantul Regency

	Thua Thien-Hue, Vietnam		Parangtritis, Indonesia	
Type of protected area	Coastal forests and sandy soils		Coastal dune ecosystem with active barchan dunes	
Elements protected	Villages		Villages	
	Infrastructure		Infrastructure	
	Agriculture activities		Agriculture activities	
	Industry			
Purpose of protection	Protection from		Protection from	
	Sand encroachment		Tsunamis	
	Sand movement		Coastal erosion	
	Landslides			
Criteria for	Size of area behind forest		Type of dune	
protection	>100 ha	≤100 ha	Barchan dune	Other dune type
Protection status	High priority area, sand extraction and use forbidden	Lower priority area, sand extraction and use allowed	High priority area, settlements and other infrastructure development forbidden	Lower priority area, limited agricultural activities, settlements, and tourism facilities allowed

development. In this context, coastal zones are considered as multi-use areas that have to be managed based on their natural characteristics and potential.

At the provincial and regency level, regulations No 2-2010 about regional planning of Yogyakarta Special District and No 4-2011 about spatial planning in Bantul Regency are relevant for the use and management of the sand dune ecosystem. These regulations define how the coastal sand dunes of Parangtritis can be used for tourism and agricultural purposes without neglecting its conservation status.

The aim of both regulations is to create sustainable coastal environments by regulating land use activities, implementing conservation and rehabilitation measures, as well as reducing risks from natural hazards and geo-processes. In the regional plan of Yogyakarta Special District, the dunes of Parangtritis and the surrounding coastal area are designated as a strategic area. The spatial plan of Bantul Regency specifies this strategic area with respect to the development of science, technology, and research. Moreover, it defines a conservation status for the barchan dunes of Parangtritis and restricts the use of these dunes (Table 18.5).

18.4.4.3 Ritoque, Chile

In Chile, CDS play an important role for the protection against tsunamis, and recently in some areas the Ministry of Interior (Ministerio del Medio Ambiente, Corporacion Nacional Forestal, CONAF) invested in the creation of coastal dune

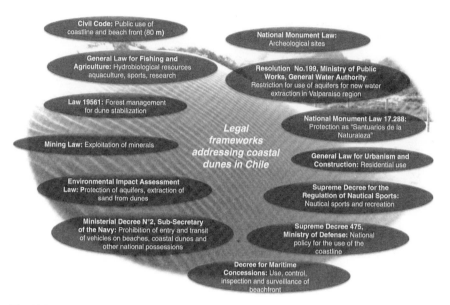

Fig. 18.8 Legal frameworks in Chile that directly or indirectly address coastal dunes (Compiled by the authors)

parks, where native vegetation is restored. However, there is an elevated tsunami risk outside the designated areas as well, as exemplified by the dunes of Ritoque, which are not integrated in the set of coastal dune parks. In Chile, coastal dunes are addressed in various national laws, as shown in Fig. 18.8, with different ministries, state agencies, local authorities, and stakeholders involved in implementing the laws. This makes it difficult to develop and coordinate an integrated planning approach to prevent the unsustainable use and degradation of CDS. The example of Chile shows, on the one hand, the government's willingness to make investments that protect and restore CDS. On the other hand, there is still a need to establish an appropriate and effective management system that maximizes both the coastal protection and the various benefits provided by CDS.

18.5   Discussion

The three case studies show that coastal dunes provide protection against coastal hazards such as storms and tsunamis and, at the same time, offer many other benefits. However, degradation and destruction of CDS reduce their protection function and have negative impacts on other ES, habitat function, as well as biodiversity and geodiversity. In recent years, the importance of coastal ecosystems as buffers against natural hazards has gained more attention in the scientific community, and coastal dunes have been incorporated in ecosystem-based approaches (Yasuda et al. 2012; Acevedo 2013; Hettiarachchi et al. 2013, Senhoury et.al., Chap. 19). However, particularly in developing and emerging countries, coastal dunes are still under-researched ecosystems compared for instance to mangroves (see Emerton et al., Chap. 2; Friess and Thompson, Chap. 4; van Wesenbeeck et al., Chap. 8; David et al., Chap. 20). Among others, there is a lack of reliable data on the efficiency of coastal dunes as buffers against various coastal hazards. Moreover, there are only a few studies on the status of coastal dunes and the provision of ecosystem services (Mendoza-Gonzales et al. 2012), and CDS monitoring systems are the exception.

The schemes we developed for assessing ES and the degradation status could therefore serve as a model for other tropical and subtropical CDS. The advantage is that the indicators are relatively simple and the underlying data easy to access, so that the schemes could also be used in data-poor environments. Nevertheless, we see our rating system, which is based on expert evaluations, as an initial step to set up an indicator system to assess the contribution that CDS make to risk reduction and adaptation. In future research, a set of more robust indicators needs to be developed to compare ES and the degradation status of CDS in a more quantitative way. This indicator set will then also allow for monitoring dune system dynamics.

It will be essential to actively involve international dune experts, local stakeholders, and communities in the development of such indicator sets. Local knowledge will allow for further improving the indicators and creating a database for tropical and subtropical CDS. At the same time, community and stakeholder participation will raise awareness of the importance of coastal dunes. In this regard, we see our work as motivation for future research. Further, we want to emphasize the urgency of actions to stop further loss and degradation of tropical and subtropical coastal dunes.

Based on our three case studies, we show that there are various drivers of dune loss and degradation. CDS are often perceived as obstacles for coastal development or merely as suppliers of raw materials, but not as valuable geo-ecosystems that provide risk reduction services. We assume that the environmental cost of dune loss and degradation will in the long term far exceed the short-term benefits of sand or mineral sale. Even worse, sand removal will put coastal communities at higher risk to natural hazards. We therefore want to stimulate further research on the costs and benefits of various management options regarding CDS. To date, there are very few studies on the economic value of coastal ecosystems and CDS in particular (Mendoza-Gonzales et al. 2012). Moreover, the need and potential for restoration should be taken into account. To this end, the ReDune index (Lithgow et al. 2014) provides a methodology to assess priorities for foredune restoration measures that can be applied by planners and decision makers. It makes use of a weighted checklist with 36 indicators to identify the most urgent need for restoration.

In all three cases one must, however, acknowledge current government efforts to better manage coastal ecosystems. Of the three countries researched, currently the legal framework for dune management in Chile is the most detailed. But even such a complex legal framework can lead to uncertainties and conflicts of interest when the elements of the framework are not yet mutually harmonized. For urban environments, Pauchard et al. (2013) pointed out that adequate planning instruments that consider biodiversity and ES are lacking. For instance, the General Law on Urban Planning and Construction (LGUC) in Chile regulates residential and industrial uses, constructions, and the location of public facilities, while the nomination of protected zones is under the Public National System of Protected Areas (SNASPE). The partly overlapping legal competences result in different sectoral planning approaches for the same area that hinder an integrated management of CDS. Based on the three case studies, we postulate that the same lack of adequate planning tools exists for rural and peri-rural environments and we emphasize the need to harmonize legal frameworks for CDS.

In general, there is growing awareness among governments and other decision-makers of the importance of coastal dunes for disaster risk reduction (DRR) and climate change adaptation (CCA). Thus, in Chile, the Sustainable Planning Programs (PREs) have been implemented by the Government in the aftermath of the earthquake and tsunami of 27 February 2010. Among other things, the programs consider the role of coastal forests and dunes as buffer zones. This must certainly be seen as a step in the right direction. Moreover, there are promising ecosystem-based initiatives, such as a dune restoration project in Puerto Saavedra, in the province of Araucanía, which explicitly aims at reforesting degraded dunes as a protection against tsunami impacts (Acevedo 2013).

Also in Indonesia and Vietnam, several efforts have been made to better manage coastal dunes by introducing specific regulations (Table 18.5). However, in both cases the regulations are very specific and could be improved by considering the multiple benefits that CDS provide. Furthermore, the legal regulations focus only on the core zones of the CDS and could be better integrated into more comprehensive landscape approaches that consider interactions with adjacent terrestrial and marine ecosystems. Therefore, we recommend zoning schemes for coastal dunes that define areas of different protection and utilization status. In Parangtritis, first steps have been taken to define such areas, but the growing population and land use pressures hamper implementation and control. This is a crucial point also in many other developing and emerging countries with dynamic coastal development. It is equally difficult to simply adopt integrated dune management strategies and programs even in industrial countries, although some recent efforts are being documented, such as the European Dune Network under the Coastal & Marine Union (EUCC) or the integrated management policies and rehabilitation techniques for coastal dunes in Queensland and New South Wales, Australia (Gold Coast City Council 2007; NSW Department of Land and Water Conservation 2001).

18.6 Conclusions

CDS play a crucial role for DRR and CCA and provide various essential ES. However, particularly in developing countries of the tropics and subtropics, CDS are severely threatened by coastal development, insufficient coverage of protected areas, and a lack of awareness of the services provided by this ecosystem. With our study we provide an easily applicable and consistent methodology for the assessment of ES and degradation status of CDS for policymakers, planners, and decision makers, which could be used in data-poor environments. Together with the ReDune index, this method serves as a basis for the evaluation of the protection function, geo-ecological value, conservation need, and restoration potential of CDS. In future studies, the methodology can be further developed by using more quantitative indicators and involving international dunes experts to improve the data base. We further recommend valuating and monetizing ES of CDS in order to estimate the costs and benefits of ecosystem-based measures compared to other solutions.

One option to make CDS more recognized is to establish coastal dune area networks. A starting point could be to connect all coastal dune areas under national protected area systems. Moreover, CDS should be given more attention in Integrated Coastal Zone Management (ICZM) approaches to coordinate the different sectoral policies including DRR and CCA. This requires further improvement and harmonization of legal frameworks to better manage CDS. Finally, community-based monitoring systems should be implemetend to strengthen the role of local communities and stakeholders in the management of coastal dunes.

Acknowledgements We would like to thank Dr. Marco Cisternas, Dr. Eduardo Salgado, Dr. Leonardo Vera, Dr. Dwi Nurwihastuti and MSc Satria Gasa for their kind support and provision of materials. This work was supported by the Center for Natural Resources and Development (CNRD), the German Federal Ministry for Economic Cooperation and Development (BMZ), the German Academic Exchange Service (DAAD), as well as the research project 'Land-use and climate change interactions in the Vu Gia Thu Bon River Basin, Central Vietnam (LUCCi)', funded by the German Federal Ministry of Education and Research (BMBF).

References

Acevedo R (2013) Bosques en las dunas: una barrera natural contra maremotos. Tendencias, La Tercera, 10 August 2013. http://diario.latercera.com/2013/08/10/01/contenido/tendencias/26-143713-9-bosques-en-las-dunas-una-barrera-natural-contra-maremotos.shtml. Accessed 16 Dec 2014

Adger WN, Hughes TP, Folke C, Carpenter SR, Rockström J (2005) Social-ecological resilience to coastal disasters. Science 309(5737):1036–1039

Alonso I, Weston K, Gregg R, Morecroft M (2012) Carbon storage by habitat – Review of the evidence of the impacts of management decisions and condition on carbon stores and sources. Natural England Research Reports NERR043

Bambaradeniya CNB, Perera MSJ, Samarawickrama VAMPK (2006) A rapid assessment of post-tsunami environmental dynamics in relation to coastal zone rehabilitation and development activities in Hambanthota District of southern Sri Lanka. IUCN Sri Lanka Occasional Paper No. 10, Colombo

Bantul Regency (2014a) Precipitation data 2009–2011. http://www.bantulkab.go.id/datapokok/0407_pola_curah_hujan.html. Accessed 17 Dec 2014

Bantul Regeny (2014b) Population data 2012. http://www.bantulkab.go.id/datapokok/0501_
kepadatan_penduduk_geografis.html. Accessed: 17 Dec 2014

Barbier EB, Hacker SD, Kennedy C et al (2011) The value of estuarine and coastal ecosystem
services. Ecol Monogr 81(2):169–193

Bhalla RS (2007) Do bio-shields affect tsunami inundation? Curr Sci 93:831–833

Brown S, Nicholls RJ, Woodroffe CD et al (2013) Sea-level rise impacts and responses: a global
perspective. In: Finkl CW (ed) Coastal hazards, vol 6, Coastal research library. Springer,
Dordrecht/New York

Carter RWG (1991) Near-future sea level impacts on coastal dune landscapes. Landsc Ecol
6:29–39

Castro C (1987) Transformaciones geomorfológicas recientes y degradación de las dunas de
Ritoque. Revista de Geografía Norte Grande 14:3–13

Chin H (2008) Evaluating the economic developing potential of sandy areas of Thua Thien-Hue
province (Vietnamese version)

CONAF (1979) Proyecto control de dunas (SERPLAC) V región

Dahm J, Jenks G, Bergin D (2005) Community-based dune management for the mitigation of
coastal hazards and climate change effects: a guide for local authorities

Darwin C (1839) Naturalist's voyage round the world – the voyage of the Beagle. Reprint 1913,
available online at Project Gutenberg. http://www.gutenberg.org/files/3704/3704-h/3704-h.
htm#chxii

De Silva M, Premachandra SPU (1998) An ecological study of the sand dune vegetation of the
Ruhuna National Park, Sri Lanka. J South Asian Nat Hist 3(2):173–192

Doody JP (2013) Sand dune conservation, management and restoration. Springer, Dordrecht

Dura T, Cisternas M, Horton BP et al (2015) Coastal evidence for Holocene subduction-zone
earthquakes and tsunamis in central Chile. Quat Sci Rev 113:93–111

French PW (2001) Coastal defences: processes, problems and solutions. Routledge, London

Fritz HM, Blount C (2007) Thematic paper: role of forests and trees in protecting coastal areas
against cyclones; Chapter 2: Protection from cyclones. In: Braatz S, Fortuna S, Broadhead J,
Leslie R (eds) Coastal protection in the aftermath of the Indian Ocean tsunami: what role for
forests and trees? Proceedings of the regional technical workshop, Khao Lak, 28–31 August
2006, RAP Publication (FAO), No 207/07

General Statistical Office (2013) Year book. Statistical publishing house, Hanoi

GFDRR, The World Bank (2010) Weathering the storm: options for disaster risk financing in
Vietnam

Gold Coast City Council (2007) Planning scheme policies: policy 15 – Management of coastal
dune areas

Gómez-Pina G (2002) Sand dune management problems and techniques, Spain. J Coast Res
36:325–332

Gonsalves J, Mohan P (eds) (2012) Strengthening resilience in post-disaster situations: stories,
experience, and lessons from South Asia. International Development Research Centre, Academic Foundation, New Delhi

Graham M (1824) Diario de su residencia en Chile (1822) y de su viaje al Brasil (1823). San
Martin, Cochrane, O' Higgins. Editorial – América, Madrid, Biblioteca Ayacucho, reprint
1916

Gray M (2004) Geodiversity: valuing and conserving abiotic nature. Wiley, Chichester

Grootjans AP, Dullo BS, Kooijman AM et al (2013) Restoration of dune vegetation in the
Netherlands. In: Martínez ML, Gallego-Fernández JB, Hesp PA (eds) Restoration of coastal
dunes. Springer, Berlin/New York

Harada K, Imamura F (2005) Effects of coastal forest on tsunami hazard mitigation – a preliminary
investigation. In: Satake K (ed) Tsunamis: case studies and recent developments. Springer,
Dordrecht, pp 279–292

Heslenfeld P, Jungerius PD, Klijn JA (2004) European coastal dunes: ecological values, threats, opportunities and policy development. In: Martínez ML, Psuty NP (eds) Coastal dunes: ecology and conservation. Springer, Berlin/Heidelberg

Hettiarachchi SSL, Samarawickrama SP, Fernando HJS et al (2013) Investigating the performance of coastal ecosystems for hazard mitigation. In: Renaud FG, Sudmeier-Rieux K, Estrella M (eds) The role of ecosystems for disaster risk reduction. United Nations University Press, Tokyo/New York/Paris

IUCN (2005) A report on the terrestrial assessment of tsunami impacts on the coastal environment in Rekawa, Ussangoda and Kalametiya (RUK) area of southern Sri Lanka

Kallesøe MF, Bambaradeniya C, Iftikhar UA et al (2008) Linking coastal ecosystems and human well-being: learning from conceptual frameworks and empirical results. Ecosystems and Livelihoods Group, Asia, IUCN, Colombo

Lithgow D, Martínez ML, Gallego-Fernández JB et al (2013) Linking restoration ecology with coastal dune restoration. Geomorphology 199:214–224

Lithgow D, Martínez ML, Gallego-Fernández JB (2014) The "ReDune" index (Restoration of coastal Dunes Index) to assess the need and viability of coastal dune restoration. Ecol Indic 49:178–187

Liu PL-F, Lynett P, Fernando H et al (2005) Observations by the international tsunami survey team in Sri Lanka. Science 308:1595

Lotze HK, Lenihan HS, Bourque BJ et al (2006) Depletion, degradation, and recovery potential of estuaries and coastal seas. Science 312:1806–1809

MA – Millennium Ecosystem Assessment (2005) Ecosystems and human well-being, vol 1, Current state and trends, findings of the condition and trends working group. Island press, Washington, DC/Covelo/London

Martínez ML (2008) Dunas costeras. Investigación y ciencia 383:26–35

Martínez ML, Psuty NP (eds) (2007) Coastal dunes – ecology and conservation, vol 171, Ecological studies. Springer, Berlin

Martínez ML, Maun MA, Psuty NP (2004) The fragility and conservation of the world's coastal dunes: geomorphological, ecological and socioeconomic perspectives. In: Martínez ML, Psuty NP (eds) Coastal dunes. Springer, Berlin/Heidelberg

Mascarenhas A, Jayakumar S (2008) An environmental perspective of the post-tsunami scenario along the coast of Tamil Nadu, India: role of sand dunes and forests. J Environ Manag 89:24–34

Massone MM (1980) Nuevas consideraciones en torno al complejo Aconcagua. Revista Chilena de Antropología 0(3):75–85

Mendoza-Gonzales G, Martínez ML, Lithgow D et al (2012) Land use change and its effects on the value of ecosystem services along the coast of the Gulf of Mexico. Ecol Econ 82:23–32

Miththapala S (2008) Seagrasses and sand dunes, Coastal ecosystems series. Ecosystems and Livelihoods Group Asia, IUCN, Colombo

MONRE (2014) Typhoon data base. http://www.thoitietnguyhiem.net/CSDLBao/CSDLbao.aspx. Accessed 17 Dec 2014

Moreno T, Gibbons W (eds) (2007) The geology of Chile. The Geological Society of London, London

Moreno-Casasola P (2004) A case study of conservation and management of tropical sand dune systems: La Mancha–El Llano. In: Martínez ML, Psuty NP (eds) Coastal dunes. Springer, Berlin/Heidelberg

Naeem S, Chapin FS, Costanza R et al (1999) Biodiversity and ecosystem functioning: maintaining natural life support processes. Issues Ecol 4:1–14

Nhuong TV, Luu LT, Tu TQ et al (2002) Vietnam shrimp farming review. Individual partner report for the project: policy research for sustainable shrimp farming in Asia. European Commission INCO-DEV Project PORESSFA No. IC4-2001-10042, CEMARE University of Portsmouth UK and RIA1, Bac Ninh

NSW Department of Land and Water Conservation (2001) Coastal dune management: a manual of coastal dune management and rehabilitation techniques, coastal unit. DLWC, Newcastle

Nurwihastuti DW, Sartohadi J, Mardiatno D et al (2014) Understanding of earthquake damage pattern through geomorphological approach: A case study of 2006 Earthquake in Bantul, Yogyakarta, Indonesia. World J Eng Technol 2(3b):61–70

Pauchard A, Barbosa O, Maira J et al (2013) Regional assessment of Latin America: rapid urban development and social economic inequity threaten biodiversity hotspots. In: Elmqvist T, Fragkias M, Goodness J et al (eds) Urbanization, biodiversity and ecosystem services: challenges and opportunities – a global assessment. Springer, New York

Peduzzi P (2014) Sand, rarer than one thinks. UNEP Global Environmental Alert ServiceNehren_et-al.docx (GEAS), Thematic focus: ecosystem management, environmental governance, resource efficiency

Phan TGT, Nguyen VH (2006) Cost-benefit analysis for coastal sand shrimp farming in Vietnam. Nong-Lam University, Faculty of Economics, Ho Chi Minh City

Prasetya GS (2007) The role of coastal forest and trees in combating coastal erosion. In: Braatz S, Fortuna S, Broadhead J, Leslie R (eds) Coastal protection in the aftermath of the Indian Ocean tsunami: what role for forests and trees? FAO, Bangkok

Saye SE, Pye K (2007) Implications of sea level rise for coastal dune habitat conservation in Wales, UK. J Coast Conserv 11:31–52

Schlacher TA, Jones AR, Dugan JE et al (2014) Open-coast sandy beaches and coastal dunes. In: Lockwood JL, Maslo B (eds) Coastal conservation. Cambridge University Press, Cambridge, MA

Spalding MD, Ruffo S, Lacambra C et al (2014) The role of ecosystems in coastal protection: adapting to climate change and coastal hazards. Ocean Coast Manag 90:50–57

Sridhar KR, Bhagya B (2007) Coastal sand dune vegetation: a potential source of food, fodder and pharmaceuticals. Livest Res Rural Dev 19/6. http://www.lrrd.org/lrrd19/6/srid19084.htm. Accessed 16 Dec 2014

Sterling EJ, Hurley MM, Minh LD (2006) Central Vietnam and the Truong Son Range: from wet mountains to dry forests. In: Vietnam: a natural history. Yale University Press, New Haven

Sudmeier-Rieux K, Masundire H, Rizvi A, Rietbergen S (eds) (2006) Ecosystems, livelihoods and disasters, an integrated approach to disaster risk management, vol 4, Ecosystem management series. IUCN, Gland

Sunarto, Marfai MA, Mardiatno D (2010) Multirisk assessment of disasters in Parangtritis coastal area. Gadjah Mada University Press, Yogyakarta

Takagi H, Esteban M, Ngyuen DT (2014) Introduction: coastal disasters and climate change in Vietnam. In: Ngyuen DT, Takagi H, Esteban M (eds) Coastal disasters and climate change in Vietnam – engineering and planning perspectives. Elsevier, London

Takle ES, Chen T-C, Wu X (2007) Protection from wind and salt. In: Braatz S, Fortuna S, Broadhead J, Leslie R (eds) Coastal protection in the aftermath of the Indian Ocean tsunami: what role for forests and trees? FAO, Bangkok

TEEB – The Economics of Ecosystems and Biodiversity (2014) Ecosystem Services. http://www.teebweb.org/resources/ecosystem-services. Accessed 16 Dec 2014

Thao ND, Takagi H, Esteban M (eds) (2014) Coastal disasters and climate change in Vietnam. Elsevier, London/Waltham

Toral Ibáñez M, Vita Alonso A, Cogollor Herreros G (1980) Dinámica superficial del campo de dunas de Ritoque. Boletin Técnico N° 60. Facultad de Ciencias Forestales, Universidad de Chile, Departamento de Silvicultura

UN Atlas of the Oceans (2014) Human settlements on the coast. http://www.oceansatlas.org/servlet/CDSServlet?status=ND0xODc3JjY9ZW4mMzM9KiYzNz1rb3M. Accessed 16 Dec 2014

Van der Meulen F, Bakker TWM, Houston JA (2004) The costs of our coasts: examples of dynamic dune management from Western Europe. In: Martínez ML, Psuty NP (eds) Coastal dunes: ecology and conservation. Springer, Heidelberg

Waycott M, Duarte CM, Carruthers TJB et al (2009) Accelerating loss of seagrasses across the globe threatens coastal ecosystems. PNAS 106:12377–12381

Wolanski E (2007) Protective functions of coastal forests and trees against natural hazards. In: Braatz S, Fortuna S, Broadhead J, Leslie R (eds) Coastal protection in the aftermath of the Indian Ocean tsunami: what role for forests and trees? FAO, Bangkok

Wolf Eigenherr B (2008) Estudio comparativo de la vegetacion de los campos de dunas de ritoque y la chepica (32° 49′ – 33° 28′ s), V Región de Valparaíso, Chile. Seminario de Investigación presentado al Instituto de Geografía de la Pontificia Universidad Católica de Chile

World Risk Report (2014) United Nations University – Institute for Environment and Human Security (UNU-EHS) and Alliance Development Works

Yamagata M (2006) Inland Sa Huynh culture along the Thu Bon river valley in Central Vietnam. In: Bakus EA, Glover IC, Sharrock PD (eds) Uncovering Southeast Asia's past: selected papers from the 10th international conference of the European Association of Southeast Asian Archaeologists. National University of Singapore Press, Singapore

Yasuda S, Tanaka N, Yagisawa J (2012) Effects of the coastal forests, sea embankment and sand dune on reducing washout region of houses at the tsunami caused by the Great East Japan Earthquake. Proceedings of the International Symposium on Advances in Civil and Environmental Engineering Practices for Sustainable Development – ACEPS:197–204

Zhang G, Song J, Yang J, Liu X (2006) Performance of mortar and concrete made with a fine aggregate of desert sand. Build Environ 41:1478–1481

Chapter 19
Managing Flood Risks Using Nature-Based Solutions in Nouakchott, Mauritania

Ahmed Senhoury, Abdeljelil Niang, Bachir Diouf, and Yves-François Thomas

Abstract Whether or not exacerbated by climate change, flood risks are becoming more frequent in the capital city of Nouakchott in Mauritania. Flooding in Nouakchott is due to a combination of both natural factors and human activities. The extreme fragility of the barrier beach that protects the city from the sea, the accelerated exploitation and inadequate infrastructure built along the coast have made this barrier beach highly vulnerable to wave action, exposing the city to a high risk of flooding. Flooding is further exacerbated by rising groundwater levels in several neighborhoods of the city. Cartographic analysis of flood risk indicated that socio-economic impacts associated with floods could be high. In the case of sea water intrusion, up to 30 % of the city could be potentially submerged. This would directly affect nearly 300,000 people and entail high risks of casualties. Associated economic losses due to flooding could be as high as USD 7 billion (Senhoury, Aménagements portuaires et urbanisation accelerée des côtes basses sableuses d'Afrique de l'Ouest dans un contexte de pejoration climatique, cas du littoral de Nouakchott (Mauritanie). Thesis state, University of Dakar, April 29, 2014, 157 pp, 2014).

The following measures based on nature-based approaches are recommended to tackle flood risks in Nouakchott:

- Restore and consolidate the barrier beach through reforestation of degraded areas;
- Put in place an appropriate drainage system for rain and marine waters and a sewage sanitation system;

A. Senhoury (✉)
University of Sciences, technology and medecine, Nouakchott, Mauritania
e-mail: senhoury@hotmail.com

A. Niang
Umm Al Qura University, Makkah, Saudi Arabia

B. Diouf
Cheikh Anta Diop University, Dakar, Senegal

Yves-François Thomas
CNRS-Meudon, Paris, France

© Springer International Publishing Switzerland 2016
F.G. Renaud et al. (eds.), *Ecosystem-Based Disaster Risk Reduction and Adaptation in Practice*, Advances in Natural and Technological Hazards Research 42,
DOI 10.1007/978-3-319-43633-3_19

- Optimize a solution to safeguard the harbor of Nouakchott; and
- Transform wetlands created by the permanent flooding of low-lying areas in the city into urban protected areas.

Keywords Flood risks • Nouakchott • Coastal risk • Hydrodynamics • Cartographic analysis • Socio-economic impacts • Nature-based solutions

19.1 Introduction

The increasing losses and costs associated with natural hazard events in recent years have highlighted the high vulnerability of contemporary cities to disasters. Like many coastal cities across the world, Nouakchott, the capital city of Mauritania, is facing serious risks of flooding. Located behind a narrow dune belt, in area largely under sea level, this city is both exposed and vulnerable to heavy rains as well as episodic increases in the sea level (Senhoury 2000; GRESARC 2006).

The city of Nouakchott is home to one third of the country's population and the country's key economic infrastructure. Protection of the city is therefore a major concern for decision makers and residents. Protection requires detailed knowledge of the flood risks confronting the city and the identification of measures for safeguarding its vulnerable areas.

Over the last years, the amount of rainfall recorded in the city has reached approximately 100 mm, which is sufficient for stagnant waters to become a nuisance and, in several neighbourhoods of the city, a disaster. Indeed, the geological nature of the soil and high groundwater level makes soil uanble to absorb, even low, rainfall. The lack of rainwater drainage and sewage disposal system results in foul-smelling waters overflowing from septic tanks and stagnating for weeks, thereby affecting city residents' well-being and comfort and disrupting economic activities in the capital.

In addition to flood risk linked to heavy rainfall, Nouakchott is also threatened by coastal flooding due to a combination of natural factors, including the ecological fragility of the coastal dune belt and the weak difference in level and presence of sebkha[1] grounds, and various anthropogenic activities, namely uncontrolled urban planning, building of infrastructure on the beach, and destruction of plant cover and mining of sand dunes for construction materials.

The general objective of this paper is to characterise the vulnerability of Nouakchott to flooding linked to coastal/marine hazards as well as the accumulation of rainwater and groundwater discharge. It will then outline solutions to mitigate flood risks. Many results discussed in this paper are from a Phd Thesis supported in 2014 at the University of Dakar (Senhoury 2014).

---

[1] flat-bottomed depression, usually flooded, where salty soils limit the presence of vegetation.

19.1.1 Nouakchott's Geography

Located in West Africa, Mauritania lies between latitudes 15° and 27° North, covers a surface area of 1,030,700 km^2 (Senhoury 2000), and has over 650 km of coastline bordering the Atlantic Ocean. Nouakchott, the capital city, is located towards latitude 18°07 North and longitude 17° West (Fig. 19.1).

Established in 1957, Nouakchott is on the oceanic front of the Sahara, at the edge of a low and narrow coastal plain, known as the *Aftout Es Sahli*.[2] Less than 5 km away from the seafront, the city is connected to the Atlantic Ocean through a narrow sand belt and has the following geographical sea-to-inland profile:

- A relatively narrow and thinly wooded coastal belt, with an average width of 150 m and an average altitude of 6 m, which provides the city its sole protection against flooding linked to coastal hazards;
- A vast flood depression with a locally variable altitude of 1–4 m;
- Large continental dune belts.

Although a young city, Nouakchott has already experienced accelerated population growth, starting with 1800 people in 1957 and growing to almost 800,000 inhabitants, as per the last census conducted in 2013 (ONS 2013). Its spatial organisation is characterised by a radial development and by a predominantly horizontal dispersed housing. The city grew as plots of land were attributed to residents, and new subdivisions or neighbourhoods created, while previously-established subdivisions were not fully developed. This expansion was made worse by the spontaneous and uncontrolled settlement of new migrants in the city's outskirts or in pockets of the existing urban fabric on undeveloped sites. As a result, many outlying areas of the city are spreading to the sensitive and lowest areas in the coastal zone.

19.1.2 The Climatic and Hydrodynamic Conditions

The high temperature, scarce rainfall, wind intensity and the wave regime in Nouakchott are elements that promote sediment mobility, especially on coastal dunes, and thus exacerbate coastal flood risk (Elmoustapha et al. 2007).

---

[2]Aftout es Sahli is a coastal lagoon, whose width varies from 3 to 7 km from the sea and the continental dunes in the east, and extends over nearly 275 km from Saint Louis (Senegal) to Nouakchott.

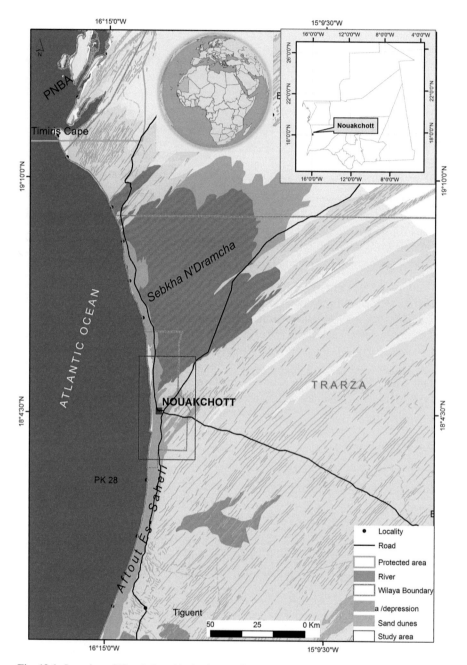

Fig. 19.1 Location of Nouakchott (Author's own figure)

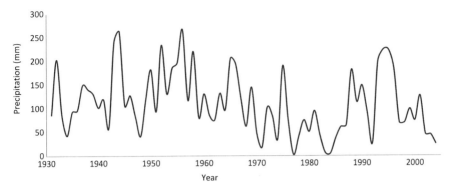

Fig. 19.2 Annual rainfall in Nouakchott between 1930 and 2004 (Data from the weather station in Nouakchott, cited in Senhoury 2014)

19.1.2.1 Average Annual Temperature

The average annual air temperature in Mauritania is between 20 and 30°C. However, temperatures above 40°C are commonly observed (Barusseau 1985). In Nouakchott, the average temperature is between 19 and 33 °C (Dubief 1963).

19.1.2.2 Rainfall

Rainfall data were obtained from the meteorological station of Nouakchott, covering the period 1930–2004. The evolution of annual precipitation over this period (Senhoury 2014) shows irregular rainfall in Nouakchott (Fig. 19.2). Rainfall variation is high, from a minimum of 5 mm recorded in 1984 to a maximum peak of 241 mm recorded in 1945.

The irregularities of precipitation observed in recent decades indicates a probable connection with climate change. This hypothesis requires further study, especially in the context of Nouakchott, where a correlation between rainfall and flood risk has been established (Senhoury 2014).

19.1.2.3 Wind Regime

Two wind regimes are active in this region: the rain-bearing Atlantic monsoon and the dry, North-Northeast trade winds linked to the Azores and Sahara region of high atmospheric pressure. Wind speeds and directions at 11 m above sea level have been continuously measured since 1975 by the Port de l'Amitié Authority (Ould Mohameden 1995). The most frequent winds are from the East-Northeast to West-Northwest sectors (81.7 % of the observations). Winds from the North-Northwest to the Northwest sectors represent 39.0 % of annual wind conditions, while winds from the North and North-Northeast sectors are present 21.0 % of the time. Wind speeds exceeding 13.9 m/s occur 1.1 % of the time, while speeds under 3.2 m/s are observed 6.3 % of the time.

19.1.2.4 Tidal Regime

Tides propagate from the north to the south along the Mauritanian coast, and are mainly semi-diurnal at Nouakchott. The tidal regime is microtidal, with ranges attaining 2 m during high spring-tide conditions (Ould Mohameden 1995).

Due to this rather small tidal range, tide-induced currents are not significant along the coast of Mauritania, which is nevertheless affected by a major oceanic circulation, the Canary current which originates from the northern Atlantic. A branch of this current veers westwards at CapeBlanc to the north of Mauritania, forming the North Equatorial current along the coasts of Mauritania and Senegal. The speed of this permanent current, orientated southward is about 0.2 m/s (Hebrard 1973).

19.1.2.5 Wave Regime

Between 1975 and 1982, wave measurements were carried out off Nouakchott wharf at a water depth of 9.5 m during a feasibility study prior to the construction of Port de l'Amitié harbour. The distribution of the mean wave heights and periods for the year 1976 has been analysed (Ould El Moustapha et al. 2007). Waves from the northwest and west-northwest are the most frequent, representing respectively 46.2 % and 23.6 % of the observations. Wave periods[3] are rather small; values smaller than 4 s and between 5 and 6 s represent 33.6 % and 38.9 %, respectively, of the observations. Periods exceed nine seconds only 15.7 % of the time. Mean wave heights are between 0.8 and 2.0 m 80 % of the time, and only rarely exceed 2.0 m. Maximum wave heights are associated with waves from the West and West-Northwest sectors.

19.1.2.6 Storm Regime

A statistical analysis of storms was performed in a previous work (Senhoury 2014) using the re-analysis method. The databases covered in this work are the data from the European Center for Medium-Range Weather Forecasts (ECMWF). The measuring point is located in the sea in front of Nouakchott (longitude: $-16.75°$ latitude: $18.00°$). These data cover the period 1957–2013. The processing of data identified 39 severe storms on the coast of Nouakchott. The results show that

---

[3]The time which separates two crests of successive waves.

although their frequency is not high in Nouakchott, severe storms are long in duration and last for several hundred hours.

Moreover, the frequency of these storms seems to be growing since the 1990s. The results show that storms have occurred frequently three times a year since 1990, while this rate has never exceeded two storms per year before 1990.

19.1.2.7 Nouakchott's Flood History

Several flood incidents have been recorded on the Nouakchott site, including the following key events:

- In 1950, the coastal plain of *Aftout Es Saheli* was flooded as a result of exceptional increases in the level of the Senegal River. That same year, following torrential rains, the Senegal River waters ran into the Atlantic Oean through *the Chott Boul* and the mouth located south of Saint-Louis, Senegal, and flowed as far as Nouakchott through *Aftout Es Saheli*. The floods destroyed the only neighbourhood standing in those days, which was rebuilt on the same site by the then colonial administration. Similar flooding events had been recorded in 1890 and 1932.
- On three separate occasions, in February 1987, August 1992 and December 1997, waves driven by violent storms crossed the coastal belt towards the direction of Nouakchott, causing the failure of the coastal belt in several places and moderate damages.
- Rainfalls recorded in 2013 (about 100 mm), although high but not exceptional, resulted in disastrous flooding in almost all the neighbourhoods of the city (Fig. 19.3). The disaster served as additional warning and a call for planned action with long-term solutions.
- In addition to flooding by rains, groundwater is also surfacing more frequently in several areas of the city, which exacerbates flooding episodes. Most of these affected areas are urbanized but are characterized by poor urban planning and therefore are at a higher risk of flooding.

19.2 Main Causes of the Vulnerability of Nouakchott

Nouakchott is highly exposed and vulnerable to flood risks, which are a result of a combination of underlying factors, including:

- Climate extremes, linked to climate change and climate variability (Niang 2014);
- Weakening and subsidence of the coastal dune belt due to poor urban planning and pressures from human activities (e.g. port facilities, construction of various

Fig. 19.3 Flood in Nouakchott following rains in 2013 (Source: enhaut; reproduced with permission)

buildings on public areas on the beach, construction of roads and terraces that prevent water run-off from permeating into the ground, construction of houses on wetlands, and from natural forcing (e.g. exceptional tides, waves, storms, etc.);
• Lack of a rainwater and sewage disposal system.

This section provides detailed information on the main factors that contribute to flood risks in Nouakchott.

19.2.1 Severe Climate Conditions

As extreme events become more recurrent and storms more intense, climate change appears as a likely assumption. Increased frequency of west winds clocking at more than 10 m/s, and the greater amplitude of swells, compound the threats faced by low-lying coastal areas. Flood risk is made worse with the likely rise in the average sea level.

19.2.2   An Artificialisation of the Coastline, Paying Little Attention to Its Fragile Balance

Natural events, however, do not solely account for the environmental deterioration of the coastline located in the vicinity of Nouakchott. Human activities that ignored the environmental dynamics of the coastline have largely contributed to disrupting a naturally fragile balance.

In this regard, the construction of the Nouakchott's port, known as *Port de l'Amitié,* in 1979, which was carried out without a prior environmental impact assessment, has strongly disturbed the hydrodynamic and sedimentary functioning of Nouakchott's coastline and the evolution of the coastal stretch. As a result, the following has been observed (Figs. 19.4 and 19.5):

- Severe siltation of the coast, north of the port, which has already caused the decommissioning of Nouakchott's wharf, and this process is threatening to spread, in the short term, to the port's basin through a detour around the far west side of the port embankment. The latter threat will remain, although current works to expand the embankment might delay the siltation process by two to three decades;
- Significant erosion south of the current port facilities, which has already prompted the adoption of safeguard measures, such as use of spurdikes and containment dikes (Fig. 19.5). This erosion is the cause of the marked destruction of the shorefront dune over several kilometres south of the port. As a result of the degradation of the coastal dune, the *Aftout Es Saheli* plain and neighbourhoods in the southern part of the city have become more prone to sea water incursions.

The constructed protection measures do not always have positive effects. Indeed, these protection measures were constructed in an *ad hoc* basis, and their negative environmental impacts have seldom been considered.

Moreover, while the *Port de l'Amitié* is by far the primary cause of the negative evolution of Nouakchott's coastline, other anthropogenic activities have also contributed significantly to undermining the coastal system, namely:

(a) **Sand removal** from the coastal dune for construction purposes is one of the main causes of weakening the coastal dune, and is at the root of the breaches seen, from which *Aftout Es Sahli* has suffered sea flooding on several occasions. Sand removal is now fully prohibited.

(b) **The construction of buildings on the coastal dune**, where in 2005 there were already five hotels and several industrial and/or trading infrastructures (factories, fish market, etc.), has also weakened this coastal dune. Built infrastructure exert pressure on and weaken the dune belt, the only form of protection against coastal flooding in some areas of the city.

Fig. 19.4 Aerial view of the site of Nouakchott's port in 1980 (Reference : 80_Mau_42_155 IGN France, reproduced with permission)

(c) **Recreational activities**, such as frequent and unregulated car stunts (a recreational, sporting activity), destroy pioneer plants, prevent the types of sediment accumulation that develop with these plants, flatten dune ridges and make it impossible to seal breaches along the dune belt. This motorised traffic is exacerbated by the anarchical trampling of spectators that come in increasing numbers to watch the car stunts.

(d) **Uncontrolled grazing by animals which consume dune vegetation and the use of dune vegetation as firewood by some households.** This vegetation, which is supposed to fix the sand dunes in place, has already been degraded by drought in the past decades.

Fig. 19.5 Aerial view of Nouakchott's port in 1991 (Reference 1991_Mau_12_150 IGN France, reproduced with permission)

19.2.3 A Brackish Water Table with Continuously Rising Water Levels

A large part of Nouakchott was built in a low-lying area whose altitude is lower than the sea level. In many parts of the city, the water table is surfacing and its level is directly related to that of the ocean. In addition to being a receptacle for marine and rain waters, the water table may threaten the city of Nouakchott in case of outcropping and degrades habitat conditions, even without flooding episodes.

Since the commissioning of Nouakchott's system of safe water supply from the Senegal River (*Aftout Es Sahli* project), water distribution in the capital city has increased from 60,000 m^3 to 170,000 m^3 daily and may reach 226,000 m^3 daily in 2030 (MHET 2014). This significant increase has taken place without an appropriate system for sanitation and wastewater treatment. Hence, a network of individual septic tanks and cesspools partly contributes to the refilling of the water table. According to current estimates, 90 % of waste waters and discharge flow directly into the water table.

When the water table is high, even a small amount of rainfall cannot infiltrate into the already saturated *sebkha* ground, which results in flooding. In addition to the flood risk, people directly exposed to polluted waters face other health risks. The rapid development of housing projects in floodprone, low-lying areas has only exacerbated these risks.

In this regard, it is crucial to take into account the functioning of the water table, when considering global sea level rise, climate extremes, the coastal regimes (tides, waves, etc), rainfall and wastewater linked to accelerated urbanization.

19.3 Review of Nouakchott Flood Risks

The topographic survey of Nouakchott completed in 2006 by the GRESARC research group (GRESARC 2006) was used to map out flood-prone areas in the city. The study showed that about one third of the city's urbanised areas are located at extremely low topographic levels, and hence prone to flooding.

Simulations have been conducted (Fig. 19.6) to help understand the extent of flooding in Nouakchott, for instance in case of several large breaches in the dune belt following a strong storm, or in the case of rising groundwater levels together with heavy rainfall. To develop this map, it was mainly assumed that the extreme sea level reaches between 1.0 and 1.4 m, or that the height of rainwater that puddles are not absorbed into the ground due to watertable saturation is approximately the same values.

Based on the study, the inhabited areas in the city most prone to flooding are essentially:

• The western part of the *Tevragh Zeina* urbanised area;
• Almost all the inhabited areas of the *Sebkha* and *El Mina-nord* neighbourhoods;
• The majority of homes in *Riyad*.

The eastern end of the *Téyaret* and *Ksar* neighbourhoods as well as the central and north-central part of *Dar Naïm* are equally located at a point below the extreme sea level of 1 m, but are separated from the sea by higher, elevated zones. These areas are less exposed to risks of coastal flooding. However, these disctricts are also particularly vulnerable to inland flooding triggered by heavy rains.

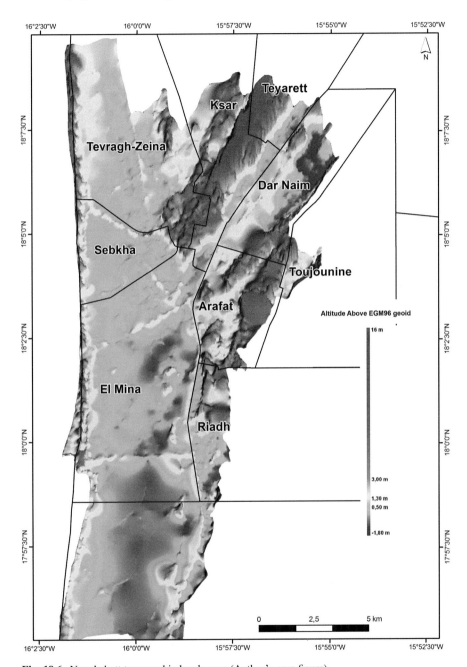

Fig. 19.6 Nouakchott topographic levels map (Author's own figure)

19.4   Assessment of the Socio-economic Impacts of Possible Flood Incidents

The socio-economic impact assessment of possibe flood incidents at Nouakchott was undertaken in 2014 by Senhoury (2014). This assessment was based on the study of flood-exposed areas and available data of the population and infrastructrure distribution. A Geographic Information System (GIS) was used which included the mapping of land-use in Nouakchott, the topography made by GRESARC in 2006 and the National Statistics Office's (ONS) statistics on population distribution. Using the ArcGis software to operate the GIS system helped to calculate flood risk areas, the length of tarred (cemented) roads, and the number of threatened socio-economic infrastructures.

The findings presented in this paper are those of a scenario whereby the extreme sea level is 1.4 m. The inundation considered in this regard supposes that one of the following events could occur:

– Appearance of several large breaches associated with a storm whose duration makes it possible for a significant amount of seawater to spill over;
– Rise in the level of Nouakchott's brackish water table to reach the extreme sea level, with possible significant rainfalls.

The flood map of Nouakchott indicating flood-prone areas, buildings and infrastructures impacted under this scenario is provided in Fig. 19.7.

The flood-prone population of Nouakchott was determined based on the size of the population residing in the city in 2011, estimated at 727,000 inhabitants. The calculations show that 38 % of the population face flood risks, representing 273,000 people that will have to be displaced in the event of flooding by sea. The bulk of potentially-affected population live in low-income neighborhoods of Nouakchott and are predominantly poor.

Calculations made with the ArcGis also determined for each commune of Nouakchott the size of flooded areas together with their economic values, based on 2005 data on the extent of urbanization. Results showed that an area of more than 10,400 hectares, including 8200 urbanised hectares, is likely to be submerged, i.e. about 30 % of the city. Risks in terms of human losses are considerable, especially when people are not informed nor trained in disaster management. The civil protection departments do not seem to have the means to handle flooding incidents of such a magnitude (MPEM 2005).

By assuming that unit costs of housing are close to market prices, and depending on whether homes are located in higher- or low-income areas, the material values of threatened areas are estimated at a cost of more than USD 7 billion. With regard to tarred roads, the simulation area covered a network of 371 km in 2005. In the event of coastal flooding, 189 km would be destroyed, representing a value of USD 31 million. Other critical infrastructures, including schools, hospitals, mosques and markets, that provide key social services would also be threatened. Table 19.1 shows the number of basic infrastructures under threat in case of flooding at 1.4 m.

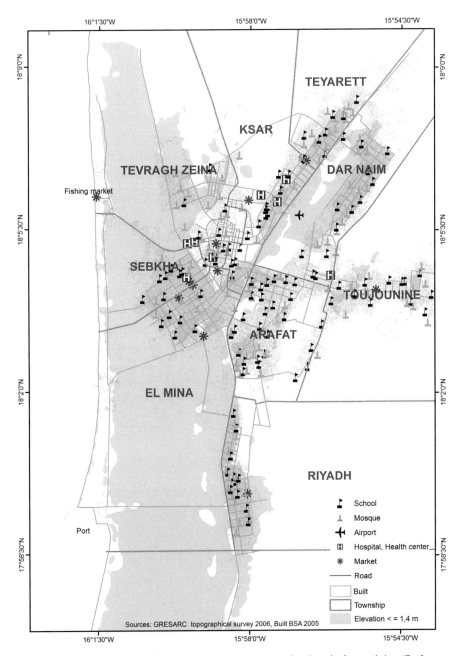

Fig. 19.7 Submersion map of Nouakchott for an extreme sea level equivalent to 1.4 m (Senhoury 2014, reproduced with permission)

Table 19.1 Threatened basic infrastructures (Senhoury 2014)

Type of infrastructure/ service	Total number of infrastructures existing in Nouakchott in 2005	Number of threatened infrastructures
Schools	107	45
Public hospitals	8	2
Mosques	112	60
Markets	11	6

It should be noted that strategic economic infrastructures are also threatened by flooding. Because of the limited information on the economic values of such infrastructures, only a limited indicative list of key threatened properties comprises:

- Part of Nouakchott's airport;
- Nouakchott's port;
- Nouakchott's wharf; fish market; and
- Two important cement factories

19.5 Proposed Corrective and Adaptive Measures

Given the flood risks facing the city of Nouakchott and the scope of their environmental and socio-economic impacts, mitigation and/or adaptation measures are required. These must include both preventive and corrective measures.

19.5.1 Preventive Measures

The preventive actions proposed consist of adopting flood-sensitive land-use measures and regularly monitoring the current evolution of the coastline, as follows:

- Develop and implement urban town planning maps and schemes that firmly prohibit construction works in areas likely to be submerged or flooded, i.e. low-lying areas of Sebkha (with an altitude between −1 and +1 m), the ridge of the dune belt and along the beach between the dune and the coastline;
- Stop any activities that threaten the coastal dune belt; it is urgent to maintain the prohibition of sand removal for construction purposes, provide options for sand collection, control car traffic along the fragile segments of the coastal belt and prohibit animal grazing on the plant cover that holds the dune belt intact;
- Put in place an early warning system, which could trigger an immediate response to a potential disaster or a serious anthropogenic stress. This system would ensure monitoring of weather, and its hydrodynamic consequences as well as the morphological monitoring of the beach and dune belt. The system would also help monitor the expected increase of the groundwater level in relation to climate drivers (sea level, rains, etc.).

19.5.2   Corrective Measures

The proposed corrective measures are based on previous studies (IRC-Consultant 2008; Senhoury 2014). They aim to mitigate flood risks through the optimum utilization, when possible, of available natural resources. The choice of nature-based approaches is dictated by two reasons. They help to correct the current malfunctioning of the coast of Nouakchott and improves the protection of its ecosystems and ecosystem services that contribute to flood risk reduction. Moreover, the engineering approaches have a relatively high cost. To this end, the following measures are recommended:

(a) Establishing a drainage and water treatment system

In order to reduce flood risks in the city because of both the rainy season and groundwater discharge, it is recommended to accelerate the establishment of a drainage system for sewage and excess run-off by building a wastewater collection network and a treatment station. The treated wastewater can be utilized for gardening and reforestation of the coastal dune. This will help to collect and redirect waters collected and promote its rational and sustainable use in order to reduce overexploitation of potable water in an arid city.

The Governments of China and Mauritania have signed in December 2014 an agreement for the provision of USD 32 million to construct a sanitation system in Nouakchott. The planned project includes the construction of modern rainwater and wastewater drainage networks. Feasibility studies suggest building two separate drainage systems, i.e. a rainwater network and a wastewater network. The proposed drainage system should take into account basic data on wastewater production in Nouakchott, currently estimated at 82,000 m^3 and the assumption of erratic rainfall varying from 5 to 241 mm.

(b) Reinforcement of the coastal dune

Low-lying areas are found in Nouakchott's coastal dune belt, which make these areas particularly susceptible to breaches and seawater intrusion. A priority action to protect the city from coastal flooding should therefore consist in plugging (i.e. repairing) the breaches and, more generally, in reinforcing the dune belt.

The optimal functioning of the coastal belt does not require that it remains strictly stable in its current position but that controls are undertaken for the structure to play its role with some degree of mobility.

According to IRC-Consultant's 2008 estimates, the optimal height required for the reinforcement of the coastal dune should be at least 6 m above the sea level (IRC-Consultant 2008). Needless to say, the most effective belts are those with a large width. From the outset, a width/height ratio of 20 seems to be appropriate, representing a minimal width of 120 m and a minimal height of 6 m.

Two complementary techniques for reinforcing Nouakchott's coastal dune area are recommended, i.e. mechanical techniques and biological techniques. In the context of Nouakchott, these techniques complement each other and offer the advantage of incurring minimal environmental impacts.

Fig. 19.8 Experience of sand dune stabilisation using Typha australis stalks on the dune between the "Plage des pêcheurs (Fishermen's Beach)" and the wharf (Author's own figure)

Across the entire dune area facing the city, north of the port facilities, the dune belt is where major sand movement takes place. Setting up shelterbelts to protect dunes may result in the quick accretion of the dune. The mechanical technique of using windbreak hedges could therefore be effective in this case. For socio-economic and ecological reasons, it is preferable to use windbreak hedges made of local plant materials, because they are low cost and accessible. Indeed, these locally-developed techniques have been proven to be efficient in several areas of the coastal dune, based on pilot projects. The most replicable example is from a pilot study conducted in 2005 which used the stalks of an invasive plant species known as *Typha australis* to build windbreaks or shelterbelts on dunes (Fig. 19.8). This invasive plant is a threat to the ecosystem of the Senegal River, south of Mauritania; hence, using the stalks of this species to make shelterbelts is one way of eliminating it.

The mechanical reinforcement of the dune belt contributes to the development of natural vegetation. It would be interesting, however, to supplement it with biological measures that support re-vegetation, which are in turn conducive to the accumulation of sand and its mobility reduction. When a satisfactory profile is achieved, certain plant species must be introduced in the dune, primarily in the rear portion. It is recommended to choose local bushy species, particularly *Nitraria retusa, Calotropis* and *Tamarix,* as they are quite adapted and highly resistant to dryness. Planting should take place in the area located between the *Port de l'Amitié* and the *Plage des Pêcheurs*, stretching 12 km long. The size of the back dune to be planted is estimated at 240 hectares (200 m wide), with a density of 200 seedlings per hectare, representing a total of 48,000 seedlings (MPEM 2005). Treated areas must be protected by a perimeter fence to ensure constant care and surveillance.

(c) Preservation of Nouakchott's port

The alarming deterioration of the coastline induced by the construction of Nouakchott's port facilities is considerable, which as a result has become the cause of major flood risks for the city. Consequently, optimal solutions should be found to manage the coastline's evolution that allows both the preservatin of the *Port de l'Amitié* and erosion reduction in its southern part.

This is why digital simulations on various options for the development of the *Port de l'Amitié* were conducted in the framework of a partnership with the GRESARC research group of the University of Caen on behalf of the consultancy firm IRC-Consultant (IRC-Consultant 2008). The UNIBEST model (Delft Hydraulics 1994) was used to make these simulations.

The results show that for the port of Nouakchott, the restoration of the sedimentary transit through by-passing seems to be an ideal solution in terms of sediment balance. The by-passing is a system which allows to restore artificially the sediment movement along a coast (by pumping or transportation in trucks). This solution has double effects. It would avoid siltation in the port by reducing accretion in the northern part and, at the same time, help to replenish the beach located south of the port. Yet, this solution appears costly owing to the significant amount of sediments for which an artificial transit is required given the intensive drift along the littoral.

(d) Creation of protected urban areas in Nouakchott

This adaptation measure is about transforming wetlands created by the permanent flooding of low-lying areas in the city into urban protected areas and using them for recreation such as bird watching as well as for excess water retention. A pilot project of this nature is under consideration in the centre of Nouakchott. In 2014, the Commune of Tevragh Zeina proposed this pilot action and is actively seeking support from national and international partners (Commune de Tevragh Zeina 2014).

The idea of this project is to harness an urban wetland created at the centre of the city, where several incidents of flooding caused by rainwaters or groundwater discharge have occurred. Indeed, surface sealing combined with increased construction and waterlogging have exacerbated rainwater stagnation in the lowest points of the city. These ponds, whose numbers grow year after year, fill with water during the rainy season and remain water-logged for an increasingly longer period, often throughout the year. A number of these urban wetlands, such as the site located in the *Ambassadeurs* neighbourhood in the centre of Nouakchott, has become home to specific vegetation and is increasingly visited by water birds (with more than 65 identified species). The site has become a real biodiversity hotspot in town.

The establishment of a protected urban area on that location boils down to enclosing and developing the site into an area for flood mitigation, birdwatching, recreational activities and promoting environmental education Such pilot projects should be duplicated and implemented in other parts of the city.

19.6 Conclusion

The city of Nouakchott provides a case study to understand the risks of flooding in coastal cities and the drivers of these risks (see also van Wesenbeeck et. al., Chap. 8; Nehren et. al., Chap. 18; David et. al., Chap. 20). The review has highlighted that Nouakchott remains vulnerable to serious flood risks of different origins, such as coastal flooding, rainfall accumulation and groundwater discharge. This vulnerability is the result of a marked deterioration of the environment within and outside the city. In fact, more than natural hazards that threaten the city (i.e. sea level rise, increased storm frequency, etc.), coastal degradation linked to unsustainable human activities have been instrumental in disrupting a fragile ecological balance and resulting in increased disaster risk to city dwellers.

Several contributing factors to the risks of flooding in Nouakchott have been considered, inluding natural factors (fragility of the coastal dune belt, topographic low of some areas) and various anthropogenic activities (uncontrolled urban planning, construction of infrastructures, destruction of plant cover and removal of construction materials from the coastal dune). This is compounded by the lack of a run-off drainage and sewage disposal system, particularly as groundwater tables are no longer capable of absorbing excess runoff, which increases flood risks for Nouakchott.

The digital terrain model, developed for mapping floods risks in Nouakchott, was used to highlight areas with topographic levels lower than the average sea level. Findings from the cartographic review of these risks suggest that with or without climate change, Nouakchott still remains subject to risks of flooding, if not major submergence. In the event of seawater intrusion, nearly 30 % of the city would be submerged to an extent far beyond recurrent floodings caused by rainwaters. Port and airport facilities, almost 200 km of tarred roads and many public infrastructures, such as health centres, universities and schools, would be affected. Under such circumstances, economic losses might reach the equivalent of US$ 7 billion.

In order to mitigate risks identified and/or adapt the coastline to the corresponding flood risks, there is a need to consider both preventive and corrective solutions. Corrective measures recommended are largely based on ecological approaches, either through optimising local natural resources (treatment and re-use of wastewaters and rainwaters for gardening) or implementing soft mechanical and biological techniques (revegetation-based reinforcement of the coastal dune).

References

Barusseau JP (1985) Evolution de la ligne de rivage en République Islamique de Mauritanie. UNESCO, Division of Marine Sciences – Marine Sedimentology Laboratory, University of Perpignan, Study reports, février 1985, 104 pp

Commune de Tevragh Zeina (2014) Projet Nature à Tevragh Zeina. Press release, 4 pp

Delft Hydraulics (1994) Unibest, a software suite for simulation of sediment transport processes and related morphodynamics of beaches profiles and coastline evolution. Theoretical reference document. H 454, octobre 1994, 40 pp

Dubief J (1963) Le climat du Sahara. Memory Occasional Papers, Inst. Rech. Sah., Université Alger.T.1, 312 pp

Elmoustapha AO, Levoy F, Monfort O, Koutitonsky VG (2007) A numerical forecast of shoreline evolution after harbour construction in Nouakchott, Mauritania. J Coast Res 23(6):1409–1417

GRESARC (2006) Cartographie des risques littoraux de Nouakchott. Study report. MPEM & UICN. 63 pp

Hebrard L (1973) Contribution à l'étude géologique du quaternaire du littoral mauritanien entre Nouakchott et Nouadhibou. Participation à l'étude des désertifications du Sahara. Thesis, University of Lyon, 483 pp

IRC-Consultant (2008) Revue de l'état des risques d'inondation de la ville de Nouakchott. Preliminary report, Volume 1: main report. Department of Environment, Mauritania, 64 pp

MHET (2014) Réactualisation de l'Etude du Plan Directeur de l'assainissement de la ville de Nouakchott. Study report, version draft.Ministry of water, energy and TICs, Mauritania, 83 pp

MPEM (2005) Rapport sur les risques d'inondation de la ville de Nouakchott. Study report, 30 pp

Niang AJ (2014) Resilience against climate change: case of Nouakchott City. Geo-Eco-Trop 38 (1):155–168

ONS (2013) Etat d'avancement des activités du Recensement Général de la Population et de l'Habitat 2013. Study report, 10 pp

Ould Mohameden A (1995) Aménagement et évolution du littoral, Apports de la télédétection et de la modélisation mathématique: cas du port de Nouakchott. Nice, France: University of Nice, PhD thesis, 149 pp

Senhoury A (2000) Influence d'un ouvrage portuaire sur l'équilibre d'un littoral soumis à un fort transit sédimentaire, l'exemple du port de Nouakchott (Mauritanie). Thesis, University of Caen, Basse-Normandie, December 15, 2000, 162 pp

Senhoury A (2014) Aménagements portuaires et urbanisation accelerée des côtes basses sableuses d'Afrique de l'Ouest dans un contexte de pejoration climatique, cas du littoral de Nouakchott (Mauritanie). Thesis state, University of Dakar, April 29, 2014, 157 pp

Chapter 20
Assessing the Application Potential of Selected Ecosystem-Based, Low-Regret Coastal Protection Measures

C. Gabriel David, Nannina Schulz, and Torsten Schlurmann

Abstract Climate change and subsequent processes triggered by climate change demand novel assessments and protection schemes in coastal environments, as frequency and intensity of extreme events as well as mean sea water levels are expected to rise. Most often, conventional coastal engineering approaches are solely built for protection purposes, but often come with negative side-effects to the coastal environment and communities. During the last decade, new concepts in coastal engineering have started emerging. Several technical measures with an ecosystem-based design have been developed and, in some places, already implemented over the last decade. These low-regret measures, for instance green belts, coir fibers and porous submerged structures, reveal their full potential as stand-alone coastal protection or when used in combination with each other. They are believed – and in some cases documented – to be a better alternative or potential complement to conventional "hard" coastal engineering protection. Concrete examples are taken from the densely populated coastal area of Jakarta Utara (North Jakarta) and the National Capital Integrated Coastal Development (NCICD), showing benefits and further opportunities, but also challenges for applied low-regret coastal protection measures and ecosystem-based disaster risk reduction. An assessment of the application potential of three "soft" protection measures is given and discussed.

Keywords Soft coastal protection measures • Ecosystem-based and Low Regret Adaptation Management (ELRAM)

C.G. David (✉) • N. Schulz • T. Schlurmann
Franzius Institute for Hydraulics, Estuarine and Coastal Engineering, Leibniz University of Hanover, Nienburger Straße 4, 30167 Hannover, Germany
e-mail: david@fi.uni-hannover.de

© Springer International Publishing Switzerland 2016
F.G. Renaud et al. (eds.), *Ecosystem-Based Disaster Risk Reduction and Adaptation in Practice*, Advances in Natural and Technological Hazards Research 42,
DOI 10.1007/978-3-319-43633-3_20

20.1 Motivation to Develop New Low-Regret Coastal Protection Systems

Coasts and estuaries are increasingly exposed to rising sea levels, varying extreme weather and climate events. The likely impacts are due, on the one hand, to gradual processes (e.g. sea-level rise, coastal erosion, salt intrusion in estuarine systems and nearshore morphological changes) and, on the other hand, to single extreme events (e.g. storms and storm surges), together with increasing threats and human pressure.

Recently, increased attention has been paid to manage risks of extreme events through the "Special Report on Managing the Risks of Extreme Events and Disasters to Advance Climate Change Adaptation" published in 2012 by the Intergovernmental Panel on Climate Change (IPCC). The report contains a diverse portfolio of innovative options of "low-regret" adaptation measures for coastal protection. Protection measures are characterised as being "low-regret", if they *yield benefits regardless of the climate scenario but are not cost-free* (Wilby and Keenan 2012 pg. 1), or if they are *beneficial regardless of climate change impacts. Moreover, these measures will improve the adaptability of* the system *to the natural variability in climate patterns* (Bou-Zeid and El Fadel 2002 pg. 1). The fifth IPCC report (IPCC 2012 pg. 16), defines "low-regret" measures, as *measures that provide benefits under current climate and a range of future climate change scenarios. [...] They have the potential to offer benefits now and lay the foundation for addressing projected changes. Many of these low-regrets strategies produce co-benefits, help address other development goals, such as improvements in livelihoods, human well-being, and biodiversity conservation, and help minimize the scope for maladaptation.*

Such low-regret protection measures provide coastal protection by dissipating wave energy and, additionally, support local coastal ecosystems and supply eco-system services. Coastal ecosystems such as tidal marshes, mangroves, dunes and coral or shellfish reefs generate almost 40 % of all ecosystem services on our planet, with these being about twice the GDP of the world population (Costanza et al. 1997). Preserving these ecosystem services can provide additional monetary benefits in the range of $4.3–$20.2 trillion/year (Costanza et al. 2014), adding up the cost-benefit of "low-regret" adaptation measures – apart from being more sustainable and positive towards the environment itself.

Present requirements for a sustainable future are no longer met solely by conventional coastal engineering approaches. A paradigm shift from building *in* nature towards building *with* nature is necessary (De Vriend and van Koningsveld 2012). The answer is to recreate or preserve and improve existing ecosystems with a combination of socio-economic and marine ecosystem disciplines and establish an "Ecosystem-based and Low-Regret Adaptation Management" (ELRAM).

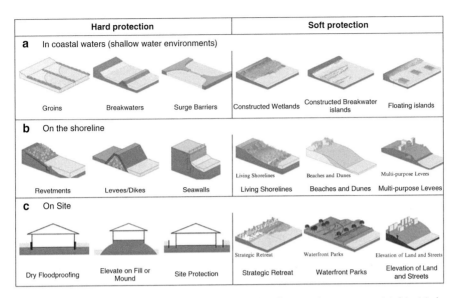

Fig. 20.1 Different types of hard protection (*left*) and soft protection measures (*right*): (**a**). in coastal waters; (**b**). on the shoreline; (**c**). and on-site (Bloomberg and Burden 2013; reproduced with permission)

20.1.1   State of the Art

In the past, coastal defence measures have been built mainly from the perspective of engineering and structural defence disciplines, but in order to enhance sustainability and long-term benefits of coastal protection measures, a more efficient integration of socio-economic and ecological principles and knowledge is required. Novel approaches attempt to recreate, preserve and improve existing ecosystems with a combination of socio-economic and marine ecosystem disciplines and to establish an ecosystem-based and low-regret adaptation management in coastal zones. Responses to coastal hazards therefore need to consider a broader range of solutions, which will demand transdisciplinary competences, knowledge of multiple coastal issues, and their links to ecosystems and society. In this context, typical examples of traditional, engineering-type, hard coastal protection measures are depicted on the left side and a selection of ecosystem-based, "low-regret" soft measures on the right side of Fig. 20.1. These measures are classified into categories whether they can be applied near-shore, onshore or on-site.

Traditional hard protection systems solely protect coastal areas from coastal hazards, but do not bring additional benefits, nor adapt to increasing future threats. They typically require continual and sometimes costly maintenance and need to be adapted, i.e. upgraded, to cope with sea level rise from climate change. They may also lead to or aggravate adverse effects on morphology, hydrodynamics, sediment transport, and nutrient budgets and impair the local economy (Cheong et al. 2013). "Soft", ecosystem-based protection systems also face these challenges, but concurrently attempt to improve local environmental settings, most often enhance structure and functioning of ecosystems and their services and to protect the location from typical coastal hazards.

Soft protection systems are not easy to implement in practice yet. The precondition of the environment and the geographic location decide whether different ecosystem-based approaches are feasible. An additional challenge is to meet security standards and the practical application when used for infrastructural and human defence. Ecosystem-based coastal defence requires more space than conventional structures, but for highly urbanized seaside cities, space is limited. In this case, conventional hard protection or a combination of hard and soft protection measures is often the only practical approach. When space between the sea and urbanized areas on the coast increases, efficiency and effectiveness of ecosystem-based flood defence increases likewise (Temmerman et al. 2013).

As of today, the performance and efficiency of created ecosystems as flood defence systems are still to a significant degree uncertain, because only a few long-term studies exist (Temmerman et al. 2013). However, instead of pursuing one particular strategy or defence for one specific hazard, coastal adaptation measures have to be dynamic, versatile and flexible to face climate change efficiently. Also, a proper design considers the influence for society, ecosystems and engineering among each other (see Fig. 20.2, Cheong et al. 2013). Involving all shareholders in planning and decision-making creates a higher acceptance for the process and thus reduces adverse political, financial and infrastructural effects, as Cheong et al. (2013) point out in order to define a novel disciplinary approach of so-called Ecological Engineering depicted in Fig. 20.2.

20.2 Possible Measures and Their Application

Soft coastal protection methods range from strategies such as integrated coastal zone management (ICZM) and early warning systems (EWS) to engineered (infra) structures. ICZM is already part of several national laws around the world (i.e. European Union, the United States of America, Australia, Sri Lanka, etc.), and early warning systems can be found for example in reference to tsunami hazards along the coastal stretches in the whole Indian Ocean (Taubenböck et al. 2013). This chapter, however, takes a closer look at and critically assesses three types of ecological engineering approaches, with a focus on the application in Southeast Asia.

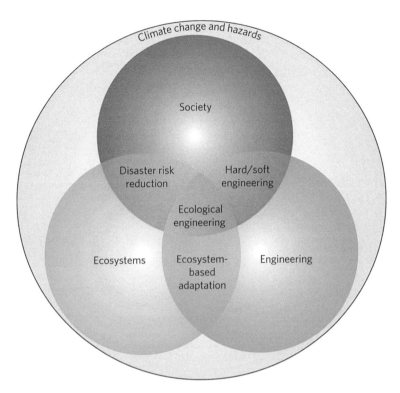

Fig. 20.2 Ecological Engineering – Combining strategies for coastal adaptation (Cheong et al. 2013; reproduced with permission)

20.3 Mangroves

Mangrove biomes consist of different tree and shrub types, which are adapted to the conditions of the coasts and estuaries in the tidal range of areas with (sub)tropical climate (Vedharajan and Gross 2007). Mangrove trees stand on their roots, which are mostly covered with water during high tide, but are exposed to air during low tide (see Fig. 20.2). They are not only an important part of the ecosystem and contribute to support the livelihoods for adjacent coastal communities, but they also serve as natural coastal protection and as a typical type of green belt.

The mangroves grow on muddy, wet and intermittently submerged beds (Fig. 20.3). Their dense root network traps organic compounds from upstream, thus creating a hypoxic or anoxic environment. In order to guarantee oxygen supply, many mangrove plants have developed aerial roots or pneumatophores. Mangrove roots are typically above ground and water level. They form a large root block which withstands wave attack and dissipates wave energy. However, the level of resistance varies depending on the root type.

Fig. 20.3 Natural mangrove in Senegal (By Wetlands International in Spalding et al. 2014; reproduced with permission)

20.3.1 Protection Potential

Several publications and reports based on field observations, physical and numerical modelling document the protection potential of mangroves in coastal areas (see also Renaud et al., Chap. 1; Friess and Thompson, Chap. 4; van Wesenbeeck et al., Chap. 8). The trunks and branches of mangrove trees serve as a barrier for wind and swell waves (McIvor et al. 2012). They reduce current velocity, flood depth and impact due to waves, floods and high winds (Hiraishi and Harada 2003; Teo et al. 2009; GIZ 2011; Lacambra et al. 2013). Mangroves can therefore protect humans, infrastructure and agricultural land against natural hazards such as storms, typhoons, and tidal waves. The level of protection, however, depends on the specific characteristics of local mangrove forests, such as tree species, age, condition, planting density and of course height and trunk diameter, as well as submergence of plants (Mazda et al. 2006; Augustin et al. 2009; Bao 2011; Hashim et al. 2013). Hashim et al. (2013) state that wave attenuating factors are not fully understood as of today, but present approaches take several variables regarding tree dimensions and planting properties into account (Mazda et al. 1997b; Quartel et al. 2007; Mendez and Losada 2004; Augustin et al. 2009; Bao 2011; Guannel et al. 2015). McIvor et al. (2012) find a proper design approach for mangrove restoration by selecting two models: a regression model by Bao (2011) and a numerical case study by Narayan et al. (2010), based on a modified version of the numerical phase averaged model Simulating WAves Nearshore (SWAN), presented in Suzuki et al. (2012). A summary with quantitative effects of mangroves on water levels and wave heights is given in Table 20.1.

Table 20.1 Summary of mangrove protection potential in different publications

Source	Effect	Condition
Mazda et al. (1997a)	Significant decrease for offshore wind waves	Width: 1500 m belt,
		Species: *Kandelia Candel*
		mature (5–6 years)
Quartel et al. (2007)	Wave height reduction between 5 and 7.5 times larger than plain seabed	
Tuyen and Hung (2009)	80 % wave height reduction	Width: 200 m or twice the wavelength
		Densely planted
McIvor et al. (2012)	5–50 cm of peak water level per kilometer mangrove forest	Planted in over large areas
	Wind and swell waves reduction greater than 75 %	1 km width of mangrove

Mazda et al. (2006) examined several plant species in a mangrove biome and found different influences on wave height among the genera. A reduction in wave height is found for genera with pneumatophores, for example any plants from the *Rhizophora* species. Compared to other mangrove types, *Rhizophora spp.* are the most favourable for wave attenuation (Hadi et al. 2013; Kathiresan and Rajendran 2005; Mazda et al. 2006; Tanaka et al. 2007). The dense and strongly connected roots also inhibit coastal erosion as well as accumulate and build up sediments, thus serving indirectly as a coastal protection measure (Gedan et al. 2011). The occurrence of *Rhizophora* species, however, differs for certain areas; while they grow seaward around India, they are found more landward in Southeast Asia (Kathiresan and Rajendran 2005).

Mangroves will not prosper on sandy soils with low humus content as well as low freshwater runoff or high salt concentrations. They lose stability when they are uprooted or bent. Uprooting is influenced by local bathymetric and geographical characteristics as well as the soil properties (Strusinska-Correia et al. 2013). Yanagisawa et al. (2009) found a correlation between survival rates after tsunamis and stem diameter of *Rhizophora* trees. Once a tree is uprooted in storm or flood events, it will become a dangerous debris itself. Nonetheless, a mangrove forest can completely regenerate after destruction within 15–30 years (EJF 2006).

In summary, in terms of coastal protection, mangroves can be restored together with conventional coastal protection measures and decrease the impact on conventional protection structures, thus reducing the dimensions required for such structures (Tuyen and Hung 2009). Mangroves shelter the hinterland from coastal hazards as stand-alone measures if sufficient space is provided (Harada et al. 2002; Barbier et al. 2013).

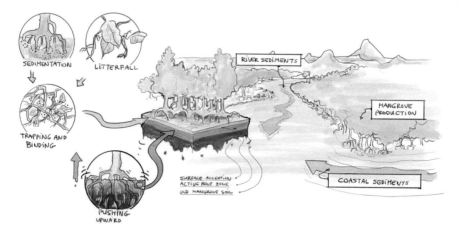

Fig. 20.4 Principle of sediment accumulation and filter capability of mangroves (Reproduced with permission from Spalding et al. 2014)

20.3.2 Ecological Benefit

Ecologically, conserving and restoring mangrove trees is important to reduce effects of climate change. They are regarded as important carbon sinks and thus contribute effectively to minimise greenhouse effects. One hectare of mangrove forest can extract approximately 1.5 tonnes of carbon per year from the atmosphere. For Indonesia 1.82 million ha of mangrove forest are available, leading to an equivalent reduction of exhaust gases from more than 5 million cars (GIZ 2011).

Additionally, the sediments beneath the mangrove trees keep another 700 tonnes of carbon per hectare by retaining the alluvial river sediments and tree leaves from upstream (see Fig. 20.4). The material is held back in the roots of mangrove trees and becomes solved organic matter in the tree's bed. Therefore, mangrove forests literally serve as carbon filters for (river) water. Mangroves also remove contaminants from the ocean and waterways (GIZ 2011). However, a clearance or die-off of a mangrove area will again release the stored pollutants.

By accumulating sediments and building upward, mangrove trees can adapt to changes in water levels, which is a decisive feature to keep pace with sea level rise. However, the plants can only respond to a limited, moderate rise of water levels (Vedharajan and Gross 2007).

In addition to coastal protection, mangroves secure the livelihood of the surrounding population by providing services such as firewood, medicines, fibres, dye and even food. They also serve as habitat and breeding ground for many fishes and animals, including shrimps, birds and marine mammals (GIZ 2011). Vedharajan and Gross (2007) estimated the economic value of mangrove forests to be around 7000 Euros per hectare, calculating without carbon storage.

A successful implementation with mangroves as sediment trap was made by the Research and Development Centre for Marine and Fisheries Technology, Jakarta (KKP-P3TKP) on the coast of Kamal Muara across from the Kamal Pantai Fish

Fig. 20.5 Extracted coir fibers, the raw material for coir products such as geotextiles, © David 2014

market in North Jakarta (Andayani et al. 2013). Before their installation, the seaward bank of the adjacent shrimp ponds were at risk of breaking due to high erosion rates. The team of KKP-P3TKP installed geotubes at the Kali Kamal river mouth to decrease flow velocity locally and planted mangrove trees on the lee-side of the tubes, which stabilized the banks successfully.

20.4   Natural Coir Fiber Geotextiles

Coir fiber geotextiles are a cost-effective, biodegradable and sustainable building material, following the paradigm of building with nature. They stabilize high slopes and banks of soil structures by ecological measures. Coir geotextiles support initial soil consolidation and protect early cultivation of vegetation, and consequently strengthens banks or dykes, eliminating the need for a permanent and persistent synthetic solution (i.e. hard revetment).

Coir is a seed fiber and a waste product of the food industry. The fibers are lignocellulosic[1] fibers, which are gained from the cortex or husk of the coconut fruit (see Fig. 20.5). The coir fiber has low cellulose but high lignin content. Low cellulose content is in general responsible for poor mechanical properties, while increased lignin content stiffens and toughens the fiber (Silva et al. 1999) and leads

---

[1]Lignocellulose: Cell walls of wooden plants, which consist of hemicellulose, cellulose and lignin.

to higher resistance to weathering, fungi and bacteria (Carus et al. 2008). Coir fibers undergo biological degradation, but at a much lower rate compared to other natural fibers (Silva et al. 1999; Lekha 2004; Lekha and Kavitha 2006; Carus et al. 2008). Coir contains the highest microfibrillar angle among typical natural fibers, which is a property of the microstructure of wooden fibers. The microfibrillar angle describes the orientation of the helical windings of fibers against the longitudinal cell axis and is critical factor for physical and mechanical resistance of wood (Cave and Walker 1994), where a high microfibrillar angle leads to a high tensile strength (Miller et al. 1998; Silva et al. 1999).

There is no universally valid lifetime for coir fibers, as the endurance of natural (geo-) textiles depend on a wide variety of environmental factors on site, but there are studies regarding their strength and site-specific lifetime: On the one hand, there are quantified measures (i.e. Balan and Venkatappa Rao 1996; Miller et al. 1998; Lekha 2004; Marques et al. 2014) measuring a loss of tensile strength of 55 % to almost 80 % after 6–7 months. On the other hand, there are measures by experience: Rajagopal and Ramakrishna (2009) give their coconut-based geotextiles a total life span of 4 years. Miller et al. (1998) speak of a 7-year design lifetime, which they also recognized as a commonly-used value in the early applications of coir fiber geotextiles. However, a reduction of tensile strength does not include a reduction of other stability attributes. The initial strength will decrease rapidly after the first few months, but decay rates will then be much lower.

20.4.1 Application

A test field investigated by the authors of coir-geotextiles for coastal engineering is located on Tabanan Beach, south shore of Bali Island, Indonesia. Tabanan is located northwest of Denpasar, and the areas surrounding the beach are experiencing rapid changes in land use from rice fields as well as from tourism-related infrastructures. The south shore of Bali is prone to wave attack and erosion. The test site is bounded by two volcanic rock peninsulas with groyne-like features, reducing long-shore sediment transport. Yet wave attack in storm conditions still causes considerable cross-shore sediment transport and thus erosion. The beach, however, will be a considerable economic asset due to the anticipated increase in tourism. Therefore, a compromise of beach preservation and coastal protection has to be found and, as of today, two approaches are being considered by the local government:

- Hard-structured concrete seawalls on the eastern part of the test site (Fig. 20.6). The seawall protects a parking lot and nearby lodges and homes. Other seawalls under similar conditions protect the area behind the structure, but have led to or increased erosion problems on the beach.
- The coir geotextile protected dune head is located further north than the hard-structure. The dune base is secured by bamboo sticks. Behind the sticks are two

Fig. 20.6 Hard-structure seawall in Tabanan, Bali © David 2014

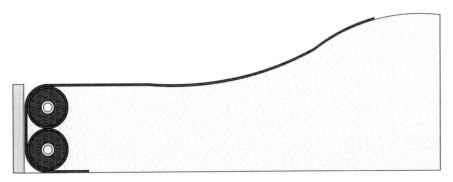

Fig. 20.7 Schematic sketch of coir fiber geotextile installation. On the *left* is the bamboo stick support, followed by two coir rolls and the dune body, which together with the rolls is wrapped in a geotextile layer

coir rolls and the dune body. The rolls and the dune body are wrapped by coir geotextiles. The installation can be seen in Figs. 20.7 and 20.8. After being installed, vetiver grass is planted on the dune to reinforce the soil.

As both approaches are just being implemented at the time of writing, a direct comparison is not immediately possible.

In practice, coir fibers are used as temporary alternatives to non-degradable synthetic geotextiles to prevent erosion and deformation, support bank stability, facilitate consolidation processes and drainage, with the aim to reinforce soil structures (Miller et al. 1998; Lekha and Kavitha 2006). Lekha and Kavitha (2006) for example use natural coir products in wetland areas, as water-permeable filter, preventing early structural failure during the consolidation process. Rajagopal and Ramakrishna (2009) and Subaida et al. (2009) studied coir textile application for rural roads. Miller et al. (1998) and Lekha (2004) studied the influence of coir geotextiles with regard to bank stability and erosion prevention. Both suggest a hybrid method of geotextiles with subsequent planting of, for

Fig. 20.8 Pictures of the on-site coir geotextile installation: (**a**). shows the cross-sectional view on the textiles; (**b**). shows the bamboo wall and the installation of the top coir layer © David 2014

example beach grass or vetiver. The geotextile supports the initial stability of the bank or dyke structure while the coir netting provides shelter for rain impact. After germinating, the seeds will grow through the coir netting, while the degrading natural coir material turns from covering shelter to nutrient supplier. Coir has also been successfully applied outside of Southeast Asia, for instance in Europe (Schurholz 1991) and India (Balan and Venkatappa Rao 1996).

Faruk et al. (2012) mentioned their lack of consistency of fiber properties and processing conditions and the fibers' sensitivity to weathering as major drawbacks of coir fibers. The latter condition, however, can be improved. A test by Miller et al. (1998) proves that degradation rates due to weathering (e.g. UV radiation, moisture and temperature) under temperate continental climate conditions led to tensile strength decrease to only <1 % for longitudinal loads and 9 % for transverse loads of the initial strength after 6 months. This confirms an influence of weathering on fiber degradation, but in comparison to degradation rates by soil-fiber interactions, coir fiber degradation rates are rather low (see Miller et al. 1998; Balan and Venkatappa Rao 1996; Lekha 2004; Rajagopal and Ramakrishna 2009; Marques et al. 2014).

20.5 Submerged Protection Structures

Most erosion problems encountered are the consequence of anthropogenic re-working of shorelines and the interference with natural sediment fluxes and alteration of sediment budgets (e.g. groins interrupting longshore sediment transport). In natural conditions, storm events subtract sediment material offshore, while swell waves steadily and slowly nourishes the beach again during low energy periods (Silvester and Hsu 1997; USACE 2002; Komar 1976). This process requires a natural margin for the coast to move between swell and storm seasons. Nevertheless, given that ocean view and beach connection are attractive for domestic or touristic properties, seaside constructions are mostly located closely to the coastline or directly on the shoreline and thus disturb the natural margin of sediment movements. Likewise, poorly-managed coastal development, deviation of freshwater streams and river damming or intense aquaculture (e.g. shrimp cultivation) aggravate erosion problems. Two potential solutions are artificial reefs or permeable submerged breakwaters. They reduce cross- and longshore erosion by reducing wave energy to overcome adverse erosion effects and recreate a natural operating space for sediment transport (e.g. Burcharth et al. 2007). If properly designed, they will protect the shoreline as well as support and conserve the coastal ecosystem by mimicking a natural near-shore habitat.

20.5.1 Low-Crested and Submerged Breakwaters

Submerged breakwaters are a special type of the traditional engineering breakwater measure, which are classified as hard protection measures, for example for port protection. Emerged breakwaters are a simple bar aiming to reduce wave heights. Sedimentation or scouring as a result of the breakwater is almost unavoidable and leads to high dredging costs, but calm water is the top priority for port efficiency and safety, thus justifying these expenses (Burcharth et al. 2007).

If aesthetic aspects of the protection measure have to be considered as well, for example on tourist beaches, submerged breakwater solutions become more attractive. Conventional, impermeable submerged breakwaters are equal in construction principles and material to normal breakwaters, only with their berm underneath the water level. If planned properly, they can be designed to decrease long-shore sediment transport by bending oblique incident waves by refraction and to reduce cross-shore sediment transport. Reduced cross-shore sediment transport then initiates tombolo or salient formation.

Submerged breakwaters affect waves by dissipation, transmission and reflection (e.g. Oumeraci et al. 2001). Energy dissipation is induced by wave breaking, friction or other non-linear interactions (i.e. Mason and Keulegan 1944; Roeber et al. 2010), leading to decreased wave height and changed wave form (Habel 2001; Bleck and Omeraci 2004). Wave attenuation depends primarily on the reef water depth and its width relative to the wave length (e.g. Oumeraci et al. 2001) and performs differently for changing sea water levels. Lowe et al. (2005) also indicate bottom friction as important factor for wave attenuation, but its importance varies depending on studies (Thornton and Guza 1983; Young 1989; Massel and Gourlay 2000). Non-linear interactions are also induced by submerged structures and transfer wave energy into higher harmonics or lower periods.

Wave energy depends on wave height (H) and wave period (T), thus decreasing if wave heights or wave periods decrease. This attenuation leads to decreased sediment transportation (Oumeraci et al. 2001; UNEP 2010), which can be further enhanced by a combination of artificial reefs, beach nourishment and groynes (Bleck and Omeraci 2004; Schlurmann et al. 2003). Wave energy transfer shows also significant effect on the longshore morphodynamics[2] (e.g. Hsu and Evans 1989; Gonzalez and Medina 2001; Cánovas and Medina 2012).

Transmission can be described by the wave transmission coefficient $K_t = H_T/H_i$ with H_T as transmitted wave height and H_i as incident wave height. Dattatri et al. (1978) studied several submerged breakwater types and their ratio of crest submergences (d_S) to water depth (d), to derive performance characteristics for the transmission coefficient. They found that for $d_S/d = 0.4$, the transmission coefficient can be in the order of 75–95 %. Habel (2001) introduced a concept of 2–4 layers submerged filter modules with K_t over 70–80 % of initial wave energy. Arnouil (2008) compiled several design criteria, for successful submerged breakwater design. The result distinguishes between tombolo and salient formation. Tombolos are created by accumulated sand on the lee-side of the breakwater, which attach the structure to the coast. Salients do not reach to the breakwater and allow further longshore sediment transport. The latter are favoured as mentioned by Chasten et al. (1993) and USACE code EM 1110-2-1617 (USACE 2002); otherwise, downstream beaches will be cut off from sand supply and will most likely erode (faster).

Reflection of the incident waves is described by the reflection coefficient $K_r = H_R/H_i$ with H_R as reflected wave height. Wave reflection influences neighbouring structures and scouring around the structure itself for reflection coefficients larger than 25 % (Omeraci et al. 2001). Therefore, a small reflection coefficient is favourable. Scouring can also occur on each end of multiple submerged breakwaters with low permeability, which require gaps between each other to maintain water circulation. However, these gaps can create rip-channels with high currents, leading to erosion.

Lower reflection and locally accumulated backflow coefficients as well as increased dissipation and transmission can be achieved by permeable structures e.g. on filter elements (Omeraci et al. 2001), plugged block modules (Habel 2001) or artificial reefs.

20.5.2 Artificial Reefs

A reef is per definition a strip, bar or ridge of seabed material rising shortly beneath the water surface. Natural reefs can consist of rock and sand as well as coral or algae. Artificial reefs exist in several forms and types with a wide range of complexity and sophistication. Building materials vary from recycled natural material to more elaborate structures, for instance piled up sand containers made out of geotextiles or concrete elements. First experiences with artificial reefs where

---

[2]Morphodynamics: The dynamic interaction of seabed material with hydrodynamic processes as waves, tides and currents, which leads to erosion or sedimentation.

made in Japan and date back to the 1950s, while research started to focus on the topic in the 1960s (i.e. Carlisle et al. 1964).

Artificial reefs are a combination of submerged breakwaters and natural reefs. They are designed as an artificial submerged structure which mimics the protection potential and ecological benefits of a natural reef (UNEP 2010; Goreau and Trench 2012). Artificial reefs can provide multiple services in addition to coastal protection. For example, Mendonca et al. (2012) use numerical models to design an artificial near-shore reef for erosion protection and improve surfing conditions. Furthermore, the artificial reef will decrease flow velocities and attenuate higher waves, which improves swimming conditions and safety. Near-shore reefs also attract divers, as the reefs are used as shelter by multiple fish and marine animals. Altogether, these factors can potentially increase local tourism (Bleck and Omeraci 2004).

Reefs also serve as shelter and habitat for algae and small fish, which then again attract bigger fishes. Reefs thus increase the ecosystem resilience (i.e. counteract against ecosystem damage) and improve marine biodiversity (Pickering and Whitmarsh 1997). Whitmarsh et al. (2008) present a positive cost-benefit calculation for artificial reefs due to an increased fish occurrence and determine their monetary value for the local fish industry.

Hence, permeable and porous submerged breakwaters like a bar-type artificial reef are favourable, if sufficiently resistant against wave attack. They combine the protection potential of submerged breakwaters with the ecological benefit and ecosystem services of reef habitats, by mimicking the properties of natural reefs and thereby providing shelter and habitat for smaller fish and plants, thus attracting larger fish and ultimately increasing biodiversity. Under proper environmental circumstances, artificial reefs can potentially attract corals and develop into a coral reef. Martin et al. (2005) investigated submerged breakwaters on European shores (i.e. Spain, Italy and UK) and noticed an improvement of local ecosystem conditions in the form of increased abundance of fish and other species, or the diversity of living organisms. Wehkamp and Fischer (2013) conducted a 3-year study in front of Helgoland (Germany) in the North Sea, proving the positive effects of submerged concrete-made tetrapods as a fish nursery ground. A significant increase of fish and juvenile fish abundance surrounding the structures indicates the structures are suitable as a fish nursery ground (Wehkamp and Fischer 2013). Moreover, proper submerged breakwater configuration is used to manipulate wave parameters to improve local surfing conditions (Black and Mead 2009).

However, Burcharth et al. (2007) also mention constrains, which have to be considered. First of all, there has to be a legal basis for submerged breakwaters in the policy and legislation for coastal protection and sea defence. Bathymetry, a negative influence of artificial structures on vulnerable neighbouring coasts (e.g. by altered sedimentation transport) and the availability of proper building material pose physical limitations for artificial reefs. Finally, Burcharth et al. (2007) encourage a thorough on-site inspection, aiming to identify sensitive sites of historic, natural or environmental value, which could be affected adversely by an artificial interference.

20.5.3 Application

Reef Balls™ are typical concrete modules for artificial reefs. They are offered by
the Reef Ball Foundation (RBF 2014), an international, non-profit environmental
NGO. Reef Balls™ are hemispherical fabrics with porous side walls and a hollow
body. Fifteen to fourty circular holes in the mantle ensure permeability for sediment
and enhances marine habitat (see Fig. 20.9). The void mimics reef refuge for fishes
and provides shelter for sea dwellers, or with smaller holes it can be filled with mud
and serves as a mangrove flower pot (RBF 2014). The concrete Reef Ball™ units
are usually produced on-site. RBF (2014) refers to American Society for Testing
and Materials (ASTM) standards for the concrete, but individual solutions with
alternative local materials have also been successfully applied.

The first use of Reef Balls™ for wave attenuation and coastal protection was in
1998 in the Dominican Republic (Harris et al. 2004), and since then many other
projects have been realized (i.e. USACE 2005; Arnouil 2008). Cesar (1996)
estimates the monetary potential for Reef Balls™ in Indonesia of up to 1 million
USD/km for highly developed areas and around 50,000 USD/km for moderately
populated areas. Recently, a local Indonesian design has been developed and
published (Akhwady 2012). It is suitable for Indonesian coastal areas with a price
adapted to the Indonesian market. The modules are bottle-shaped porous-hollow
cylindrical pieces (see Fig. 20.10) made from concrete according to Indonesian SNI
7394:2008 standard with a yield strength of 24 MPa. The "Bottle Reef" is 1 m tall,
the body has a diameter of 0.9 m, the neck of 0.7 m and the side walls are perforated
with 12 pores. The analysis of a 2D physical model showed improved K_t compared
to the Reef Ball™ (Akhwady 2012). Additionally, Bottle Reef can be arranged to
minimize gaps between the units and reduce wave energy further, but compared to
other reef units, such as the A-Jack, Tetrapod and Cube, Bottle Reef units have the
lowest armor stability (Akhwady 2012).

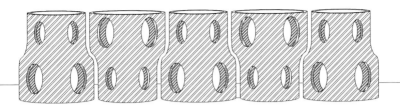

Fig. 20.10 A series of bottle reef units

Both Reef Balls™ and Bottle Reef units were installed at White Sand Beach (Pasir Putih) in Situbondo in Northeast Java Island, separated by approximately 200 m of each other. Although the former was intended as an artificial reef and was thus approximately 1.5 m deeper, they have not been damaged by winter storm events and were just slightly displaced, while the Bottle Reef units were completely scattered and several units broke in the first year of field testing. Akhwady (2012) and his research team at the Research and Development Centre for Marine and Fisheries Technology, Jakarta (KKP-P3TKP) and Sepuluh November Institute of Technology, Surabaya (ITS) could not access sufficient funding, making private investment necessary to build the Bottle Reef modules at Pasir Putih Situbondo. Consequently, the units were smaller and of lower quality concrete components, thus light-weight and of weak resistance. Moreover, the budget did not allow monitoring after installation, so that there is no precise data when exactly or under which conditions the Bottle Reef installation failed. Local communities reported a severe storm in early 2014, which most probably was cyclone Gillian and which could have damaged the Bottle Reef.

The added-value and net benefit of artificial reefs or submerged breakwaters is still debated. Without any doubt wave attenuation, erosion protection and improvement of the marine environment are verified in physical, numerical and practical studies, but the application of such systems as an engineering tool is not yet established and doubts with regards to their effectiveness still exist (Scyphers et al. 2011). Baine (2001) and Ranasinghe and Turner (2006) published reviews on the performance of submerged breakwaters and evaluated their feasibility based on literature and their own experiments. The two reviews reached differing conclusions, but both assumed that submerged breakwaters bring benefits to several aspects of the coastal environment. Ranasinghe and Turner (2006) saw the major constraint in poorly understood theoretical principles, which would be necessary to deliver proper combinations of design parameters for a successful artificial reef design. Baine (2001) gave an overview of many reef applications and recommends their application, but suggest sound planning and managing. Arnouil (2008) explicitly mentions design parameters for successful implementation of submerged breakwaters.

Another comprehensive overview of "low-crested coastal structures" (LCS) is given in Burcharth et al. (2007). They mention several considerations when building LCS and mention conceptual designs, as well as design parameters. Burcharth et al. (2007) outline a positive effect on sedimentation of submerged (porous) breakwaters, but also mention stagnant water of poor quality on the lee-side of the structures. Additionally they mention the complicated interaction between waves, water levels, currents and sediment transport, making long-term predictions of morphological changes difficult.

Other studies show good results of applied reef bars and submerged breakwaters, but on-site conditions and requirements on the structure have to allow for successful construction and operation (Dean et al. 1997; Habel 2001; Harris et al. 2004). Further examination will gradually close existing knowledge gaps and lead to more practical experience like the Reef Ball Foundation, as well as development of further technologies like the Bottle Reef by Akhwady (2012). The Bottle Reef is still a prototype model and in need of further investigation. To ensure comparable results and create an alternative to more sophisticated externally-developed methods such as the Reef Ball™, proper funding is essential.

Future research must contribute to understand the underlying, complex physical processes, identify further possibilities, as well as diminish the limitations and constrains of submerged breakwaters. Future results must aim to outline their low-regret characteristics and thus promote a more confident implementation of permeable submerged structures in coastal protection.

20.6 Missed Opportunity? Low-Regret Solutions Not Considered in Jakarta's Coastal Defense

With 250 million inhabitants, some 17,500 islands, 35,000 km of coastline and a high exposure to natural hazards (among others, a high probability of earthquakes and tsunamis), the Republic of Indonesia is particularly vulnerable to coastal hazards. Jakarta is directly located on the shores of the Java Sea and is crossed by 13 rivers draining into the ocean. Moreover, ground water pumping and subsequently land subsidence have increased drastically in the last decades, resulting in 10.3 % of Jakarta's land area lying below sea level, which will double by 2030 and triple by 2050 according to models by Irzal (2013). Flooding is literally a daily problem for residents during the rainy season, and hazard events as well as exposure will further increase in the future as a result of anticipated sea level rise and more extreme precipitation (IPCC 2012).

Figures 20.11 and 20.12 illustrate these problems for Northwest Jakarta and North Jakarta, close to the airport and harbour, respectively. The maps were coupled with Shuttle Radar Topography Mission (SRTM) data, describing the area's elevation. The figures show the present coastline as a white, dashed line and the altered coastline in red, if the mean sea level (MSL) rose by 50 cm.

The disaster risk and need for coastal protection in North Jakarta have become obvious and are addressed by the upcoming National Capital Integrated Coastal Development (NCICD 2014) plan. The plan presents three solutions to tackle

Fig. 20.11 Static flooding simulation of North Jakarta, port area (Source: Google Maps; topographic data: SRTM)

Fig. 20.12 Static flooding simulation of North Jakarta, airport area (Source: Google Maps; topographic data: SRTM)

Jakarta's flooding problem (NCICD 2014): either abandoning the urban agglomerations (retreat); onshore protection; or seawall protection. The current design chooses the latter option and plans for construction of a dike allowing for traffic, as well as reclaimed offshore islands similar to the Palm Islands in Dubai. These

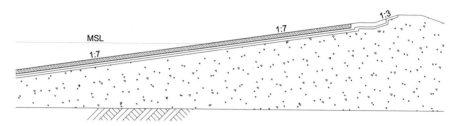

Fig. 20.13 Cross section of the land reclamation, according to the design of NCICD (2014)

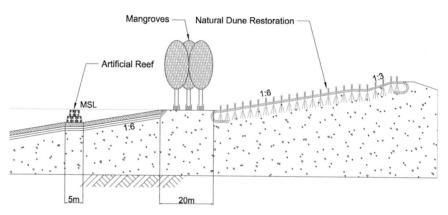

Fig. 20.14 Cross section of a possible ecosystem based design of the land reclamation

islands will have the shape of Indonesia's heraldic symbol, the *Garuda* and will be created in the bay of Jakarta. The Jakarta barrier will extend along Jakarta Bay and connect the northern parts of east and west Jakarta, thus aiming to reduce the city's problem of congestion and urbanisation. However, the islands are man-made dikes, coming at cost of putting in place a proper ICZM. The plan misses an opportunity to use modern ecosystem-based disaster risk reduction (Eco-DRR) approaches, which could offer sustainability in an ecological sense, but also yield other benefits.

NCICD (2014) creates further living space and tries to improve congestion and thus tackles its primary objectives, but the implementation misses a contemporary and modern response to sustainable ecological and social demands. The technical implementation of the land reclamation in the master plan of NCICD (2014) is shown in Fig. 20.13. The seabed is at −17.16 m and goes up to +7.7 m at the top of the dike. The bank slope is 1:7, while the top of the dike has a slope of 1:3. Figure 20.14 shows an alternative concept, which enhances the initial design with low-regret measures. It proposes a 1:6 slope, which is typical for grass covered sea dykes on the German North Sea coast (EAK 2002 2007). Below mean sea level (MSL), the soil composition of the system in Fig. 20.14 is equal to the design in NCICD (2014). In shallower water, a berm with artificial reef elements is installed, decreasing wave energy. At MSL, another berm follows, hosting a mangrove belt on muddy soil. The mangroves trap sediments, attenuate waves, and reinforce the soil. The upper part of the alternative protection concept applies coir geotextiles coupled with vetiver grass on a 1:6 and 1:3 slope. The low-regret approach in Fig. 20.14 uses a 6 m berm for the artificial reef elements, which is needed for a three-row Reef Ball™ setup with a diameter of 1.5 m, 1.6 m (Harris 2002) or 1.83 m per unit Reef Ball™ and a 20 m wide berm for the mangroves, allowing for wave energy dissipation of 50–70 % (Vo-Luong and Massel 2006, 2008). In total, the bank is about 4 m longer (about 3 % more than the initial design) and requires about 10 % more soil material for construction than the initial design.

Before the final design is published, the different concepts and ideas should have been available to all relevant stakeholders. NCICD (2014) analyses the social impact of the planned sea wall to the community and presents steps to mitigate problems for low-income households and local fishery. However, the master plan misses the opportunity to involve such stakeholders into decision making and consider their expertise and interests in the master plan. Similar as in Rosenzweig et al. (2011), stakeholders and experts should discuss on the basis of a portfolio of adaptation measures, leading to a final design. A broader stakeholder involvement encourages identification and sense of ownership and thus acceptance. It also helps to define competences and jurisdiction among stakeholders in a legal framework, which play a crucial role in reducing disaster risks and which supports resilience significantly (IFRC 2014).

A concept which contains ecosystem-based and local, low-regret measures could have made the master plan a benchmark project for Eco-DRR. However, it misses the opportunity to include local knowledge and apply innovative protection measures that build with nature.

20.7 Conclusions

Conventional engineering approaches do not preserve existing ecosystems or support the livelihood of the coastal communities. The integration of socio-economic and marine ecosystem-based approaches into engineering approaches allow for more sophisticated and profitable Ecosystem-based and Low-Regret Adaptation Management (ELRAM), which not only protects the environment from severe hazardous events, but also supports local coastal ecosystems and provides ecosystem services. There are numerous ELRAM-approaches, of which mangroves, natural geotextiles or submerged reefs are just a few examples. Mangroves have proven their efficiency for wave attenuation and sedimentation. However, they grow mainly on muddy soils. Natural coir geotextiles, as an example of the available techniques for dune restoration, initially protect the dunes and support both freshly planted and juvenile dune vegetation until it develops full protection potential. Both mangroves and dunes require physical space to efficiently protect the coastal area. An offshore option for limited space are submerged porous breakwaters, which also could act as artificial reefs if properly designed.

There are a wide variety of methods for coastal protection, of which none is a panacea for Eco-DRR. Mostly they work best in combination with each other (Temmerman et al. 2013). Such combined designs will for many purposes potentially outperform designs, which are based solely on conventional engineering approaches in the future. To increase confidence towards ecosystem-based measures, research must point out the potential of ELRAM-approaches, but also identify limitations and constraints on their suitability. However, the lack of practical experience and knowledge about long-term performance and efficiency prevents a broader application of ELRAM-approaches. Also, the lack of practical guidance, standard of practices and design guidelines hamper its use. To promote the paradigm shift of building *with* nature and initiate more confident and widespread implementation, knowledge gaps must be filled and scientific findings must lead to practical recommendations and offer design patterns for a successful implementation.

Acknowledgment The authors appreaciate the support of the German Federal Ministry of Education and Research (BMBF), who funded this work within the TWIN-SEA project.

References

Akhwady R (2012) Kinerja terumbu buatan silinder berongga (bottle reeftm) sebagai pemecah gelombang ambang terbenam. Ph.D. dissertation, Insititute of Tenth November Surabaya Indonesia (in Indonesian language)

Andayani A, Soejarwo PA, Indriasari VY (2013) Uji Coba Pemasangan Karung Geotekstil Memanjang (KGM) di Pantai Kamal Muara. (Trial test of longitudinal geotextile tube installation in Kmal Muara coastal area). Unpublished report. Pusat Pengkajian dan Perekayasaan Teknologi Kelautan dan Perikanan – BADAN LITBANGKP-KKP (in Indonesian language)

Arnouil DS (2008) Shoreline response for a Reef Ball™ submerged breakwater system offshore of Grand Cayman Island. Master thesis, Florida Institute of Technology, Melbourne, Florida

Augustin LN, Irish JL, Lynett P (2009) Laboratory and numerical studies of wave damping by emergent and near-emergent wetland vegetation. Coast Eng 56:332–340

Baine MSP (2001) Artificial Reefs: a review of their design, application, management and per-formance. Ocean Coast Manag 44(3–4):241–259. doi:10.1016/S0964-5691(01)00048-5

Balan K, Venkatappa Rao G (1996) Erosion control with natural geotextiles. In: Rao GV, Banergee K (eds) Environmental geotechnology with geosynthetics. The Asian Society for Environmental Geotechnology and CBIP, New Delhi, pp 317–325

Bao TQ (2011) Effect of mangrove forest structures on wave attenuation in coastal Vietnam. Oceanologia 53(3):807–818

Barbier EB, Georgiou IY, Enchelmeyer B, Reed DJ (2013) The value of wetlands in protecting southeast Louisiana from hurricane storm surges. PLoS One 8(3), e58715. doi:10.1371/journal.pone.0058715

Black K, Mead S (2009) Design of surfing reefs. Reef J 1(1):177–191

Bleck M, Omeraci H (2004) Analytical model for wave transmission at artificial reefs. In: Proceedings 29th international conference on Coastal Engineering (ICCE), ASCE, Vol 1, Lisbon, pp 269–281

Bloomberg MR, Burden AM (2013) Coastal climate resilience – urban waterfront adaptive strategies. Department of City Planning, The City of New York

Bou-Zeid E, El-Fadel M (2002) Climate change and water resources in Lebanon and the Middle East. J Water Resour Plan Manag 128(5):343–355

Burcharth HF, Hawkins SJ, Zanuttigh B, Lamberti A (2007) Environmental design guidelines for low crested coastal structures. Elsevier, Amsterdam

Cánovas V, Medina R (2012) A long-term equilibrium beach planform model for coastal work design. Coast Eng Proc 1(33) doi:http://dx.doi.org/10.9753/icce.v33.sediment.43

Carlisle JR, Turner CH, Ebert EE (1964) Artificial habitat in the marine environment. California Department of Fish and Game Fish Bulletin 124, 93 p

Carus M, Gahle C, Pendarovski C et al (2008) Studie zur Markt- und Konkurrenzsituation bei Naturfasern und Naturfaser-Werkstoffen (Deutschland und EU). Gülzower Fachgespräche 26, Fachagentur Nachwachsende Rohstoffe e.V. (Hrsg.), Gülzow 2008, S. 126

Cave ID, Walker JFC (1994) Stiffness of wood in Farown plantation softwood: the influence of microfibril angle. For Prod J 44(5):43–48

Cesar H (1996) Economic analysis of Indonesian coral reefs. World Bank Environment Department Paper Department, Environmentally Sustainable Development Vice Presidency. The World Bank, New York

Chasten MA, Rosati JD, McCormick JW, Randall RE (1993) Engineering design guidance for detached breakwaters as shoreline stabilization structures, Technical report CERC-93-19. US Army Corps of Engineers Waterways Experiment Station, Vicksburg

Cheong SM, Silliman B, Wong PP et al (2013) Coastal adaptation with ecological engineering. Nat Clim Chang 3(9):787–791

Costanza R, d'Arge R, de Groot R et al (1997) The value of the world's ecosystem services and natural capital. Nature 387:253–260

Costanza R, de Groot R, Sutton P (2014) Changes in the global value of ecosystem services. Glob Environ Chang 26:152–158. doi:10.1016/j.gloenvcha.2014.04.002

Dattatri J, Raman H, Shankar N (1978) Performance characteristics of submerged breakwaters. Coast Eng Proc 1(16). doi:http://dx.doi.org/10.9753/icce.v16.%p

De Vriend HJ, van Koningsveld M (2012) Building with nature: thinking, acting and interacting differently. EcoShape, Building with Nature, Dordrecht

Dean RG, Chen R, Browder AE (1997) Full scale monitoring study of a submerged breakwater, Palm Beach, Florida, USA. Coast Eng 29(3–4):291–315. ISSN 0378–3839

EAK 2002 (2007) Empfehlungen für die Ausführung von Küstenschutzbauwerken – Korrigierte Ausgabe 2007. Recommendations for coastal protection works – Ammended Edition 2007 Herausgeber (ed) Kuratorium für Forschung im Küsteningenieurwesen. Die Küste, Heft 65

EJF (2006) Mangroves: nature's defence against Tsunamis – a report on the impact of mangrove loss and shrimp farm development on coastal defences. Environmental Justice Foundation, London

Faruk O, Bledzki AK, Fink HP, Sain M (2012) Biocomposites reinforced with natural fibers: 2000–2010. Prog Polym Sci 37(11):1552–1596

Gedan KB, Kirwan ML, Wolanski E et al (2011) The present and future role of coastal wetland vegetation in protecting shorelines: Answering re-cent challenges to the paradigm. Clim Chang 106:7–29

GIZ (2011) Mangroves. Deutsche Gesellschaft für Internationale Zusammenarbeit (GIZ) GmbH

Gonzalez M, Medina R (2001) On the application of static equilibrium bay formulations to natural and man-made beaches. J Coast Eng 43:209–225

Goreau TJ, Trench RK (2012) Innovative methods of marine ecosystem restoration. CRC Press, Boca Raton, pp 5–10

Guannel G, Ruggiero P, Faries J et al (2015) Integrated modeling framework to quantify the coastal protection services supplied by vegetation. J Geophys Res Oceans 120:324–345. doi:10.1002/2014JC009821

Habel R (2001) Künstliche Riffe zur Wellendämpfung. Berlin (Deutschland, Bundesrepublik). Dissertation, Mensch und Buch Verlag, Berlin 2001, 128 S.; TU Berlin

Hadi S, Latief H, Muliddin M (2013) Analysis of surface wave attenuation in mangrove forest. Proc ITB Eng Sci 35:89–108

Harada K, Imamura F, Hiraishi T (2002) Experimental study on the effect in reducing tsunami by the coastal permeable structures. In: Proceedings of the 12th International Offshore and Polar Engineering Conference. Kita-Kyushu, Japan, May 26–31

Harris LH (2002) Submerged reef structures for habitat enhancement and shoreline erosion abatement. U.S. Army Corps of Engineers Coastal & Hydraulic Engineering Technical Note (CHETN), Vicksburg

Harris LE, Turk G, Mead S (2004) Combined recreational amenities and coastal erosion protection using submerged breakwaters for shoreline stabilization. Beach Preservation Technology 2004, FSBPA

Hashim AM, Catherine SMP, Takaijudin H (2013) Effectiveness of mangrove forests in surface wave attenuation: a review. Res J Appl Sci Eng Technol 5(18):4483–4488

Hiraishi T, Harada K (2003) Greenbelt tsunami prevention in south-pacific region. Rep Port Airport Res Inst 42:1–23

Hsu JRC, Evans C (1989) Parabolic bay shapes and applications. Proc Inst Civ Eng (Part 2) 87:557–570. Thomas Telford, London

IFRC (2014) Viet Nam: country case study report – how law and regulation support disaster risk reduction. From: IFRC-UNDP Series on Legal Frameworks to support Disaster Risk Reduction, Geneva

IPCC (2012) In: Field CB, Barros V, Stocker TF, Qin D, Dokken DJ, Ebi KL, Mastrandrea MD, Mach KJ, Plattner G-K, Allen SK, Tignor M, Midgley PM (eds) Managing the risks of extreme events and disasters to advance climate change adaptation. A special report of working groups I and II of the Intergovernmental Panel on Climate Change. Cambridge University Press, Cambridge, UK/New York, 582 pp

Irzal F (2013) Grey solutions for urban water management: Jakarta case. Presented on "C40 workshop on climate adaptation and risk assessment" Rotterdam, the Netherlands June 3–6, 2013. http://www.deltacities.com/documents/presentations/04%20Jakarta_CDC%20website.pdf

Kathiresan K, Rajendran N (2005) Coastal mangrove forests mitigated tsunami. Estuar Coast Shelf Sci 65:601–606

Komar PD (1976) Beach processes and sedimentation. Prentice-Hall, Inc., Englewood Cliffs

Lacambra C, Friess DA, Spencer T, Möller I (2013) Bioshields: mangrove ecosystems as resilient natural coastal defenses. In: Fabrice R, Rieux KS, Marisol E (eds) The role of ecosystems in disaster risk reduction. United Nations University Press, Tokyo, pp 82–108

Lekha KR (2004) Field instrumentation and monitoring of soil erosion in coir geotextile stabilized slopes: a case study. Geotext Geomembr 22:399–413

Lekha KR, Kavitha V (2006) Coir geotextile reinforced clay dykes for drainage of low-lying areas. Geotext Geomembr 24(1):38–51

Lowe RJ, Falter JL, Bandet MD et al (2005) Spectral wave dissipation over a barrier reef. J Geophys Res 110:C04001. doi:10.1029/2004JC002711

Marques AR, de Oliveira Patrício PS, dos Santos FS et al (2014) Effects of the climatic conditions of the southeastern Brazil on degradation the fibers of coir-geotextile: evaluation of mechanical and structural properties. Geotext Geomembr 42(1):76–82

Martin D, Bertasi F, Colangelo MA (2005) Ecological impact of coastal defence structures on sediment and mobile fauna: evaluating and forecasting consequences of unavoidable modifications of native habitats. Coast Eng 52(10–11):1027–1051. http://dx.doi.org/10.1016/j.coastaleng.2005.09.006

Mason MA, Keulegan CH (1944) A wave method for determining depths over bottom discontinuities, U.S. Army Beach Erosion Board. Tech. Memo. 5. Beach Erosion Board, Washington, DC, 29 pp

Massel SR, Gourlay MR (2000) On the modelling of wave breaking and set-up on coral reefs. Coast Eng 39:1–27

Mazda Y, Magi M, Kogo M, Hong PN (1997a) Mangrove as a coastal protection from waves in the Tong King delta, Vietnam. Mangrove Salt Marshes 1:127–135

Mazda Y, Wolanski E, King B et al (1997b) Drag force due to vegetation in mangrove swamps. Mangrove Salt Marshes 1:193–199

Mazda Y, Magi M, Ikeda Y et al (2006) Wave reduction in a mangrove forest dominated by Sonneratia sp. Wetlands Ecol Manag 14:365–378

McIvor AL, Möller I, Spencer T, Spalding M (2012) Reduction of wind and swell waves by mangroves. Natural coastal protection series: report 1. Cambridge coastal research unit working paper 40. Published by The Nature Conservancy and Wetlands International. 27 p. ISSN 2050–7941

Mendez FM, Losada IJ (2004) An empirical model to estimate the propagation of random breaking and non-breaking waves over vegetation fields. Coast Eng 51:103–118

Mendonca A, Fortes CJ, Capitao R et al (2012) Hydrodynamics around an artificial surfing reef at Leirosa, Portugal. J Waterw Port Coast Ocean Eng 138:226–235

Miller DE, Hoitsma TR, White DJ (1998) Degradation rates of woven coir fabric under field conditions. In: Hayes DF (ed) Engineering approaches to ecosystem restoration. Proceedings of the 1998 wetlands engineering and river restoration conference, Denver, Colorado, March 22–27, 1998. American Society of Civil Engineers, Reston, pp 266–271

Narayan S, Suzuki T, Stive MJF, Verhagen HJ, Ursem WNJ, Ranasinghe R (2010) On the effectiveness of mangroves in attenuating cyclone-induced waves. In: Proceedings of the international conference on coastal engineering 32 (no page numbers). URL: http://journals.tdl.org/ICCE/article/view/1250

NCICD (2014) Draft master plan. National Capital Integrated Coastal Development Project, Jakarta

Oumeraci H, Clauss GF, Habel R, Koether G (2001) Unterwasserfiltersysteme zur Wellendämpfung. Abschlussbericht zum BMBF-Vorhaben "Unterwasserfiltersysteme zur Wellendämpfung" (in German)

Pickering H, Whitmarsh D (1997) Artificial reefs and fisheries exploitation: a review of the 'attraction versus production' debate, the influence of design and its significance for policy. Fish Res 31:39–59

Quartel S, Kroon A, Augustinus PGEF et al (2007) Wave attenuation in coastal mangroves in the Red River delta. Vietnam J Asian Earth Sci 29(4):576–584

Rajagopal A, Ramakrishna S (2009) Coir Geotextiles as separation and filtration layer for low intensity road bases. Indian Geotechnical Conference (IGC-2009), Guntur, India. Vol II, pp 941–946

Ranasinghe R, Turner IL (2006) Shoreline Response to submerged structures: a review. Coast Eng 53:65–79. http://dx.doi.org/10.1016/j.coastaleng.2005.08.003

RBF (The Reefball Foundation) (2014) Webpage: http://www.reefball.org/. Accessed 20 Nov 2014

Roeber V, Cheung KF, Kobayashi MH (2010) Shock-capturing Boussinesq-type model for nearshore wave processes. Coast Eng 57(4):407–423

Rosenzweig C, Solecki WD, Blake R et al (2011) Developing coastal adaptation to climate change in the New York City infrastructure-shed: process, approach, tools, and strategies. Clim Chang 106:93–127. doi:10.1007/s10584-010-0002-8

Schlurmann T, Bleck M, Oumeraci H (2003) Wave transformation at artificial reefs described by the Hilbert-Huang Transformation. In: Proceedings of the 28th International Conference on Coastal Engineering (ICCE2002). American Society of Civil Engineers (ASCE) 2:1791–1803

Schurholz H (1991) Use of woven coir geotextiles in Europe. Coir XXXV(2):18–25

Scyphers SB, Powers SP, Heck KL Jr, Byron D (2011) Oyster reefs as natural breakwaters mitigate shoreline loss and facilitate fisheries. PLoS One 6(8), e22396. doi:10.1371/journal.pone.0022396

Silva GG, De Souza DA, Machado JC, Hourston DJ (1999) Mechanical and thermal characterization of native brazilian coir fiber. J Appl Polym Sci 76(7):1197–1206

Silvester R, Hsu JRC (1997) Coastal stabilization. World Scientific Publ. Co, Singapore, 578 pp. (Reprint of Silvester and Hsu, 1993)

Spalding M, McIvor A, Tonneijck FH et al (2014) Mangroves for coastal defence. Guidelines for coastal managers & policy makers. Wetlands International and The Nature Conservancy, Wageningen, 42 p

Strusinska-Correia A, Husrin S, Oumeraci H (2013) Tsunami damping by mangrove forests: a laboratory study using parameterized trees. Nat Hazards Earth Syst Sci 13:483–503

Subaida EA, Chandrakaran S, Sankar N (2009) Laboratory performance of unpaved roads reinforced with woven coir geotextiles. Geotext Geomembr 27:204–210

Suzuki T, Zijlema M, Burger B et al (2012) Wave dissipation by vegetation with layer schematization in SWAN. Coast Eng 59(1):64–71. http://dx.doi.org/10.1016/j.coastaleng.2011.07.006

Tanaka N, Sasaki Y, Mowjood MIM et al (2007) Coastal vegetation structures and their functions in tsunami protection: experience of the re-cent Indian Ocean tsunami. Landsc Ecol Eng 3:33–45

Taubenböck H, Goseberg N, Lämmel G et al (2013) Risk reduction at the "Last-Mile": an attempt to turn science into action by the example of Padang. Nat Hazard 65:915–945

Temmerman S, Meire P, Bouma TJ et al (2013) Ecosystem-based coastal defence in the face of global change. Nature 504(12):79–83. doi:10.1038/nature12859

Teo FY, Falconer RA, Lin B (2009) Modelling effects of mangroves on tsunamis. Water Manag 162:3–12

Thornton EB, Guza RT (1983) Transformation of wave height distribution. J Geophys Res 88:5925–5938

Tuyen NB, Hung HV (2009) An experimental study on wave reduction efficiency of mangrove forests. In: Proceeding of the 5th International Conference on Asian Pacific Coasts (APAC2009). Nanyang Technological University (NTU), Oct 13–16, Singapore, 4:336–343

UNEP (2010) Linking ecosystems to risk and vulnerability reduction. The case of Jamaica. UNEP, Geneva. See also: http://postconflict.unep.ch/publications/RiVAMP.pdf

USACE-U.S. Army Corps of Engineers (2002) Coastal engineering manual. Engineer manual 1110-2-1100. U.S. Army Corps of Engineers, Washington, DC. (in 6 volumes)

USACE-U.S. Army Corps of Engineers (2005) National Erosion Control Development and Demonstration Program (Section 227) Miami Beach, Florida

Vedharajan B, Gross O (2007) MANGREEN – Mangrove ecology and restoration in India. Report 2006/07 (DEEPWAVE Report 09/07), 40 p

Vo-Luong P, Massel SR (2006) Experiments on wave motion and suspended sediment concentration at Nang Hai, Can Gio mangrove forest, Southern Vietnam. Oceanologia 48(1):23–40

Vo-Luong P, Massel S (2008) Energy dissipation in non-uniform mangrove forests of arbitrary depth. J Mar Syst 74(1–2):603–622

Wehkamp S, Fischer P (2013) Crustaceans and fish abundances and species at and around artificially introduced tetrapod fields in the southern North Sea, 2013. Alfred Wegener Institute for Polar and Marine Research – Biological Institute Helgoland. doi:10.1594/PANGAEA. 821916

Whitmarsh D, Santos MN, Ramos J, Monteiro CC (2008) Marine habitat modification through artificial reefs off the Algarve (southern Portugal): an economic analysis of the fisheries and the prospects for management. Ocean Coast Manag 51:463–468

Wilby RL, Keenan R (2012) Adapting to flood risk under climate change. Prog Phys Geogr 36 (3):348–378

Yanagisawa H, Koshimura S, Goto K et al (2009) The reduction effects of mangrove forest on a tsunami based on field surveys at Pakarang Cape, Thailand and numerical analysis. Estuar Coast Shelf Sci 81:27–37

Young IR (1989) Wave transformation over coral reefs. J Geophys Res 94:9779–9789

Chapter 21
Risk Perception for Participatory Ecosystem-Based Adaptation to Climate Change in the Mata Atlântica of Rio de Janeiro State, Brazil

Wolfram Lange, Christian Pirzer, Lea Dünow, and Anja Schelchen

Abstract A perception analysis is an important approach for developing adequate sensitization activities and increasing the participation of local populations in ecosystem-based disaster risk reduction (Eco-DRR) and ecosystem-based adaptation (EbA). These concepts have great potential in the study area, the mountain region of Rio de Janeiro state, where a disaster in 2011 showed once more that landslides, mudslides and floods are recurrent. Although degradation of the natural ecosystems is one of the main reasons for the high vulnerability of the local population, ecosystem-based measures to reduce disaster risks and to adapt to climate change are still uncommon. Valuing the benefits of nature through community-based adaptation measures is one promising approach to reduce landscape and ecosystem degradation and vulnerability, but a high level of community awareness is needed to generate their active participation in protecting and restoring ecosystems. To analyze the degree of awareness and the reasons for the barriers to participation, a perception analysis was conducted based on collected quantitative and qualitative data. Results show that people (a) have a high perception of their vulnerability, but (b) have poor knowledge about the relation between risks and ecosystem services, (c) do not feel responsible for participating, and (d) do not see possibilities for a better engagement in disaster risk reduction and climate change adaptation. We conclude that these three gaps (b, c and d) need to be addressed as a main component of a sensitization concept for Eco-DRR and EbA in the region.

W. Lange (✉) • A. Schelchen
Centre for Rural Development, Humboldt-Universität zu Berlin, Germany
e-mail: w.lange@gmx.net

C. Pirzer
Endeva, Berlin, Germany

L. Dünow
AMBERO Consulting Gesellschaft mbH, Kronberg im Taunus, Germany

© Springer International Publishing Switzerland 2016
F.G. Renaud et al. (eds.), *Ecosystem-Based Disaster Risk Reduction and Adaptation in Practice*, Advances in Natural and Technological Hazards Research 42,
DOI 10.1007/978-3-319-43633-3_21

Keywords Brazil • Risk • Perception • Vulnerability • Adaptation • Climate change • Ecosystem services • Sensitization • Ecosystem-based adaptation • Extreme weather event

21.1 Introduction

21.1.1 Objectives

Our main objective is to show a perception analysis that provides a deeper under-standing of the awareness and knowledge to enhance local residents' participation in disaster risk management. The interventions recommended are limited exclu-sively to sensitization activities and exclude other possible interventions such as financial transfers. We analyze the perception of the population in four areas in the municipality of Teresópolis (Rio de Janeiro state) with regard to disaster risks associated with extreme climate events. The perception analysis results identified the drivers and barriers preventing residents from participating more actively and effectively in DRR and CCA. This can serve as a basis for developing a sensitiza-tion strategy to enhance local population's participation in disaster risk reduction.

The study was undertaken by a team from the German Center for Rural Devel-opment (SLE) of the Humboldt University of Berlin within the project "Biodiver-sity and Climate Change in the Mata Atlântica" which is being implemented by the Brazilian Ministry of Environment (Ministério do Meio Ambiente do Brasil – MMA) with technical support by Deutsche Gesellschaft für Internationale Zusammenarbeit GmbH (German Cooperation for Sustainable Development – GIZ) and financial support by KfW Entwicklungsbank (German Development Bank).

21.1.2 Disaster Risks and Climate Change in Rio de Janeiro State

Disaster risks associated with extreme events such as torrential rainfall are preva-lent in the country's southeast and southern regions. In these areas, disasters caused by natural hazards, namely landslides, mudslides and floods, are recurrent, and their intensity and strength are likely to increase due to climate change (PBMC 2013a). In the Região Serrana (mountain region) of Rio de Janeiro state, yet another disaster occurred in January 2011, with severe consequences (Fig. 21.1). Torrential rains caused landslides, mudslides and floods that killed more than 900 people and made over 35,000 people homeless. This event put Brazil in third place among countries most affected by catastrophes in 2011 (Guha-Sapir et al. 2012). The World Risk Report of 2012 (Brodbeck 2012) focused on one issue closely linked to many disasters: degradation of the environment. In the case of Brazil, this relationship

Fig. 21.1 Area in Teresópolis affected by the disaster in 2011, 2 years later: flooded area with abandoned houses in the foreground and still visible landslides in the background (Photo: W. Lange)

between environmental degradation and disasters is evident, as changes to nature by humans contribute significantly to the intensity and impacts of disasters.

The mountain region of Teresópolis is part of the Mata Atlântica biome, the third largest Brazilian vegetation complex and a global biodiversity hotspot. Once encompassing 3500 km along the Brazilian coast and covering 1.0–1.5 million km^2 (Galindo Leal and Gusmão-Câmara 2003), only between 11 and 16 % of the original forest cover remains today, found mainly in small fragments (Ribeiro et al. 2009). The Mata Atlântica suffered greatly from exploration for development and large areas of land were converted to different land uses (e.g. agriculture and urban settlements) through various development cycles. Until the nineteenth century, the area was one of the main coffee-producing regions in Brazil. Later, it gained importance for intensive agricultural activities (e.g. producing vegetables such as lettuce, tomato and onions), which remains the main livelihood of the rural population, but also one of the main drivers of environmental degradation (Nehren et al. 2009). Consequently, most of the ecosystems have been destroyed or degraded through industrial and urban expansion, as well as by intensification of land use for agricultural and grazing purposes (Smyth and Royle 2000). However, nowadays, deforestation has decreased considerably (Fundação SOS Mata Atlântica and Instituto Nacional de Pesquisas Espaciais 2014).

Furthermore, unplanned occupation of slopes and areas close to rivers and streams are other risk factors (Nehren et al. 2009). Once the vegetation cover is removed, these areas become more susceptible to landslides, mudslides and flooding, leading to high risks for residents. Degraded ecosystems cannot efficiently carry out their functions for risk reduction, such as maintaining slope stability, flood control, and the balance of regional climate (Renaud et al. 2013). The existing efforts for ecosystem protection and restoration cannot cope with the high level of degradation in the region. Additionally, the rugged topography and the vulnerable geology (i.e. the basement underneath the soil is of sliding granitic) and soils (i.e. comprised of silty weathering mantle) increase the risks of mudslides, landslides, and floods (CEPED-UFSC 2011; DRM-RJ 2012).

Extreme meteorological events are the main trigger factor for landslides and floods (Fernandes et al. 2004) and are not new to the region. Due to climate change, their frequency and intensity have increased since the middle of the twentieth century (Marengo 2008). Projections for the study region indicate that rainfall will increase by 20 % until 2070 and by up to 30 % until 2100 (PBMC 2013a). Consequently, climate change will further increase disaster risks and vulnerability (PBMC 2013b) in an already fragile region. Hence, there is an urgent need for efficient measures to adapt to climate change and reduce disaster risk.

21.1.3   DRR and CCA in Rio de Janeiro State

Aware of the challenges the country will face due to climate change, the Federal Government of Brazil passed a law establishing the National Politics of Climate Change in 2009. The law was followed by several sectorial plans that focused on mitigation and reduction of carbon emissions. Adaptation to climate change was given more priority in 2013 when an inter-ministerial working group, headed by the Ministry of Environment, was established to develop a National Plan of Adaptation until 2015. In the policy framework of Rio de Janeiro state (SEA 2012), activities related to climate change primarily focus on reducing greenhouse gas emissions. Adaptation is mentioned in the State Plan of Climate Change, but concrete measures and activities for adaptation still have to be defined and implemented.

After the disaster in 2011, major efforts were made to increase the state's capacity for DRR, but measures mainly focused on engineered infrastructure. Engineered structures such as dredging, dams, embankment restoration and other hydraulic-engineering solutions to control flooding, as well as slope stabilization measures such as use of concrete walls to prevent landslides, were implemented in many parts of the state. These activities involved various actors like the State Government of Rio de Janeiro, the municipalities and the communities (CEPED-UFSC 2011). However, most of these engineered measures are relatively expensive, and do not address the underlying risk factors. Implementation of disaster prevention activities should therefore be supplemented by improving long-term planning procedures including ecosystem-based solutions. Better coordination between the different stakeholders working with adaptation to climate change is also needed, because the activities have failed to integrate with each other.

21.1.4 Effective DRR and CCA Based on Ecosystem Services

An emerging approach to DRR and CCA is based on the sustainable use and management of ecosystem services. Ecosystem-based adaptation (EbA) seeks to integrate the use of ecosystem services and biodiversity into an overall strategy to help people adapt to the adverse impacts of climate change (Colls et al. 2009). Ecosystem-based disaster risk reduction (Eco-DRR) consists in a similar approach that supports healthy, well-managed ecosystems to act as natural infrastructure, reducing physical exposure to many hazards and increasing the socio-economic resilience of people and communities to disasters (Sudmeier-Rieux and Ash 2009). Both approaches have more in common than they are different. Both EbA and Eco-DRR aim to reduce vulnerability to disaster and climate risks by focusing on the sustainable use, management, conservation and restoration of ecosystems (UNEP 2015). The main differences between the two approaches are that EbA is a long-term approach, as the impacts of climate change tend to increase in decades, while Eco-DRR mostly addresses current and recurring hazards. They both address climate- or water-related hazards, although Eco-DRR would also consider other types of hazards, including geological hazards (e.g. earthquakes, tsunamis) as well as technological hazards (e.g. oil spills impacting on coastal and marine ecosystems). EbA may take into account uncertainties associated with long-term climatic changes, and seek to adapt to the impacts of climate change in vulnerable development sectors, such as agriculture, while Eco-DRR could also be applied to tackle non-climate induced hazards such as tsunamis, landslides, avalanches and rockfall (Venton 2008).

Despite their great potential in the study region, ecosystem-based solutions, such as the conservation (responsible and sustainable use of nature without reducing the services of ecosystems) and restoration (return to its original state) of ecosystems, have been under-estimated by public policymakers, both in the past and present. DRR in Brazil is still mainly focused on the short-term activities of disaster preparedness and response rather than on long-term disaster prevention and mitigation strategies. Compared to technological and engineered infrastructure measures (also commonly referred to as "grey infrastructure"), ecosystem-based solutions (or "green infrastructure") can provide several co-benefits such as carbon sequestration, climate regulation, and water security (Renaud et al. 2013).

In Brazil, a key policy instrument based on an ecosystem-based approach is the permanent preservation area (Área de Preservação Permanente or APP). Often located on steep slopes and around rivers, APPs are protected by the National Forest Code (Brasil 2012) due to their environmental functions. The high risk of landslides and flooding is explicitly mentioned as one of the reasons why those areas must not be developed. Unfortunately, the lack of law enforcement and land use pressures, such as construction and agricultural development, means the APP instrument is often ineffective. This was apparent in the 2011 disaster where most of the damage occurred in APPs, because people had settled in these areas (SBF 2011). Other activities and measures that are already being implemented, but need more effort, include reforestation, river restoration, and alternative land-use systems such as agroforestry or silvopastoral systems. These solutions consider landscapes and ecosystems as a holistic system and aim to create long-term effects to achieve disaster resilience and support sustainable development. Mainstreaming ecosystem services for DRR and CCA should be improved at all governmental levels as part of the long-term planning process.

Participation by local government and especially local communities in DRR and CCA is vital (Allen 2006). The protection and restoration of ecosystems do not work without the involvement of local people, and their benefits are usually better demonstrated over the medium or long term (Colls et al. 2009).

Participation of the local population is specifically important in Teresópolis, as the human-induced degradation of ecosystem services and unsustainable land use increases disaster risks. One approach of integrating people in EbA or Eco-DRR activities is community-based adaptation (CbA) (IIED 2009; Care 2010). CbA involves participatory processes that increase the local population's awareness and whose primary objective is to improve the capacity of local communities to adapt to climate change (Care 2010). It also aims to protect and sustain ecosystems, not only for livelihoods of people, but also to reduce disaster risks. Participation in CbA primarily depends on the knowledge, needs, and priorities of the local people directly concerned (IIED 2009). CbA can be integrated with Eco-DRR and EbA measures, because it directly engages people affected by disaster risks and supports them to implement activities within their own environment.

21.1.5 Importance of Perception for Participation

In the context of CbA, the perception of local residents affected by disasters or living in high-risk areas plays a significant role (see also Harmáčková et al. Chap. 5; Takeuchi et al. Chap. 14; Fedele et al., Chap. 23). Understanding their perception is important for motivating individuals to actively avoid, mitigate, and reduce risks (Wachinger et al. 2013). Perception may vary depending on several factors such as the type and context of risk, socialization, biases, and social context and is influenced by knowledge, experience, values, attitudes, and emotions.

An analysis of the local population's perception allows us to identify how people deal with disaster risks and climate change. These insights assist in the development of a sensitization strategy which considers drivers and barriers, in order to foster enhanced participation in Eco-DRR and EbA activities. For Teresópolis, community participation is analyzed in the context of the protection and restoration of the Mata Atlântica.

21.1.6   Four Key Factors for Active Participation

According to the psychometric paradigm of risk perception, which assumes that the perception of risk leads to a specific behavior (Grothmann and Patt 2005), the response of people to natural hazards is influenced by risk perception, their judgments, and preferences (Slovic and Weber 2002). Information can lead to behavioral change through adequate educational campaigns (Madajewicz et al. 2007), but the way risk information is formulated has an effect on judgments and vice-versa (Plapp 2003). Both the way information about risks is formulated and the information's availability are extremely important for behavioral response. The individual's decision to act is also determined by how he or she interprets the given information based on previous experiences (Plapp 2003). Depending on the knowledge of alternatives, a decision about whether to take action or not can be made. In order to have the possibility to choose between different alternatives, they have to be available so that possible consequences can be considered (Weber 1997).

We can therefore conclude that perception of risk depends greatly on experience, knowledge and judgment. It has an influence on behavior to take up risk-reducing activities and is one of the factors which could serve as a barrier or driver for enhanced participation. Therefore, our analysis consists of four main dimensions:

(a) People's perception of their own **vulnerability** to disasters, considering exposure, sensitivity, and adaptive capacity as the factors that define vulnerability (IPCC 2001). Exposure is defined as the degree to which a system is exposed to climate-related threats, such as construction of housing on steep slopes which are threatened by landslides, mudslides and floods. Sensitivity is the degree to which the system is affected by the threats, such as direct and indirect damage caused by landslides, mudslides and floods (Messner et al. 2006). Adaptive capacity is the ability of the system to respond successfully to the threat (IPCC 2001; Mytanz 2013), which means for example the knowledge about effective measures to protect and restore ecosystems as well as the financial resources and capacity to implement these measures. If people do not perceive that they are exposed and sensitive to risks, they do not necessarily see the need to act and protect themselves. Furthermore, when they feel overwhelmed and incapable of adapting to risks, there is a high probability that they will not act on those risks.

(b) People's **knowledge** of the relationship between ecosystems and natural hazards: In the context of Teresópolis, this refers to people's knowledge of how ecosystem services can contribute to reducing disaster risks and how ecosystems can be protected and restored.

(c) People's attitudes with regard to their own contribution towards ecosystem protection and restoration: This component refers to people's perception of their **responsibility** to contribute to the protection and restoration of ecosystems. The main premise is that people accept and assume responsibility for these measures, because ecosystem-based measures will only work if everyone participates over the long term.

Fig. 21.2 Four dimensions
as basis for better
participation (Own graphic)

(d) People's perception of **possibilities** and available alternatives to engage more actively in ecosystem-based measures. The main premise is that people must recognize the available opportunities for action in order for them to act and contribute to specific activities.

These four dimensions serve as an analytic scheme to identify barriers in the perception of people that prevent them from participating more actively and effectively in Eco-DRR and EbA (Fig. 21.2). In order to develop more specific recommendations for an adequate sensitization strategy at local level, it has to be taken into account which mass media is used and what are the experiences with environment sensitization activities that have already been realized in the study region.

21.2   Methodology

Using a multi-criteria approach, four geographical areas of the municipality of Teresópolis, Caleme, Granja Guarani, Santa Rita and Vieira, were selected for the study. The selection was based on a list of ten potential areas that had been defined in conjunction with a representative of the environmental department at Teresópolis city hall, where relevant data on the characteristics had been collected in a pre-study. All four selected areas were determined to have a high potential for conservation and/or restoration of ecosystems to reduce disaster risks. To obtain an accurate representation of the whole municipality of Teresópolis, both urban and rural areas as well as those affected and unaffected by the tragedy of 2011 were selected, resulting in a selection of one urban-affected (Caleme), one urban-unaffected (Granja Guarani), and two rural-affected (Vieira and Santa Rita) areas

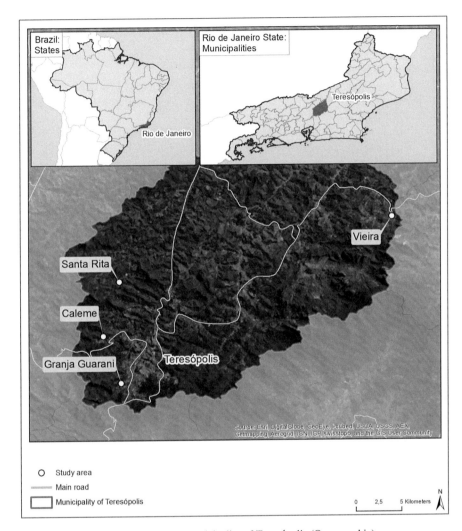

Fig. 21.3 Research areas within the municipality of Teresópolis (Own graphic)

(all rural areas had been affected by the 2011 events) (Fig. 21.3). Additionally, attention was paid to socio-economic criteria (i.e. age and income) to ensure the sample covered the socio-economic characteristics of the entire municipal area.

Three main empirical methods were used for the data collection: semi-structured questionnaires as the main quantitative data collection tool, and focus group discussions and qualitative interviews with local key actors as additional sources of information. The qualitative empirical instruments aimed to enhance the results obtained from the semi-structured questionnaires and the triangulation of quantitative and qualitative methods allowed a higher validity and more sophisticated interpretation of the results (Flick 2011).

All empirical instruments were pretested and adapted accordingly, before they were applied in the research areas to ensure that appropriate terminology was used and that there were no leading or biased questions. A psychologist in the research team evaluated the research methods to guarantee that they were locally-sensitive, in view of the 2011 tragedy. Additionally, during the training of the researchers, special emphasis was placed on possible negative impacts of the research to avoid harm to the participants (Flick 2007).

During the data collection phase, all research participants were informed briefly about the research topic in order not to distort the results. All agreed to participate, and their anonymity was assured. All methods aimed to gather data regarding the four key factors as well as additional information about existing environmental education activities and the population's extent of using media such as television, newspaper and the internet. The methods have been developed by the research group based on existing literature (Plapp 2003; Mytanz 2013).

The intention was to obtain representative and individual data in order to observe possible variations in people's perceptions. As a unit of analysis, individuals older than 16 years were selected, assuming that perceptions are individually determined (Slovic 1992). Therefore, the semi-structured questionnaire was the principal research method, containing 41 questions (including 17 questions about vulnerability, 9 regarding knowledge of ecosystem services and disasters, 3 on responsibility and 2 about possibilities). Sample questions are given in the results section.[1]

Within the four selected communities, a systematic random sample based on households was applied in order to guarantee that each unit of analysis had the same probability to be chosen for the semi-structured interviews. A monitoring system was used to ensure representation by gender and income groups in equal proportions to their composition in the population of each area according to the census (IBGE 2010). Whenever a discrepancy was detected, the sampling strategy was changed in order to specifically reach under-represented groups in the remaining households. A total of 271 semi-structured, face-to-face interviews in Portuguese were conducted (67 in Caleme, 89 in Granja Guarani, 62 in Santa Rita, 53 in Vieira), which is 14.8 % of the total population as per the demographic census in the four sample districts (IBGE 2010). This sample size assures representative results with a level of confidence of 90 % and a single size error of 5 %. The four research areas were equally represented in the sample.

In addition, focus groups were held to obtain deeper qualitative information on the research issues. Residents older than 16 years were invited to participate. A convenience sample of individuals available and willing to participate in the study was chosen (Collins et al. 2007). Their selection was supported by a "gatekeeper", a socially engaged and well-known person in the community, to guarantee an adequate composition of the group. There was one focus group discussion in each area.

---

[1]All questions and results presented in this article are translations from Portuguese into English. For a list of all questions and results from the semi-structured questionnaire in Portuguese please see Lange et al. (2013).

Different participatory methods, such as community mapping and small discussion groups, were used within the focus group to obtain as much information as possible on the different research dimensions. For example, community mapping assessed the perception of disaster risks and understanding of its causes including ecosystem degradation. The focus group methods were facilitated and moderated by two people in Portuguese, and at least three researchers observed the discussion, taking notes on pre-prepared observation sheets. The communities were informed of the results by an information leaflet after the main results were analyzed.

Furthermore, 19 qualitative problem-centered interviews based on an open questionnaire with 27 questions were conducted with local key actors (e.g. representatives of the community association, local NGO leaders, school directors, priests, etc.). The main criteria for their selection were their social engagement in community activities, especially after the 2011 tragedy, and their knowledge about the structure of the community and its problems. In each community, a "gatekeeper" was identified, who supported the research group to make contact with other experts. This was especially important to create a trustworthy and open atmosphere during the interviews. The main objective of the qualitative interviews was to understand the perception of key protagonists in the communities and their assumptions about the local population's perception of risks. Following the principles of qualitative research, interviews were conducted until saturation level of data was achieved.

The quantitative data obtained from the semi-structured questionnaires were analyzed by quantitative content analysis (Mayring 2010) with the support of the statistical program SPSS. There were categorical and open questions in the questionnaires. The open questions were quantified before the analysis. The quantification of the answers was done by both concept-driven coding, to take into account the results of preliminary research, and open, data-driven coding, to ensure the capture of additional data and phenomenon which were considered of explanatory value for the research (Gibbs 2007). Both the focus group discussions and the qualitative interviews with local key actors were analyzed by qualitative content analysis (Mayring 2010). We used a step model of deductive category application. This means that we worked with coding rules for each category previously developed, determining exactly under what circumstances a text passage can be coded with a category (Mayring 2000). The software ATLAS.ti was used for the analysis.

The indices of perception of vulnerability, exposure, sensitivity and adaptive capacity presented in the results chapter of the article were calculated based on quantitative data from the semi-structured questionnaires. While the index of perception of vulnerability is calculated by adding the perception of exposure and sensitivity and subtracting the perception of adaptive capacity (GIZ 2013), each of these three indices is composed of multiple questions from the questionnaires.[2]

All three empirical research methods were designed with the intention that the methods would also be used in future research on the same and similar topics.

---

[2]For the exact composition of each index please see Lange et al. (2013).

Detailed information on all methods, including questionnaires, focus group instructions and interview guidelines are provided in the original study in the form of a perception analysis toolkit (Lange et al. 2013).

To ensure the reliability of the methods and the replicability in other research contexts on disasters, the authors aimed to develop research methods which could be adapted to different disasters and to different levels of affected populations.

21.3 Results

21.3.1 Perception of Risks and Adaptation Measures in Teresópolis

The results show that three main barriers are impeding the local population from more actively engaging in Eco-DRR and EbA activities. First, while the population has some knowledge of the importance of environmental protection, only a few people have a wider understanding of the role of ecosystem services in reducing disaster risks. Second, although a major part of the population feels responsible for conserving ecosystems, they do not feel responsible for undertaking ecosystem restoration activities. Lastly, a great number of people perceive that the greatest obstacle to better value ecosystems is the lack of resources and options, as well as the lack of opportunities to obtain resources and options. Our results also suggest that people in Teresópolis already have a relatively high perception of their vulnerability to disasters and believe disaster risk will increase in the future. Hence, this important condition for increasing participation in Eco-DRR and EbA is already met. Sensitization activities should therefore primarily focus on the other three barriers: knowledge, perception of responsibility and perception of possibilities.

The following discussion presents the most important insights for each of the four dimensions. This differentiated look at the results is especially important when it comes to developing appropriate sensitization activities and ways to engage the local population in the reduction of disaster risks.

21.3.2 Perception of Vulnerability with Respect to Disaster Risks

Our analysis of the perception of vulnerability shows that people in Teresópolis have a relatively high awareness of their exposure and sensitivity and perceive their capacity to adapt to disaster risks as relatively low. This means that most people know that they live in risk-prone areas and realize that they are insufficiently protected against potential disasters. But they are typically not aware of the possibilities to better protect themselves from disasters.

A vivid example that illustrates people's level of awareness of their exposure is that 86 % of the population stated that either landslides, mudslides or floods, the most prevalent hazards in the mountain area of Teresópolis, would pose a direct risk to their life. This high percentage can be explained by the devastating impact of the tragedy of 2011 and the resulting heightened local awareness of disasters. Focus group discussions and interviews with local key actors confirmed that after the events of 2011 the perception of exposure to hazards increased significantly within the community – especially in those areas worst affected. Almost every resident knows at least one family member who was directly harmed by the disaster in 2011.

Whilst the tragedy increased local awareness, it also made it hard for people to differentiate between levels of risk across different locations. It even led to an overestimation of risks. For example, more than half of the population (58.5 %) believe that every area in the region has equal probability or risk of experiencing a disaster, not taking into account that some areas could have higher levels of risk than others. This means that, although the awareness of risk is relatively high, many people are not capable of making differentiated judgments about the actual levels of risk of an area.

For the perception of sensitivity and adaptive capacity, the results are similar, although less distinct. More than 70 % of the population perceive that disasters can cause serious damage to their own lives as well as to their livestock and other assets. In the case of a disaster, most people feel helpless and incapable of adequately protecting themselves. One-third (34 %) do not even have a basic idea of what they could do to prevent or mitigate the impacts of disasters, and some do not believe actions can be taken at all to protect against disasters. This is mainly due to the lack of knowledge of adequate measures for disaster risk reduction and adaptation, and the perception of a lack of financial resources to take up those measures.

Histogrammes in Fig. 21.4 shows the distribution among the local population of the perception of (a) vulnerability and factors that define it, (b) exposure, (c) sensitivity and (d) adaptive capacity. While the index of perception of vulnerability is a combination of the other three indices,[3] each of these indices is derived from different questions[4] from the semi-structured interviews. Each histogram shows the distribution of perception from a value of 0 (very low perception) to a value of 1 (very high perception).

The histograms show that there is a relatively high average perception of vulnerability of 0.62 with a relatively low standard deviation of 0.19 – that is, the

---

[3]The formula used to calculate the index of perception of vulnerability is: perception of vulnerability = perception of exposure + perception of sensitivity – perception of adaptive capacity (GIZ 2013). To generate a value between 0 and 1 for the vulnerability index, we used the formula "1 – ((Nmax -Nx)/(Nmax-Nmin))" (UNDP 1990).

[4]For more information on the composition of each index, including the sets of questions, please see Lange et al. (2013).

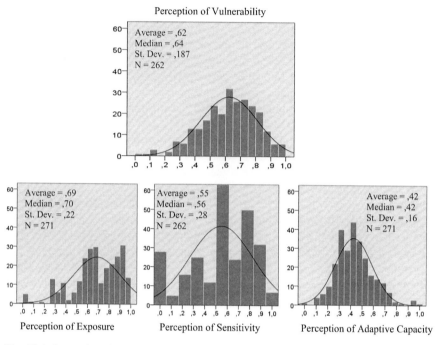

Fig. 21.4 Perception of vulnerability and its constituent dimensions (Own graphic)

perception of vulnerability deviates quite homogeneously around the average value. For the perception of exposure, this differs slightly. The perception is higher with an average of 0.69 and also deviates more (0.22) with two clusters around the values 0.6 and 0.9. This indicates that one part of the population already has a very high perception of exposure, while the perception of exposure by the rest of the population is far lower. Similar characteristics apply to the perception of sensitivity (high standard deviation of 0.28), although the general perception is lower with an average value of 0.55. The perception of adaptive capacity is generally low with an average of 0.42, deviating homogeneously around this value.

It is also crucial to understand how people perceive their future vulnerability. This knowledge is of special relevance for the planning and design of projects and strategies, especially in the context of Eco-DRR and EbA, because ecosystem-based measures need some time until they can fully yield DRR or CCA benefits and are therefore often undertaken over the medium- or long-term (Renaud et al. 2013).

Of the people surveyed, 76 % perceived an increase in frequency and magnitude of rainfall during the last 10 years, and 80 % stated that a future increase would have more negative than positive effects on their livelihoods. Also, 68 % of the population believed that landslides (compared to 63 % for floods) had increased during the last 10 years in frequency and/or magnitude.

These results were confirmed in all four focus groups, especially in the rural areas. Here, people talked comprehensively about the increase in rainfall, landslides, mudslides and floods and more generally about the increase in disaster risks. Although the results are probably also influenced by the 2011 tragedy, people perceive that disaster risks increased during the past few years and, based on this trend, expect that their exposure is likely to increase in the future.

All these results suggest that the perception of vulnerability, as one important driver for enhanced participation in DRR and CCA, already exists in Teresópolis. Sensitization activities should therefore not focus on further increasing awareness of vulnerability. Instead, efforts should focus on more active engagements with the local population to discuss effective measures for DRR and CCA, and enhance local understanding of exposure to disaster risks. A high perception of vulnerability does not automatically mean that people act or even choose adequate measures to reduce risks and/or recognize the value of ecosystems for DRR and CCA. Although disasters could affect everyone, there are certainly areas of high risk and areas of medium or low risk. Finding optimal solutions for each area, on a case-by-case basis, is crucial and requires a well-informed and differentiated perception of risks – especially the perception of exposure to risks. People will need this deeper understanding to choose between risk reduction measures that are most appropriate to their context.

21.3.3 Perception of Ecosystem Services and Their Functions to Reduce Disaster Risks

The effectiveness of Eco-DRR and EbA measures depends highly on the acceptance – and ideally the participation – of the local population (IIED 2009), who need to have a good understanding of the risk reduction functions of ecosystems (Borrini-Feyerabend et al. 2007; King and Marfai 2008). For Teresópolis, our results suggest that even though the population has some knowledge about the importance of ecosystem protection, only a few people have a deeper understanding of the role of ecosystem services in the reduction of disaster risks.

Over 50 % of those interviewed explicitly indicated activities in the area of ecosystem degradation as the main cause for the increase of landslides, mudslides and floods in the past 10 years. This basic knowledge of the causality between ecosystem degradation and risk was also confirmed in the focus group discussions. Participants talked about the severe effects of deforestation on increasing the prevalence of landslides, mudslides and floods. However, in addition to recognizing that ecosystem degradation increases disaster risks, people also need to understand how intact ecosystems and the services they provide can contribute to reducing risks, and how ecosystem restoration could therefore be viewed as a key measure for implementing DRR or CCA. In this regard, protected area management can play an essential role.

According to a study on the role of protected areas for risk reduction in the mountain region of Teresópolis, the regions most affected by the catastrophe in 2011 were in APPs (SBF 2011). APPs with intact vegetation were significantly less impacted by the disaster than APPs with degraded vegetation (SBF 2011). This is primarily due to the slope stabilization and water regulation functions of the ecosystems in APPs with intact vegetation (SBF 2011).

When asked how the forests in APPs benefit the population, most interviewees mentioned that forests conserve biodiversity by protecting animals and produce fresh air and clean water. Only 12 % spontaneously mentioned that forests can also contribute to reducing the risk of disasters. When specifically asked to elaborate on the risk reduction functions of forests in the focus groups, some participants noted that the roots of the trees would stabilize the soil and thus prevent landmasses from sliding. However, other risk reduction functions of a forest, such as its water absorption capacity and associated flood reduction services, were not mentioned at all.

When asked about the functions of protected areas (including APPs) in general, results were similar. Although 72 % of the people surveyed could name at least one protected area in their region, only 5 % perceived that protected areas could contribute to reducing disaster risks. Instead, most people understood the functions of protected areas to be mainly for the conservation and restoration of nature in general or for the protection of animals. When directly asked if protected areas could contribute to the reduction of disaster risks, 65 % of the people interviewed agreed. However, only one third (35 %) could afterwards explain at least one concrete example or mechanism how protected areas could reduce disaster risks. These results confirm the observations made during the focus group discussions. Although many participants perceived a certain connection between ecosystem degradation and disaster risks, and some even knew that ecosystems were important to reduce disaster risks, very few people understood the mechanisms by which ecosystems can reduce risks.

This crucial observation shows that people in Teresópolis have a basic understanding but no sophisticated knowledge on the relationship between ecosystems and risk reduction. The limited knowledge is one important factor preventing people to value ecosystem services and participate in risk reduction measures. People need to know that effective conservation and restoration of their surrounding ecosystems can protect them against landslides, mudslides and floods, and they need to know in which areas restoration is especially crucial. If they are not aware of these relationships, why should they be motivated to invest their scarce time to participate in risk reduction measures?

Lack of knowledge about the ecosystem's risk reduction functions is the first barrier to the population's enhanced participation in Eco-DRR and EbA measures identified in our study. As a consequence, sensitization activities in the region should focus on increasing knowledge about the role of local ecosystem services and their functions in reducing disaster risks.

21.3.4   Perception of Responsibility to Protect and Restore Ecosystems

As mentioned above, for Eco-DRR and EbA measures to be effective, the local population should accept and support the measures and ideally, be actively involved (IIED 2009). Another factor that can enhance participation is the population's sense of responsibility to contribute to Eco-DRR and EbA. If people do not have this sense of responsibility, there is little reason why they should be intrinsically motivated to participate (Wachinger et al. 2013). Therefore, we analyzed the local population's perception of its own responsibility to conserve and restore ecosystems. We compared people's perceived role of the government in risk reduction with the perception of the responsibilities people see for their community and themselves (see Fig. 21.5).

While almost half the people interviewed feel responsible for contributing to the mitigation of disasters by means of conservation of nature, people generally consider the government to be responsible for traditional DRR measures, such as engineered infrastructure (37.3 %). Also, more than one third of the people claimed that the government should force people to leave high risk areas and provide social housing in safe areas. The restoration of ecosystems, by contrast, is neither perceived as a responsibility of the government (5.5 %) nor of the community (5.2 %). This might be explained by the relatively limited knowledge of the effectiveness of ecosystem restoration for risk reduction, as shown above.

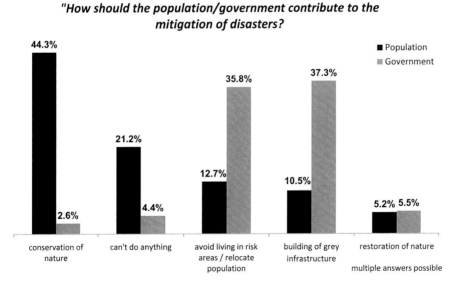

Fig. 21.5 Perception of responsibility of the population and of the government (Own graphic). The figure only shows answers that are relevant for EbA and Eco-DRR measures. Other answers, e.g. on governmental inspection or social organization, are excluded to simplify the illustration. For the complete set of answers please see Lange et al. (2013)

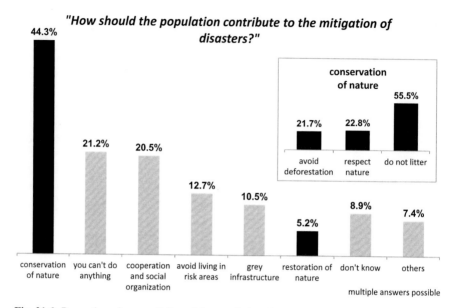

Fig. 21.6 Perception of responsibility of the population (Own graphic)

When looking in more detail at the responsibilities people see for their community and themselves (Fig. 21.6), we find that in the area of conservation of ecosystems, those interviewed feel responsible for taking up avoidance behavior (e.g. avoid deforestation, do not litter) rather than undertaking the active implementation of activities (e.g. remove litter from rivers or forests or to actively educate other people).

Similar observations were made by local community representatives who stressed that after the tragedy of 2011, the main behavioral change in the community was to avoid throwing solid waste into the rivers. Although this activity can certainly reduce the risk of floods (Jha et al. 2012), it cannot effectively reduce the risks of landslides and other hazards. Active ecosystem restoration is also crucial, particularly in the area of Teresópolis where a lot of environmental degradation has already taken place (Smyth and Royle 2000).

At the same time, 21 % of the people think that they cannot do anything and another 9 % have no idea how to contribute to risk reduction. This means that almost one third of the population do not feel capable of contributing to any DRR measures.

High local awareness of proper waste disposal, however, shows that environmental education could have a significant impact on people's perceptions. Since 2011, there have been various campaigns for proper waste disposal in the Teresópolis area which, according to our interviews, have been well-accepted by the local community.

Lack of the sense of responsibility of the population to effectively conserve, and especially to restore ecosystem services, is the second barrier to enhanced participation in EbA and Eco-DRR identified in our study. Sensitization activities should therefore also focus on increasing the perception of responsibility among the local population.

Results of our study also suggest that the perception of responsibility is strongly connected to knowledge of the functions of ecosystem services. If people have limited knowledge of the effectiveness of certain measures (e.g. reforestation), it seems less likely that they will feel responsible for implementing them. Therefore, enhancing local knowledge of the effectiveness of ecosystem services for DRR and CCA (the first barrier identified earlier in this paper) seems to be a first step to increasing people's sense of responsibility for implementing effective Eco-DRR or EbA measures.

21.3.5 Perception of Possibilities to Conserve and Restore Ecosystems

For people living in low and medium risk areas, the likelihood and the impact of disasters can often be significantly reduced by conserving and restoring the surrounding ecosystems (Nehren et al. 2014). These measures are especially important because in Teresópolis, according to our interviews with local key actors, resettlement capacities are limited and often associated with negative consequences, especially for the most vulnerable and marginalized. Resettled communities run the risk of losing their valuable social networks and supporting infrastructure.

As shown above, a major part of the population is aware of their vulnerability to disasters, and some people also know about the importance of ecosystems for reducing disaster risks. Also, a segment of the population values ecosystems. Why then do so few people actively participate in the protection and restoration of ecosystems?

As shown in Fig. 21.7, more than one third of the people surveyed in Teresópolis (37 %) perceived a lack of willingness, awareness or sensitization as important factors preventing them from more actively contributing to the protection and restoration of ecosystems. This confirms the above results and the barriers identified.

While only 18 % mentioned a lack of education and information during the interviews, this topic was extensively discussed in the focus groups. Participants identified the lack of knowledge of how to collectively undertake the protection and restoration of ecosystems as equally important as the lack of willingness and awareness. A key outcome of the focus discussions was that people do not feel capable of guiding or instructing others on how to undertake Eco-DRR or EbA activities. Hence, lack of knowledge or training on implementation of Eco-DRR and EbA measures seems to be a major reason why people do not engage more actively in Eco-DRR and EbA.

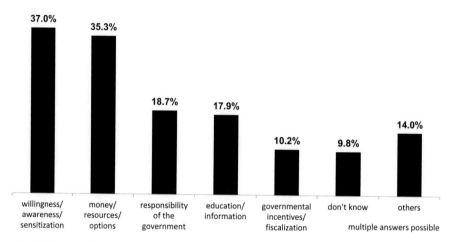

Fig. 21.7 Perception of obstacles to contribute to conservation and restoration of ecosystems (Own graphic)

In addition to knowledge on how to effectively conserve and restore ecosystems, people need to feel they have the resources and capability to do so. Figure 21.7 also shows that more than one third of the population (35.3 %) sees the lack of money and resources and the lack of opportunities to obtain such resources as another great obstacle. Several interviewees also mentioned that the lack of money would result in a lack of time for investing in Eco-DRR and EbA activities, because people would need to work extra hours to earn their income. Thus, particularly time-consuming activities like planting trees during weekends would simply not be an option for them. Lack of both resources and knowledge of how to obtain resources is the third gap identified in this study. Sensitization activities should therefore also focus on increasing knowledge of the local population on how and where to obtain financial resources and options to invest in ecosystem-based measures.

21.4 Conclusion

Ecosystem-based measures such as reforestation and protected area management have great potential to reduce disaster risks and adapt to climate change in the mountain area of Teresópolis (Nehren et al. 2014). In many cases, they can complement or even substitute hard engineering measures. However, to be effective, policymakers must design measures sensibly and involve the local community in planning and implementation (Allen 2006).

An analysis of the perceptions of local residents allows us to identify the drivers and barriers for effective participation in EbA and Eco-DRR. The perception analysis is a tool, especially designed to obtain a better understanding of people's awareness of their vulnerability, knowledge of the relationships between environment and disaster risk, perception of self-responsibility and possibilities for action.

Based on the results and additional research regarding environmental education and relevant stakeholders (Lange et al. 2013), a locally adapted sensitization and communication strategy could be developed. In the municipality of Teresópolis, the three barriers identified through our perception analysis need to be addressed. They should constitute the main content of a sensitization strategy and should also be used as a basis for elaborating educational and/or informational materials.

Instead of implementing top-down approaches, perception analyses provide a feasible way to adapt Eco-DRR and EbA strategies to different local needs, thereby increasing their effectiveness and efficiency. A perception analysis, therefore, plays an important role in advising policymakers from national, state, municipal, and local levels on how to implement more effective, ecosystem-based measures for DRR and CCA with strong involvement of the local communities.

It is important to emphasize that a perception analysis serves to complement Eco-DRR and EbA measures at local level. To engage the population and increase their participation, it is especially important to align certain measures bottom-up, although government institutions are foremost responsible for creating the conditions for local participation, such as providing relevant information about where and how people can contribute effectively. Furthermore, whilst the proposed Eco-DRR or EbA measures might be cost-effective, financial support for their implementation is crucial, such as providing a fund for local communities or even non-governmental organizations to undertake such activities.

We also suggest that public decision makers and other stakeholders in the region use the results and recommendations of our study to develop adequate intervention strategies to increase public participation in disaster risk management. Our methods and data could then be used as a baseline study to measure the impact of their actions.

In addition to the conclusions of our case study, we recommend that the elaborated methods (toolkit)[5] that were successfully applied in this analysis be used by other researchers in similar research areas in order to allow for future comparison of results and to develop bottom-up activities. We strongly recommend that the methodological toolkit is applied, tested and adapted by other researchers in various contexts of disasters in order to contribute to the development of reliable methods for measuring perception of ecosystem services for disaster risk reduction and adaptation. We believe that Eco-DRR and EbA measures can be successfully implemented and will contribute to effective risk reduction strategies.

---

[5]For more information on the toolkit please see Lange et al. (2013).

References

Allen K (2006) Community-based disaster preparedness and climate adaptation: local capacity-building in the Philippines. Disasters 30(1):81–101

Borrini-Feyerabend G, Pimbert M, Farvar T et al (2007) Sharing power: a global guide to collaborative management of natural resources. Earthscan, London

Brasil (2012) Código Florestal: Lei n° 12.651 de 25 de maio de 2012. Presidência da República, Brasília. http://www.planalto.gov.br/ccivil_03/_ato2011-2014/2012/lei/l12651.htm

Brodbeck N (ed) (2012) World risk report. Bündnis Entwicklung Hilft, Berlin

Care (2010) Community based adaptation toolkit. Digital toolkit. Care. http://www.careclimatechange.org/files/toolkit/CARE_Integration_Toolkit.pdf. Accessed 10 Oct 2014

CEPED-UFSC (2011) Atlas Brasileiro de Desastres Naturais: 1991–2010. Volume Rio de Janeiro. Universidade Federal de Santa Catarina, Florianópolis

Collins KMT, Onwuegbuzie AJ, Jiao QG (2007) A mixed methods investigation of mixed methods sampling designs in social and health science research. J Mixed Methods Res 1(3):267–294

Colls A, Ash N, Ikkala N (2009) Ecosystem-based Adaptation: a natural response to climate change. IUCN, Gland. 16 pp

DRM-RJ (2012) Diagnóstico sobre o risco a escorregamentos no estado do Rio de Janeiro e plano de contingencia para atuaçao do NADE/DRM-RJ no período de dezembro de 2011 a abril de 2012. Governo do Estado do Rio de Janeiro, Rio de Janeiro

Fernandes NF, Guimaraes RF, Gomes RAT et al (2004) Topographic controls of landslides in Rio de Janeiro: field evidence and modeling. CATENA 55(2):163–181

Flick U (2007) Designing qualitative research. SAGE, London

Flick U (2011) Triangulation: Eine Einführung. Reihe Qualitative Sozialforschung. VS Verlag für Sozialwissenschaften, Wiesbaden

Fundação SOS Mata Atlântica, Instituto Nacional de Pesquisas Espaciais (2014) Atlas dos remanescentes florestais da Mata Atlântica, período 2012–2013. São Paulo. http://mapas.sosma.org.br/site_media/download/atlas_2012-2013_relatorio_tecnico_2014.pdf. Accessed 10 Oct 2014

Galindo Leal C, Gusmão-Câmara I (2003) The Atlantic Forest of South America: biodiversity status, threats, and outlook, vol 1, State of the Hotspots. Island Press, Washington, DC

Gibbs GR (2007) Analyzing qualitative data. In: Flick U (ed) The SAGE qualitative research kit. SAGE, Los Angeles

GIZ (2013) Assessing and monitoring climate vulnerability. Deutsche Gesellschaft für Internationale Zusammenarbeit, Eschborn

Grothmann T, Patt A (2005) Adaptive capacity and human cognition: the process of individual adaptation to climate change. Glob Environ Chang 15:199–213

Guha-Sapir D, Vos F, Below R, Ponserre S (2012) Annual disaster statistical review 2011 – the numbers and trends. Université catholique de Louvain, Brussels

IBGE (Instituto Brasileiro de Geografia e Estatística) (2010) Censo Demográfico 2010. http://censo2010.ibge.gov.br. Accessed 10 Oct 2014

IIED (2009) Participatory learning and actions. Community-based adaptation to climate change. IIED, London

IPCC (2001) 3rd Assessment Report, Working Group II, Appendix I, www.grida.no/publications/other/ipcc_tar/. Accessed 10 Oct 2014

Jha AK, Bloch R, Lamond J (2012) Cities and flooding: a guide to integrated urban flood risk management for the 21st century. The World Bank, Washington, DC. http://www.gfdrr.org/gfdrr/sites/gfdrr.org/files/urbanfloods/pdf/Cities%20and%20Flooding%20Guidebook.pdf. Accessed 10 Oct 2014

King L, Marfai M (2008) Coastal flood management in Semarang, Indonesia. Environ Geol 55(7):1507–1518

Lange W, Cavalcante L, Dünow L et al (2013) HumaNatureza^2 = Proteção Mútua. Percepção de riscos e adaptação à mudança climática baseada nos ecossistemas na Mata Atlântica, Brasil.

Schriftenreihe des Seminars für Ländliche Entwicklung 255 (SLE). Humboldt-Universität zu Berlin, Landwirtschaftlich-Gärtnerische Fakultät, Berlin. http://edoc.hu-berlin.de/series/sle/255/PDF/255.pdf. Accessed 11 Nov 2014

Madajewicz M, Pfaff A, van Geen A et al (2007) Can information alone change behavior? Response to arsenic contamination of groundwater in Bangladesh. J Dev Econ 84:731–754

Marengo JA (2008) Água e Mudanças Climáticas. Estudos Avançados 22(63):83–96

Mayring P (2000) Qualitative content analysis. Forum Qual Soc Res 1(2)

Mayring P (2010) Qualitative Inhaltsanalyse: Grundlagen und Techniken. Beltz, Weinheim

Messner F et al (2006) Flood damage, vulnerability and risk perception – challenges for flood damage research. In: Schanze J, Zeman E, Marsalek J (eds) Flood risk management: hazards, vulnerability and mitigation measures, vol 67, NATO Science Series. IV. Earth and Environmental Sciences. Springer, Dordrecht, pp 149–167

Mytanz C (2013) Indicators for local and regional vulnerability assessment in rural Cameroon. Climate Protection Programme for Developing Countries. Eschborn

Nehren U, Alfonso de Nehren S, Heinrich J (2009) Forest fragmentation in the Serra dos Órgãos: historical and landscape ecological implications. In: Gaese H et al (eds) Biodiversity and land use systems in the fragmented Mata Atlântica of Rio de Janeiro. Cuvillier, Göttingen, pp 39–64

Nehren U, Sudmeier-Rieux K, Sandholz S et al (eds) (2014) The ecosystem-based disaster risk reduction. Case study and exercise source book. UNEP and CNRD, Geneva

PBMC (2013a) Contribuição do Grupo de Trabalho 1 ao Primeiro Relatório de Avaliação Nacional do Painel Brasileiro de Mudanças Climáticas. Sumário Executivo GT1. PBMC, Rio de Janeiro

PBMC (2013b) Contribuição do Grupo de Trabalho 2 ao Primeiro Relatório de Avaliação Nacional do Painel Brasileiro de Mudanças Climáticas. Sumário Executivo do GT2. PBMC, Rio de Janeiro

Plapp S (2003) Wahrnehmung von Risiken aus Naturkatastrophen. Eine empirische Untersuchung in sechs gefährdeten Gebieten Süd- und Westdeutschlands. http://digbib.ubka.uni-karlsruhe.de/volltexte/3542003

Renaud FG, Sudmeier-Rieux K, Estrella M (2013) The relevance of ecosystems for disaster risk reduction. In: Renaud F, Sudmeier K, Estrella M (eds) The role of ecosystems in disaster risk reduction. UNU Press, Tokyo, pp 3–25

Ribeiro MC, Metzger JP, Martensen AC et al (2009) The Brazilian Atlantic forest: how much is left, and how is the remaining forest distributed? Implications for conservation. Biol Conserv 142(6):1141–1153

SBF (2011) Áreas de Preservaçao Permanente e Unidades de Conservaçao x Áreas de Risco. O que uma coisa tem a ver com a outra?. Série Biodiversidade 41. Ministério do Meio Ambiente, Brasília

SEA (2012) Plano Estadual sobre Mudança do Clima. Governo do Estado do Rio de Janeiro, Rio de Janeiro. http://download.rj.gov.br/documentos/10112/1312221/DLFE-56319.pdf/planoEstadualmudclima.pdf. Accessed 11 Oct 2014

Slovic P (1992) Perception of risk: reflections on the psychometric paradigm. In: Krimsky S et al (eds) Social theories of risk. Praeger, New York, pp 117–152

Slovic P, Weber EU (2002) Perception of risk posed by extreme events. Working paper. Columbia University, New York. http://www.rff.org/Documents/Events/Workshops%20and%20Conferences/Climate%20Change%20and%20Extreme%20Events/slovic%20extreme%20events%20final%20geneva.pdf. Accessed 15 Oct 2014

Smyth C, Royle S (2000) Urban landslide hazards: incidence and causative factors in Niterói, Rio de Janeiro State, Brazil. Appl Geogr 20:95–117

Sudmeier-Rieux K, Ash N (2009) Environmental guidance note for disaster risk reduction: healthy ecosystems for human security, Revised Edition. IUCN, Gland

UNDP (1990) Human development report 1990. The concept of measurement of human development. United Nations Development Programme, New York/Oxford

UNEP (2015): Promoting ecosystems for disaster risk reduction and climate change adaptation: opportunities for integration. Discussion paper. http://www.unep.org/disastersandconflicts/por tals/155/publications/EcoDRR_Discussion_paper_web.pdf. Accessed 10 Sept 2015

Venton P (2008) Linking climate change adaptation and disaster risk reduction. Tearfund, Teddington. http://www.preventionweb.net/files/3007_CCAandDRRweb.pdf Accessed 10 Oct 2014

Wachinger G, Renn O, Begg C, Kuhlicke C (2013) The risk perception paradox – implications for governance and communication of natural hazards. Risk Anal 33(6):1049–1065

Weber EU (1997) Perception and expectation of climate change: precondition for economic and technological adaptation. In: Bazerman M, Messick D, Tenbrunsel A, Wade-Benzoni K (eds) Psychological perspectives to environmental and ethical issues in management. Jossey-Bass, San Francisco, pp 314–341

Chapter 22
Strategies for Reducing Deforestation and Disaster Risk: Lessons from Garhwal Himalaya, India

Shalini Dhyani and Deepak Dhyani

Abstract Forest ecosystem services are significant for local communities, especially for mountain communities dependent on natural resources. This chapter examines the contribution of forests to local communities dwelling at various elevations (from 1400 to 2800 m.a.s.l.) in Upper Kedarnath Valley of Garhwal, India. It is based on a study which provides an overview of common fodder extraction practices in the region and their impact on disaster risk. The research pointed to exceptional variations in temperature, snowfall and rainfall intensity that were reported in the past three decades. According to local communities, during this period deforestation and forest degradation were the result of land conversion, construction of hydropower dams, and increased biomass extraction particularly for firewood and fodder production as well as extraction of forest products. Extreme climate events and disasters are closely linked to these forest cover changes. The research showed that livestock per household, individuals per household involved in fodder harvesting, and the altitude of the village are important factors affecting forest health, or forest degradation patterns, respectively. The study provides an overview of impact of climate variabilities and forest degradation on local communities. Fodder banks are discussed as a nature-based (or ecosystem-based) solution that can address forest degradation in the Indian Himalayan Region and neighboring mountain countries. The approach is based on the principles of 'community and ecosystem management' to provide an alternative for fodder resources to local communities. Efforts from this practical experience reflect the need of proactive planning to enhance adaptive capacities of mountain communities in

S. Dhyani (✉)
National Environmental Engineering Research Institute (CSIR-NEERI), Nehru Marg,
Nagpur 440020, Maharashtra, India
e-mail: shalini3006@gmail.com

D. Dhyani
Society for Conserving Planet and Life (COPAL), Srinagar Garhwal 246174,
Uttarakhand, India
e-mail: drddhyani@gmail.com

© Springer International Publishing Switzerland 2016
F.G. Renaud et al. (eds.), *Ecosystem-Based Disaster Risk Reduction and Adaptation in Practice*, Advances in Natural and Technological Hazards Research 42,
DOI 10.1007/978-3-319-43633-3_22

507

India and South Asia in general. This study is intended to enable more effective targeting of forest management interventions to reconcile the goals of poverty reduction and forest conservation.

Keywords Garhwal • Disaster risk • Natural solutions • Fodder bank • Community participation

22.1   Introduction

Global attention to forest ecosystem services has increased because of their function in providing benefits important to rural livelihoods and reducing climate-related vulnerabilities in many developing nations (Kalaba et al. 2013). The physical and social nature of Garhwal, which is part of Uttarakhand State in the Indian Himalayan Region (IHR), makes the region and its people extremely vulnerable to natural and man-made hazards, in particular landslides, floods, droughts, earthquakes, and epidemics. Inhabitants of the region often lack the capacity to cope with these extreme climate events and disasters. Disaster preparedness is crucial for areas like Garhwal where resources for relief, recovery, and reconstruction are limited and disasters cannot be predicted or fully prevented. The "Himalayan Tsunami" that occurred in Kedarnath valley in June, 2013 due to continuous heavy rains, frequent cloud burst incidents and glacial lake outburst resulted in massive floods (Das 2013). On 16th and 17th of June 2013, the Upper Kedarnath valley experienced a cloud burst, coupled with a glacial lake outbreak along with torrential rains. This caused sudden flash floods associated with landslides and earth flows in the valley both upstream and downstream of the valley. From 14th to 17th June 2013, in just 3 days, the entire Garhwal including the study area received heavy rainfall, which was about 375 % more than the standard rainfall that falls during the normal monsoon in the region (Satendra et al. 2014). Following the cloud burst, the Chorabari Lake (3800 m.a.s.l.) above the Kedarnath shrine collapsed resulting in a flash flood in the valley. Millions of tons of debris and rocks were carried downstream by these flash floods. This resulted in washing out of human settlements in Kedarnath, Rambara and many small villages downstream, and the loss of more than 20,000 lives including tourists. It devastated agriculture and forest lands both upstream and downstream, and submerged villages and towns, generating fear among local communities (Das et al. 2013). The Kedarnath disaster left the entire country contemplating the failure of the National Disaster Management Authority, unorganized tourism in sensitive valleys and enhanced deforestation and degradation at high altitudes.

Inaccessibility of the remote mountain areas and impoverished local communities make these communities dependent on forest resources in form of biomass and other life-sustaining products (Singh et al. 1998). Livelihoods of mountain communities in Garhwal depend on biomass harvested from forests, forest-dependent traditional farming practices and forest-dependent livestock rearing. In Garhwal, production is mainly subsistence-based and systems are influenced by traditions and indigenous knowledge. Women are mainly involved in collection of biomass (firewood, fodder, leaf litter and wild edibles) from forests for various household needs. Women spend a lot of their daily time and energy in extracting bio resources and this greatly adds to their drudgery. Hence, women are considered the backbone of the economy of Garhwal Himalayas. Enhanced pressures of local communities on forest resources for a variety of biomass demands (fuel, fodder, leaf litter, timber, wild edibles, etc.) has increased the degradation of forests. Livelihoods in the lower Himalayas are agriculture-based; in the middle Himalayas they are agriculture- and livestock-based whereas, in the higher Himalayas they are livestock-based. Quality fodder is considered vital to livestock productivity, and thus it becomes imperative to understand how to produce sustainable fodder resources from pastures and forests without deterioration of the ecosystems through appropriate management solutions (Rubanza et al. 2006).

Traditional and indigenous knowledge can guide suitable utilization, management, and conservation of forests to reduce disaster risk in mountain areas, when it is used adequately. Cost effective, nature-based solutions, such as Ecosystem-based Disaster Risk Reduction (Eco-DRR) with active community participation can help to reduce the pressure from forests and at the same time reduce disaster risks. These innovative but traditional preventive measures can promote effective strategies to develop alternative resources, build resilience and reduce the frequency and severity of disasters (Sudmeier-Rieux et al. 2013). Eco-DRR can be an effective approach in planning and implementing disaster risk reduction measures for sustaining Himalayan ecosystems and human well-being. Forest and natural resource management are therefore pragmatic approaches for reducing disaster risk. Community Based Adaptation (CBA) is an approach which takes into account the priorities of communities, their knowledge, and capacities to empower people to plan and cope with the impacts of increasing climate-related vulnerabilities and disaster risks (Reid et al. 2009). The objectives of reducing pressure on forests cannot be successful without being based on equitable participation of local communities in land management decisions, land-use trade-offs and long-term goal setting. We therefore undertook a study on the natural and man-made pressures in six villages in Upper Kedarnath valley, Rudraprayag district of Garhwal, where we implemented two innovative solutions: first the Fodder bank (FB) and secondly, the Livelihood Resource Center (LRC). Both of them were explored as more sustainable approaches to withstanding long periods of fodder shortage and alternative livelihoods for reducing pressure from forests.

The purpose of this chapter is to understand the extent to which resource extraction is accelerating the pace of forest degradation in the Himalayas under the current scenario of climate change coupled with enhanced developmental activities. The study also sheds light on the performance and the extent to which two cost effective nature-based solutions (FB and LRC), which contributed to reducing short and long term pressures on forests, can also contribute to reducing disaster risks.

It has three main objectives:

I. Analyse local livelihood activities and resource extraction from forests in the selected villages with a focus on women;
II. Assess perceived climatic changes by local residents;
III. Evaluate and discuss the performance of a Fodder bank (FB) and a Livelihood Resource Center (LRC) as nature-based solutions.

22.2 Study Area and Methods

22.2.1 Study Area

The study area of Upper Kedarnath valley is situated in Rudraprayag district of Garhwal in Uttarakhand state of India (Fig. 22.1). One among the *Panch Kedars*, the ancient Hindu shrine of Lord *Shiva viz. Kedarnath* is situated in the valley. The study area is characterised by a series of verdant valleys and hill ranges. The Mandakini River, one of the major tributaries of the Ganga River flows between the high mountains of Kedarnath valley. There are six villages located in the Upper Kedar valley: Tausi (~2800 m.a.s.l.), Triyuginarayan (~2600 m.a.s.l.), Kongarh

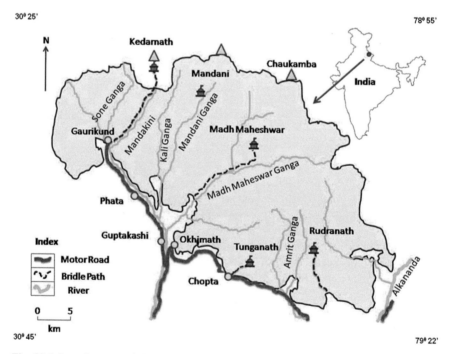

Fig. 22.1 Location map of the study villages (*red stars*) of Garhwal, in Upper Kedar Valley Uttarakhand, India. The map shows the boundaries of Kedarnath WLS

(~2200 m.a.s.l.), Nyalsu (~2000 m.a.s.l.), Shersi (~1800 m.a.s.l.) and Maikhanda (~1400 m.a.s.l.). These villages were selected for the study as they represent an elevation gradient with variations in biomass production and forest extraction practices. All villages comprise forests that are protected under different legal categories (Tausi and Triyuginarayan: Kedarnath Wildlife Sanctuary; Kongarh and Maikhanda: reserve/revenue forests; Nyalsu and Shersi: community forests). Maikhanda village was selected for developing a FB and a LRC.

Eighty percent of the annual rainfall occurs during the monsoon season between June-September, with >200 mm/month in July, August and September. The main valleys are fully exposed to the summer monsoon. Temperatures are highest in June, July and August prior to the monsoon. The highest and lowest temperatures recorded during the study period (January–December, 2010) are 28 °C and just below 0 °C, respectively (Misra 2010). The higher reaches of the valley are snow-bound for about 3 months (December, January and February).

The valleys and slopes of the hills are covered by moist temperate broad leaved forests (Champion and Seth 1968; Roy et al. 2015). Dominating tree species in the forests around the villages are the ring cupped oak (*Quercus glauca*), white oak (*Q. leucotrichophora*), green oak (*Q. floribunda*), brown oak (*Q. semecarpifolia*), West Himalayan fir (*Abies pindrow*), West Himalayan spruce (*Picea smithiana*), Himalayan yew (*Taxas baccata*), oval leaved lyonia (*Lyonia ovalifolia*), and rhodo-dendron (*Rhododendron arboreum*). The forests of the valley provide refuge to some rare and threatened wildlife, as for example the Himalayan musk deer (*Moschus chrysogaster*), Himalayan snow leopard (*Panthera uncia*), Indian leopard (*Panthera pardus fusca*), Himalayan tahr (*Hemitragus jemlahicus*), blue sheep/bharal (*Pseudois nayaur*), and Himalayan monal (*Lophophorus impejanus*) (Misra 2010).

All the studied villages together comprise 735 households and 2879 individuals (Census of India 2011) and are part of the Garhwali community. Additionally, there is an immigrant Nepalese population of about 5000–8000 individuals that visits the valley for occasional employment opportunities as household help, porters and daily wagers during the Kedarnath pilgrimage season (from May–October).

22.2.1.1 Socio-Economic Conditions of the Villages

The average family size of sampled households was recorded at six persons/house-hold, which was higher than the regional average of five persons/household (International Institute for Population Sciences (IIPS) and Macro International 2007). Keeping animals is critical for these households as it is an important source of income and wealth in the agriculture-based economy of the study area. Average livestock possession is 3–5 animals per household, usually including a cow, two bullocks or a buffalo. A few rich families also have bigger livestock holdings, including two cows, two bullocks and a few sheep. Exceptions are temporary migrant livestock holders, with large sheep herds as livestock rearing is their primary occupation. As they practice transhumance, they do not stall-feed their animals but opt for open grazing and hence, were not involved in the study. Per capita landholding was estimated at 0.59 hectare of rainfed land. Nearly 15–20 crops (e.g., traditional cereals, millets, vegetables, pulses, oil yielding, condiments, spices and medicinal plants) are grown by the farmers, however, only a few are grown on a large scale. Traditionally, Farm Yard Manure (FYM) prepared from cattle dung is used in agriculture without any chemical inputs. Well-maintained orchards of Malta fruit *(Citrus sinensis* variety*)*, elephant ear fig *(Ficus auriculata)* and English walnut *(Juglans regia)* are observed in the villages. Farm area and cropping intensity have changed over the last two decades (Misra et al. 2008) from traditional crops, *i.e.* finger millet *(Eleusine coracana)* and amaranth *(Amaranthus frumentaceus)* to cash crops such as potato *(Solanum tuberosum)*. Cultivation of traditional crops such as Indian barnyard millet *(Echinocloa frumentocea)*, soy bean *(Glycine max)*, foxtail millet *(Setaria italica)*, proso millet *(Panicum miliaceum)* and pearl millet *(Pennisetum typhoides)* have been largely abandoned. The area under potato and kidney bean has also increased in the last three decades, adding more pressure on forests. The earlier sustainable harvesting of fodder and other biomass by local communities from the forests is now becoming unsustainable because of increasing human and livestock populations and also because of increasing market demands for various forest-based products (Singh et al. 1998; Misra 2010). Agriculture and animal husbandry along with tourism related jobs are the main sources of income in the valley.

22.2.2 Methodology

A detailed review of forest and land use policies and archival reports related to resource rights, management and rural development available at various institutes and government departments was carried out to understand the to understand the socio-ecological system related to forest resources that included rights and resources locals have on the forests. Detailed data about demography, settlement patterns and local communities were collected from the Office of Revenue Department, Ukhimath Block, District Rudraprayag. Resource rights on forests were also verified from the village institutions' records and elected representatives of the village institutions. Primary data were collected by using different tools such as semi-structured questionnaires followed by group discussions and household interviews as well as guided field walks, and field observations. Local communities interviewed during the study were mostly illiterate. They were composed of farmers, housewives, and people belonging to the poorer segments of society who live at the frontline of changes in the environment across the region. Interviews were conducted in the local vernacular languages (Garhwali and Hindi), in which the researchers were conversant. The interviews included several sections covering livelihood activities and resource extraction from nearby forests and are facing serious impacts of climate-related variabilities and vulnerabilities. A seasonal calendar was established to understand when resources were removed from forests in the study villages. In-depth interviews focused on the tendency of change in weather conditions including temperature, rainfall and extreme events in recent years (1980s–2012 based on household members recollection), the impacts of change in weather conditions, and increased extreme events on components of water and forest resources. In these surveys, timeline and historical recall methods were used in order to identify climate extreme events and their variability over time, frequency and intensity. After gathering and organising information and data related to demography and historical records, a survey was conducted based on a semi-structured questionnaire. Stratification for the sampling was based on altitude and distance of the villages from the forests. Respondents selected for in-depth interviews were those who have rich experiences (based on the detailed responses during questionnaire survey and focussed group discussions). These respondents were mostly elderly respondents with a memory of climate conditions that spans three decades. This approach was applied to obtain information on impacts of climate-related variabilities and vulnerabilities on their day-to-day life in the last three decades (1980s onwards). This included changes in local weather conditions such as rise in temperature, shift and change in rainfall, snowfall patterns; occurrence of extreme events, change in water sources, etc.

The Questionnaire Had Two Main Modules

(A) Information on Climate Change (CC) and impact of CC according to local communities:

1. Weather conditions: change in daily weather conditions and reasons behind these.
2. Water: shift and change in rainfall and snowfall patterns. Lack of water: drying up of natural water sources.
3. Extreme climate events (storm; cloud burst; flash flood; landslide): increase/decrease; how much and which year; why and which flood extreme; before and after the year of changes.
4. Impact and probable reasons of extreme events.
5. Shortage of biomass from forests: impact or no impact on locals especially women as they are directly involved in collection from forests (depend on local perception).
6. Result of shortage in forest bio-resources especially fodder.
7. Options and opportunities: nature based cost effective site specific solutions.

(B) Information on innovative applications and approaches

8. Options: alternatives to fodder resources; alternative livelihoods; diversifying income generation activities, etc.

In order to better understand the changes in climate conditions during the last three decades (1980 onwards), we conducted household surveys of all households harvesting fodder from the forests: 25–50 % of households in each village (according to Census of India 2011). A total of 15, 50, 15, 50, 21 and 35 households interviews were completed from Tausi, Triyuginarayan, Kongarh, Nyalsu, Shersi and Maikhanda villages, respectively (Table 22.1). In addition, at least one individual from each household was interviewed, representing a sample size >20 % as recommended by Adhikari et al. (2004).

Table 22.1 People's perceptions regarding change in local weather conditions in past 30 years (1979–2013) in the study area

Village	Total HH[a]	Sampled HH[a]	% response of informants					
			Shift in snowfall pattern	Increase in day temp.	Irregular rainfall pattern	Decrease in total rain days/year	Increased frequency of storms
Shersi	48	21	90	65	78	45	23
Nyalsu	201	50	95	62	77	56	31
Kongarh	25	15	89	58	79	42	22
Triyuginarayan	195	50	65	49	67	39	34
Tausi	23	15	45	28	62	41	14
Maikhanda	55	35	25	88	90	48	60

[a]HH = Household Numbers

Community dependence on the forests for fodder was estimated based on the frequency of forest forays by individuals from each household per day and was cross-checked by actual field visits. Round the year fodder harvesting activities were divided into two seasons: summer and monsoon, *i.e.* when plenty of green fodder is available and winter (lean period for fodder in the study area), *i.e.* when only dry grasses are available and harvested by local communities. Prominent fodder plants harvested from forests by local communities were identified during the survey by using regional floral keys (Gaur 1999). The quantity of fodder collected by each sample household was calculated over 24 h by adopting a weight survey method (Mitchell 1979; Martin 1995).

22.3   Results

22.3.1   Climate-Related Disasters and Vulnerabilities

As one of the main objectives of this study was to understand local inhabitants' perceptions of climate change and variability, the information value of local communities was considered important. Across the region, almost all respondents identified significant changes to their environment and basic resources options. Master trainers were invited to demonstrate and generate awareness about innovative nature-based solutions and approaches. During our questionnaire survey, local respondents reported their observations and experiences of drastic changes in local weather in the last 15 years (since 2000). Ninety-four percent of the respondents mentioned the increase in local day temperatures in their valley over the past two decades (since 1990s) as a major reason explaining the drying up of natural water bodies in their villages. Observations reported by our survey also included frequent fluctuations in day temperature during the last three decades (since mid-1980s). According to survey respondents, there have been frequent changes in rainfall and snowfall patterns in the past 10 years (post 2000) (Table 22.1). Rainfall in the valley used to be distributed throughout the year, but nowadays rainfall has decreased to a few months (July, August, September) resulting in water shortages that were an uncommon situation 30 years ago.

Increased frequency of storms in the last decade (since 2000) was also experienced and reported by local respondents during our questionnaire survey (Table 22.1). Respondents reported extreme climate events and climate-related vulnerabilities being at their peak during 2011–2013 when heavy rains started even before the rainy season (June/July onwards), leading to the Kedarnath disaster described in the introduction.

22.3.2   Fodder, Drudgery and Nature-Based Solutions

Forest dependent agriculture and animal husbandry is a prime source of livelihood for over 70 % of local communities in the study area and these are directly affected by climate-related extreme events (Misra et al. 2008). Communities use a large number of plants for their diverse needs that grow extensively in the forests of the region (Misra et al. 2009). We observed daily, monthly, seasonal and annual resource extraction from forests, allowing us to estimatelocal communities' demands and dependence on different types of biomass (Table 22.2). The scarcity of fodder in hills is observed not only because of declining forest areas but also due to a rapidly growing population. Unavailability of green forage during lean winter periods has always been a serious issue in Garhwal that has added to the drudgery of women and enhanced forest degradation (Fig. 22.1). The problem has also led to (1) resource rights and extraction conflicts with nearby village *Van Panchayats* (community forests); and (2) increased women and children health ailments as well as malnutrition and improper education of female children. A large variety of tree shrubs, herbs, grass and agricultural by-products are used for livestock fodder. The major part (62.2 %) of the fodder is extracted from forests while the remaining fodder (37.8 %) is derived from agro-forestry systems, low altitude pastures, wastelands and alpine areas (Singh et al. 1998). More than 50 preferred fodder plants were identified during our primary field survey. Preferred fodder species (based on palatability and lactation enhancing) are white oak (*Quercus leucotrichophora*), brown oak (*Q. semecarpifolia*), green oak (*Q. floribunda*), ring cupped oak (*Q. glauca*), Indian maple (*Acer caesium*), horse chestnut (*Aesculus indica*), dudhilo (*Ficus nemoralis*), punjab fig (*F. palmata*), syanru (*Debregeasia salicifolia*), wild Himalayan cherry (*Prunus cerasoides*), hornbeam (*Carpinus viminea*), and Indian cranberry (*Vibrnum mullaha*). In earlier times, livestock were left to graze in the forests. Animals sought out their own food and were only assembled for milking and to protect them from wild animals. Nowadays, cattle are generally stall fed due to a ban imposed on open grazing (Misra et al. 2009). Nonetheless, a few buffaloes, sheep and goats are also left for grazing in nearby forests, alpine and *kharaks* (small pastures in oak forests) with few herder families of these villages (Dhyani et al. 2011).

Women spend a lot of time and energy in procuring fodder, adding to their everyday hard work and despondent life and life style. According to local respondents and our personal field observations, fodder collection is a frequent household activity other than fuelwood, with at least one female from each household involved in this activity. Involvement of males in fodder collection was almost negligina jable. The total green fodder collection reported during summer and monsoon months was highest for Shersi village (84 ± 6.23 kg/household/day) and lowest for Maikhanda village (64.4 ± 3.60 kg/household/day) (Table 22.3). Similarly, the highest total dry fodder collection during lean winter months was again recorded for Shersi village (80.4 ± 5.11 kg/household/day) and the lowest for Maikhanda village (62.4 ± 1.66 kg/household/day) (Table 22.3).

Table 22.2 Indigenous resource use pattern in the villages of the study area

Resource type	Extraction process	Species	Frequency	Consumption calendar	[a]Average
Fuelwood	Felling, lopping, collecting	Q.leucotrichophora Prunus cerasoides, Lyonia ovalifolia, Pyrus pashia, Alnus nepalensis	Once in 2–3 days	Entire year	35.00 Kg/2–3 day
Fodder	Lopping, chopping	Q.leucotrichophora Myrica esculenta, Debregeasia salicifolia	Twice a day	February–May and August–September	86.8 ± 10.5–157.5 ± 17.5 Kg/day
NTFP and wild edibles	Lopping, chopping, uprooting	Myrica esculenta, Rhododendron arboreum, Diplazium esculentum, Paeonia emoddi	2–3 times a week	March–June	1.5 Kg/household/week
Leaf Litter	Collecting	Leaves of Alnus nepalensis, Aesculus indica and Acer caesium	Twice a day	November–April	45 Kg/household/day

[a]Consumption is averaged for all households and all sites

Table 22.3 Amount of fodder collected from March–October (Summer + Monsoon) and amount of winter fodder from November–February (Winter)

Village name	(Summer + Monsoon)		(Winter fodder)	
	Amount Kg/HH/day	Amount Kg/HH/month	Amount Kg/HH/day	Amount Kg/HH/month
Tausi	84 ± 3.6	2506 ± 108.8	74 ± 4.9	2230 ± 148.3
Triyuginarayan	84 ± 6.2	2520 ± 186.8	67 ± 3.5	2016 ± 107.5
Kongarh	82 ± 5.3	2457 ± 159.5	74 ± 4.9	2232 ± 148.0
Nyalsu	83 ± 4.2	2478 ± 126.8	73 ± 3.9	2196 ± 117.1
Shersi	84 ± 6.2	2522 ± 187.8	80 ± 5.1	2412 ± 153.4
Maikhanda	64 ± 3.6	1932 ± 108.1	63 ± 1.6	1872 ± 49.8

Where *HH* household, *Kg* kilogram

Livestock size, distance of forests from villages, individuals involved per household as well as altitude of the villages influence the collection of fodder in these villages. The closest distance of all of the villages to forests was from Shersi village (1.5 km) and forays to forests by households were most frequent from this village when compared to all other studied village (Table 22.2 and Table 22.4). Collection during winter months (November–February) was comparatively lower. Women overcome long walking distances and climb rocks and hills to collect less nutritious dry fodder. According to personal observations and information collected during questionnaire surveys, accidents were frequent among women of the studied villages during winter fodder collection. Alternatives to travelling long distances for fodder collection and availability of fodder during lean winter periods are considered a key priority for local communities.

Wastelands, or unproductive grazing land occupy 2,940,000 ha in Uttarakhand state of IHR (Singh et al. 2012). As agriculture is the main source of income in Garhwal, large areas of forests and highlands were initially taken up for cultivation. Gradually these areas were left barren as people started migrating to other places because of better opportunities. Cultivable waste land has a major share of land area (98 %) in the state which is further divided into waste and fallow land (Singh et al. 2012). Cultivable wastelands offer an outstanding opportunity to reduce pressure from forests and women drudgery as they can be developed to satisfy biomass demand. Planting fast growing, high biomass yielding and nutritious plants not only increases fodder availability options but also reduces soil erosion and frequency of landslide events. Developing FBs on wastelands can provide a natural solution to reducing forest degradation, soil erosion and women drudgery. The basic idea behind this approach is to ensure native biodiversity conservation, restoration of degraded wastelands and providing nutritious fodder resources to local communities in their village vicinities (Fig. 22.1). FBs help to reduce pressure on forests, allow surviving lean fodder periods, and also help to reduce the distance frequently travelled by women to collect fodder. FBs can also help in generating awareness about better livestock feeding, livestock health, improved milk and meat yield by feeding nutritious fodder for better monetary income to local communities (Dhyani and Maikhuri 2012). The authors were involved in strengthening the

Table 22.4 Forest resource type, sources, availability status and dependency on forests of the study area

Settlements	Altitude (m.a.s.l.)	Resource type	Forest category[a]	Resource dependency [b, c]	Bioresource availability based on distance from village (km)
Maikhanda	1400	Fuelwood, fodder, leaf litter	RF	Medium	2–3
Shersi	1800	Fuelwood, fodder, leaf litter, timber and NTFPs	CF	High	1.5
Nyalsu	2000	Fuelwood, fodder, leaf litter, timber and NTFPs	RF	Medium	4–5
Kongarh	2200	Fuelwood, fodder, leaf litter, timber and NTFPs	RF	Medium	4–5
Triyuginarayan	2600	Fuelwood, fodder, leaf litter, timber and NTFPs	PF	Medium-High	2–3
Tausi	2800	Fuelwood, fodder, leaf litter, timber and NTFPs	PF	High	1.5

[a]*CF* community forest, *PF* protected forest, *RF* reserve forest
[b]High = >1 to = 3 km; Medium = 3 to < 5 km; Low = >5 km
[c]Production > Consumption = low;   Production = Consumption = Medium,   and   Production < Consumption = High

approach by introducing these tested fodder species in village cropland bunds to reduce soil erosion from croplands, a common phenomenon in hill agriculture.

22.3.3   Performance of the Fodder Banks (FBs)

An integrated ecosystem approach was adopted and was coupled with a community participation approach to develop an alternative to unsustainable fodder harvesting practices in the vicinity of Maikhanda village. Based on the approach, Maikhanda village in the valley with a majority of poor and scheduled (historically disadvantaged people in India) families was chosen for developing a FB model to fulfill fodder demands for lean periods. A FB was developed over 5 years (2009–2014) by introducing fast growing, high biomass yielding and nutritious species of fodder in

Fig. 22.2 (**a**) Woman carrying harvested green fodder; (**b**) Stored dry winter fodder; (**c**) Plantation at FB in Maikhanda village; (**d**) Local meetings for FB; (**e**) FB training programme; (**f**) LRC training programme

the community wasteland of Maikhanda village (Fig. 22.2). The choice was based on the willingness of local communities to provide their village community land to develop a FB and a small agriculture land for developing a fodder nursery. Meetings with *Mahila Mangal Dals* (Women welfare Groups *Custodians of Forest for Conservation and Sustainable Harvesting*) were held before and during the execution of each activity i.e., land preparation, fencing, pits digging, species selection, fodder plantation, etc. *Mahila Mangal Dals* were identified to take care of the site and engage in the rotational harvesting of fodder. To understand the benefits of the FB, livestock milk yields and reduced forays to forests were taken as indicators.

The FB was designed to develop fast growing and high biomass yielding nutritious fodder resources (both indigenous as well as introduced). The indigenous fodder plants were suggested by local communities based on their needs, their knowledge about species with regard to enhanced lactation, better nutrition and long life. Introduced fodder varieties were selected based on the altitude and climate suitability after detailed discussions with practitioners and scientists. This exercise was carried out to maximise the benefits and reduce the chances of detrimental invasion of plants. Women were trained in scientific propagation, plantation, multiplication of plants, and sustainable harvesting of resources. Planting was carried out twice a year on wastelands (a critical aspect of this activity that fully justifies the Eco-DRR approach), once during monsoon and other during spring. Indigenous grass species included multipurpose Ringal bamboo (*Chimnobambusa falcata, Thamnocalamus spathiflorus, Arundinaria spp.*). Tree species included among others alder (*Alnus nepalensis*), ring cupped oak (*Quercus glauca*), white oak (*Quercus leucotrichophora*), dudili (*Ficus nemoralis*), fig (*Ficus auriculata, Ficus subincisa*), and syanru (*Debregeasia salicifolia*). Introduced tree species included the European nettle tree (*Celtis australis*), mulberry (*Morus alba*), orchid tree (*Bauhinia variegata*) and grasses like napier (*Pennisetum purpureum*), Bermuda grass (*Cynodon plectostachyus*), orchard grass (*Dactylis glomerata*), and buffalo grass (*Panicum maximum*). All the seedlings were mass propagated in FB nurseries following scientific norms and were successfully planted by local communities in FBs. An assessment of survival percentages of plant species in FBs planted from 2009 to 2014 was carried out after every 4 months of planting (Dhyani et al. 2013). Regular capacity building and training workshops were carried out every year at FB sites to build capacity and transfer of knowledge to local communities. A large quantity of fodder plant material was also distributed free of cost among local communities during these onsite capacity building workshops. More than 93 women belonging to the same number of households of Maikhanda village cluster reported continuous harvesting and stall feeding of napier grass (*Pennisetum purpureum*) to their milk yielding animals from June 2010 onwards. During the first phase of this programme (2009–2012) women of Maikhanda village reported reduced visits to forests from 30 days to 10–15-days/month. The number of female beneficiaries who are introducing fast growing, high biomass yielding species in their cropland bunds increased by 10–15 women every 6 months.

Lactation yield of the animals is a direct indicator of the nutritional quality of the fodder. "*For the women of Kedarnath valley, this FB has brought the innovation that the solution to seasonal fodder deficit and milk yield, lies not only with the breed of cattle, but also with growing smart grass (fast growing, high biomass yielding, nutritious grasses), now being promoted*" said Birulal, native of Maikhanda village involved development of the FB. The smart grass growing technique is fast growing, high biomass yielding, nutritious napier grass (*Pennisetum purpureum*) and other such fodder grasses. Local women were also questioned about the effect of this new feed on the lactation yield of their milk yielding animals. Women of the village informed experiencing drastic changes in lactation yields of their animals after just 2–3 stall feeds (even with as little as 25–50 ml/yield) and regular stall feeding. In 2013, after 4 years of development, the model was handed over to *Mahila Mangal Dal* (women welfare groups). Since 2013, FB is managed by active local community involvement and cooperation of Maikhanda village (Dhyani 2012; Dhyani et al. 2013). Regular bimonthly monitoring of the FB is carried out to understand the sustainability of harvesting by visiting the site and regular interactions with local communities. Local communities of Maikhanda village have readily adopted the entire concept of using ecosystem and community-based approaches for rehabilitating the remaining wastelands and common lands of the village. Established fodder nursery fulfills seedling requirements of the village and livelihoods of two families. The main obstacle in implementing the ecosystem and community-based approach was developing trust among local communities about the initiative. This was achieved after a year, when local communities experienced benefits and results of the FB and they also started planting fodder species in their cropland bunds along with regular activity of developing FBs on village wastelands.

22.3.4 Performance of the Livelihood Resource Center (LRC)

The LRC can help in promoting the role of livelihoods in initiating forest conservation post-FB. A survey was undertaken in order to understand the effectiveness of the LRC and the potential of this approach in reducing pressure on forests by involving forest dependent poorer segments of the community and other key stakeholders. Local communities were interviewed regarding their understanding about the linkages between conservation and livelihoods to reduce pressure on natural forests and options for *ex-situ* conservation of economically important plants. More than 68 % respondents of Maikhanda village followed by 61 %, 59 %, 56 %, 53 %, 51 % from Shersi, Tausi, Triyuginarayan, Kongarh and Nyalsu, respectively responded positively for linking conservation with livelihoods to ensure long-term reduction of pressure on forests. According to respondents from all these villages, most of the governmental and also non-governmental projects do not attract the attention of villagers when monetary benefits are low or negligible, a major reason explaining the failure of such projects.

The approach involves quarterly capacity building and training programmes with the aim to improve the understanding of local communities of sustainable resource harvesting from nearby forests. LRC also involves training local communities for innovative livelihood options. Master trainers can demonstrate and generate awareness about innovative livelihood options such as off-season organic vegetable cultivation, bamboo made handicrafts, nursery raised important indigenous plants, dry flower arrangements, and value added agricultural product development. Successful entrepreneurs can also be involved to share their success stories to motivate local communities. LRC explores locally available cost-effective natural resources for enhancing livelihood options for local communities. This approach was to link FB to market to generate alternative economic opportunities for local communities.

Establishment of a LRC was undertaken through the active involvement of local communities and the village *Panchayat*. This allowed integrating livelihood and conservation goals to strengthen Community Conservation Initiatives (CSI). The LRC was developed in Maikhanda village based on suggestions from respondents during the personal interviews. The overall goal of the programme was to generate awareness about the role of FB and LRC in forest conservation, and sustainable utilization of natural resources by innovative livelihood options for local communities. LRC is a long-term forest conservation and management approach that involves forest dependent marginalised local communities and other key stakeholders. LRC explores locally available cost-effective natural resources for enhancing livelihood options for local communities. The emphasis was on regular and innovative capacity-building programmes that involved livelihood options that are neither dependent nor supported by overharvesting of forest resources. Impact of each and every capacity building programme among local communities was regularly monitored. Some households of the village were motivated to grow Ringal bamboo (*Chimnobambusa falcata, Thamnocalamus spathiflorus, Arundinaria spp.*) and local fig trees (*Ficus nemoralis, F. auriculata and F. subincisa*) in their cropland bunds for raw material production instead of harvesting these resources from forests. One household developed a plant nursery to increase the availability of planting material in the villages. Vegetable cultivation from indigenous seeds was promoted as an important practice among women. Incentives in the form of plant cutters, vegetable and fodder seeds and handicraft knives were used to motivate local communities. The next step in strengthening the entire approach to reducing pressure from forests is to establish market links between FBs and LRCs. This is still under process and more results will be available as we progress with this research.

22.4 Discussion

The present study reflects the concerns and difficulties related to increasing pressures on forest resources and climate variabilities in the Indian Himalayan Region and how they are having direct impacts on life and livelihoods of local communities. Increased fluctuations in monthly and annual temperature, rainfall and snowfall have a serious impact on forest dependent agriculture and on natural water resources that are vital for supporting rural life in remote villages of Garhwal. Responses received from the communities are important considering some extreme climate events and disasters (cloud bursts, landslides, glacial lake outbursts) that have been more frequent in the last few years. *"Cloudbursts, landslides and flash floods have become an annual affair in the valley. The monsoon has been bringing*

with it such massive losses of lives, property, crops and infrastructure since, 2005
the development clock had been set back by a decade. Things are much, much worse
every passing year. With many highways being damaged, bridges being washed
away, electricity and phone networks down, several ravaged places continue to be
marooned every years" told Harsh Prakash Semwal resident of Shersi village
during one of personal interviews. Some regional studies have also supported the
opinion of local communities collected in this study, reporting average day temperatures increasing by 0.75 °C per decade in the Himalayas during the last
100 years (Singh 2007). Frequent heavy rainfall events are causing landslides as
well as wash away agriculture land near natural streams and the river Mandakini.
The authors observed these phenomena throughout the study period. Areas affected
by landslides recover very slowly, and they often do not return to pre-landslide
conditions. The authors have personally observed few major landslides in the
region, with scars still evident more than two decades after the event. These
problems partially find their source in the fact that local communities and policy
planners do not factor in scientific knowledge and do not consider ecosystem or
nature based approaches when managing the resources.

In the last few decades the Government of India has also considered the severity
of extreme climate events as a serious issue for the country. Climate variabilities
and vulnerabilities are having a very profound impact on sensitive Himalayan
ecosystems. Sustaining Himalayan ecosystems has been considered an important
core area in National Action Plan for Climate Change (NAPCC 2008) as Himalayan
ecosystems are not only vital for communities dwelling in these states but also for
all those that are located downstream. Increasing population, agriculture activities
and combined upcoming hydroelectric projects have accelerated the pace of land
use changes in these areas in last few decades. These, coupled with climate-related
variabilities, are causing significant damages to natural forest ecosystems and the
socio-economic environment for local communities. Results clearly reflect that
increasing population and changes in resource requirements of local communities
(because of increasing agriculture activities and livestock holding) has added clear
and significant pressure on nearby forests.

Sources of income generation of local communities in the study area have
always been dependent on seasonal tourism activities (major Hindu shrine of
Kedarnath located in the valley), forest-based agriculture and selling forest based
Non Timber Forest Products (NTFPs). About 77.4 % of the total population of
Garhwal is rural and dwells in geographically inaccessible areas and have low
connectivity with other areas of the country (Singh et al. 2010). This inaccessibility
of the area and underprivileged socio-economic status of local communities is
responsible for the total dependence of the local communities on forest areas for
bio-resource supplies. Increased land use pressure has enhanced human interferences that have significantly affected resource availability in the Himalayas. This is
even more so considering the short- and long-term effects of climate change that are
perceived by local communities of the region. Climate and environmental changes
coupled with human interferences are modifying vegetation cover in the hilly
terrain of Garhwal Himalaya.

Transformation in the traditional lifestyle due to changing socio-economic conditions is also largely responsible to changes in resource use in Garhwal Himalaya. This has also been discussed by other authors and they also consider that forests in the Himalayas have undergone extensive degradation. Particularly in the last two centuries there has been extensive degradation of forested land, which has produced a mosaic of natural and managed ecosystems in the Indian landscape (Singh 1991; Roy et al. 2015). It is clear that this continuous depletion of the resource base is eroding living standards as well as ecosystem stability at a very large scale. Livelihoods in the higher Himalayas are more linked to livestock than agriculture, unlike lower and middle altitude villages of Himalayas. The study looks into one of the major forest resource harvesting activity of fodder in villages of Garhwal and also other parts of IHR. This activity leads to rapid removal of leaf, seedling and herb biomass from the forest floor for stall feeding livestock. At the same time, the activity enhances women drudgery due to lack of fodder options in village vicinity. In the long run continuous removal of grasses from forests is also leading to removal of seedlings affecting forest regeneration.

The study highlighted a clear knowledge gap in potential impacts of climate change on Himalayan ecosystems and a lack of ecosystem management approaches (requiring human capacity and management skills). Issues are very local but in the long run they are affecting the ecosystems at a large scale and amplifying communities dwelling in these areas. Nature-based solutions or Eco-DRR approaches were perceived to be required for large scale application to address this as well as similar problems. The FB can be considered a nature-based solution to allow rehabilitating wastelands and its potential to full fodder demands of locals during lean periods. Linking it with LRC establishment can ensure the sustainability of FB in the long run. The approach was considered innovative with the potential to provide not only economic, environmental but also social benefits to local communities, as recognized by various national and international platforms "International Centre for Integrated Mountain Development" (ICIMOD, Nepal); South Asian Association for Regional Cooperation (SAARC, Bhutan) and International Union for Conservation of Nature (IUCN) where this approach was presented for discussion. Ecosystem-based Adaptation (EbA) and Eco-DRR approaches are decentralized, cost effective, proven solutions with multiple benefits (Convention on Biological Diversity 2009). We need to first understand that the role of restoration of forests is not only to provide protection from soil erosion and landslides, but also to provide increased opportunities for local communities for improving their livelihoods and make use of several ecosystem services such as carbon sequestration and cultural services.

Land, one of the most important resources of Indian Himalayan Region (IHR) is under continuous stress and it is being rapidly converted into wasteland. The problem of wasteland in IHR differs from that of other areas in the country. The well-balanced land-vegetation associations in the study area have disrupted due to a variety of reasons. The wasteland problem has received less attention in the context of the degradation of the Himalayan environment. There is an urgent need of reclaiming and rehabilitating these wastelands by using appropriate cost-effective nature-based solutions and practices.

Our study provides instances and lessons that have been learnt; cultivable wastelands provide potential land options to local communities for growing fodder resources using FB and LRC approaches that can provide surplus fodder for local communities and also reduce pressure from forests. Moreover, development planning in the Himalayas needs to identify the potential for harnessing ecosystem services while addressing vulnerabilities linked to ecosystem degradation. Therefore, special efforts will be needed to foster social engineering to give due share to the socially excluded by developing and planning the approach involving local communities that many times also includes marginalized populations. Other important lessons learnt were developing a sense of belonging among local communities especially women of the Himalayas. Women in Garhwal part of IHR are considered to be the backbone of the economy and custodians of forest conservation. They have helped generating significant social, economic and cultural co-benefits for all. A clear message from the discussed nature-based solutions approach can be taken that by enhancing cross-sectoral joint ventures local communities can proactively develop and rapidly learn to promote integrated approaches to adaptation and disaster risk reduction.

22.5   Conclusions

The long-term mission to bring the region to acceptable disaster risk levels has just begun. The region is among the most disaster-prone in the world if one considers the number and severity of disasters, casualties, and impact on the national economy (Pradhan 2007). There seems to be no practical alternative to biomass extraction from forests as a source of basic energy until the socio-economic conditions of people living at subsistence level is improved. Afforestation with ecologically as well as socio-economically practical plants will not only fulfill the biomass demands but will also reduce forest degradation. A better understanding of local knowledge practices can help identifying what is important and can be promoted at local and regional level. Building on local knowledge and practices that capitalize on local strengths can reduce dependencies on external aid, too. Only through strong commitment, hard work, and joint efforts of using nature-based solutions this situation can be improved. More efforts are desirable to take the entire region into a mode of sustainable preparedness, resting on empowered communities, with proper inclusion of socially marginalized and vulnerable groups.

Acknowledgements The authors thank their field Assistant in Maikhanda village Sri. Birulal and all our informants of Upper Kedar valley, who wanted to share their traditional knowledge with us. Authors wish to thank the Director, CSIR-National Environmental Engineering Research Institute (NEERI), Nagpur and Director, Society for Conserving Planet and Life (COPAL) for encouragement. Thanks also to Dr. Karen Sudmeier-Rieux, Dr. Udo Nehren and Dr. Fabrice Renaud for their constructive comments and suggestions to substantially improve the manuscript. Financial support for the work from Department of Science and Technology (SYSP scheme) and Rufford Small Grants Programme, UK is thankfully acknowledged.

References

Adhikari B, Di FS, Lovett JC (2004) Household characteristics and forest dependency: evidence from common property forest management in Nepal. Ecol Econ 48(2):245–257

Census of India (2011) Village wise household schedule for district Rudraprayag, Uttarakhand. Government of India. Available via http://www.censusindia.gov.in/2011-Schedule/Shedules/English_HH_Side_B_NT.pdf. Accessed 24 Mar 2014

Champion HG, Seth SK (1968) A revised survey of the forest type of India. Manager of Publications, Government of India, New Delhi

Convention on Biological Diversity (2009) Secretariat of the convention on biological diversity. Year in Review 2008. Montreal, 68 pp

Das PK (2013) 'The Himalayan Tsunami' – Cloudburst, Flash Flood & death toll: a geographical postmortem. IOSR J Environ Sci Toxic Food Tech 7(2):33–45

Das S, Ashrit R, Moncrieff MW (2013) Simulation of a Himalayan cloudburst event. National Center for Medium Range Weather Forecasting, National Center for Atmospheric Research, Boulder

Dhyani S (2012) Strengthening fodder resources through fodder banks to reduce drudgery of rural women in Upper Kedar valley, Uttarakhand. In: Chettri N, Sherchan U, Chaudhary S et al (eds) India mountain biodiversity conservation and management selected examples of good practices and lessons learned from the Hindu Kush Himalayan region. International Centre for Integrated Mountain Development, Kathmandu, pp 52–54

Dhyani S, Maikhuri RK (2012) Fodder Banks can reduce women drudgery and anthropogenic pressure from forests of Central Himalaya. Curr Sci 103(7):763

Dhyani S, Maikhuri RK, Dhyani D (2011) Energy budget of fodder harvesting pattern along the altitudinal gradient in Garhwal Himalaya, India. J Biomass Bioenergy 35(5):1823–1832

Dhyani S, Maikhuri RK, Dhyani D (2013) Utility of fodder banks for reducing women drudgery and anthropogenic pressure from forests of Central Himalaya. Natl Acad Sci Lett 36 (4):453–460

Gaur RD (1999) Flora of the district Garhwal, Northwest Himalaya (with ethnobotanical notes). Trans Media, Srinagar Garhwal

International Institute for Population Sciences (IIPS), Macro International (2007) National Family Health Survey (NFHS-3), 2005–06: India: Volume I. IIPS, Mumbai

Kalaba FK, Quinn CH, Dougill AJ (2013) Contribution of forest provisioning ecosystem services to rural livelihoods in the Miombo woodlands of Zambia. Popul Environ 35:159–182

Martin GJ (1995) Ethnobotany a methods manual. Earthscan Publishers Ltd., London

Misra S (2010) Impact of natural and man-made disturbances on vegetation structure and diversity in Guptakashi range of Kedarnath forest division, Uttarakhand. Ph.D. thesis. Forest Research Institute University, Dehradun, India

Misra S, Dhyani D, Maikhuri RK (2008) Sequestering carbon through indigenous agriculture practices. LEISA India 10(4):21–22

Misra S, Maikhuri RK, Dhyani D et al (2009) Assessment of traditional rights, local interferences and natural resource management issues in Garhwal part of Indian Himalayan region. Int J Sustain Dev World Ecol 16(6):404–416

Mitchell R (1979) An analysis of Indian agroecosystem. Interprint, New Delhi

National Action Plan on Climate Change (2008) Available at http://pmindia.nic.in/Pg01-52.pdf. Assessed 11 Oct 2014

Pradhan BK (2007) Disaster preparedness for natural hazards: current status in Nepal. Mats GE, Greta MR, Joyce MM et al (eds) Published by International Centre for Integrated Mountain Development G.P.O. Box 3226, Kathmandu, pp 67

Reid H, Alam M, Berger R et al (2009) Community-based adaptation to climate change: an overview. Particip Learn Action 60:13

Roy PS, Roy A, Joshi PK et al (2015) Development of decadal (1985–1995–2005) land use and land cover database for India. Remote Sens 7(3):2401–2430

Rubanza CDK, Shem MN, Ichinohe T et al (2006) Biomass production and nutritive potential of conserved forages in silvopastoral traditional fodder banks (Ngitiri) of Meatu District of Tanzania. Asian-Aust J Anim Sci 19(7):978–983

Satendra KJ, Kumar A, Naik VK, AVSM KC (2014) India disaster report 2013 compiled by National Institute of Disaster Management, pp 124

Singh SP (1991) Structure and function of low and high altitude grazing land ecosystems and the impact of the livestock component in the Central Himalaya. Final technical report, Department of Environment, Goverment of India, New Delhi

Singh SP (2007) Climate change an overview. In: Selvasekarapandian S, Pandit GG, Ramachandran TV et al (eds) Fifteenth National Symposium on Environment (NSE-15). Mitigation of pollutants for clean environment BARC Macmillan India Ltd., Mumbai, p 24–29

Singh JS, Singh SP, Ram J (1998) Fodder and fuelwood resources of Central Himalaya: problems and solutions. Report submitted for study group on fuel and fodder. Planning Commission, Government of India, New Delhi

Singh G, Rawat GS, Verma D (2010) Comparative study of fuelwood consumption by villagers and seasonal "Dhaba owners" in the tourist affected regions of Western Himalaya, India. Energy Policy 38:1895–1899

Singh C, Dadhwal KS, Dhiman RC et al (2012) Management of degraded boulder riverbed lands through Paulownia based silvipastoral systems in Doon valley. Indian Forester 138(3):243–247

Sudmeier-Rieux K, Ash N, Murti R (2013) Environmental guidance note for disaster risk reduction: healthy ecosystems for human security and climate change adaptation, 2013 edn. IUCN, Gland, iii + 34 pp. First printed in 2009 as Environmental Guidance Note for Disaster Risk Reduction: Healthy Ecosystems for Human Security

Chapter 23
Ecosystem-Based Strategies for Community Resilience to Climate Variability in Indonesia

Giacomo Fedele, Febrina Desrianti, Adi Gangga, Florie Chazarin, Houria Djoudi, and Bruno Locatelli

Abstract Rural communities have long been using ecosystems to sustain their livelihoods, especially in times of disasters when forests act as safety nets and natural buffers. However, it is less clear how climate variability influences changes in land uses, and their implications for human well-being. We examined how forests and trees can reduce human vulnerability by affecting the three components of vulnerability: exposure, sensitivity, and adaptive capacity. A total of 24 focus group discussions and 256 household surveys were conducted in two smallholder-dominated rural landscapes in Indonesia, which were affected by floods, drought and disease outbreaks. Our results suggest that forests and trees are important in supporting community resilience and decreasing their vulnerabilities to climate-related stresses in different ways. The role of trees varied according to the type of ecosystem service, whether provisioning or regulating, in relation to the phase of the climatic hazard, either in the pre-disaster phase or in the post-disaster recovery phase. It is therefore important to distinguish between these elements when ana-lyzing people's responses to climatic variability in order to fully capture the contribution of forests and trees to reducing people's vulnerability. Landscape spatial characteristics, environmental degradation and community awareness of climate variability are crucial because if their linkages are recognized, local people

G. Fedele (✉)
CIFOR, Center for International Forestry Research, Bogor, Indonesia

CIRAD, Centre de Coopération Internationale en Recherche Agronomique pour Le Développement, UPR Forêts et Sociétés, Montpellier, France
e-mail: greengiac@gmail.com

F. Desrianti • A. Gangga • H. Djoudi
CIFOR, Center for International Forestry Research, Bogor, Indonesia

F. Chazarin
CIFOR, Center for International Forestry Research, Lima, Peru

B. Locatelli
CIRAD, Centre de Coopération Internationale en Recherche Agronomique pour Le Développement, UPR Forêts et Sociétés, Montpellier, France

CIFOR, Center for International Forestry Research, Lima, Peru

© Springer International Publishing Switzerland 2016   529
F.G. Renaud et al. (eds.), *Ecosystem-Based Disaster Risk Reduction and Adaptation in Practice*, Advances in Natural and Technological Hazards Research 42,
DOI 10.1007/978-3-319-43633-3_23

can actively manage natural resources to increase their resilience. Interventions related to forests and trees should take into consideration these aspects to make ecosystem services a valuable option for an integrated strategy to reduce disaster risks and climate-related vulnerabilities.

Keywords Climate variability • Climate change adaptation • Ecosystem services • Ecosystem-based adaptation • Natural resource management • Social-ecological systems • Social vulnerability

23.1 Introduction

Societies have long been using and managing ecosystems for subsistence, livelihoods (Shackleton and Shackleton 2004) and protection against risks caused by fluctuations in rainfall and temperature (CBD 2009). In times of extreme weather events, the literature has often identified forests as important safety nets and natural buffers that help reduce people's vulnerability by providing food, drinkable water, shelter and regulation of ecological processes (e.g. McSweeney 2004; Angelsen and Wunder 2003). In many parts of the world, natural systems and resources are a critical asset for local communities because they provide the foundations to respond to extreme weather events or disasters, especially if other technological options are limited (Sudmeier-Rieux et al. 2006; Roberts et al. 2011). Such dependency on natural resources, however, can also make rural populations prone to social and economic vulnerabilities, which a changing climate can exacerbate (IPCC 2014).

The effects of climate variability are already visible in many parts of the world, where people have been experiencing a general increase in extreme high temperatures, in drying trends, and in the number of heavy precipitation events (IPCC 2014). According to the Intergovernmental Panel on Climate Change (2014), climate variability refers to fluctuations in the means of climatic parameters such as those mentioned above, and can appear as unusual events and changes that occur within relatively short timeframes (seasons or years). If changes in variability are persistent for an extended period such as decades or longer, it can suggest that a change in climate has occurred (IPCC 2014). The effects of climate variability, due to either subtle shifts or more extreme events, directly impact poor people's lives. It has been predicted that the effects of climate variability will cause a decline in agricultural yields, reduce access to water, increase the severity of damages to assets in flood-prone areas, and increase vulnerability to human and non-human diseases (e.g. vector-borne diseases or pest species) among other impacts (IPCC 2014). Rural areas are particularly at risk from the impacts of climate variability due to their underlying vulnerabilities related to geographic situations, limited financial and technological means, and the sensitivity of their livelihoods to weather conditions, which can turn a hazard event into a disaster.

Healthy, diverse and well-managed ecosystems are able to resist, absorb and recover from unwanted changes and risks (CBD 2000; Gunderson and Holling

2002). Community management decisions can change the type, magnitude, distribution and relative mix of services that ecosystems provide, which in turn can reduce or increase a community's vulnerability to adverse climate (Rodríguez et al. 2006). Adaptation strategies based on ecosystems can complement and sometimes substitute other approaches involving hard infrastructure, technological solutions or capacity building (CBD 2009; Raudsepp-Hearne et al. 2010). In this way, communities can respond to the challenges posed by climatic variability, while also generating additional positive environmental, social, economic and cultural benefits, making sustainable ecosystem management a cost effective and suitable option for community climate change adaptation (see also Lange et al., Chap. 21).

Although research on adaptation to climate-related stress based on ecosystems is relatively new, there is an increasing recognition of the role of ecosystems in response strategies to climate change (Doswald and Osti 2011; Jones et al. 2012; Pramova et al. 2012). Several guiding principles have been developed by international organizations (CBD 2000; UNEP 2012; GIZ 2013; UNFCCC 2013; EU 2009) and practitioners (e.g. BirdLife International, International Union for Conservation of Nature and Natural Resources, World Wide Fund for Nature, in Heath et al. 2009; Colls et al. 2009; Andrade Pérez et al. 2010). However, regarding scientific knowledge on climate change and variability, few studies have focused on aspects related to human adaptation at the local level (Tompkins et al. 2013; IPCC 2012), in rural areas (IPCC 2014) within forested landscapes (IUFRO 2009). In addition, the recent IPCC 5th Assessment Report (IPCC 2014) indicated that more research was needed to better understand how climate variability influences changes in land use, which in turn can affect the provision of ecosystem services relevant for people's well-being. Especially lacking is quantitative evidence of the effects of management practices and landscape configurations (including forest types) on benefits to climate change adaptation (Harvey et al. 2013).

Indonesia has one of the largest areas of tropical forest in the world, which is rapidly disappearing (FAO 2010) and is among the top five countries most frequently hit by natural disasters (EM-DAT 2013). In this study, we examined the benefits provided by ecosystems in reducing local community vulnerability to climate variability in two smallholder-dominated rural landscapes in Indonesia, where households experience floods, drought and diseases outbreaks. In particular, we assessed the roles that forests and trees play in helping communities reduce their exposure and sensitivity, and increase their adaptive capacity to climate variability and decrease disaster risks (IPCC 2014). The chapter is organized into five sections: research background, study sites and methodology, results, discussion and conclusion. The results section is divided into three parts: (i) a description of the exposure to climatic variability and their impacts on local people's lives, (ii) the sensitivity of the socio-ecological systems, and (iii) household response strategies. At the end of each section, we focus on the results related to forests and trees. The discussion focuses on: (i) the role of forests and trees in reducing the potential impact of disasters (exposure and sensitivity) and (ii) their role in strengthening local people's response strategies (adaptive capacities).

23.2   Methods

23.2.1   Study Sites and Selection Criteria

Our four study sites were located in the provinces of West Kalimantan and Central Java (see Fig. 23.1). Criteria for site selection encompassed the communities' exposure to recent severe weather events and a diversity of forest conditions (low to high levels of degradation) and population density (low to high levels). In West Kalimantan, we selected two villages (Nanga Jemah and Tubang Jaya) characterized by low population density and low forest degradation compared to the two villages (Selopuro and Sendangsari) in Central Java. We also chose the villages according to their vegetation cover to allow further comparison (Table 23.1).

In West Kalimantan, the villages of Nanga Jemah and Tubang Jaya, in Boyan Tanjung Sub-district, Kapuas Hulu District, are located on the banks of the Boyan River. The Boyan River flows through the foothills (100–500 m a.s.l) of the Muller-Schwaner Mountain Range (PPSP 2013), in which most of the remaining dipterocarp forests of Kalimantan are found (MacKinnon 1996). Local livelihoods are centered on artisanal gold mining, agriculture and harvesting forest products such as the Borneo ironwood (*Eusideroxylon zwageri – belian*), and *gaharu* (from the heartwood of *Aquilaria* spp. infected by a fungi). The main agricultural crops include upland rice, maize, cassava and sweet potato. The agricultural land is dotted amongst rubber plantations, secondary forests, and natural forests, which provide additional income. Other livelihood activities include animal husbandry, fishing in rivers or growing fish in ponds, and hunting.

In Central Java, the villages of Selopuro and Sendangsari, in Batuwarno Sub-district, Wonogiri District, are located in the karst and limestone foothills of the southern part of the Thousand Mountains (*Pegunungan Seribu*) where the

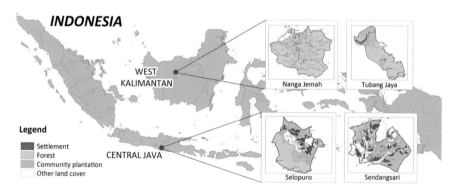

Fig. 23.1 The location of the study sites: two villages in West Kalimantan Province and two in Central Java Province. In each province one village with more forest and tree cover (Nanga Jemah in West Kalimantan and Selopuro in Central Java) and one with less (Tubang Jaya in West Kalimantan and Sendangsari in Central Java) were selected (*green* areas in the map). (Source: participatory mapping)

Table 23.1 Socio-economic and environmental characteristics of the four study villages in the provinces of West Kalimantan and Central Java

	West Kalimantan		Central Java	
	Nanga Jemah	Tubang Jaya	Selopuro	Sendangsari
Tree cover [% village territory]	98.2 %	97.5 %	75.5 %	64.2 %
Village plantations [% village territory]	4.2 %	25.6 %	39.5 %	62.0 %
Population density [households/km²]	0.7	4.1	65.3	72.6
Main livelihoods (in order of importance)	Rubber farmer, gold miner, farmer	Gold miner, rubber farmer, farmer	Farmer, cattle farmer, construction labour	Farmer, cattle farmer, farm labour
Climate-related event (in order of importance)	Flood, drought, human disease	Flood, drought, human disease	Drought, plants disease	Drought, plants disease, flood
Households affected by the most important climatic event	37 %	52 %	100 %	100 %

Bengawan Solo River originates (Surono et al. 1992). Both villages border pine monoculture forests and mixed species forest that are owned and managed by Perum Perhutani, a state-owned company. Other trees in the landscape include white albizia (*Falcataria moluccana – sengon laut*), teak (*Tectona grandis – jati*), and mahogany (*Swietenia macrophylla – mahoni*) growing on private land, along fields or on dry land (*tegalan*). In 2004, Selopuro's planted 'community forests' received the Indonesia Ecolabel Institute (LEI) certification. The main livelihoods in Selopuro and Sendangsari are in agriculture, mostly rice, corn and soybean as well as income from occasional off-farm jobs. Laborers help either in the villages during field preparations for seeding, weeding or at harvesting time, or temporarily migrate to cities to work as construction workers or merchants. Most of the population raises livestock to support their income, mostly cows and goats.

23.2.2 Research Methods

Quantitative and qualitative participatory methods were combined to gather information on interactions between the social and ecological systems that help people to adapt to the adverse effects of climate change. To guide our data collection and analysis, we used the Sustainable Livelihoods Framework, which considers five capitals (natural, physical, financial, human and social) as the basis of local livelihood choices (DFID 1999). We took a closer look at the natural capital, which include the resource stocks (e.g. land, water, or forests) as well as the

ecosystem services (e.g. soil stabilization, pest control, water regulation and purification). We complemented the framework by considering additional elements
such as 'knowledge and information', including learning from experience, 'innovative ways of thinking', 'forward looking governance structure', and 'access and
entitlements' as suggested by the Africa Climate Change Resilience Alliance (Jones
et al. 2010). According to the Sustainable Livelihoods Framework, the availability
and control of assets under constraints of policies, regulations and vulnerabilities
influence local people's ability to achieve livelihood outcomes such as food security. We broke down the concept of vulnerability in its defining three components:
exposure, sensitivity, and adaptive capacity, following the most-widely used definition from the IPCC Fourth Assessment Report (IPCC 2007) in order to tackle
these distinct aspects for a better overall understanding of the salient issues. A
reduction of climate vulnerability can be achieved by a combination of measures
that reduce the exposure and the sensitivity of social-ecological systems or enhance
their adaptive capacity, which in turn improves their resilience to climate hazards.
We used Folke's (2006) definition of resilience as the capacity of social-ecological
systems to cope with, adapt to, and retain essential structures, processes, and
feedbacks and learn to live with uncertainty and surprise (such as climate variability). We distinguished people's responses to climatic events between coping and
adaptive strategies. Coping strategies refer to short-term actions aimed at meeting
immediate needs and are always reactive, whereas adaptive strategies take into
consideration long-term perspectives and possible future changes, which can be
either reactive or anticipatory (IPCC 2012).

We conducted 24 focus group discussions (FGD) using different participatory
rural appraisal techniques, and 256 household surveys selected through stratified
random sampling. The participants in the focus group discussions were selected by
taking into consideration different areas of expertise, sources of livelihoods, and
gender, as well as geographical representation within the village. Five to seven
FGDs were conducted per village (more FGDs in the larger villages in West
Kalimantan), through which we explored the dependencies of community livelihoods on natural resources and climate as well as their interactions. The household
surveys were conducted with a representative sample size according to the equation
of Arkin and Colton (1963) at a 95 % confidence level and a ±10 % relative error
limit. The survey aimed at obtaining specific information on assets, damages and
response strategies of local people affected by the consequences of climate variability. For quantitative and qualitative analysis, we coded and categorized local
people's answers. The major themes that emerged were then analyzed in more
detail, comparing trends in percentages of people and strategies between sites.

In order to better understand climate variability at the village level, we used
satellite data of the Tropical Rainfall Measuring Mission (TRMM). The TRMM
estimations combine microwaves and infrared technologies that are calibrated
against ground based monthly rain gauge totals to produce 3-hourly precipitation
information at a spatial resolution of 0.25° latitude/longitude (or approximately
25 km). We used monthly average precipitation anomalies in order to reveal
unusual trends. Anomalies represent the deviation from the mean and were

calculated by subtracting long-term climatological monthly trends from observed data. This dataset (Huffman et al. 2010) was chosen because of its finer estimations compared to other satellite information when we checked against ground measurements in areas nearby our study villages.

23.3 Results

23.3.1 Exposure to Climate Variability and Their Impacts

Participants in our focus group discussions identified several climatic events that severely affected their productive activities or assets in the last 10 years. They suffered from multiple climate-related events such as floods, drought, and disease outbreaks (see Table 23.1). In West Kalimantan participants highlighted, among the most severe climate-related events, the recent floods of December–January (2012/13 and 2013/14), the chikungunya disease (viral disease transmitted by infected mosquitoes) of 2010 (a year's duration) as well as the droughts of 2012 and 2014. The main climatic events identified by households in Central Java were the dry periods in 2002, 2011, 2012, the plant disease outbreaks (*Patah leher* or "rotten neck" a rice leaves blast disease most likely caused by the fungus *Pyricularia oryzae*) in 2010 and 2013, as well as the heavy rains of 2008 and 2010.

The information gathered in the focus group discussions were compared with monthly precipitation anomalies calculated from TRMM satellite data (Fig. 23.2). There is a good match between perception and extreme weather events reported by local people and satellite estimates for rainfall in all study sites. In West Kalimantan, the floods local people reported corresponded with precipitation anomalies of up to +200 mm/month. In Central Java, dry periods were identified in the same year climatic data showed a below-average rainfall (around −75 mm/month). Diseases

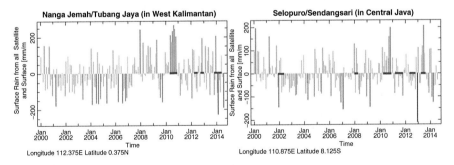

Fig. 23.2 Monthly precipitation anomalies in Nanga Jemah and Tubang Jaya, West Kalimantan and Selopuro and Sendangsari in Central Java. The *darker* the color the higher the precipitation anomaly in ± mm/month compared to the average. The *red* thick bars on the X-axis indicate the occurrence of climate-related events as identified by the communities (see text) (Source: TRMM 3B-42 ver. 6, Huffman et al. (2010))

that affected humans (caused by a vector-borne virus) and crops (due to a rice fungal pathogen) have a good overlap with particularly wet periods as estimated by satellite data. This could be explained by the fact that both these kinds of human and rice plant diseases spread easily in a wet environment (Ditsuwan et al. 2011 and Iglesias and Rosenzweig 2007 respectively).

According to the villagers, their primary difficulty in preparing for future climate-related risks was the increased unpredictability of weather in the past few years, with noticeable changes in precipitation intensity (Sendangsari), frequency (Tubang Jaya), or both (Nanga Jemah and Selopuro). No remarkable change in the seasonality of livelihood activities compared to 10 years ago was identified by farmers, who preferred to continue following traditional practices. However, time shifts and adjustments for some agricultural practices were reported by several farmers. For example, in West Kalimantan slash and burn for cultivating upland rice, spraying herbicides and picking fruit all had to be delayed due to rain. Villagers in West Kalimantan also indicated that they were still using traditional practices to predict seasonal changes based on their observations of natural phenomena such as the flowering of fruit trees, insect behaviors and cloud shapes. Approximately 38 % of people in Nanga Jemah and 15 % in Tubang Jaya were aware of imminent floods or drought using traditional knowledge. However, according to some villagers, traditional predictions have now become less reliable. Regardless of the source, external or traditional knowledge, the majority of people (around 60 %) thought there was insufficient information available, especially in more remote areas such as the villages in West Kalimantan.

People affected by drought reported losses for on- and off-farm activities, as well as changes in food, water and health conditions (Table 23.2). Most of the impacts caused by drought were related to a decrease in quantity and quality of products harvested compared to the normal situation, either because of damage (farm activities) or impaired access (off-farm). The impacts that caused the biggest loss in well-being were, in order of importance: decreases in agricultural production in all locations (66 % of people in West Kalimantan and 90 % in Central Java), followed by clean water access in West Kalimantan (28–46 %) and higher food prices in Central Java (39–45 %). Maize and vegetables were the cultivated lands most severely impacted by drought in terms of losses in productivity, followed by rice whose yields were halved in all villages. In West Kalimantan, transportation was severely disrupted due to low water level in the river, which subsequently effected Non-Timber Forest Products (NTFPs) harvests. Several other activities that depended on water also were discontinued, causing the loss of job opportunities for workers in gold mining, farming or construction in all locations.

Harvest failures can also affect demand and supply and thus influence market prices. Fewer households reported an increase in food prices in West Kalimantan (less than 10 %) compared to Central Java (around 40 %), which could imply more people in Central Java were not able to cover their needs from their own production and had to buy extra supplies. At the same time, only people in Central Java, especially the village with less forest, suffered from food shortages. This could indicate that alternative sources of food are less abundant in places with less forest.

Table 23.2 Impacts of drought on main livelihood activities, water and food in West Kalimantan and Central Java

Impact on resources/activities/ people	West Kalimantan						Central Java					
	Nanga Jemah			Tubang Jaya			Selopuro			Sendangsari		
	% people n = 50	Quantity	Quality[a]	% people n = 50	Quantity	Quality	% people n = 78	Quantity	Quality	% people n = 77	Quantity	Quality
Agriculture	66			66			91			88		
Rice	54	−52 %	(—)	28	−55 %	(—)	70	−58 %	(—)	84	−56 %	(—)
Maize	8	−75 %	=	2	−50 %	(—)	47	−77 %	(—)	13	−61 %	(—)
Vegetables	48	−74 %	(—)	30	−70 %	(———)	1	−40 %	(—)	2	−75 %	(—)
Forest related												
Timber	21	−69 %		4	−63 %		0	0 %		0	0 %	
NTFPs	20	−88 %		22	−69 %		0	0 %		0	0 %	
Rubber	74	−42 %		86	−33 %		N/A	N/A		N/A	N/A	
Off farm												
Gold mining	8	−50 days		34	−38 days		N/A	N/A		N/A	N/A	
Labor	N/A	N/A		N/A	N/A		20	−3 months		14	−4 months	
Health												
Unsafe water	28	30 days		46	34 days		16	93 days		27	93 days	
Sickeness	22	3 days		14	3 days		3	1 day		6	2 days	
Food shortages	0	0 days		0	0 days		4	0 days		16	30 days	
Food prices	8	+0.56 USD[b]		2	0 USD		39	+0.11 USD		45	+0.12 USD	

Note: The percentages indicate the proportion of people who experienced such consequences and the average change in quantity and quality compared to normal times

[a]Quality changes: = no decrease nor increase, (−/+) slight decrease/increase,, (− −/++) moderate decrease/increase, (− − −/+++) major decrease/increase. N/A no household undertaking these activities

[b]USD rate on 28 November 2014 was USD 1 = IDR 12,197.

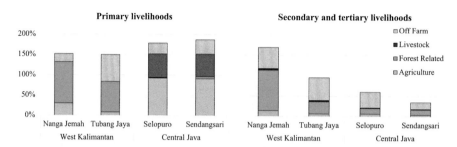

Fig. 23.3 Diversification of livelihoods in the study sites in West Kalimantan and Central Java according to the percentage of people involved (values more than 100 % because of multiple activities) (Source: Data from household survey (Nanga Jemah and Tubang Jaya N = 50; Selopuro N = 79; Sendangsari N = 77)). Note: For more details on the activities included in each livelihood category see Table 23.2

23.3.2   Livelihoods and Their Sensitivity to Climate Variability

The communities in the four study villages were mostly rural smallholder farmers characterized by their diversity of livelihood sources and dependency on natural resources. Several household decisions on productive activities were taken according to weather conditions, relying on favorable temperature and rainfall for agriculture or forest related activities. Such dependencies demonstrate the tight relationships between the social and natural systems in these landscapes.

In West Kalimantan, most of the households in the two villages used forests and trees for their livelihoods (lumbermen, rubber farmers and NTFP collectors), while in Central Java the majority were involved in agriculture and animal husbandry. The respondents identified both agriculture and forest related activities as being sensitive to climate variability. In addition to their main source of livelihood, people in all study villages had a range of activities to supplement their income. In Central Java, they were mostly off-farm such as construction work, temporary migration, and animal husbandry. In West Kalimantan, forest related works include cutting and transporting trees and collecting NTFPs (e.g. rubber tapping, birds and mammals, *gaharu* or agarwood). Interestingly, the diversification of livelihoods decreased with decreasing forest and tree cover (see Fig. 23.3).

Although the people interviewed were generally not able to elaborate on the reasons why disasters were happening, around one quarter of the affected households (and in Tubang Jaya more than half) linked the occurrence of disasters with environmental conditions, in particular environmental degradation. In all study sites, trees and forests were highly valued for decreasing the impact of extreme weather events; they were considered 'very important' or 'important' in helping to prevent severe drought and floods. There was a gradual increase in the recognition of these benefits as vegetation cover decreased (from 41 % in Nanga Jemah, 55 % in Tubang Jaya, 75 % in Selopuro to 82 % in Sendangsari).

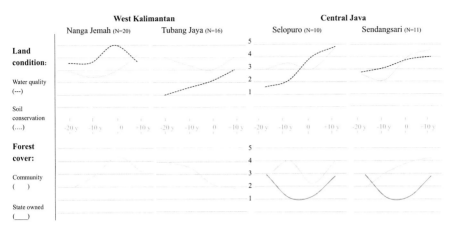

Fig. 23.4 Local perceptions of water and soil quality, and changes in forest conditions over time from 20 years ago to 10 years in the future (Source: Focus group discussions where participants scored their satisfaction with the condition of water, soil and forests on a scale (Y axis) from "very satisfied" (5) to "very unsatisfied" (1))

Villagers in all locations also associated water issues with environmental conditions. For example, in Central Java, villagers recalled that water sources started to decrease 15 years ago when semi-natural forests were replaced with a pine monoculture plantation. Villagers' satisfaction with water-related regulating services such as water quality and soil conservation, followed similar trends as forest cover (see Fig. 23.4). They perceived water quality corresponded with changes in forest cover. In one case, the village with the least forest in West Kalimantan, said the increase in their water quality was not related to a change in forest cover, but to the construction of water wells supported by the government. In Central Java water availability increased due to a similar program. This highlights the need to consider technological innovation when assessing sensitivity to climate variability.

While community dependence on environmental factors can increase their sensitivity to climatic variability, forests and trees offer several opportunities to reduce the associated risks. Communities plan according to landscape characteristics to reduce the risk of being severely affected by disasters. For example, the risks of floods were considered in the selection of new locations for housing as well as for productive activities, when opening forests for rubber plantations, agriculture, or building new fish ponds. Similar concerns were taken into account when building new houses or making renovations. People decided on the locations and the height of the house poles based on their experience of the highest water level previously reached and predictions. In both sites in West Kalimantan, entire hamlets relocated to safer places further away from the river to avoid flooding. In the last 20 years Tubang Jaya moved four times and Nanga Jemah once. In Central Java some agricultural fields were abandoned or converted to other land use to avoid wildlife (monkeys and boars) damage to crops and low productivity, and also because of forest expansion and reduced human capital (aging population and migration). This was mostly dry land near forest margins that were cultivated once, or occasionally twice, a year with red rice (an early maturing species that is more drought resistant), corn, soybeans, cassava, and trees on the edges.

People in all study locations used trees in order to protect or restore watershed services and reduce potential future impacts. Slopes were stabilized against erosion by planting trees on the hills surrounding the villages and by building terraces with the help of government programs in Central Java (1973–1975). In West Kalimantan, to reduce riverbank erosion, villagers planted and maintained durian and other trees, coconut and palms, along the river. Formal and informal regulations were also established at the village level to ban logging and maintain trees in strategic locations such as hilltops or along rivers.

In Central Java in the late 70s, a farmer planted teak on his land with such success that the practice spread. The farmer explained that he started because of the better opportunities for this type of land. Trees required less attention compared to crops especially in such dry areas. Trees also offered more flexibility since they can be used whenever needed, and are more profitable in the market. In Selopuro, there is now an organized group with official representatives in each sub-village, and together they have agreed on regulations governing the management of trees in their area. Currently, all hamlets have planted teak, mahogany, and white albizia in their gardens. Not only have these trees provided alternative incomes, but they have also helped bring water to the surface. Households in the surroundings now no longer experience as severe shortages of clean water during dry seasons as before. According to some farmers, they were also able to extend the planting season and share the water among multiple users.

23.3.3 Adaptive Strategies in Response to Climate Variability

Households in the study villages have been experiencing the impacts of climate variability and have been devising a variety of strategies to respond (Table 23.3). For response strategies, we do not distinguish between villages with different levels of tree and forest cover as the difference in households' numbers was only a maximum of ±7 %. On average the main climatic event in each village resulted in around four strategies adopted per household (±0.1) for responses to floods in West Kalimantan and drought in Central Java, while for secondary events (i.e. drought in West Kalimantan and plant diseases in Central Java), there were 1.8 strategies adopted per household (±0.1 depending on the village). Nanga Jemah had a slightly lower average than the others (1.3 strategy/household). The strategies were categorized according to the livelihood capital the household used to overcome the difficulty (means), which should not be confused with sectors affected by climatic stress (target). We also considered the level at which they are implemented: actions that are taken at the individual or household level (spontaneously or autonomously) and those that were taken more collectively at the village level (often government/policy supported practices).

Table 23.3 Summary of household response strategies to drought and floods in the study sites in West Kalimantan and Central Java. The numbers indicate the percentage of the total activities adopted and are only those used by more than 5 % of households

Capital			Response strategies to climate variability			
			West Kalimantan (Flood N = 367; drought N = 160)		**Central Java** (Flood N = 9; drought N = 620)	
Natural	**Drought**	9 %	Water seeds and crops	20 %	Pump or drain ground-water for agriculture	
		5 %	Store clean water or find alternative sources (e.g. river) for household consumption	16 %	Diversification of crop and species	
				6 %	Selling yields, livestock, timber	
				15 %	Substitute livestock fodder for leaves	
					Expand crops in areas near water	
					Plant trees and use fuel wood	
	Flood		Relocate house/crops to higher places		Plant trees to avoid erosion	
			Plant trees to avoid landslide			
			Store clean water			
			Change seed variety;			
			Harvest forest products (timber, NTFPs)			
Physical	**Drought**	20 %	Fertilize rubber trees	7 %	Stem river or build water channels	
		7 %	Use pesticides and fertilizer on crops		Dig a well or pipe water to the house	
					Use irrigation system for crops	
	Flood	12 %	Moving assets to higher place			
		12 %	Install net in fish pond			
		6 %	Change equipment in rubber			
			Collection or in wood transportation			
			Elevate house			
Social	**Drought**		Temproary shelter with family/neighbors		Clean water assistance from government	
	Flood		Provide help in cleaning/ recovery		Consult government agricultural experts	

(continued)

Table 23.3 (continued)

Capital			Response strategies to climate variability		
			West Kalimantan (Flood N = 367; drought N = 160)		**Central Java** (Flood N = 9; drought N = 620)
Financial	**Drought**		Save money in preparation		Buy drinking water and rice
					Buy gasoline for waterpump
	Flood				Borrow money from social groups or neighbors
Human	**Drought**	9 %	Stop activities (e.g. mining, logging)	6 %	Stop or reduce farming
		4 %	Change transportation arrangements		Manage food supply and change diet
		24 %	Change rubber harvest timing		Find new job opportunities (migration)
					Change timing of planting and harvesting
	Flood	8 %	Preventive collection and storage		Change harvest timing
		5 %	Collect lost items		Cleaning the field
			Stop activties (gold mining, rubber, logging)		Planting management
			Monitor river flow		
			Maintain house and clean environment		

In West Kalimantan, during or after floods, local communities focused on the recovery of their main livelihoods and immediate needs (around 60 % of the strategies were for short term benefits). Most of their actions used physical capital or existing infrastructure to protect valuable goods, for example to secure the harvested rubber (19 %), move household assets into roof spaces or rafters (12 %), or protect fishponds with nets (12 %). In addition, few households discontinued the harvesting of rubber, cutting trees and gold mining (<5 %). In case of drought, half of the people adopted strategies related to rubber harvesting (fertilizing, reducing tapping, and changing equipment).

In Central Java in times of drought, 70 % of the strategies reported in the household surveys were for short-term benefits in response to the consequences of drought. Local people used natural capital to address issues related to water harvesting and management (20 %), substituted livestock fodder for leaves (15 %), and sold timber or fuel wood (6 %). Around 6 % of the participants adopted more long-term strategies of species diversification, and changed from paddy to other cash crops (locally known as *palawija*) or seed varieties of paddy and soybean. In

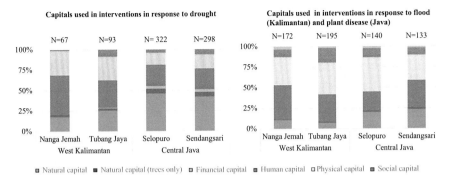

Fig. 23.5 Capital used by households to respond to drought and floods or plant diseases in the study villages in West Kalimantan and Central Java. N indicates the total number of interventions undertaken by the households interviewed. (Source: household surveys)

addition, 6 % of the local people avoided possible loss or damage due to a lack of water by stopping or reducing the number of species and/or the amount planted during the driest months, mostly soybeans and peanuts.

Although several response strategies to floods and drought were employed, trees and forests represented only 2–6 % of the total interventions used. People in places with less forest cover used more of their natural capital and trees in response to climate-related events (see Fig. 23.5).

Several collective actions that provide benefits for adaptation were found at the village or sub-village level, which typically involved shared means (people or land). Forests are ideal for collective actions as they are often held under communal or state tenure. Their resources are available due to free access or traditional regulations. In the four sites, a common response strategy was to extend agriculture or collection of natural resources wherever possible, especially in communal or more risky areas (insecure use rights or exposed to extreme weather). For example, farmers started planting on riverbanks (Nanga Jemah, Tubang Jaya and Sendangsari), at the edge of water reservoirs as the level decreased (in Selopuro) or on hilltops (in villages in West Kalimantan). In Central Java, when certain resources become limited due to drought, people entered the perimeters of the Perum Perhutani to collect leaves, fuel wood, and grass, and to use water resources.

In the focus group discussions, the participants identified several local rules aimed at the sustainable use of forest resources and to maintain vegetation cover. For example, in the Central Java community forest, for each tree felled, 10 must be planted if space and conditions allow. In West Kalimantan, in Nanga Jemah no more than three trees may be felled at the same time, while in Tubang Jaya a village rule bans the felling of 'primary' forest for cultivation and prescribes the use of secondary regrowth instead. Government supported programs for raising awareness and technical advice have been implemented, such as preventive interventions (terraces, wells and water harvesting systems, and reforestation) and household assistance (seeds, rice, and water tanks).

23.4   Discussion

23.4.1   Using Trees to Reduce Exposure and Sensitivity

The four study sites in West Kalimantan and Central Java have all experienced intra-seasonal variations in precipitation. Vogel (2000) wrote that rainfall has been regarded as the most significant climate parameter affecting human activities. In Indonesia, agricultural production is strongly influenced by annual and inter-annual variations in precipitation, where the Austral-Asia monsoon and El Nino-Southern Oscillation (ENSO) dynamics play an important role (Naylor et al. 2007). Having regularly experienced drought, floods and disease outbreaks, the people interviewed were well aware of the climatic variability and associated risks. Other studies have found that farmers recognize subtle changes in climate (Maddison 2007; Thomas et al. 2007; Gbetibouo 2008; Kalinda 2011; Amdu et al. 2013; Boissière et al. 2013). Household experiences of recent climatic variability showed clear agreement with satellite estimations of anomalies.

Farmers decide when and what crop varieties to plant based on their prediction of precipitation trying to reduce the risk of crop failure by diversifying income opportunities. Fluctuations in precipitation and temperatures can quickly lead to shorter or more unpredictable periods during which the risks of losses increase dramatically. Because the study villages are predominantly dependent on rainfed agriculture or forest products for their livelihoods, which are highly prone to damage due to climate-related events, they can be defined to be climate-sensitive resource dependent (Adger 1999). In addition, some farmers decided to cultivate their land even in unfavorable conditions, often leading to low crop yields or harvest failure during the driest months. As a result, local communities are pursuing a range of livelihood activities to spread the risk associated with crop losses.

Livelihood diversification helps reducing vulnerability, unless several activities are affected by climate variability. Expanding livelihoods opportunities with less climate-sensitive activities mitigate climate-related risks by helping before (*ex-ante*) or to cope later (*ex-post*) (Godoy et al. 1998; Lanjouw 1999; Adger 2006). Both strategies eventually help families to smooth income fluctuations given the seasonality of agricultural production (Kant et al. 1996; Paavola 2008). Villages with more abundant vegetation had a larger range of income opportunities. However, even though forests and trees contribute to broadening adaptation options, they can be themselves affected by climatic events (i.e. prevent access to forest resources or impairing the delivery of ecosystem services), putting people who rely heavily on forest resources more at risk. For example, some of the new livelihoods, such as harvesting rubber, raising livestock or fish, gold mining or farm labor, also remain sensitive to climate variability. Therefore, promoting diversified livelihoods should focus on alternatives that are less climate dependent, especially in areas where people's activities are based on natural resources, which although diversified could still be sensitive to climate variability.

All households affected by climatic events implicitly or explicitly recognized the importance of natural capital, including forests and trees, in regulating the intensity of natural disturbances. This is central to adaptation as it enables communities to actively use their natural capital. They can then take advantage of the services delivered by ecosystems and their physical protection together with geographic features in reducing climate-related risks. Past experience and future projections were part of the rationale in the selection of locations for productive activities or housing settlements. This is particularly valid for West Kalimantan, probably due to the nature of the main disaster (flood), but also due to the fewer constraints in land availability compared to Central Java. Land availability and financial resources are, however, the main reasons for delaying or not taking action. In these cases, such activities were simply discontinued. Other studies on perceptions of climate variability point out that farmers are more likely to adapt if they can perceive the changes in the climate (Maddison 2007; Simelton et al. 2013).

High awareness on the linkages between the effects of climate variability and environment conditions helped communities not only to locate their property and economic activities in less risky places, but also allowed them to actively reduce future impacts through landscape interventions. Exploiting spatial diversity in the landscape to improve livelihood outcomes has been seen as a possible strategy that people can use to spread the risks associated with climate variability (Eakin 2000). In fact, several communities tried to maintain or enhance land characteristics of interest in strategic places, such as trees on hilltops or along rivers to prevent erosion and regulate water run-off, when considering climate-related risks. These interventions were mostly collective and often involved formal or informal regulations. Interestingly, in places with less forested areas, people's strategies to respond to climate-related risks were more based on planting trees or harvesting tree parts. This makes ecosystem restoration and reforestation plans that are already recognized and accepted locally viable options. In these cases, supporting existing collective efforts and organizations would help communities reach a scale that provides visible benefits and ensures continuity.

People took action to protect, increase or manage trees with particular attention to the positive benefits of maintaining and regulating water availability, as shown by the household survey. This is the case for new teak plantation in Central Java where most environments are already degraded, as well as in more pristine forests of West Kalimantan, where existing tree cover is kept in specific areas. Villagers' considerations are in line with studies that recognize the important role of forested landscapes in regulating watershed processes (Pattanayak and Kramer 2001; Ilstedt et al. 2007; Rodríguez Osuna et al. 2014). On the other hand, findings from Bosch and Hewlet (1982) highlighted that the afforestation of former grassland with pine not only reduces annual stream flow but also reduces the dry season flow, which can decrease water availability for agricultural purposes. Vincent et al. (1995) estimated that an increase in coniferous species could proportionally reduce annual water yields. Furthermore, deciduous tree species (e.g. teak in our study site) were found to typically generate less evapotranspiration than evergreen and thus help diminishing negative effects on the water balance (Ellison et al. 2012; Wattenbach et al. 2007). The findings highlight the importance of selecting appropriate species and locations when planning changes in tree cover over large areas, as well as the role that local experiences related to land management can play in informing such initiatives. In this way, it is possible to provide additional adaptation benefits while minimizing unwanted side-effects in surrounding areas.

23.4.2　Using Trees to Strengthen Response Strategies and Adaptive Capacities

Few families adopted strategies based on the use of forests and trees to respond to climatic shocks, as shown by the household surveys. However, several response strategies targeting forest related activities are used where livelihoods depend on them as shown in West Kalimantan. These strategies mostly involve changes in land management practices. In Central Java, local people use tree leaves as fodder and sell timber, but more as a last resort after having sold other assets such as livestock. The limited use of forest products is in partial contrast with the forest safety nets or their natural insurance role that has been observed elsewhere (floods in East Kalimantan, Indonesia: Liswanti et al. 2011; floods and diseases in Peru: Takasaki et al. 2004; floods and drought in Malawi: Fisher et al. 2010; storms, flooding and plant and animal diseases in Vietnam: Völker and Waibel 2010). At the same time, however, in other places an increase in the use of forest products, because of climatic events, was not observed (floods and drought in Papua, Indonesia: Boissière et al. 2013; environmental shocks around the world: Wunder et al. 2014; hurricane in Honduras: McSweeney 2004).

Provisioning services of forests are often used in reactive (*ex-post*) strategies, whereas anticipatory (*ex-ante*) strategies rely more on regulating services. We argue that the importance of trees and forests for reducing human vulnerability can be described more clearly by specifying the type of ecosystem service provided in relation to the particular phase of the climatic hazard, whether before or after the impacts of the climatic events materialize (i.e. phases of disaster risk management). People in the study villages valued regulating services from forested ecosystems for their function in preventing or reducing possible impact caused by climatic hazards, especially for their role in regulating water and soil processes. The provisioning services of forests were mostly used in reactive strategies after the occurrence of a climatic hazard. People harvested trees' parts to substitute sources of income or food (e.g. they sold timber or firewood, or used leaves as fodder for animals). In addition, distinguishing the type of ecosystem services and people's response strategies would help ensure that the full potential of forested ecosystems is accounted for when comparing and selecting the most cost-effective options to reduce climate-related vulnerabilities.

There could be some methodological caveats that underestimate the role of forests. People who live near forests, and utilize them regularly, might not consider unusual to undertake additional forest activities in relation to a climatic event and thus such activities may go unreported. Another explanation could be that regulating services of ecosystems were not specifically taken into account in previous research or the focus was on reactive strategies. Furthermore, several forest products require time to harvest and process before being used. Their harvest access can also be interrupted due to the climatic event, and therefore less suited in case of urgent need. Their benefits might also be evident later on. Moreover, it remains challenging for researchers and communities alike to clearly identify and quantify these benefits. This is probably related to the intrinsic differences in the services; regulating services are more abstract and easier to demonstrate qualitatively, do not immediately display changes in use, and provide collective benefits, but do not require direct access in order for people to benefit. On the contrary, provisioning services are more tangible, easier to measure quantitatively, stocks are depleted by use, and usually specific individuals who control the resources gain the benefits.

23.5   Conclusion

This chapter revealed that smallholders in the four communities in Indonesia living in areas with different vegetation covers and changes are actively engaging in several strategies related to the use of forests and trees to respond to the adverse impacts of climate variability such as drought, floods and disease outbreaks. However, these strategies constitute a limited contribution to overall vulnerability reduction when considered alongside the variety of measures taken by the study communities. Most of the people responded to the climatic hazards adopting technological solutions (e.g. pumping water, developing irrigation systems and protective systems, and changing to more modern equipment), increasing the use of agricultural inputs (fertilizers, pesticides, and seeds varieties), adjusting agricultural management practices and crop species rotation, and seeking external help through social networks or government agencies. The role of forests and trees is particularly important as part of community *ex-ante* strategies to better prepare for and reduce potential damages (i.e. decrease exposure and sensitivity). However, the role as a coping and recovery mechanism is more limited and few people rely on forests and trees during or immediately after hardship situations by using forest products (e.g. selling timber or NTFPs).

Local communities living in areas with fewer trees, which tend to have a more degraded environment and be closer to the ecosystem thresholds for sustaining ecological functions, have experienced more changes and seem to value more and be more involved in managing the remaining vegetation. In these areas, in order to still be able to benefit from ecosystem services, especially those related to water regulation and provision and soil stabilization, people have to actively influence their natural capital. On the other hand, in areas with more preserved forests people can benefit more passively without having to develop particular actions that affect them. In addition, in villages with more forests, livelihoods are more diversified, suggesting that they have more available alternatives to replace a temporary loss of income due to climatic events. Nevertheless, several natural resource dependent activities are also highly sensitive to climatic variability making them a double-edged sword. Therefore, in areas where people's activities depend on natural resources, efforts to promote livelihood diversification should focus on alternatives that are less climate-sensitive in order to mitigate climate-related risks.

In future research or development interventions, it is crucial to understand the complex linkages between forest cover and human vulnerability by considering the whole local context and temporal dimension. There is a need to explicitly distinguish the support of forests according to the timing, *ex-post* vs. *ex-ante* adaptation respectively, and the nature of the service, regulating vs. provisioning services. This would help take into account the full benefits provided by forested ecosystems, in particular for reducing and mitigating climate-related risks through water regulation and provision, and soil stabilization. The role of ecosystems regulating services is not always fully taken into account or could be easily underestimated when comparing possible adaptation interventions. Nonetheless, it is an essential part of the safety net function of forests. In addition, other factors greatly influence community vulnerability, such as alternatives related to other capitals including technological development, the awareness and experience with the event, the ecosystems' conditions, in particular threshold effects of tree cover degradation on ecosystem services. Furthermore, people's lives are impacted by multiple and interconnected disturbances that can be slow or sudden in nature, such as subtle shifts in climate or extreme weather events. However, focusing on the effects of and common solutions to climate variability (such as ecosystems management) rather than the differences in time-frame of their occurrence (long/short or sudden/gradual impacts) would help the development of comprehensive strategies to reduce people's vulnerability and increase their resilience that span across sectors and disciplines (e.g. disaster risk reduction and climate change adaptation).

Community awareness regarding climate variability and environmental degradation is crucial. If these linkages are recognized, it encourages people to actively manage their environment and natural resources, which could be an entry point for ecosystem-based interventions. Furthermore, for adaptation it would help to identify priority spots where there is a strong demand from local users for ecosystem services that can support the reduction of climate-related risks. Favorable spatial land characteristics that influence ecosystem services relevant for strengthening people's adaptation, especially regarding the regulation of water and soil processes, should be identified and carefully evaluated, and future changes planned together with local communities. These are prerequisites that make ecosystem services a valuable option for an integrated strategy to reduce disaster risk and climate-related vulnerabilities that suit existing community contexts and needs and thus are more likely to be successful.

Acknowledgement The authors would like to thank the local communities of Nanga Jemah, Tubang Jaya, Selopuro, Sendangsari for their participation in the research activities. We also thank Nyimas Wardah for her help in the study preparations and the fieldwork in West Kalimantan and Central Java; Siti Nurika Sulistiani and Tutup Kuncoro for their help in Central Java data collection; Serge Rafanoharana for his support in producing the maps; Glen Mulcahy, two anonymous reviewers, and the book editors for their valuable comments and reviews. This research was carried out by the Center for International Forestry Research (CIFOR) as part of the CGIAR Research Program on Forests, Trees and Agroforestry. It received financial support from the Australian Agency for International Development (AusAID) under the Agreement 63650.

References

Adger N (1999) Social vulnerability to climate change and extremes in coastal Vietnam. World Dev 27(2):249–269

Adger N (2006) Vulnerability. Glob Environ Chang 16:268–281. doi:10.1016/j.gloenvcha.2006.02.006

Amdu B, Ayehu A, Deressa A (2013) Farmers' perception and adaptive capacity to climate change and variability in the upper catchment of Blue Nile, Ethiopia. African Technology Policy Studies Network, ATPS working paper no 77

Andrade Pérez A, Fernández Herrera B, Cazzolla Gatti R (eds) (2010) Building resilience to climate change: ecosystem-based adaptation and lessons from the field. IUCN, Gland, 164 pp

Angelsen A, Wunder S (2003) Exploring the forest – poverty link: key concepts, issues and research implications. CIFOR occasional paper 40

Arkin H, Colton R (1963) Table for statistics. Barnes and Noble Publication: Barnes & Noble, New York

Boissière M, Locatelli B, Sheil D et al (2013) Local perceptions of climate variability and change in tropical forests of Papua, Indonesia. Ecol Soc 18(4):13. http://dx.doi.org/10.5751/ES-05822-180413

Bosch JM, Hewlet JD (1982) A review of cacthment experiments to determine the effect of vegetation changes on water yield and evapotranspiration. J Hydrol 55:3–23

CBD- Secretariat of the Convention on Biological Diversity (2000) Report of the fifth meeting of the subsidiary body on scientific, technical and technological advice, Kenya

CBD- Secretariat of the Convention on Biological Diversity (2009) Connecting biodiversity and climate change mitigation and adaptation: report of the second ad hoc technical expert group on biodiversity and climate change. Technical series no. 41, Montreal

Colls A, Ash N, Ikkala N (2009) Ecosystem-based adaptation: a natural response to climate change. IUCN, Gland

DFID- Department for International Development of the UK (1999) Sustainable Livelihoods Guidance Sheets, Numbers 1–8. Department for International Development, London. Also available on www.livelihoods.org

Ditsuwan T, Liabsuetrakul T, Chongsuvivatwong V et al (2011) Assessing the spreading patterns of dengue infection and chikungunya fever outbreaks in lower southern Thailand using a geographic information system. Ann Epidemiol 21(4):253–261

Doswald N, Osti M (2011) Ecosystem-based approaches to adaptation and mitigation: good practice examples and lessons learned in Europe. The BfN-Skripten, Bonn

Eakin H (2000) Smallholder maize production and climatic risk: a case study from Mexico. Clim Chang 45:19–36

Ellison D, Futter N, Bishop K (2012) On the forest cover-water yield debate: from demand to supply-side thinking. Glob Chang Biol 18:806–820. doi:10.1111/j.1365-2486.2011.02589.x

EM-DAT International Disaster Database (2013) Université Catholique de Louvain, Brussels, Belgium. www.emdat.be. Accessed 28 Oct 2014

EU- Commission of the Europian Communities (2009) Adapting to climate change: towards a European framework for action. Commission of the European Communities, Brussels

FAO (2010) Global forest resources assessment 2010. Food and Agriculture Organization of the United Nations', Rome

Fisher M, Chaudhury M, McCusker B (2010) Do forests help rural households adapt to climate variability? Evidence from southern Malawi. World Dev 38(9):1241–1250. doi:10.1016/j.worlddev.2010.03.005

Folke C (2006) Resilience: the emergence of a perspective for social–ecological systems analyses. Glob Environ Chang 16:253–267

Gbetibouo GA (2008) How can African agriculture adapt to climate change? Insights from Ethiopia and South Africa: understanding farmers' perceptions and adaptations to climate change and variability. The case of the Limpopo Basin, South Africa. IFPRI Research Brief 15–8

GIZ, Göhler D, Müller F, Mytanz C et al (2013) Ecosystem-based Adaptation (EbA). In: Environment and climate change. Deutsche Gesellschaft für. Internationale Zusammenarbeit (GIZ) GmbH

Godoy R, Jacobson M, Wilkie D (1998) Strategies of rain-forest dwellers against misfortunes: the Tsimane' Indians of Bolivia. Ethnology 37(1):55–69

Gunderson LH, Holling CS (2002) Panarchy: understanding transformations in human and natural systems. Island Press, Washington, DC

Harvey CA, Chacón M, Donatti CI et al (2013) Climate-smart landscapes: opportunities and challenges for integrating adaptation and mitigation in tropical agriculture. Conserv Lett 7 (2):77–90

Heath M, Phillips J, Munroe R et al (eds) (2009) Partners with nature: how healthy ecosystems are helping the world's most vulnerable adapt to climate change. BirdLife International, Cambridge, UK

Huffman GJ, Adler RF, Bolvin DT et al (2010) The TRMM Multi-satellite Precipitation Analysis (TMPA). In: Hossain F, Gebremichael M (eds) Satellite rainfall applications for surface hydrology. Springer, Dordrecht. ISBN: 978-90-481-2914-0, p 3–22

Iglesias A, Rosenzweig C (2007) Climate and pest outbreaks. Encycl Pest Manag 2:87–89

Ilstedt U, Malmer A, Verbeeten E et al (2007) The effect of afforestation on water infiltration in the tropics: a systematic review and meta-analysis. For Ecol Manag 251:45–51. doi:10.1016/j.foreco.2007.06.014

IPCC (2007) Climate Change 2007: impacts, adaptation and vulnerability. contribution of working group II to the fourth assessment report of the intergovernmental panel on climate change. Parry ML, Canziani OF, Palutikof JP, van der Linden PJ, Hanson CE (eds). Cambridge University Press, Cambridge, UK

IPCC (2012) Summary for policymakers. In: Managing the risks of extreme events and disasters to advance climate change adaptation. A special report of working groups I and II of the intergovernmental panel on climate change. Field CB, Barros V, Stocker TF et al (eds) Cambridge University Press, Cambridge, UK/New York

IPCC (2014) Climate change 2014: impacts, adaptation, and vulnerability. contribution of working group II to the fifth assessment report of the intergovernmental panel on cimate change. Field CB, Barros VR, Mach KJ et al Cambridge University Press, New York

IUFRO (2009) Adaptation of forest and people to climate change – a global assessment report. Seppälä R, Buck A, Katila P (eds) IUFRO world series vol 22, Helsinki

Jones L, Ludi E, Levine S (2010) Towards a characterisation of adaptive capacity: a framework for analysing adaptive capacity at the local level. Overseas Development Institute. Background Note, December 2010

Jones HP, Hole DG, Zavaleta ES (2012) Harnessing nature to help people adapt to climate change. Nat Clim Chang 2:504–509

Kalinda T (2011) Multiple shocks and risk management strategies among rural households in Zambia's Mazabuka District. J Sustain Dev 7(5):52–67. doi:10.5539/jsd.v7n5p52

Kant S, Nautiyal JC, Berry RA et al (1996) Forest and economic welfare. J Econ Stud 23(2):31–43

Lanjouw P (1999) Rural nonagricultural employment and poverty in Ecuador. Econ Dev Cult Chang 48(1):91–122

Liswanti N, Sheil D, Basuki I et al (2011) Falling back on forests: how forest-dwelling people cope with catastrophe in a changing landscape. Int For Rev 13(4):442–455

MacKinnon K (1996) The ecology of Kalimantan. Oxford University Press, New York

Maddison DJ (2007) The perception of and adaptation to climate change in Africa. Research working paper of World Bank Policy, no 4308

McSweeney K (2004) Natural insurance, forest access, and compounded misfortune: forest resources in smallholder coping strategies before and after Hurricane Mitch, northeastern Honduras. World Dev 33(9):1453–1471. doi:10.1016/j.worlddev.2004.10.008

Naylor RL, Battisti DS, Vimont DJ et al (2007) Assessing risks of climate variability and climate change for Indonesian rice agriculture. Proc Natl Acad Sci 104(19):7752–7757. doi:10.1073/pnas.0701825104

Paavola J (2008) Livelihoods, vulnerability and adaptation to climate change in Morogoro, Tanzania. Environ Sci Pol 11:642–654. doi:10.1016/j.envsci.2008.06.002

Pattanayak SK, Kramer RA (2001) Worth of watersheds: a producer surplus approach for valuing drought mitigation in eastern Indonesia. Environ Dev Econ 6:123–146

Percepatan Pembangunan Sanitasi Pembangunan [PPSP] (2013) Buku Putih Sanitasi. Kabupaten Kapuas Hulu, Indonesia

Pramova E, Locatelli B, Brockhaus M et al (2012) Ecosystem services in the national adaptation programmes of action. Clim Pol 12(4):393–409. http://dx.doi.org/10.1080/14693062.2011.647848

Raudsepp-Hearne C, Peterson GD, Tengö M et al (2010) Untangling the environmentalist's paradox: why is human well-being increasing as ecosystem services degrade? Bioscience 60:576–589

Roberts D, Boon R, Diederichs N et al (2011) Exploring ecosystem-based adaptation in Durban, South Africa: "learning-by-doing" at the local government coal face. Environ Urban 24 (1):167–195. doi:10.1177/0956247811431412

Rodríguez JP, Beard TD, Bennett GS (2006) Trade-offs across space, time, and ecosystem services. Ecol Soc 11(1):28

Rodríguez Osuna V, Börner J, Nehren U et al (2014) Priority areas for watershed service conservation in the Guapi-Macacu region of Rio de Janeiro, Atlantic Forest, Brazil. Ecol Process 3:16

Shackleton S, Shackleton C (2004) Everyday -resources are valuable enough for community-based natural resource management programme support: evidence from South Africa. In: Fabricius C, Koch E (eds) Rights, resources and rural development: community-based natural resource management in southern Africa. Earthscan, London

Simelton E, Quinn CH, Batisani N et al (2013) Is rainfall really changing? farmers' perceptions, meteorological data, and policy implications. Clim Dev 5(2):123–138. doi:10.1080/17565529.2012.751893

Sudmeier-Rieux K, Masundire H, Rizvi A, Rietbergen S (eds) (2006) Ecosystems, livelihoods and disasters: an integrated approach to disaster risk management. IUCN, Gland/Cambridge, UK

Surono BT, Sudarno I, Wiryosujono S (1992) Geology of the Surakarta Giritontro Quadrangles, Java. Geological Research and Development Center, Bandung

Takasaki Y, Barham B, Coomes O (2004) Risk coping strategies in tropical forests: floods, illnesses, and resource extraction. Environ Dev Econ 9:203–224. doi:10.1017/S1355770X03001232

Thomas DS, Twyman C, Osbahr H et al (2007) Adaptation to climate change and variability: farmer responses to intra-seasonal precipitation trends in South Africa. Clim Chang 83:301–322. doi:10.1007/s10584-006-9205-4

Tompkins EL, Mensah A, King L et al (2013) An investigation of the evidence of benefits from climate compatible development. Centre for Climate Change Economics and Policy. Working paper no 124. Sustainability Research Institute, University of Leeds, UK, 2013

UNEP (2012) Ecosystem-based adaptation guidance: moving from principles to practice. Travers A, Elrick C, Kay R et al (eds) United Nations Environment Programme Working Document 2012

United Nations Framework Convention on Climate Change [UNFCCC] (2013) National inventory submissions. UNFCCC, Bonn

Vincent JR, Kaosa-ard M, Worachai L et al (1995) The economics of watershed protection: a case study of Mae Taeng River, Thailand. TDRI and HIID, Bangkok/Cambridge, MA. Policy Brief

Vogel C (2000) Usable science: an assessment of long-term seasonal forecasts amongst farmers in rural areas of Sourth Africa. S Afr Geogr J 82(2):107–116

Völker M, Waibel H (2010) Do rural households extract more forest products in times of crisis? evidence from the mountainous uplands of Vietnam. Forest Policy Econ 12(6):407–414. doi:10.1016/j.forpol.2010.03.001

Wattenbach M, Zebisch M, Hattermann F et al (2007) Hydrological impact assessment of afforestation and change in tree-species composition–a regional case study for the Federal State of Brandenburg (Germany). J Hydrol 346(1):1–17

Wunder S, Börner J, Shively G (2014) Safety nets, gap filling and forests: a global-comparative perspective. World Dev 64(1):S29–S42. http://dx.doi.org/10.1016/j.worlddev.2014.03.005

Chapter 24
Defining New Pathways for Ecosystem-Based Disaster Risk Reduction and Adaptation in the Post-2015 Sustainable Development Agenda

Marisol Estrella, Fabrice G. Renaud, Karen Sudmeier-Rieux, and Udo Nehren

Abstract This chapter seeks to articulate future directions in the field of Eco-DRR/CCA, in the context of the new post-2015 sustainable development agenda. It synthesises the experiences featured in this book and highlights the key challenges and opportunities in advancing Eco-DRR/CCA approaches. Four main themes are discussed: demonstrating the economic evidence of Eco-DRR/CCA; decision-making tools for Eco-DRR/CCA; innovative institutional arrangements and policies for mainstreaming Eco-DRR/CCA; and research gaps. The major global policy agreements in 2015 are examined for their relevance in promoting Eco-DRR/CCA implementation in countries. Finally, the authors reflect on a new agenda for Eco-DRR/CCA and outline some of the key elements required to significantly advance and scale-up Eco-DRR/CCA implementation globally.

M. Estrella
Post-Conflict and Disaster Management Branch, United Nations Environment Programme, Geneva, Switzerland

F.G. Renaud (✉)
United Nations University Institute for Environment and Human Security (UNU-EHS), Platz der Vereinten Nationen 1, 53113 Bonn, Germany
e-mail: renaud@ehs.unu.edu

K. Sudmeier-Rieux
Commission on Ecosystem Management International Union for Conservation of Nature, Gland, Switzerland

U. Nehren
Institute for Technology and Resources Management in the Tropics and Subtropics (ITT), TH Köln, University of Applied Sciences, Köln, Germany

© Springer International Publishing Switzerland 2016
F.G. Renaud et al. (eds.), *Ecosystem-Based Disaster Risk Reduction and Adaptation in Practice*, Advances in Natural and Technological Hazards Research 42,
DOI 10.1007/978-3-319-43633-3_24

24.1 Introduction

As this book was being finalised, the United Nations Conference on Climate Change had just concluded in December 2015 in Paris, attended by 151 Heads of State and Government, making it the largest gathering of world leaders. As the Paris outcomes are assessed and the post-2015 era begins, we stand before a unique crossroad which could potentially reshape development on a global scale. With increasing disaster and climate change risks, growing populations and expanding economies, the imperative to chart more sustainable and resilient development pathways – globally, nationally and locally – is more critical now than ever.

This concluding chapter synthesises experiences featured in this current book and highlights the emerging challenges and opportunities in advancing ecosystem-based approaches to disaster risk reduction and climate change adaptation (Eco-DRR/CCA). What clearly emerges from the chapters in this book is the importance of tackling multiple development challenges through integrated approaches to ecosystem management, disaster risk reduction (DRR) and actions on climate change.

24.2 The Post-2015 Sustainable Development Agenda: Opportunities for Ecosystem-Based Disaster Risk Reduction and Climate Change Adaptation

In 2015, the international community reached consensus on three major global policy agreements and charted the post-2015 sustainable development agenda. The Sendai Framework for Disaster Risk Reduction (SFDRR) was adopted at the Third World Conference on Disaster Risk Reduction in Sendai, Japan, in March 2015, and was subsequently endorsed by the United Nations (UN) General Assembly in May 2015. The SFDRR provides the global framework for DRR actions for the next 15 years (2015–2030), and will in effect supersede the Hyogo Framework for Action (2005–2015). In October 2015, the UN General Assembly approved the 17 Sustainable Development Goals (SDGs), which will guide national and local development agendas until 2030 and take over from the Millennium Development Goals. The UN Convention to Combat Desertification (UNCCD) also convened its 12th Conference of the Parties (CoP) in October 2015 and reached a landmark agreement to achieve land degradation neutrality by 2030. Finally, in December 2015, the 21st CoP of the UN Framework Convention on Climate Change adopted the Paris Agreement on Climate Change and resulted in firmer commitments to reduce carbon emissions globally as well as key principles for climate change adaptation (CCA). An important thread running through all global policy agreements in 2015 is a clear recognition of the role that ecosystems play in safeguarding and enabling development gains, and in building resilience against disasters and climate change.

A number of international environmental conventions are also closely in line with the post-2015 global policy agreements recently adopted and have endorsed key decisions of major relevance to Eco-DRR/CCA. Firstly, during its 12th CoP in October 2014, the UN Convention on Biological Diversity (CBD 2014) adopted Decision XII/20 which encourages Governments and other relevant organisations to promote and implement ecosystem-based approaches to climate change related activities and DRR. Although the CBD has long championed ecosystem-based approaches to CCA, DecisionXII/20 for the first time explicitly links biodiversity conservation and ecosystem-based approaches to DRR and climate change mitigation and adaptation. Secondly, the Ramsar Convention on Wetlands of International Importance also recently adopted Resolution 13 during its 12th CoP in June 2015, which calls for integrating DRR in wetlands management. Thirdly, the World Heritage Convention (WHC 2007) agreed at its 31st session as early as 2007 to develop and implement its "Strategy for Risk Reduction at World Heritage Properties", which includes both cultural and natural heritage sites around the world. WHC has developed resource materials on integrating DRR in the management of World Heritage Sites.[1]

Other multi-lateral environmental fora have also addressed the issue of climate change and disaster risks, which further encourage countries to implement integrated ecosystem management, DRR and CCA approaches. For example, at the last World Parks Congress, convened by the International Union for the Conservation of Nature (IUCN), in Australia, in November 2014, the "Promise of Sydney" emerged as the main outcome and outlined a 10-year agenda for investing in protected areas for sustainable development and responding to global challenges, including climate change and disaster risks. With over 6000 participants from over 170 countries represented, the 2014 World Parks Congress for the first time focused on human security and DRR, with clear recommendations for integrating DRR and ecosystem management in protected areas.[2]

Countries around the world will clearly be seeking ways to implement the post-2015 global policy agreements and commitments. Table 24.1 outlines each of the major post-2015 global policy agreements as well as key decisions adopted by multi-lateral environmental agreements and their relevance for advancing implementation of Eco-DRR/CCA (see also Fig. 1.1 discussed in Chap. 1). It further describes key provisions related to Eco-DRR/CCA that have been adopted and the implications for countries and communities when these provisions are indeed implemented.

There is great momentum for ecosystem-based approaches to DRR and CCA in other policy fora, including at regional, national and sub-national levels, which will further drive implementation of Eco-DRR/CCA in countries. The European Commission has developed an EU research and innovation agenda on the topic (European Commission 2015) in the context of its Horizon 2020 research

---

[1]http://whc.unesco.org/en/disaster-risk-reduction/
[2]http://worldparkscongress.org/downloads/approaches/Stream4.pdf

Table 24.1 Key provisions on Eco-DRR/CCA in the post-2015 global policy agreements

Policy agreement	Summary description	Number of signatory parties (Member States)	Key provisions on ecosystem-based DRR/CCA in the agreement	National-level instruments/mechanisms for implementing the agreement and scope for promoting eco-DRR/CCA through these instruments
Sendai Framework on Disaster Risk Reduction – SFDRR(2015–2030)[a]	The SFDRR is the global framework on Disaster Risk Reduction, with seven targets and four Priorities for Action. It seeks to prevent new and reduce existing disaster risk through the mainstreaming of disaster risk reduction across all development sectors, programmes and policies. While the SFDRR is a voluntary, non-binding agreement, it calls for an all-of-society engagement, with governments having the primary role of reducing disaster risk	187 Member States adopted the SFDRR at the 3rd World Conference on DRR in March 2015. The SFDRR was subsequently endorsed at the UN General Assembly's 69th Session in June 2015	The SFDRR recognizes ecosystem degradation as a driver of risk as well as the environmental impacts of disasters. A new milestone is that the sustainable management of ecosystems is recognized as a key measure for building resilience to disasters. The role of ecosystems will need to be taken into account in disaster risk assessments (Priority Action 1), strengthening risk governance (Priority Action 2) and investments in disaster resilience (Priority Action 3). The SFDRR also calls for greater collaboration between institutions and stakeholders from other sectors, including from the biodiversity and environment sectors. It calls for	Countries will develop their national and local DRR strategies and plans. A focus on strengthening environmental resilience and ecosystem-based approaches could be featured in national and local DRR strategies, with targets and indicators developed as appropriate. Eco-DRR/CCA should also be mainstreamed across sectoral development plans and strategies (e.g. environment, water, agriculture, rural and urban land-use planning, etc.) (see also Sustainable Development Goals, below)

Sustainable Development Goals – SDGs (2015–2030), also the 2030 Sustainable Development Agenda[b]	With a total of 17 goals and 169 targets, the SDGs focus on three main areas: (i) eradication of poverty; (ii) protecting the planet from degradation, while ensuring that economic, social and technological progress occurs in harmony with nature; and (iii) promoting universal peace and just and inclusive societies. While the SDGs are not legally binding, governments are expected to take ownership and establish national frameworks for the achievement of the 17 Goals	193 UN Member States have endorsed the SDGs, at the UN General Assembly's 70th Session in September 2015	A major pillar of the SDGs is protecting the planet from degradation, including through sustainable consumption and production, sustainably managing its natural resources and taking urgent action on climate change	ecosystem-based approaches to be implemented in transboundary cooperation for shared resources, such as within river basins and shared coastlines
			Sustainable ecosystem management is explicitly addressed under targets of Goals 2, 6, 11, 14 and 15, and with reference to curbing environmental degradation under Goal 8	193 countries will develop their respective National Sustainable Development Strategies or National SDG Frameworks, and countries are tasked to develop corresponding indicators for each of the targets listed under the SDGs

(continued)

Table 24.1 (continued)

Policy agreement	Summary description	Number of signatory parties (Member States)	Key provisions on ecosystem-based DRR/CCA in the agreement	National-level instruments/mechanisms for implementing the agreement and scope for promoting eco-DRR/CCA through these instruments
			DRR (and resilience) is explicitly mentioned under targets of Goals 1, 2, 4, 9, 11, 13 and 15, with reference to strengthening resilience under Goal 14	
			Climate change (and climate extremes) is explicitly mentioned under targets of Goals 1, 2, 11, 13, with reference to ocean acidification under Goal 14 and climate-related hazards under Goal 15	
			Maximum inter-linkages between ecosystems, DRR and climate change actions can therefore be supported under **Goal 2** (End hunger, achieve food security and	To maximize integration between ecosystems, DRR and climate change, efforts can focus initially on Goals 2, 6, 11, 13, 14 and 15 where there are already strong linkages, as

			improved nutrition and promote sustainable agriculture), **Goal 6** (Ensure availability and sustainable management of water and sanitation for all); **Goal 11** (Make cities and human settlements inclusive, safe, resilient and sustainable), **Goal 13** (Take urgent action to combat climate change and its impacts), **Goal 14** (Conserve and sustainably use the oceans, seas and marine resources for sustainable development) and **Goal 15** (Protect, restore and promote sustainable use of terrestrial ecosystems, sustainably manage forests, combat desertification, and halt and reverse land degradation and halt biodiversity loss)	expressed through their respective targets: Goal 2 – Target 2.4, Goal 6 – Targets 6.5, 6.6, Goal 11 – Targets 11.4, 11.a, Goal 13 – Targets 13.1-3, Goal 14 – Target 14.2, Goal 15 – Target 15.1, 15.3
UNFCCC 21st Conference of the Parties – Paris Agreement on Climate Change[c]	The Paris Agreement seeks to significantly scale-up climate actions and deal more comprehensively with climate change impacts to safeguard development and eliminate poverty. Countries committed to hold the global average temperature to well below 2°C above pre-industrial levels (and to pursue	195 Countries adopted the Paris Agreement, which is legally binding	The Paris Agreement recognises protecting the integrity of ecosystems and biodiversity for both climate change mitigation and adaptation actions. It specifically lays out principles of adaptation that takes ecosystems into consideration. It also calls for integrating adaptation into relevant environmental	National Adaptation Planning (NAP) enables all developing and least developed country (LDC) Parties to assess their vulnerabilities, to mainstream climate change risks and to address adaptation

(continued)

Table 24.1 (continued)

Policy agreement	Summary description	Number of signatory parties (Member States)	Key provisions on ecosystem-based DRR/CCA in the agreement	National-level instruments/mechanisms for implementing the agreement and scope for promoting eco-DRR/CCA through these instruments
	efforts to limit the increase to 1.5°C. It specifically aims to "significantly reduce the risks and impacts of climate change and foster climate resilience"		policies and actions, where appropriate, as well as for building resilience of ecosystems through sustainable management of natural resources	
	Countries have also agreed to a global goal for adaptation that considers enhancing adaptive capacity, strengthening resilience and reducing vulnerability to climate change		It further recognises the importance of reducing the loss and damage associated with climate change impacts, including extreme events and slow onset events, and the role of sustainable development in reducing risk of loss and damage. Within the Warsaw International Mechanism for Loss and Damage associated with Climate Change, it calls for early warning systems, preparedness, comprehensive risk assessments and management as well as a range of insurance solutions	In decision 1/CP.20 the COP also invited all Parties to consider communicating their efforts in adaptation planning or consider including an adaptation component in their Intended Nationally Determined Contributions (INDCs). Efforts should focus on ensuring that NAPs and INDCs incorporate the key adaptation principles set out in the Paris Agreement, which include taking into account building ecosystem resilience in adaptation and protecting the integrity of ecosystems
				Eco-DRR/CCA projects and technical assistance can be supported through:
				Green Climate Fund
				Climate Technology Centre and Network (CTCN)

Other Multi-lateral Environmental Agreements with direct references to the post-2015 sustainable development agenda

Convention on Biological Diversity (CBD), 12th Conference of the Parties, Decision XII/20[d]	The CBD recognised for the first time in international law that the conservation of biological diversity is a universal concern for humankind and is integral to sustainable development. It covers all ecosystems, species, and genetic resources. The CBD is a legally-binding agreement	Under the CBD, there are 168 signatory Member States	Decision XII/20 on Biodiversity and Climate Change and DRR recognises that while biodiversity and ecosystems are vulnerable to climate change, the conservation and sustainable use of biodiversity and restoration of ecosystems can play a significant role in climate change mitigation and adaptation, combating desertification and disaster risk reduction. It calls on governments and other relevant organisations to promote Eco-DRR/CCA approaches and integrate these into their respective policies and programmes	168 countries are obligated to develop, implement and regularly review their National Biodiversity Strategic Action Plans (NBSAPs), which should take into account Decision XII/20 and integrate DRR and climate change actions in their respective NBSAPs
			Decision XII/20 supports implementation of the Aichi Targets, specifically Target 15: By 2020, ecosystem resilience and the contribution of biodiversity to carbon stocks has been enhanced, through conservation and restoration, including restoration of at least 15% of degraded	CBD signatory Member States can leverage Decision XII/20 to advocate for a stronger role for biodiversity conservation and ecosystem-based approaches in local and national DRR strategies as well as in National Adaptation Plans (NAPs)

(continued)

Table 24.1 (continued)

Policy agreement	Summary description	Number of signatory parties (Member States)	Key provisions on ecosystem-based DRR/CCA in the agreement	National-level instruments/mechanisms for implementing the agreement and scope for promoting eco-DRR/CCA through these instruments
Ramsar Convention (or formally the Convention on Wetlands of International Importance), 12th Conference of the Parties, Resolution 13[e]	The Ramsar Convention provides the framework for national action and international cooperation for the conservation and wise use of wetlands and their resources. It is a non-binding agreement	There are 169 signatory Member States, also referred to as Contracting Parties	Resolution 13 on wetlands and disaster risk reduction strongly encourages countries to mainstream disaster risk reduction measures in wetland management plans, especially Ramsar Sites, which integrate the principles of ecosystem-based management and adaptation against natural hazards and accelerated sea level rise. It further calls for integration of DRR in all relevant policies, action plans and programmes	169 countries can promote Eco-DRR/CCA approaches through wetland management plans, which cover 2,231 Ramsar-designated sites with a total surface cover of 214,936,005 ha
	The Convention uses a broad definition of wetlands. It includes all lakes and rivers, underground aquifers, swamps and		It further calls on countries to integrate ecosystem management related considerations, in particular relating to wetland and	Ramsar Contracting Parties can leverage Resolution 13 to promote Eco-DRR/CCA in wetland management policies and plans (in both Ramsar and

ecosystems, thereby contributing to climate change mitigation and adaptation and to combating desertification

	marshes, wet grasslands, peatlands, oases, estuaries, deltas and tidal flats, mangroves and other coastal areas, coral reefs, and all human-made sites such as fish ponds, rice paddies, reservoirs and salt pans		water management, in their national disaster risk reduction and climate change adaptation strategies	non-Ramsar wetland sites), as well as in national and local DRR and CCA strategies, plans and programmes
United Nations Convention to Combat Desertification – UNCCD , 12th Conference of the Parties[f]	The UNCCD provides the global framework for tackling the issue of land degradation and desertification. It is the only legally-binding international agreement with a focus on sustainable land management. The Convention addresses specifically the arid, semi-arid and dry sub-humid areas, known as drylands. It seeks to improve the living conditions for people in drylands, to maintain and restore land and soil productivity, and to mitigate the effects of drought	194 countries and the European Union are Contracting Parties to the UNCCD	At the 12th CoP in October 2015, Parties to the UNCCD reached a breakthrough, with the adoption of the land degradation neutrality (LDN) target. This means that Parties have agreed that the amount of healthy and productive land should stay stable starting in 2030. Parties also agreed to develop indicators for measuring progress in LDN and for enhancing land resilience to climate change and halting biodiversity loss linked to ecosystem degradation	National Action Programmes (NAPs) are the key instruments to implement the Convention, which are often supported by action programmes at sub-regional (SRAP) and regional (RAP) levels. The UNCCD urges Parties to align their action programmes, as well as other relevant implementation activities relating to the Convention, to the UNCCD's 10-Year Strategy
			Outcomes of CoP-12 strengthens implementation of the UNCCD's 10-year Strategic Plan and Framework (2008–2018), which was adopted in 2007 at CoP-8.[g] This	With the adoption of the LDN target, countries will have to formulate their own respective strategies for achieving LDN and integrate them into their NAPs

(continued)

Table 24.1 (continued)

Policy agreement	Summary description	Number of signatory parties (Member States)	Key provisions on ecosystem-based DRR/CCA in the agreement	National-level instruments/mechanisms for implementing the agreement and scope for promoting eco-DRR/CCA through these instruments
			Strategic Plan seeks to reverse and prevent land degradation and desertification, and specifically recognizes the important services provided by ecosystems, especially in dryland ecosystems, for drought mitigation and the prevention of desertification. The following strategic objectives and expected impacts outlined in this Strategic Plan have direct relevance to Eco-DRR/CCA , calling for enhanced measures on sustainable land management:	
			Objective 1 – Expected impacts: 1.1., 1.2	Countries can thus leverage the agreements reached at CoP-12 to promote the sustainable use, conservation, and restoration of ecosystems and biodiversity in the context of reducing the risk of desertification and drought
			Objective 3 – Expected impact 3.1	

[a] http://www.preventionweb.net/files/43291_sendaiframeworkfordrren.pdf
[b] https://sustainabledevelopment.un.org/sdgs
[c] https://unfccc.int/resource/docs/2015/cop21/eng/l09r01.pdf
[d] https://www.cbd.int/decision/cop/default.shtml?id=13383; For CBD's Aichi Targets, see: https://www.cbd.int/sp/targets/
[e] http://www.ramsar.org/sites/default/files/documents/library/cop12_res13_drr_e_0.pdf
[f] http://www.unccd.int/en/about-the-convention/the-bodies/the-cop/COP_12/Pages/default.aspx
[g] http://www.unccd.int/Lists/SiteDocumentLibrary/10YearStrategy/Decision%203COP8%20adoption%20of%20The%20Strategy.pdf

programme. Four goals have been identified covering sustainable urbanisation, restoration of degraded ecosystems, development of climate change adaptation and mitigation, and improving risk management and resilience for which nature-based solutions have been identified to play an important role (European Commission 2015:4). From these four goals, seven nature-based solutions for research and innovation have been identified, covering urban regeneration and well-being, coastal resilience, watershed management, sustainable use of matter and energy, enhancing the insurance value of ecosystems, and increased carbon sequestration (European Commission 2015:4). Building on the work of Sutherland et al. (2014), the document also lists 310 nature-based solutions that could be considered for enhancing the regulating services of ecosystem (including natural hazard and climate regulation).

In October 2015, the Executive Office of the Government of the United States issued a "White House Memorandum" directing U.S. government agencies to factor the value of ecosystem services into federal planning and decision-making. The memorandum specifically recognises the role ecosystems play in "enhancing the resilience of communities and ecosystems including reducing vulnerability to climate change impacts". It also mentions government efforts "to incorporate natural and nature-based infrastructure (e.g. dunes and barriers islands) to enhance storm and flood protection". Finally, it concludes that this "increased emphasis on ecosystem services to enhance resilience underscores the need for a consistent framework for incorporating ecosystem services into Federal decision making"(Executive Office of the President of the United States 2015: 3).

24.3   Emerging Issues in Eco-DRR/CCA

Overall, the future of Eco-DRR/CCA is positive, if one were to track references to such approaches in international policy agreements, as discussed above. Projects implemented with ecosystem-based DRR or ecosystem-based adaptation components are also becoming more common, not only within the environmental and disaster risk management sectors but also across other development sectors, particularly in the water and agriculture sectors.

Nonetheless, Eco-DRR/CCA still faces a long road towards being fully mainstreamed into local and national development policies, programmes and budgets (Renaud et al. 2013). Drawing from this book's chapters, we highlight key issues which describe both opportunities and challenges in further advancing Eco-DRR/CCA efforts:

- Demonstrating the economic evidence of Eco-DRR/CCA;
- Developing decision support tools for Eco-DRR/CCA to inform policy and practice;
- Promoting innovative, institutional arrangements and policies for Eco-DRR/CCA;
- Enhancing the evidence base of Eco-DRR/CCA through research and education.

24.3.1 Demonstrating the Economic Evidence of Eco-DRR/ CCA

The challenge of making the economic case for Eco-DRR/CCA is already well-recognised, and it was already highlighted in our first book volume (Renaud et al. 2013). Some progress has been made to improve understanding of this issue, by featuring it as a main theme at the International Science-Policy Workshop on Eco-DRR/CCA, co-organised among others by the Partnership for Environment and Disaster Risk Reduction (PEDRR), the Center for Natural Resources and Development (CNRD), and the Indonesian Institute of Sciences (LIPI) in Bogor, Indonesia, in June 2014.[3] However, significant gaps remain on how we can effectively make the economic case for Eco-DRR/CCA.

Under-appreciation of Eco-DRR/CCA approaches is often attributed to the lack of, or limited economic evidence, to demonstrate why investing in ecosystems offers a cost-effective means of reducing disaster risks and adapting to climate change. As a consequence, there is generally a persistent pattern of under-investment in maintaining or enhancing ecosystems and ecosystem services for DRR and CCA (Shreve and Kelman 2014).

Several economic decision-making tools are available that have potential applications for Eco-DRR/CCA, including cost-benefit analysis (CBA), replacement cost methods, cost-effectiveness analysis, as well as avoided cost (i.e. cost of losses or damage avoided) and opportunity cost analysis. Applied in the context of Eco-DRR/CCA, all of these methods, in different ways, may be utilised to assign a value to the services provided by ecosystems for DRR and/or CCA.

In this book, Emerton et al. (Chap. 2) argue the importance of accounting the *total* economic values of ecosystems as part of development and investment decisions, including valuing the protection and regulatory services provided by ecosystems which could be considerable. Recognising the full range of services that an ecosystem generates will build a stronger economic case that ecosystems provide multiple benefits and therefore have greater added value compared to hard engineering or "grey" infrastructure options.

One of the main challenges in economic valuation is the quantification of the risk reduction services (i.e. protection and regulatory services) provided by ecosystems. As pointed out by Emerton et al., while ecosystem valuation methodologies have been evolving over the last 20 years, their applications in the context of CCA and DRR still remain nascent. Few real-world applications have been documented, including case studies featured in this volume (Nehren et al. 2014; Emerton et al., Chap. 2 and Vicarelli et al., Chap. 3).

Another available case study is from Fiji (Rao et al. 2013), where cost-benefit analyses were carried out to assess climate change adaptation options for the city of

---

[3]http://pedrr.org/training/current-event/international-science-policy-workshop-bogor-2014/

Lami against sea level rise and extreme events (e.g. flooding, storm surges). The study compared "green" solutions, such as planting mangroves and replanting stream buffers, to engineering measures, such as building seawalls and increasing drainage. The study concluded that ecosystem-based measures yield a USD $19.50 benefit to cost ratio, as compared to USD $9.00 for engineering actions. Nonetheless, the study revealed that in terms of avoided (flood) damage, engineered measures provided 15–25 % greater protection than ecosystem-based measures; thus, a combination of both green and grey infrastructure was recommended for the city's coastal defence and adaptation strategy. The study further attempted to capture the full range of social, cultural and economic benefits generated by ecosystems, going beyond quantifying and monetizing the direct physical costs and benefits.

While cost-benefit analysis (CBA) is a predominant economic decision-making tool and has significant potential for prioritising Eco-DRR/CCA interventions, CBA also has its own limitations and its applications are not always rigorous or systematic, as demonstrated by Vicarelli et al. in Chap. 3. In order to improve CBA practice in the context of Eco-DRR/CCA, Vicarelli et al. recommend greater rigour and consistency in the analytical frameworks and process used, for instance in defining analytical boundaries (i.e. geographic scale and time horizons), data gathering to establish baseline conditions, setting parameters (i.e. longevity of benefits and discount rates), and undertaking peer reviews of the economic analyses generated.

It is equally important to assess the potential costs associated with ecosystem loss and degradation, as a means of securing DRR and CCA benefits. In the case of Puttalam Lagoon, Sri Lanka, Emerton et al. demonstrated in Chap. 2 that since 1992, conversion of mangroves to shrimp farming and other land uses, regarded as more productive or profitable, cost the local economy more than USD $31 million in foregone benefits, amounting to a sum more than twice as high as the income earned from land uses that replaced mangroves.

Nonetheless, Emerton et al. (Chap. 2) argue that rather than comparing grey and green approaches in opposition to each other, they should both be regarded "as part of the same economic infrastructure that is needed to deliver essential development, adaptation and disaster risk reduction services". By explicitly recognising the total economic values associated with ecosystem services, decision makers are better able to factor them into investment calculations, and develop improved policy instruments and approaches which will promote ecosystem-based solutions for DRR and CCA.

Economic analyses should also consider different scenarios for making decisions between grey and green measures for DRR and CCA. In Chap. 5, Harmáčková et al. analyse the economic costs and benefits of stakeholder-defined adaptation scenarios for the Sumava National Park in the Czech Republic and assessed their impact on ecosystem services related to DRR and CCA. A cost-benefit analysis was undertaken, quantifying the management and investment costs of each adaptation scenario as well as the benefits originating from maintaining ecosystem regulating services.

When undertaking economic analyses of Eco-DRR/CCA approaches, however, one should guard against making the assumption that policymakers and decision-makers prioritise decisions based solely or primarily on such economic figures. While generating economic evidence may be necessary for increasing the policy and budgetary priorities given to ecosystem-based DRR and CCA solutions, it is not sufficient on its own and requires a broader communication and advocacy strategy (Emerton et al., Chap. 2; Vicarelli et al., Chap. 3). In reality, most political and investment decisions are made based on multiple, competing development needs and priorities; hence, efforts must go beyond mere valuation of ecosystem services for DRR and CCA. Emerton et al. call for a greater appreciation of the drivers of peoples' economic behaviours (e.g. of policymakers as well as land-holders and resource users) in order to develop the right mix of policies, economic incentives and financing required for improved ecosystem management. In their chapter, Harmáčková et al. further highlight the importance of combining economic valuation studies with multi-stakeholder participation as well as biophysical modelling in analysing different scenarios, which create better conditions for identifying shared solutions.

Applying an economic lens also helps us better understand and manage the potential "trade-offs" of implementing Eco-DRR/CCA. It has been frequently argued and demonstrated that Eco-DRR/CCA can provide "win-win" situations because of the multiple benefits derived, such as livelihoods support, carbon sequestration, and biodiversity conservation, in addition to DRR and CCA (CBD 2009; World Bank 2010; IPCC 2012; Munang et al. 2013; Renaud et al. 2013). However, there are possible trade-offs in applying Eco-DRR/CCA measures, whereby stakeholder benefits derived from ecosystems do not necessarily yield "win-win" results for all, although one important criterion in analysing trade-offs is the time frame. For instance, Emerton et al. (Chap. 2) illustrate the case in Sri Lanka where converting mangroves into aquaculture and agriculture makes more financial sense in the short-term to local landowners rather than sustainably using and managing mangroves, given that shrimp farming, coconut farming and salt production generate higher case returns and immediate sources of income, even if these activities are proved unsustainable over the long-term or result in significant negative impacts on other groups or sectors. In this case, direct economic incentives for land holders to maintain mangroves on their land are limited or non-existent. In Chap. 5, Harmáčková et al. also show that by comparing alternative adaptation scenarios and options, there may be trade-offs when securing ecosystem services for climate change regulation, water purification and sediment retention and hydropower production. In Chap. 18, Nehren et al. highlight the potential conflicts between utilising the provisioning and regulating services of coastal sand dunes; sand extraction for construction purposes (provisioning service) may exceed or undermine the dunes' natural buffering capacity against coastal storm surges (regulating service). In this case, sand companies and the tourism sector benefit from sand extraction, at the expense of coastal communities and ultimately tax payers, who will suffer increased losses and damages from coastal storm surges over the long run. In this regard, various decision-making tools to facilitate stakeholder dialogue and consensus-building could help resolve some of these trade-offs, and are discussed further in the next section.

24.3.2 Decision-Making Tools for Eco-DRR/CCA

Over the last decade, we have seen increased interest in decision-making tools that enable greater implementation and investments in Eco-DRR/CCA measures. A whole range of decision-making tools exist, as evidenced from the literature and in this book, and it is important to differentiate these tools from each other. At the 2014 PEDRR/CNRD/LIPI workshop in Bogor, participants suggested categorising decision-making tools into three broad classes: decision support tools (DSS), assessment tools and management tools. While these types of decision-making tools may have unique purposes, the terminology is often used inter-changeably, and there is a great deal of overlap and blending between such tools, as can be inferred from this book. Nonetheless, for purposes of improving understanding of how such tools are being applied, we will refer to these three broad types of tools.

24.3.2.1 Management Tools

Eco-DRR/CCA "management tools" are more similar to implementation approaches and typically focus on aspects relevant to natural resource management or ecosystem governance. Such tools or approaches include risk-sensitive spatial planning, Integrated Coastal Zone Management (ICZM), Integrated Water Resource Management (IWRM), protected area management, integrated drought management, drylands management, and Integrated Fire Management (IFM), among others. Such tools often provide guidelines for Eco-DRR/CCA implementation (for an overview, see Estrella and Salismaa 2013). Many of these ecosystem management approaches are not new and have long been practiced in the conservation sector; however, their applications in the context of DRR and CCA have only emerged in the past decade.

In Chap. 8, van Wesenbeck et al. shed a new perspective on ICZM for flood risk reduction and consider the emerging field of practice of including coastal and fluvial ecosystems in flood defence systems. They describe an implementation approach for integrating ecosystem-based measures into coastal engineering projects. While inclusion of ecosystems may be undertaken together with traditional structural measures, such as levees and dykes, the authors argue that such integrated approaches will require new and adapted design, construction and management methods, which only a close collaboration between engineers, ecologists and experts in public administration can make possible. In Chap. 9, Kloos and Renaud discuss the importance of agro-ecosystem-based solutions in drought risk management and provide a review of several approaches in drylands management which contribute to drought risk reduction. They contend that more information is needed on the limits of applying such approaches, in order to better understand how they can be effectively integrated into Eco-DRR/CCA programmes and complemented by conventional DRR and adaptation strategies. Finally, in Chap. 12, Kumar et al. present an innovative "clustering" approach to watershed or river basin management, which adopts a landscape-scale approach to risk management by explicitly taking into account the drivers of disaster risk and community vulnerabilities, administrative and political boundaries, and ecosystem functions across multiple spatial scales. The cluster scale approach in risk reduction planning comprises smaller landscape units of communities facing similar risks. It helps bridge administrative and ecological scales for reaching more effective risk reduction outcomes and identifying suitable ecosystem-based measures.

24.3.2.2 Assessment Tools

Eco-DRR/CCA "assessment tools" generally serve to analyse and demonstrate ecosystem services that contribute to DRR and CCA. Assessed information can then be used by policymakers and planners whether to invest or prioritise Eco-DRR/CCA measures. Assessment of ecosystem services may utilise quantitative or qualitative methods, or a combination of both.

Advances in geographic information systems (GIS) have made it possible to apply spatial modelling for assessing the DRR and adaptation services generated by ecosystems. While spatial models are commonly applied in both ecological assessments and in disaster risk assessments, the practice of combining these two applications is still limited. In this book, several chapters describe various spatial tools for integrating ecosystem and disaster risk assessments. Bayani and Barthélemy in Chap. 10 demonstrate applications of InVEST (Integrated Valuation of Ecosystem Services and Tradeoffs), in countries where available data is limited. InVEST is a suite of spatial models that have been designed to assess how changes in ecosystems and land use influence the flow of natural capital, i.e. ecosystem services, including for instance coastal protection and sediment retention. By applying InVEST in Haiti and the Democratic Republic of the Congo, Bayani and Barthélemy show how ecosystem services for DRR and CCA can be visually assessed and thus help advocate for ecosystem protection and management. In Chap. 6, Whelchel and Beck also describe the use of InVEST as well as other spatial tools, such as the Coastal Resilience Tool, Climate Wizard and Coastal Defense Application, which are designed to consider and integrate ecosystem components into disaster and climate resilience planning.

Another type of assessment tool includes biophysical modelling of ecosystem services for hazard mitigation, which provides more quantitative information of ecosystem services. In Chap. 11, Dorren and Schwarz measure the effect of forests in Switzerland on slope stability and protection against shallow landslides, through an online tool called SlideforNET. One important feature of such modelling tools is that they are freely available web tools, making it easier than ever for planners and practitioners to consider ecosystem-based solutions in planning and decision-making.

However, these chapters also point out limitations in applying such assessment tools, including accessing high resolution data, factoring the uncertainties and different time horizons, and delimiting spatial scales (see also Krol et al., Chap. 7). Assessment tools also become more useful when they can generate comparative outcomes between decisions or scenarios (Whelchel and Beck, Chap. 6), but it may not always be feasible to do so. Moreover, for such tools to be fully maximised and translated into actionable decisions, they need to be applied as part of broader, multi-stakeholder decision-making and planning processes (discussed further below).

24.3.2.3 Decision Support Tools

Eco-DRR/CCA "decision support tools" generally refer to a broad suite of tools, methods and approaches that constitute a decision support system (DSS) and are designed to facilitate collaborative decision-making between alternative options. An effective DSS often entails a multi-stakeholder process that articulates different perspectives and sets prioritized actions. Decision support tools therefore provide an opportunity for comparing and deciding between alternative risk reduction and adaptation measures, including ecosystem-based measures.

Krol et al. (Chap. 7) consider the potential of applying Spatial Decision Support Systems (SDSS) which facilitate the use of geographical information for assessing risk, comparing between different risk reduction measures (including ecosystem-based measures), and most importantly, for facilitating collaborative decision-making. SDSS support the selection of the most optimal alternatives for intervention. Key features of the SDSS described include multi-stakeholder engagement (including information providers as well as end-users), developing multiple scenarios, accounting for uncertainty and identifying trade-offs between alternative ecosystem services used and other possible risk reduction interventions. Other important considerations include factoring in changing risk contexts (i.e. current and future risks) as well as multiple criteria evaluations, when comparing alternatives and identifying trade-offs. In practice, a combination of both structural and non-structural measures, of engineering and ecosystem-based measures, is applied as part of a broad suite of strategies to cope with current and future risk scenarios.

In addition to featuring a range of tools used for decision-making, Whelchel and Beck (Chap. 6) also outline several approaches that follow a "7 step-wise, planning-to-action framework" which closely characterises DSS tools, as described above. Integrated Eco-DRR/CCA approaches to coastal zone management as well as watershed management for flood risk reduction are described, applying the 7 step planning-to-action framework. In addition to the issues discussed by Krol et al. (Chap. 7), other considerations highlighted in this chapter include monitoring and assessing the effectiveness of actions or interventions taken, which should feed into an iterative planning and decision-making process. The issue of governance (i.e. mechanisms for involving those who make decisions, have influence over decisions or are impacted by the decision made) is also emphasised by the authors as an essential pre-requisite for effective multi-stakeholder participation and decision-making. The authors further recognise the difficulties of knowing which tools and approaches are best suited for different purposes and for different scales (e.g. multi-national to local community); a table presenting the different tools and their various applications is included in this chapter.

24.3.3   Innovative Institutional Arrangements and Policies for Mainstreaming Eco-DRR/CCA

While Eco-DRR/CCA projects are now increasingly more common than they were a decade ago, Eco-DRR/CCA as part of development policy, planning, budgeting and practice is not yet a common standard or mainstreamed (see Box 24.1). The issue of mainstreaming is important because it ensures that Eco-DRR/CCA projects do not become 'one-off' interventions and are systematically considered when decisions on DRR and CCA are taken. Furthermore, mainstreaming of Eco-DRR/CCA facilitates replication and up-scaling of such approaches over time and over larger geographic areas. It would further improve access to development and sectoral budgets, and thus enhance disaster and climate resilience in development allocations and investments.

> **Box 24.1. What Do We Mean by 'Mainstreaming'?**
> The terms 'mainstreaming' and 'integration' are generally used synonymously. However, integrating may simply mean incorporating i.e. concepts, while mainstreaming refers to the process of 'institutionalisation'. Mainstreaming seeks to change institutional and personal behaviours and practices, i.e. Eco-DRR/CCA becomes regarded as standard policy and institutional practice. Mainstreaming is essentially a governance process negotiated between government and non-government actors.
> *Source: PEDRR/CNRD (2013) Graduate Module on Disasters, Ecosystems, and Risk Reduction*

In this book, several chapters address the important issue of mainstreaming as a critical component of promoting, sustaining and scaling-up Eco-DRR/CCA initiatives. Experiences suggest that having an enabling policy, legal and institutional environment is needed to encourage implementation of Eco-DRR/CCA initiatives. However, given the multi-disciplinary and multi-sectoral nature of Eco-DRR/CCA, the challenge is working with existing policy, legal and institutional frameworks that *do not necessarily* support nor encourage such integrated approaches.

At present, we already see strong momentum at the global policy level to link ecosystem management, DRR and CCA, as discussed earlier in Sect. 24.2. At the same time, there is a proliferation of Eco-DRR/CCA "practice" being implemented through community-based, field level projects, as evidenced by the number of reports, publications and case studies (Doswald and Estrella 2015; see also Chap. 1). The biggest challenges in accelerating Eco-DRR/CCA implementation, therefore, are found in the "in-between" governance spaces, i.e. translating global policy commitments into national- and local-level policies and legal frameworks and developing institutional mechanisms and process that encourage implementation.

24.3.3.1 Mainstreaming Eco-DRR/CCA in the Environmental Sectors

Progress in mainstreaming Eco-DRR/CCA could be achieved if DRR and CCA were more explicitly addressed in national environmental policies and legislative frameworks, as well as in environmental strategies and programming, as called for in the SFDRR.

One promising area in this regard is protected area management. In Chap. 17, McNeely considers the role of ecosystems in protected areas that mitigate climate change and their impacts, thereby also contributing to reducing disaster risks (see also Harmáčková et al., Chap. 5). McNeely argues that the management of protected areas needs to incorporate climate change and disaster risk considerations at both site and ecosystem scales, in order to maximise ecosystem services for DRR, climate change mitigation and adaptation.

Protected area management could also become integral to post-disaster recovery and reconstruction strategies, as an opportunity to increase protection against future risks. One example is the creation of the Sanriku Fukko Reconstruction National Park in Japan, following the Great East Earthquake and Tsunami of 2011 (see Furuta and Seino, Chap. 13; Takeuchi et al., Chap. 14; McNeely, Chap. 17). Implementation guidance for incorporating DRR and CCA in protected areas is now available (UNESCO 2010; Dudley et al. 2015), and the next step will be to generate robust experiences and lessons in this field (Murti and Buyck 2014). Mainstreaming Eco-DRR/CCA in protected areas is important because it would support national-level implementation of the various multi-lateral environmental conventions that have endorsed greater integration of ecosystems management, DRR and climate change actions, in particular the CBD Decision XII/20 and the Ramsar Convention CoP-12/Resolution 13 (as discussed in Sect. 24.2).

Another promising area is in the field of Environmental Impact Assessments (EIAs) and Strategic Environmental Assessments (SEAs). The potential of EIAs and SEAs to mainstream DRR in development planning and investments has long been recognised, including in the Hyogo Framework for Action and again in the SFDRR as well as in the literature and our first book (see CDB 2004; Benson and

Twigg 2007; OECD 2010; Gupta and Nair 2013). Several country-level experiences of promoting and implementing EIA-DRR initiatives were detailed in Gupta and Nair (2013). However, few recent examples exist or have been documented on implementing SEA-DRR initiatives, such as the Integrated Strategic Environmental Assessment (ISEA) conducted by the Government of Sri Lanka in its Northern Province, in collaboration with the United Nations Development Programme (UNDP) and United Nations Environment Programme (UNEP).[4] In this example, the ISEA identified suitable areas for development investments, while taking into account environmentally-sensitive and high disaster risk areas.

Although this current book does not have a focus on EIAs, it continues to be recognised as a global priority. The SFDRR, for instance, specifically refers to EIAs in the context of promoting disaster resilient public and private investments at national and local levels (UNISDR 2015, Priority 3/c). Furthermore, at the International Conference on Disasters and Biodiversity in Sendai, Japan, in September 2014, one of the key recommendations included the importance of undertaking EIAs in post-disaster reconstruction, as a means of ensuring social and environmental safeguards to mitigate adverse environmental impacts of reconstruction that may exacerbate current and future disaster risks (Furuta and Seino, Chap. 13). In the Philippines, the Government's Environment Management Bureau (EMB) issued a special memorandum dated 11 November 2011, which produced the "EIA Technical Guidelines Incorporating Disaster Risk Reduction and Climate Change Adaptation Concerns" (Gupta and Nair 2013). Since then, the EMB in collaboration with the Asia Disaster Preparedness Center (ADPC) has produced a short training course on "Application of the Technical Guidelines for Incorporating DRR/CCA Concerns in the Philippine EIS System" (in press), which is designed to enable EIA practitioners incorporate DRR and CCA considerations in the EIA process. Many of the key challenges outlined by Gupta and Nair (2013) still remain, not least overcoming the limited collaboration between the ecosystem management and disaster management sectors. However, emergence of training courses, such as those produced by the Government of the Philippines and ADPC and the Ramsar Regional Training Center for the Americas' International Training Course on EIA and SEA (CREHO 2015), offer great opportunity for advancing practice in this area.

24.3.3.2 Mainstreaming Eco-DRR/CCA into Other Sectors

Drawing from several chapters in this book, the experiences point to clear opportunities for mainstreaming Eco-DRR/CCA into other development sectors. Given that national- and local-level planning as well as financial allocation decisions take place mostly through sectors, embedding Eco-DRR/CCA components in sectoral plans and programmes is a critical step towards successful mainstreaming.

---

[4]http://www.cea.lk/web/index.php/en/news-and-events/25-what-s-new/854-integrated-strategic-environmental-assessment-for-the-northern-province

Eco-DRR/CCA efforts appear to be well-represented in the water sector, in particular with respect to integrated watershed or water resource management (Whelchel and Beck, Chap. 6; Krol et al., Chap. 7; Kumar et al., Chap. 12). Other sectors and processes where Eco-DRR/CCA is discussed include the agricultural sector, specifically in drylands (Kloos and Renaud, Chap. 9), coastal zone management (Nehren et al., Chap. 18; David et al. Chapt 20), urban development (Sandholz, Chap. 15) and disaster risk management (Furuta and Seino, Chap. 13; Takeuchi et al., Chap. 14), including management of fire (Kieft et al., Chap. 16), flood risks (van Wesenbeeck et al., Chap. 8; Senhoury et al., Chap. 19) and community participation in DRR (Lange et al., Chap. 21).

While these sectors and planning processes represent many opportunities for mainstreaming Eco-DRR/CCA, there are also major constraints for doing so, mainly due to unclear, overlapping and contested policies, legal frameworks and institutional mandates. For instance, in Chap. 18, Nehren et al. describe the policy gaps in Chile to support an effective coastal dune management system, which could play an important role in protecting against tsunamis and coastal storms. In Chile, the management of coastal dunes are generally outside the designated coastal dune parks and other protected areas, making it difficult to ensure integrated planning and management. Due to such policy and management gaps, severe degradation of coastal sand dunes is occurring, linked to construction, tourism, and residential and commercial developments. In Chap. 15, Sandholz examines increasing disaster and climate risks in Kathmandu Valley, Nepal, resulting from rapid urbanisation and ecosystem degradation. One major stumbling block is the large number of regulations and policies on urban planning and construction that are poorly executed and regulated, without environmental considerations. These policy and institutional gaps then pose major constraints to inter-sectoral collaboration, which is essential to implementing Eco-DRR/CCA approaches (Sandholz, Chap. 15; see also Furuta and Seino, Chap. 13).

One way around policy and institutional bottlenecks may be influencing ongoing development planning and governance initiatives that serve as an 'umbrella' framework for multi-sectoral engagement. For instance, the Kathmandu Metropolitan City Risk Sensitive Land Use Plan is based on a multi-sectoral approach and includes environmental considerations (Sandholz, Chap. 15). In Japan's post-disaster reconstruction process following the 2011 Great East Earthquake, recognition of ecosystem functions for DRR has been incorporated into the 'Fundamental Plan' supporting implementation of the 'Basic Act for National Resilience Contributing to Preventing and Mitigating Disasters for Developing Resilience in the Lives of the Citizenry' (Act No.95 of 2013) (Furuta and Seino, Chap. 13).

24.3.3.3 Innovative Financing for Eco-DRR/CCA

Financial instruments that seek to encourage and incentivise investments and actions for maintaining, restoring and managing ecosystems are emerging but still in early stages of implementation. Payment for Ecosystem Services (PES) is one such instrument that has commonly been used for financing conservation and, increasingly, for carbon storage (e.g. schemes such as Reducing Emissions from Deforestation and Forest Degradation or REDD+) and for watershed or water resource management, but PES schemes are not yet commonly applied for capturing DRR or CCA benefits. In Chap. 4, Friess et al. explore the potential for applying PES schemes for mangroves in the context of DRR. However, several challenges – such as quantifying ecosystem services that contribute to DRR (e.g. wave attenuation, coastal erosion control), managing financial risks associated with climate change and natural hazards, distinguishing between service providers and users – pose major constraints and could make PES for DRR unfeasible in some contexts.

Another type of innovative financing include 'co-financing' of Eco-DRR/CCA projects. In Chap. 6, Whelchel and Beck elaborate on how national government funds in the United States can provide initial seed funding to kick-start Eco-DRR/CCA related initiatives that are also co-financed by local governments and communities. Pooling of resources from national to local levels through a competitive selection process can help provide predictable financing as well as encourage Eco-DRR/CCA innovations.

Other types of innovative financing may become more available as carbon markets mature. Wylie et al. (2016) describe and analyse such markets based on trading carbon emissions which are derived from coastal "blue carbon" ecosystems, such as seagrasses, salt marshes and mangroves. These ecosystems are known for sequestering and storing large amounts of carbon as well as providing other important services such as protection from storms and coastal erosion, support to livelihoods and other adaptation benefits. Carbon stored in coastal ecosystems can be sold as credits, which buyers then use to offset their carbon emissions, thus potentially creating mechanisms for funding and investing in ecosystem conservation projects. Wylie et al. analyse four blue carbon projects implemented in Kenya, India, Vietnam and Madagascar, which were financed by selling carbon credits on the voluntary carbon market. In contrast to other carbon financing mechanisms that require strict compliance procedures such as REDD+ and the Clean Development Mechanism (CDM), voluntary carbon markets allow for greater flexibility by utilising different voluntary standards for undertaking carbon accounting, verification and certification, and thus have lower transaction costs (Wylie et al. 2016: 78). However, the authors also point out some of the limitations and challenges of utilising voluntary markets for blue carbon projects, but they also anticipate new opportunities emerging.

24.3.3.4 Fostering Private Sector Engagements in Eco-DRR/CCA

New types of implementation arrangements in Eco-DRR/CCA increasingly involve the private sector, in particular the business sector. Private sector engagement has the potential to provide additional financing or co-financing, stimulate innovations, and create business opportunities that will encourage adoption of ecosystem-based solutions for DRR and CCA. There is clear interest from the private sector to invest in ecosystem-based approaches, in particular so-called "hybrid", or combined natural and engineered infrastructure solutions, for increasing business resilience to external shocks and other objectives (e.g. capital expense savings, reduced operating expenses; see The Joint-Industry White Paper published by the Dow Chemical Company, Shell, Swiss Re, Unilever and The Nature Conservancy in 2013). In Chap. 6, Whelchel and Beck describe the success of 'Water Funds' in catalysing integrated watershed and flood risk management. Involving large businesses, government agencies and municipalities as water users, the water funds facilitate joint investments in water resource management which also include ecosystem-based projects for flood risk reduction (see also Friess et al., Chap. 4). However, private sector experiences in applying ecosystem-based solutions for DRR or CCA are still fairly nascent, and further documentation is needed to synthesise best practices and lessons.

Global efforts to engage the private sector include the new Natural Infrastructure for Business Platform (NI4Biz), launched at the UNFCCC CoP21 in Paris by the World Business Council for Sustainable Development (WBCSD) which has approximately 200 members from private companies.[5] The NI4Biz initiative seeks to encourage private sector investments in natural infrastructure and ecosystem-based solutions to deliver on a wide range of services, including for water purification, agricultural production, soil remediation, power generation as well as for disaster risk reduction or adaptation to climate extremes. The aim is to mainstream ecosystem and natural infrastructure solutions as part of standard business practices, not only part of corporate social responsibility (which has previously been the case). UNEP, Wetlands International and The Nature Conservancy are collaborating to support WBSCD members in their implementation of the NI4Biz, for instance through compilation of case studies from the business sector, demonstration projects as well as development of a short training course geared for the business sector.

Another example is the Private Sector Alliance for Disaster Resilient Societies (ARISE), a voluntary group of more than 100 large companies and small and medium enterprises (SMEs) and convened by the United Nations Office for Disaster Risk Reduction (UNISDR). ARISE has seven work streams which allow the private sector to implement tangible projects and initiatives related to DRR. Unlike the NI4Biz platform, ARISE does not (yet) have a focus on natural infrastructure or ecosystems, but its different work streams suggest potential scope for promoting Eco-DRR/CCA, for instance through their work on developing investment metrics, benchmarking and standards, and engagements with the hotel and tourism industry.

Promising initiatives are also emerging from the insurance sector. As risk managers, risk carriers and investors, the insurance industry is increasingly being called upon to support low-carbon, disaster- and climate-resilient economies and communities (UNEP 2015). Not only are insurers looking for ways to reduce growing insured losses and business interruptions due to extreme events, they are

---

[5]For further information, see http://www.naturalinfrastructureforbusiness.org/

also actively exploring new business opportunities, products and services which could be derived from implementing Eco-DRR/CCA measures (UNEP 2014). For example, AXA Group has partnered with CARE France since 2011 to increase the resilience of coastal communities against storm surges through mangrove reforestation in Thanh Hoa province in Vietnam (UNEP 2015). While this particular partnership was undertaken as part of AXA's corporate social responsibility policy (UNEP 2015:25), the long-term aim should be that private sector investments in Eco-DRR/CCA measures become part of core business practices.

UNEP's Principles for Sustainable Insurance (PSI) initiative, the largest collaborative initiative between the UN and the insurance industry, convenes major insurance companies, insurance associations and insurance regulators and aims to scale-up commitments from the insurance sector in building resilience to disasters and climate change. For instance, AXA and Insurance Australia Group have already pledged commitments through the PSI initiative to integrate natural ecosystems, climate change and socio-economic vulnerability factors into catastrophe risk analysis and models.[6] The Nature Conservancy has also partnered with Swiss Re in developing a set of tools and approaches for quantifying risks from coastal hazards and climate change and demonstrating the cost-effectiveness of coastal ecosystems in adaptation and risk reduction (Box 24.2). Nonetheless, there is still large scope for promoting Eco-DRR/CCA across other commitments expressed by insurers, such as in developing insurance products for DRR and CCA, and investing in climate and disaster-resilient infrastructure.

Box 24.2. Partnering with Insurers: Gulf of Mexico Experience

Since 2013, The Nature Conservancy (TNC) has been working with Swiss Re in order to better understand how incorporating ecosystems and nature-based coastal defences into insurance industry models could improve the assessment of risks from natural hazards and climate extremes, as well as provide a comparative assessment of the costs and benefits between different adaptation or risk reduction measures. As one of the world's largest reinsurance companies, Swiss Re plays an important role in estimating, modelling and pricing risk.

Collaboration between TNC and Swiss Re has involved sharing models and data to demonstrate the cost effectiveness of coastal ecosystems in adaptation and risk reduction, drawing from TNC's expertise in quantifying ecosystem services of coral and oyster reefs, marshes, and mangroves with Swiss Re's methodologies and tools. They developed an open-source model that examines risks from climate change and economic growth and compares the cost-effectiveness of green and grey infrastructure solutions for reducing that risk in the Gulf of Mexico (see http://www.maps.coastalresilience.org/

(continued)

---

[6]For information: http://www.unepfi.org/psi/commitments/

However, Eco-DRR/CCA experience in the insurance sector is also still very recent, and more documented cases will be needed to draw conclusions on the sector's role in promoting and mainstreaming Eco-DRR/CCA practice. In Chap. 4, Friess et al. cite an example from Bell and Lovelock (2013) about efforts to develop a mangroves-for-DRR insurance product that aimed to protect coastal land from the impacts of storms. Bell and Lovelock (2013), however, identify a number of considerations to make ecosystem services-based insurance feasible, including calculating the DRR value of the ecosystem, clear parameters of what insured events will be covered or not, assessing the frequency and severity of weather events in a particular location in order to set the insurance premium, and a protocol of actions to be undertaken by the insurer should an insured event occur. Because of multiple requirements, Friess et al. argue that such financial mechanisms may be mostly feasible for developed countries. The authors also discuss perverse insurance incentives, for instance stemming from state subsidised or semi-private insurance schemes, that may actually perpetuate or increase disaster risks (e.g. by encouraging development in high risk areas such as coastlines).

24.3.4 Research Gaps and Opportunities

While there is already solid empirical evidence that Eco-DRR/CCA works in many contexts, knowledge gaps remain, which require action by scientific and research communities. Addressing knowledge gaps is important if we want to attract further up-take of Eco-DRR/CCA globally. For instance, the role of ecosystems in protecting people and assets during low intensity, high frequency events is relatively well-established: an example would be in terms of coastal vegetation (be it mangroves or other) protecting against coastal erosion or moderate storm surges. However, the role of these coastal ecosystems during high intensity (low frequency) events, such as tsunamis, is much more debated scientifically (see e.g. discussions and references in Lacambra et al. 2013; Renaud et al. 2013). This very issue was discussed at length after the 2004 Indian Ocean Tsunami and is still being discussed in the context of the 2011 Japan Earthquake and Tsunami. The implications are, however, important, as decision-makers need to protect their populations against both categories of hazards, but it is difficult to do so when empirical evidence is lacking. Similar conclusions can be drawn with respect to floods, whereby the role of ecosystems in the context of low intensity flood events is recognised, but further empirical evidence of that role is still required when dealing with large scale, high intensity flooding events (Bullock and Acreman 2003; van Eijk et al. 2013).

The main difficulty in establishing such empirical evidence is the fact that ecosystems and how they protect or buffer against certain hazards are locally-specific, which makes it difficult to replicate and upscale the same measures in other locations and achieve the same results. Because of the important variability of environmental and geomorphic features, what works in one place, may not work a few kilometers away (Chatenoux and Peduzzi 2007). We re-emphasise the importance of carrying out research that includes empirical work in "nested scales", meaning investigating cause-effect relationships at the local to regional (or national) scales (Estrella and Salismaa 2013). This poses important methodological challenges, although such efforts to capture ecosystem services through a landscape scale approach, e.g. linking village-level to district- and delta-level analysis, was successfully tested in the Mahanadi delta, in India (Kumar et al., Chap. 12).

We also need to better understand whether or not communities are actually interested in Eco-DRR/CCA solutions and not presume that this is the best option available for all. This requires, as for all matters linked to development, cross- and inter-disciplinary research specifically addressing this question. It is not just a matter of implementing participatory research/development activities on the ground (which is of course very important), but also to develop generic tools and models that can be adapted to local circumstances and that allow capturing people's and communities' preferences, ensuring that the perspectives of most (if not all) social groups within a community are considered. Linked to this, and as an important area of research, is the need to understand better potential unintended consequences of Eco-DRR/CCA projects. It is recognized that adaptation measures may have unintended consequences on e.g. longer term sustainability (see Eriksen et al. 2011; see also discussion in Doswald and Estrella 2015). In terms of Eco-DRR/CCA, for example, one should ensure that restoring wetlands does not increase the incidence of vector-borne diseases as wetlands provide habitat for mosquitos' eggs and larvae (Dale and Knight 2008). More generally, Eco-DRR/CCA approaches should therefore account for all potential consequences, good or bad, and research should focus on these linkages too.

Increasingly, Eco-DRR/CCA measures are being implemented together with conventional engineering measures – so-called "hybrid" approaches – in order to provide maximum protection against different types and intensities of hazard events (see van Wesenbeeck et al., Chap. 8; David et al., Chap. 20). However, it is important to keep in mind the range of other services provided by ecosystems from which local communities and society depend, beyond natural infrastructure protection against hazards. Therefore, the selection and application of ecosystem-based measures for DRR and CCA will also need to be weighed in conjunction with other local priorities, for instance with respect to livelihoods, food and water security.

From a scientific perspective, there seems to be much less knowledge (or if it exists, it has not necessarily been synthesized) on the role of ecosystems in mitigating "creeping" or "slow-onset" hazards. An example would be in the context of droughts. For instance, in the context of agro-ecosystems facing droughts, comprehensive research is still needed on how ecosystem services can increase the resilience of livelihoods, and the degree to which Eco-DRR/CCA can contribute to reducing the impacts of hazard events of varying intensities and duration (Kloos and Renaud, Chap. 9).

Another important aspect but not yet fully examined in the Eco-DRR/CCA literature is gender. Significant literature exists on the role of gender in the context of DRR (Fordham 1998; Fordham 2001; UNISDR et al. 2009; Fordham et al. 2010; UNDP 2013; among others), but much less so in the context of *ecosystem-based* DRR/CCA. Given that the roles of men and women differ with regards to natural resource management and given that men and women experience disasters and climate change impacts differently, these gender-based differences will need to be taken into account when designing and implementing Eco-DRR/CCA initiatives. Based on an initial literature review on the role of women in Eco-DRR conducted by UNEP,[7] a majority of case studies published on this topic relate to water scarcity and management of agro-ecosystems (e.g. agro-forestry, drylands cultivation). The initial set of case studies reviewed by UNEP highlight the importance of addressing women's access, use rights and ownership of key natural resources, such as water and land. However, more extensive empirical research is needed to fully understand the gender components of Eco-DRR/CCA.

Finally, ecosystems themselves are not invulnerable to the impacts of climate change and disasters; they remain under pressure from anthropogenic stressors and are being degraded at a global scale (MEA 2005). Evidence is scattered as to the impacts of disasters and climate extremes on ecosystems and their threshold capacity in the face of different disaster types and intensities. It is also difficult to understand both the short-term and longer-term impacts of disasters on ecosystems and their functionality. This poses a research question related to the metrics that need to be established in order to quantify ecosystem impacts of disasters, as well as ecosystem recovery. Efforts to assess the environmental impacts of disasters include, for instance, Post-Disaster Needs Assessments (PDNAs) and post-disaster, Rapid Environmental Assessments (REAs); however, these efforts are limited to assessing the immediate environmental impacts, primarily for the purpose of costing post-disaster environmental recovery needs, and often do not undertake quantitative assessments or analyse the long-term recovery needs of ecosystems.

---

[7]This research is still on-going.

24.4   Charting New Pathways for Eco-DRR/CCA

We would like to conclude this chapter by reflecting on the future of advancing Eco-DRR/CCA knowledge and practice. As countries start to grapple with implementing the post-2015 sustainable development agenda and the major global policy commitments, and as disaster and climate risks continue to increase, the urgency to translate these multiple policy agreements into concrete actions will be great. Ecosystem-based approaches have the potential for successfully integrating multiple priorities and delivering multiple benefits for sustainable development, disaster risk reduction and climate change mitigation and adaptation. As discussed in Chap. 1, there has been tremendous progress in the field of Eco-DRR/CCA. However, we need to continuously define new pathways of advancing Eco-DRR/ CCA knowledge and practice, in order to meet the demand from countries and communities for cost-effective, sustainable solutions that tackle the global challenges of the twenty-first century. For this, we will require concerted efforts, engaging with national and local governments, academia and the scientific community, the business sectors, and all of civil society.

24.4.1   Leveraging Scientific Knowledge to Influence Policy and Practice

We have long recognised the importance of leveraging our knowledge base on Eco-DRR/CCA to inform and influence policy and practice. However, this knowledge base is continually under pressure to provide guidance needed to apply Eco-DRR/CCA approaches effectively. Growing losses and costs associated with disasters and climate change only further increase the need for alternative measures, as structural, hard engineering interventions reach the limit of their cost-effectiveness.

In further expanding our knowledge base on Eco-DRR/CCA, there are at least two critical questions. The first question deals with *who* should be engaged in generating and legitimising this knowledge?" Just as implementation of Eco-DRR/ CCA requires multiple stakeholders and different expertise, knowledge generation in Eco-DRR/CCA also calls for a multi-stakeholder, multi-disciplinary approach. This means supporting the "co-production" of knowledge, between producers and users of knowledge, for instance fostering greater collaboration between indigenous and local communities, engineers, ecologists as well as decision makers and policy makers. An implicit assumption is mutual cooperation is essential to test, innovate and implement effective solutions (The Royal Society 2014).

The complex challenges of DRR and CCA require close interaction and cooperation between the scientific community, policymakers as well as practitioners, local stakeholders and communities. This becomes obvious, when we for instance take a look at international transdisciplinary cooperation initiatives, such as the

German-Vietnamese research project LUCCi (Land Use and Climate Change Interactions in Central Vietnam). In this project in the Vu Gia-Thu Bon river basin of Central Vietnam, researchers from various disciplines have been working hand in hand with the responsible ministries and local authorities, stakeholders, and affected communities to optimise land and water resources management, adapt to climate change and reduce disaster risk.[8] As in other transdisciplinary and applied research projects on natural resources management, close communication between the involved researchers, policymakers, stakeholders and communities during the whole project cycle is an essential requirement. This includes, among others, kick-off workshops where the research scope and targets are jointly defined, participatory on-site activities, feedback mechanisms, as well as monitoring and evaluation measures to inform decision-making. Furthermore, it is important that research outputs are disseminated not only in the scientific community, but also translated into non-technical language and made accessible to the various stakeholder groups involved and the interested public.

Knowledge platforms thus remain crucial for bringing stakeholders together, fostering learning exchanges and exploring solutions to inter-connected challenges (poverty, development, disasters and climate change). In January 2016, UNISDR organized a Science and Technology Conference to discuss how the science and technology community could best support implementation of the SFDRR. Outcomes from this Conference are still being finalised as this book went to press, but a common roadmap on future research and technology priorities is anticipated.

However, more dialogue of a similar nature needs to be encouraged and supported at regional and national levels in order to better reflect local and national priorities in knowledge generation and application. For instance, in the Asia-Pacific region, a UNISDR scientific advisory group exists comprised of key leaders from the scientific and academic community in the region which interacts regularly with regional policy bodies such as the United Nations Economic and Social Commission for Asia and the Pacific (UNESCAP). Such mechanisms should be maximised to further dialogue and knowledge on Eco-DRR/CCA.

24.4.2 Re-strategising How We Develop Capacities for Implementing Eco-DRR/CCA

With the recent surge of interest globally for Eco-DRR/CCA solutions, meeting this demand will require significant, purposeful investment in capacity development in both developed and developing countries. However, we will need to reassess *what* and *how* capacity development support should be delivered, with the aim of accelerating implementation and investments in Eco-DRR/CCA.

---

[8]http://www.lucci-vietnam.info/

Current capacity building and development support for Eco-DRR/CCA are constrained by several factors. One factor is cost. Traditionally, support for developing national and local capacities on Eco-DRR/CCA – as well as on broader DRR and climate change related topics – has been delivered mainly through in-classroom trainings or courses, which are cost-intensive in terms of bringing together multiple participants from different locations and covering facilitators' time and travel. Sustainable financing for ongoing and increasing capacity needs is a major issue. Over a given period, there are only so many people who can be trained and exposed to the emerging Eco-DRR/CCA field; hence, the target audience range will inevitably be limited, and training delivery unsustainable over the long-run. Moreover, it requires significant time investment (e.g. up to 1 week) on the part of trained participants themselves, who are generally from government, academia, and civil society organisations.

A second factor is the high turnover of human capital, especially in the public sectors as well as in civil society. Hence, once a cohort of "experts" are trained in Eco-DRR/CCA, they are unlikely to remain in a given institution or professional position for a long period of time, resulting in the loss of institutional memory and skills.

A third factor is the limited scope, in both developed and developing countries, of inter-disciplinary research and studies in academia, in addition to other challenges found in tertiary level education and professional instruction. Given that Eco-DRR/CCA is an essentially cross-disciplinary subject, advancing more integrated and applied research that cut across disciplines (e.g. geography, urban planning, ecology, engineering, economics, sociology) and overcoming academic silos will be critical. Growing knowledge and practice of Eco-DRR/CCA, especially tailored to local and country contexts and capacity needs, is possible starting with an enabling academic environment that favours inter-disciplinary learning and collaboration.

A fourth factor relates to dissemination and transfer of knowledge. While a significant amount of training and resource materials on Eco-DRR/CCA already exist, they are mostly located in international organisations and agencies rather than in national institutions. More widespread dissemination of training and resource modules need to reach the last mile, including individuals in remote locations who can adopt, implement and innovate Eco-DRR/CCA solutions. Online technology now makes this possible, but how can they become more mainstreamed and integrated as part of capacity development strategies?

These are only some of the major considerations when re-thinking how capacity development support is designed and delivered. Addressing growing capacity gaps and needs will require a scaled-up approach, which would:

– Cultivate a new generation and steady stream of government policymakers/ planners/decision makers/practitioners, with high awareness and training on Eco-DRR/CCA related topics;
– Stimulate field-based and applied research on Eco-DRR/CCA that can feed into better practice locally and at country level;

- Embed Eco-DRR/CCA knowledge in existing teaching and training curricula, by collaborating with universities as well as national and regional training institutions, which have the institutional mandate (and the regular budgets) to train on a regular basis a range of DRR-related topics, e.g. The National Institute of Disaster Management in India covers Eco-DRR/CCA components in their standard DRR trainings given to thousands of public officials each year[9];
- Encourage online teaching mechanisms, which could be combined with in-classroom teaching (see Box 24.3);
- Identify gaps in specific development sectors, such as in water, agriculture, or urban planning, where introducing Eco-DRR/CCA training materials would be beneficial.

An important aspect is linking applied trans-disciplinary research with education. Innovative teaching concepts such as inquiry-based and discovery-based learning, for instance, offer a direct pathway from problem-based thinking to solution-oriented approaches. Moreover, various forms of distance and blended learning allow not only for the integration of audio-visual media, but also the dissemination of the learning materials to a wide international audience via the internet. Web-based learning has the advantage of providing materials and translating research outputs in a way that allows global and open access to education, in particular to developing countries that often have limited access to learning facilities.

Box 24.3. Harnessing Online Technology for Scaled-Up Capacity Development on Eco-DRR/CCA

Online technology offers great opportunity for scaling up delivery of capacity building activities and overcoming human resource constraints. TH Köln, University of Applied Sciences based in Germany and the United Nations Environment Programme (UNEP) launched the first Massive Open Online Course (MOOC) on "Disasters and Ecosystems – Resilience in a Changing Climate" in January 2015 on the German MOOC platform Iversity, combining the three topics of (a) disaster risk reduction, (b) climate change adaptation, and (c) ecosystem management, which have been thus far largely discussed separately in training courses. By combining these topics, the course designers aimed at a more comprehensive approach to enhancing the resilience of communities, landscapes and ecosystems (Sudmeier-Rieux et al. 2015). The online course is based on the international post-graduate course "Disasters, Environment and Risk Reduction (Eco-DRR)" that was launched at the 7th World Environmental Education Congress in Marrakech in 2013. Available in both English and Spanish, the post-graduate course was

(continued)

---

[9]See for example: http://nidm.gov.in/PDF/modules/climate.pdf; http://nidm.gov.in/PDF/modules/Legal_new.pdf

Box 24.3 (continued)

by CUAS, UNEP and other international partners from academia and practice. While so far 50 international universities have implemented the postgraduate course in their curricula, the first MOOC intake reached more than 12,000 participants from 183 countries, of which many were practitioners and policymakers (Nehren et al. 2015). The course was presented in two parts: leadership track (3 weeks) and received a completion rate of 18 %, and the expert track (10 weeks) for participants interested in more in-depth information on Eco-DRR, which had a completion rate of 8.9 %, both relatively high figures for MOOCs. Apart from the educational aspects, the MOOC also laid the foundation for a very active community of practice that exchanges knowledge, experience and information via social media. As an additional resource of the MOOC, a case study and exercise book was developed (Nehren et al. 2014), which includes a variety examples of Eco-DRR and EbA projects around the globe. Other MOOC initiatives in this field include the UNEP MOOC on Coursera: "Pathways to Climate Change. The Case of Small Island Developing States" or the Cornell University MOOC on Environmental Education with subthemes on DRR and CCA.

Given the increased interest in Eco-DRR/CCA from the perspectives of policy, research and practice, the global demand for development of educational and capacity development programmes addressing the issue is already high. Despite some limitations, web-based learning formats such as MOOCs have the potential to overcome disciplinary boundaries and create an open learning environment that connects people from various cultural and technical backgrounds, which is essential for interdisciplinary topics. Furthermore, online education offers the possibility to efficiently transmit new knowledge from the research front to different learner groups and thereby contribute to higher education and capacity development. It is therefore hoped that the examples mentioned above will be looked upon in a few years as precursors to a well-developed field.

24.4.3 Scaling-Up Investments in Eco-DRR/CCA

Over the last decade, significant progress has been made in implementing Eco-DRR/CCA projects worldwide, in both developed and developing countries as well as in emerging economies, particularly China, India and Brazil. These trends are evidenced in the proliferation of published reports, case studies and papers on Eco-DRR/CCA since 2005.[10] Financial resources for implementing Eco-DRR/CCA have also become more available, mainly through international

---

[10]See for instance, PEDRR's Virtual Library : http://pedrr.org/activities/graduate-course/

development assistance, and particularly through climate change financial instruments (e.g. Adaptation Funds, Green Climate Fund, etc.).

However, experiences in Eco-DRR/CCA are still mostly limited to small-scale or "pilot" demonstration projects. As such, field interventions are carried out within a limited geographical area and over short timeframes (usually 2–5 years). Only very few examples exist of genuinely large-scale Eco-DRR/CCA efforts that cover wide geographical spaces over an extended time period, and that become part of national programmes or strategies (see for example: Reij et al. 2009a, b; Temmerman et al. 2013; van Eijk et al. 2013; Wehrli and Dorren 2013), and there are equally few documented cases of large-scale Eco-DRR/CCA initiatives where lessons and best practices are analysed and shared. As a result, Eco-DRR/CCA advocates and practitioners face major difficulties demonstrating *how* Eco-DRR/CCA approaches can be translated at scale and replicated.

One of the main constraints to scaling-up Eco-DRR/CCA approaches is the lack of standardised, technical guidelines for designing and using ecosystem-based measures for disaster and climate risk reduction. In part, this is a reflection of the limited integration or up-take of Eco-DRR/CCA within the engineering community (Thummarukudy, personal communication 2016). Civil, geotechnical, hydraulics and coastal engineers are invariably associated with the design and construction of "grey" infrastructure (e.g. seawalls, dykes, embankments, etc.) for managing disaster risks, which they carry out based on engineering codes of practices developed over many decades. Therefore, only conventional DRR measures are typically used by the engineering community, as they are perceived to be effective and "bankable". It thus becomes more feasible for government decision makers (and engineering companies) to obtain funding for concrete structures, e.g. for coastal protection, than if they were to consider Eco-DRR/CCA measures, such as mangrove rehabilitation, even when Eco-DRR/CCA measures are proven to be more cost-effective. This issue needs to be addressed (Ibid).

Although technical guidelines for implementing Eco-DRR/CCA measures have become much more available over the past decade, many of these guidelines have not yet been subject to rigorous testing and standardisation, nor are they readily available or implementable for all types of ecosystems and hazards. There are only very few widely-accepted implementation guidelines, for instance in the case of establishing and managing "protection forests" – known as the NaiS guidelines – in Switzerland and used by other Alpine countries, to reduce risks from mountain hazards (Wehrli and Dorren 2013). Therefore, engaging with engineers, engineering institutions and associations, would be a critical step towards advancing Eco-DRR/CCA practices. Given that every major line of engineering emerged following decades of experimentation to develop the engineering codes of practice we have today, it should be very feasible to engage with engineers to test new Eco-DRR/CCA approaches (Thummarukudy, personal communication 2016). Efforts to introduce ecological engineering into engineering colleges and curricula should therefore be pursued (Ibid).

Some of this work is already being taken up by engineers in large companies, for instance Shell which is testing applications of "living shorelines" and oyster reefs for coastal erosion control and LafargeHolcim using quarry rehabilitation to create constructed wetlands and better manage storm water and flooding.[11] With improved implementation standards for Eco-DRR/CCA measures, this would in turn help government regulatory bodies become more supportive of Eco-DRR/CCA measures, for instance by granting more permits and incentives to companies willing to invest in such measures.

In reality, there is no single, "magic bullet" for tackling the challenge of scaling-up investments in Eco-DRR/CCA. Scaling-up will require concerted efforts from all sides, including strengthening the economic case of Eco-DRR/CCA, developing effective decision support tools, addressing research and knowledge gaps, standardising implementation guidelines, engaging the private sector and raising greater awareness and creating demand for Eco-DRR/CCA approaches from "end-users", including national and local governments and local communities, as well as developing the right mix of policy and financial incentives for investing in Eco-DRR/CCA – which have been discussed in this chapter and throughout this book.

References

Asia Disaster Preparedness Center (ADPC) (in press). Short training course on "Application of the technical guidelines for incorporating DRR/CCA concerns in the Philippine EIS system"

Bell J, Lovelock CE (2013) Insuring mangrove forests for their role in mitigating coastal erosion and storm -surge: an Australian case study. Wetlands 33:279–289

Benson C, Twigg J (2007) Guidance Note 7. In: Tools for mainstreaming disaster risk reduction: guidance notes for development organizations. International Federation of Red Cross and Red Crescent Societies/ProVention Consortium, Geneva, pp 79–89

Bullock A, Acreman M (2003) The role of wetlands in the hydrological cycle. Hydrol Earth Syst Sci 7(3):253–260

Caribbean Development Bank (CDB) and Caribbean Community Secretariat (CARICOM) (2004) Sourcebook on the integration of natural hazards into the Environmental Impact Assessment (EIA) Process: NHIA-EIA Sourcebook. Cariddean Development Bank, Bridgetown. Available at https://www.caribank.org/uploads/projects-programmes/disasersclimate-change/reports-and-publications/Source%20Book5.pdf. Accessed 2 Nov 2012

Centro Regional para el Hemisferio Occidental (CREHO) (2015) International training course on environmental impact assessments and strategic environmental assessments, held on 26–30 October 2015, Panama City. (http://www.creho.org/nosotros/cursos/cursos-2015/evaluacion-de-impacto-ambiental/. Downloaded on 06 March 2016)

Chatenoux B, Peduzzi P (2007) Impacts from the 2004 Indian Ocean Tsunami: analysing the potential protecting role of environmental features. Nat Hazards 40:289–304

Convention on Biological Diversity (CBD) (2009) Connecting biodiversity and climate change mitigation and adaptation: report of the second ad-hoc technical expert group on biodiversity and climate change, CBD technical series No. 41. Secretariat of the Convention on Biological Diversity, Montreal, 126p

---

[11] See the NI4Biz Platform at: http://www.naturalinfrastructureforbusiness.org/case-studies/

Convention on Biological Diversity (CBD) (2014) COP12 Decision XII/20. Biodiversity and climate change and disaster risk reduction. Convention on Bilological Diversity, Pyeongchang

Dale PER, Knight JM (2008) Wetlands and mosquitoes: a review. Wetl Ecol Manag 16:255–276

Doswald N, Estrella M (2015) Promoting ecosystems for disaster risk reduction and climate change adaptation: opportunities for integration. Discussion Paper, United Nations Environment Programme, Geneva, 48p

Dudley N, Buyck C, Furuta N, Pedrot C, Renaud F, Sudmeier-Rieux K (2015) Protected areas as tools for disaster risk reduction. A handbook for practitioners. MOEJ and IUCN, Tokyo/Gland, 44pp

Eriksen S, Aldunce P, Bahinipati CS et al (2011) When not every response to climate change is a good one: identifying principles for sustainable adaptation. Clim Dev 3:7–20

Estrella M, Salismaa N (2013) Ecosystem-based disaster risk reduction (Eco-DRR): an overview. In: Renaud F, Sudmeier-Rieux K, Estrella M (eds) The role of ecosystems in disaster risk reduction. United Nations University Press, Tokyo, pp 26–54

European Commission (2015) Towards an EU research and innovation policy agenda for Nature-Based Solutions & Re-Naturing Cities. Final report of the Horizon 2020 Expert Group on 'Nature-Based Solutions and Re-Naturing Cities'. European Commission

Executive Office of the President of the United States (2015) Memorandum for executive departments and agencies (M-16-01) on Incorporating ecosystem services into federal decision making, issued on 7 October 2015. https://www.whitehouse.gov/sites/default/files/omb/memoranda/2016/m-16-01.pdf. Accessed Mar 2016

Fordham M (1998) Making women visible in disasters: problematising the private domain. Disasters 22(2):126–143

Fordham M (2001) Challenging boundaries: a gender perspective on early warning in disaster and environmental management, UNDAW, Environmental management and the mitigation of natural disasters: a gender perspective, Report of the Expert Group Meeting (Ankara, Turkey, November 2001), New York

Fordham M, Gupta S, Akerkar S, Scharf M (2010) Leading resilient development: grassroots women's priorities, practices and innovations. New York: GROOTS International and UNDP Gender Team, Bureau of Policy Development, New York. Accessed at http://huairou.org/sites/default/files/Leading%20Resilient%20Development%20GROOTS.pdf

Gupta A, Nair S (2013) Applying environmental impact assessments and strategic environmental impact assessments in disaster management. In: Renaud F, Sudmeier-Rieux K, Estrella M (eds) The role of ecosystems in disaster risk reduction. United Nations University Press, Tokyo, pp 416–436

Intergovernmental Panel on Climate Change (IPCC) (2012) Managing the risks of extreme events and disasters to advance climate change adaptation: special report of the intergovernmental panel on climate change. Cambridge University Press, New York. Accessed at https://www.ipcc.ch/pdf/special-reports/srex/SREX_Full_Report.pdf

International Union for Conservation of Nature World Parks Congress (IUCN) (2014) The promise of Sydney. International Union for Conservation of Nature, Sydney

Lacambra C, Friess D, Spencer T, Möller I (2013) Bioshield: mangrove ecosystems as resilience natural coastal defences. In: Renaud F, Sudmeier-Rieux K, Estrella M (eds) The role of ecosystems in disaster risk reduction. United Nations University Press, Tokyo, pp 82–108

Millennium Ecosystem Assessment (MEA) (2005) Ecosystems and human well-being: current state and trends: findings of the Condition and Trends Working Group. Island Press, Washington, DC

Munang R, Thiaw I, Alverson K, Liu J, Han Z (2013) The role of ecosystem services in climate change adaptation and disaster risk reduction. Terr Syst 5(1):47–52

Murti R, Buyck C (eds) (2014) Safe havens: protected areas for disaster risk reduction and climate change adaptation. IUCN, Gland, xii + 168 pp

Nehren U, Sudmeier-Rieux K, Sandholz S et al (ed) (2014) The ecosystem-based disaster risk reduction case study and exercise source book. CNRD/PEDRR, ISBN: 978-3-00-045844-6

Nehren U, Sudmeier-Rieux K, Sandholz S, Straub G (2015) Development and implementation of the Massive Open Online Course (MOOC) Disasters and eco-systems – resilience in a Changing Climate. Presentation at the 8th World Environmental Education Congress, Gothenburg, June 29–July 2 2015

Organization for Economic Co-operation and Development (OECD) (2010) Strategic Environmental Assessment (SEA) and Disaster Risk Reduction (DRR). DAC Network on Environment and Development Co-operation (ENVIRONET), Advisory Note. Organization for Economic Co-operation and Development, Paris. Available at http://www.oecd.org/dac/environmentanddevelopment/42201482.pdf. Accessed 2 Nov 2012

Partnership for Environment and Disaster Risk Reduction (PEDRR), Cologne University of Applied Sciences'Center for Natural Resources and Development (CNRD) (2013) Graduate module on disasters, ecosystems, and risk reduction. Partnership for Environment and Disaster Risk Reduction in collaboration with Cologne University of Applied Sciences' Center for Natural Resources and Development, Geneva

Ramsar I (2015) COP12 draft resolution X11.13 – wetlands and disaster risk reduction. Ramsar Convention on Wetlands, Punta del Este

Rao NS, Carruthers TJB, Anderson P, Sivo L, Saxby T, Durbin T, Jungblut V, Hills T, Chape S (2013) An economic analysis of ecosystem-based adaptation and engineering options for climate change adaptation in Lami Town, Republic of the Fiji Islands. A technical report by the Secretariat of the Pacific Regional Environment Programme – Apia, Samoa: SPREP 2013. Available at http://ian.umces.edu/pdfs/ian_report_392.pdf

Reij C, Tappan G, and Smale A (2009a) Agroenvironmental Transformation in the Sahel: another kind of "Green revolution". IFPRI Discussion Paper 00914, International Food Policy Research Institute, Washington, DC

Reij C, Tappan G, and Smale A (2009b) Re-greening the Sahel: farmer-led innovation in Burkina Faso and Niger. In: Speilmam DJ, Pandya0Lorcg R (eds) Millions fed: proven success in agricultural development. International Food Policy Research Institute, Washington, DC, p 53–58

Renaud FG, Sudmeier-Rieux K, Estrella M (eds) (2013) The role of ecosystems in disaster risk reduction. UNU Press, Tokyo

Shreve CM, Kelman I (2014) Does mitigation save? Reviewing cost-benefit analyses of disaster risk reduction. Int J Disaster Risk Reduct 10(part A):213–235

Sudmeier-Rieux K, Pradhan M, Nehren U, Sandholz S (2015) Improving environmental education and human security through web-based learning tools. Presentation at the 8th World Environmental Education Congress, Gothenburg, June 29–July 2 2015

Sutherland WJ, Gardner T, Bogich TL, Bradbury RB, Clothier B, Jonsson M, Ka-pos V, Lane SN, Möller I, Schroeder M, Spalding M, Spencer T, White PCL, Dicks LV (2014) Solution scanning as a key policy tool: identifying management interventions to help maintain and enhance regulating ecosystem services. Ecol Soc 19(2):3

Temmerman S, Meire P, Bouma T, Herman P, Ysebaert T, De Vriend H (2013) Ecosystem-based coastal defence in the face of global change. Nature 504(7478):79–83

The Royal Society (2014) Resilience to extreme weather. The Royal Society Science Policy Centre, London. Accessed from https://royalsociety.org/resilience

Thummarukudy M United Nations Environment Programme, Post-Conflict and Disaster Management Branch, Senior Programme Manager. Personal communication, 21 March 2016

UNDP (2013) Gender and disaster risk reduction. http://www.undp.org/content/dam/undp/library/gender/Gender%20and%20Environment/PB3-AP-Gender-and-disaster-risk-reduction.pdf. Accessed 22 Jul 2016

UNISDR, UNDP and IUCN (2009) Making disaster risk reduction gender-sensitive. http://www.unisdr.org/files/9922_MakingDisasterRiskReductionGenderSe.pdf. Accessed 22 Jul 2016

United Nations Educational Scientific and Cultural Organization World Heritage Centre (UNESCO), International Centre for the Study of the Preservation and Restoration of Cultural Property (ICCROM), International Council on Monuments and Sites (ICOMOS), International

Union for Conservation of Nature (IUCN) (2010) Managing disaster risks for world heritage. UNESCO World Heritage Centre, Paris. Available at http://whc.unesco.org/en/activities/630/

United Nations Environment Programme (UNEP) (2014) Building disaster-resilient communities and economies. United Nations Environment Programme Finance Initiative, Geneva

United Nations Environment Programme (UNEP) (2015) Collaborating for resilience: Partnerships that build disaster-resilient communities and economies. United Nations Environment Programme Finance Initiative, Geneva

United Nations Office for Disaster Risk Reduction (UNISRD) (2015) Sendai framework for disaster risk reduction 2015–2030. United Nations Office for Disaster Risk Reduction, Geneva. 18 March

van Eijk P, Baker C, Gaspire R, Kuman R (2013) Good flood, bad flood: maintaining dynamic riven basins for community resilience. In: Renaud F, Sudmeier-Rieux K, Estrella M (eds) The role of ecosystems in disaster risk reduction. United Nations University Press, Tokyo, pp 221–247

Wehrli A, Dorren L (2013) Protectuib forests: a key factor in integrated risk management in the Alps. In: Renaud F, Sudmeier-Rieux K, Estrella M (eds) The role of ecosystems in disaster risk reduction. United Nations University Press, Tokyo, pp 321–342

World Bank (2010) Convenient solutions to an inconvenient truth: ecosystem-based approaches to climate change. International Bank for Reconstruction and Development/The World Bank, Washington, DC

World Heritage Committee (2007) Decisions adopted at the 31st session of the World Heritage Committee. United Nations Educational Scientific and Cultural Organization, Christchurch, 2 July

Wylie L, Sutton-Grier A, Moore A (2016) Keys to successful blue carbon projects: lessons learned from global case studies. Mar Policy 65(2016):76–84

Index

© Springer International Publishing Switzerland 2016
F.G. Renaud et al. (eds.), *Ecosystem-Based Disaster Risk Reduction and Adaptation
in Practice*, Advances in Natural and Technological Hazards Research 42,
DOI 10.1007/978-3-319-43633-3